EMERGING INFECTIONS

Biomedical Research Reports

Volume Editor

Richard M. Krause

Fogarty International Center
National Institutes of Health
Bethesda, Maryland

ACADEMIC PRESS

San Diego   London   Boston   New York   Sydney   Tokyo   Toronto

Cover photograph; Seattle police during the 1918 influenza epidemic. Courtesy of the National Archives, Washington, D. C.

Figure 3 (chapter 6) reprinted with permission from Quinn, R. W. (1982).
Streptococcal infections. In A. S. Evans and H. A. Feldman, (eds).
"Bacterial Infections of Humans," Plenum Publishing, New York, p 525–552.
Table V (chapter 6) reprinted with permission from Holm, et al. (1989).
Volume 166. The University of Chicago Press. pp 31–37.
Figure 1 (chapter 6) reprinted with permission from Davies, J. Inactivation of antibiotics and the dissemination of resistance genes. *Science* **264**, 375–382.
Figure 2 (chapter 2) reprinted with permission from Bissonnette, L. and Roy, P. H. Characterization of the InO of Pseudomonas aeruginosa plasmid pVSI, an ancestor of integrons of multiresistance plasmids and transposons of Gram-Negative Bacteria. *J. Bacteriol.* **174**, 1248–1257.

This book is printed on acid-free paper.

Academic Press
a Harcourt Science and Technology Company
525 B Street, Suite 1900, San Diego, California 92101-4445
http://www.academicpress.com

Academic Press Limited
Harcourt Place, 32 Jamestown Road, London NW1 7BY, UK

Library of Congress Cataloging-in-Publication Data
Emerging infections / volume editor, Richard M. Krause.
        p.   cm. — (Biomedical research reports)
    Includes bibliographical references and index.
    ISBN 0-12-425931-6 (alk. paper)
    1. Epidemiology.   2. Communicable diseases.   I. Krause, Richard
  M., 1925– .   II. Series.
    [DNLM: 1. Communicable Diseases—epidemiology.   WC 100 E533 1998]
  RA651.E46   1998
  614.5—dc21
  DNLM/DLC
  for Library of Congress   97-45688
                                                                          CIP

PRINTED IN THE UNITED STATES OF AMERICA
00   01   02   03   04   05   MM   9   8   7   6   5   4   3   2   1

EMERGING
INFECTIONS

Biomedical Research Reports

Biomedical Research Reports

Series Editors

John I. Gallin
Warren G. Magnuson Clinical Center
National Institutes of Health
Bethesda, Maryland

Anthony S. Fauci
National Institute of Allergy and Infectious Diseases
National Institutes of Health
Bethesda, Maryland

CONTENTS

CONTRIBUTORS

Roy M. Anderson (23) The Wellcome Trust Centre for the Epidemiology of Infectious Disease, University of Oxford, Oxford OX1 3PS, United Kingdom

Lucy M. Bartley (301) The Wellcome Trust Centre for the Epidemiology of Infectious Disease, Department of Zoology, University of Oxford, Oxford OX1 3PS, United Kingdom

Barry R. Bloom (51) Howard Hughes Medical Institute, Department of Microbiology and Immunology, The Albert Einstein College of Medicine, Bronx, New York 10461

Julian E. Davies (239) Department of Microbiology and Immunology, University of British Columbia, Vancouver, British Columbia, Canada V6T 1Z3

Karen P. Day (463) The Wellcome Trust Centre for the Epidemiology of Infectious Disease, Department of Zoology, University of Oxford, Oxford OX1 3PS, United Kingdom

Anthony S. Fauci (327) National Institute of Allergy and Infectious Diseases, National Institutes of Health, Bethesda, Maryland 20892

Geoffrey P. Garnett (301) The Wellcome Trust Centre for the Epidemiology of Infectious Disease, Department of Zoology, University of Oxford, Oxford OX1 3PS, United Kingdom

Edward C. Holmes (301) The Wellcome Trust Centre for the Epidemiology of Infectious Disease, Department of Zoology, University of Oxford, Oxford 0X1 3PS, United Kingdom

James Hope (447) BBSRC Institute for Animal Health, Compton Laboratory, Compton, Berkshire RG20 7NN United Kingdom

William R. Jacobs, Jr. (51) Department of Microbiology and Immunology, Howard Hughes Medical Institute, The Albert Einstein College of Medicine, Bronx, New York 10461

Richard M. Krause (1, 185) Fogarty International Center, National Institutes of Health, Bethesda, Maryland 20892

Adel A. F. Mahmoud (431) Department of Medicine, Case Western Reserve University, Cleveland, Ohio; University Hospitals of Cleveland, Cleveland, Ohio 44106

Elizabeth A. McGraw (163) Institute of Molecular Evolutionary Genetics, Department of Biology, Pennsylvania State University, University Park, Pennsylvania 16802

John D. McKinney (51) Howard Hughes Medical Institute, Department of Microbiology and Immunology, The Albert Einstein College of Medicine, Bronx, New York 10461

John J. Mekalanos (147) Department of Microbiology and Molecular Genetics, Shipley Institute of Medicine, Harvard University Medical School, Boston, Massachusetts 02115

Frederick A. Murphy (375) School of Veterinary Medicine, University of California, Davis, Davis, California 95616-8734

James M. Musser (185) Section of Molecular Pathobiology, Department of Pathology, Baylor College of Medicine, Houston, Texas 77030

Neal Nathanson (365) Department of Microbiology, University of Pennsylvania Medical Center, Philadelphia, Pennsylvania 19104-6146

Stuart T. Nichol (365) Special Pathogens Branch, Division of Viral and Rickettsial Diseases, Center for Disease Control and Prevention, Atlanta, Georgia 30333

C. J. Peters (375) Special Pathogens Branch, Division of Viral and Rickettsial Diseases, Centers for Disease Control and Prevention, Atlanta, Georgia 30333

Thomas C. Quinn (327) National Institute of Allergy and Infectious Diseases, National Institutes of Health, Bethesda, Maryland 20892

Sean D. Reid (163) Institute of Molecular Evolutionary Genetics, Department of Biology, Pennsylvania State University, University Park, Pennsylvania 16802

Eric J. Rubin (147) Department of Microbiology and Molecular Genetics, Harvard Medical School, Boston, Massachusetts 02115

Allen C. Steere (219) Tufts University School of Medicine, New England Medical Center, Boston, Massachusetts 02111

Walter J. Tabachnick (411) Arthropod-borne Animal Diseases Research Laboratory, USDA, Agricultural Research Service, Laramie, Wyoming 82071

Matthew K. Waldor (147) Division of Geographic Medicine and Infectious Diseases, Tupper Research Institute, New England Medical Center, Tufts University School of Medicine, Boston, Massachusetts 02111

Vera Webb (239) Department of Microbiology and Immunology, University of British Columbia, Vancouver, British Columbia, Canada V6T 1Z3

Robert G. Webster (275) Department of Virology and Molecular Biology, St. Jude Children's Research Hospital, Memphis, Tennessee 38105

Thomas S. Whittam (163) Institute of Molecular Evolutionary Genetics, Department of Biology, Pennsylvania State University, University Park, Pennsylvania 16802

■ PREFACE

Biomedical Research Reports is introduced at a time of unprecedented advances and public interest in medical research. Our intention is to provide the reader with an in-depth analysis of subjects of interest to the biomedical researcher from the broad perspective of multiple authors, each covering a particular facet of the subject. We intentionally have chosen a broad scope for this new series to maximize flexibility in the range of topics that will be reviewed.

We will select, annually at a minimum, subjects that we believe are both timely and challenging. It is particularly appropriate to launch the series with the topic of "Emerging Infections." We are indeed fortunate that Richard M. Krause, M.D., a long-time student and authority on the subject, has agreed to edit our first volume. Dr. Krause has highlighted the leading emerging infectious diseases of this century, and in his characteristically eloquent style has packaged them into an easy to read volume. The introductory chapter by Dr. Krause, and the chapters by Roy M. Anderson, "Analytical Theory of Epidemics," and Julian E. Davies and Vera Webb, "Antibiotic Resistance in Bacteria," provide a background for understanding some of the principles of emerging infectious diseases. Influenza, dengue, AIDS, Hantavirus, Ebola virus, streptococcal diseases, Lyme disease, tuberculosis, cholerae, malaria, parasitic infections and slow virus infections were selected by Dr. Krause as examples of infectious diseases that pose enormous scientific and clinical challenges.

We will select future topics for this series based on their broad scientific, clinical and social implications. We welcome ideas for future topics from our readers, and we look forward to our interactions with Academic Press, our partner in this new series.

John I. Gallin, M.D.
Anthony S. Fauci, M.D.

REFLECTIONS AND ACKNOWLEDGMENTS

Infectious disease is and always has been part of the every day experience of life. In every generation, men of affairs have had to cope the best they could with the practical problems it presents, while priests, philosophers, and, later, scientists have had perhaps the harder task of interpreting the significance of such disease in accordance with the intellectual outlook of their time.

Sir Macfarlane Burnet

My career in basic and clinical infectious disease research began 47 years ago in 1950, when, as a medical student, I took a year's leave as a preventive medicine fellow to work on the epidemiology and natural history of streptococcal infections, acute rheumatic fever, and acute glomerulonephritis, with the late Charles H. Rammelcamp, Jr. That experience changed my life, and there followed years filled with joy at the Rockefeller University with Maclyn McCarty and Rebecca C. Lancefield, and at the National Institutes of Health. As Director of the National Institute of Allergy and Infectious Diseases my mentors in the arcane ways of government included John R. Seal and Kenneth W. Sell. From the laboratory chiefs and their colleagues I obtained a crash course that spanned the broad spectrum of infection and immunity, and I came to the view that they were opposite sides of the same coin. Not all would agree with that then or would do so now, but HIV/AIDS comes as close as anything can in Nature to illustrate that this is so.

Forty-seven years is a long time in infectious disease research. Indeed, the last half of the twentieth century has seen the advent of the wonder drugs antibiotics—the wonder, to be sure, tempered with the advent of widespread antibiotic resistance—and the astonishing success of the childhood vaccines. And finally, the re-emergence of infectious diseases occurred after what many had predicted would be a slow decline and eventual demise. Alas, that was not to be.

In my introduction to this book, I recount highlights of the historical record for these past 47 years as I have lived them, and through this prism make a few observations on the early history of infectious diseases and the efforts to confront them. The selection of the subjects included in this book and the authors has been mine. That is the only "spin" I have given to this subject, with the exception that I have suggested to the authors that they consider relevant matters pertaining to molecular and evolutionary epidemiology and population biology. I did not suggest any structure or format and left the manner of the presentation entirely up to the contributors.

I acknowledge especially the generous time and effort that the authors have devoted to their tasks. In addition, I wish to thank Louis H. Miller and Jose M. C. Roberio for many useful discussions on the intricate interactions of host, microbe, vector, and reservoir, and Oswald (Ozzie) A. Bushnell, who has graciously received me in his Honolulu home where I learned from him about the life of the Hawaiian people before and after the arrival of Captain Cook. I thank especially Anthony S. Fauci and John I. Gallin for the opportunity to edit this volume. Kerry Willis, Editor-in-Chief, Life Sciences, has been patient with the pace of my progress. I am particularly grateful to Philip E. Schambra, Director of the Fogarty International Center, who has given me the opportunity to pursue this and other efforts concerning the emergence of infectious diseases at home and abroad. Finally, my Introduction, and the chapter on the streptococcal infections with James M. Musser, would not have been brought to completion without the very able assistance of Anthony L. Davis and Susan E. Harrison.

■ INTRODUCTION

The current edition of *Emerging Infections* is out of print, and it is our hope that this paperback edition of these timely chapters will be valuable to a wider audience. The material is still sufficiently up to date such that revisions were not required.

Many factors such as changes in demography, lifestyle, and agriculture are cited as a cause of emerging infectious diseases. These factors have been reviewed extensively in this book. Considered also are the general principles of bacterial and viral genetics which influence the emergence or reemergence of microbes and the waxing and waning of epidemics as biological expressions of host-microbe associations.

Various microbial outbreaks are reported daily by ProMED and other sources of medical information. Via the Internet, we learn immediately about local outbreaks such as malaria at the Luxembourg airport, dengue in Texas, and cholera imported to New Zealand. Each of these events was a contained outbreak, but together they indicate the persistent, dynamic, and evolving nature of emerging infections.

Whether early warnings such as these also indicate a major health threat is never clear at the outset. One recent case illustrates this dilemma. As reported in *MMWR* (Centers for Disease Control and Prevention, 1999), a "previously unrecognized paramyxovirus related to, but distinct from, the Australian Hendra virus" appeared to be responsible for an outbreak of "swine fever" in southeast Asia in spring 1999. The case is described as follows:

> During September 29, 1998–April 4, 1999, 229 cases of febrile encephalitis (111 {48%} fatal) were reported to the Malaysian Ministry of Health (MOH). During March 13–19,

1999, nine cases of similar encephalitic illnesses (one fatal) and two cases of respiratory illness occurred among abattoir workers in Singapore. Tissue culture isolation identified a previously unknown infectious agent from ill patients. . . .

Cases have occurred primarily among adult men who had histories of close contact with swine. Concurrent with the human cases, illness and death occurred among swine from the same regions. Initially, Japanese encephalitis (JE) virus was considered the probable etiologic agent for this outbreak, and specimens from some patients tested positive for infection with JE virus. However, the predominance of cases in men who had close contact with swine suggested the possibility of another agent. Tissue culture isolation from central nervous system specimens at the Department of Medical Microbiology, University of Malaya, identified a previously unknown infectious agent. [In addition to other analyses and studies], additional laboratory testing . . . indicated the virus was related but not identical to the Hendra virus. . . .

Illness has been characterized by 3–14 days of fever and headache followed by drowsiness and disorientation that can progress to coma within 24–48 hours; a few patients had respiratory illness. Of the 229 case-patients, most have been men working on pig farms in Perak and Negri Sembilan. . . .

In addition to active surveillance for encephalitis cases, studies are under way to determine risk, if any, for human-to-human transmission among health-care workers and family members, to confirm the source of human infection (presumedly pigs), to define specific risk factors associated with exposures to pigs and tissues from infected animals, and to determine the case-to-infection ratio and the epidemiology of this infection in pigs. Preliminary assessment suggests that spread of the virus among states in Malaysia has occurred through transport of infected swine. . . .

To prevent further outbreaks, Malaysian authorities have banned transport of pigs within the country. Army personnel and police are enforcing this ban, and quarantined pigs are being culled within a 3-mile (5-km) perimeter around recognized outbreak areas. . . . Approximately 800,000 pigs have been killed. (Centers for Disease Control and Prevention, 1999, pp. 265–269)

Just when I thought I had completed this introduction, we were alerted to the unexpected detection of West Nile virus as the cause of encephalitis in New York, a diagnosis which had gone undetected because the tests had been performed with reagents to detect St. Louis encephalitis. Further examination revealed that at least some of the cases were not due to St. Louis encephalitis but to West Nile encephalitis virus or a variant of it. As reported in *The Washington Post* (Duke, 1999a),

the diagnostic turnabout came last week with the discovery that a mysterious spate of Bronx Zoo bird deaths was caused by a West Nile–like fever. This prompted more testing of the human encephalitis cases. In New York City, 14 confirmed cases of what previously had been called St. Louis encephalitis, including three fatal ones, were reclassified as West Nile. . . . The New York metropolitan area, and parts of Connecticut where dead crows were diagnosed, now have become laboratories for scientists tracking the course of a virus that first surfaced in Africa 60 years ago, became endemic to Asia and surfaced in Europe three years ago but never had been seen in North or south America. . . . "From a scientific point of view, this is a bombshell," said Tracey McNamara, head of pathology at the Wildlife Conservation Society at the Bronx Zoo, whose dead birds led to the diagnosis of both animal and human West Nile.

A subsequent report (Duke, 1999b) continues,

The discovery of West Nile fever here has spawned an epidemiological mystery, for scientists have no clue how the virus got to this hemisphere. The virus is endemic in parts of east Africa, Asia, and the Middle East, and it broke out as recently as three years ago in Eastern Europe. But birds generally do not migrate across the Atlantic Ocean, except for the odd occasions when birds are lost and "get up in the air currents and end up in a different continent," [Duane] Gubler [of the Centers for Disease Control and Prevention (CDC)] said. More likely . . . is that an imported bird brought the virus here, or that a human infected with it traveled to this region. In either case, with the mosquito as the vector, the virus soon spread. . . .

While birds of many kinds are believed to be carriers of the virus, its presence in crows has proved a particularly telling sentinel for the outbreak. "More crows are dying than anything else, possibly because crows are more susceptible to the disease," said Nicholas Komar, a vertebrate ecologist with the CDC. After a 1950 outbreak of West Nile virus in Egypt, studies there showed crows to be particularly susceptible to the virus, Komar said. . . .

But questions about the virus they bear will keep virologists busy for months to come. "The important thing to remember: We don't even know when this was introduced," Gubler said. "It could have been introduced last year and we just missed it in our [mosquito] surveillance. It could have been introduced in one of the southern states and introduced into New York in the spring of this year."

In his review of my earlier book, *The Restless Tide: The Persistent Challenge of the Microbial World* (Krause, 1981), Lawrence Altman noted that the "essays [in the book] are the type needed to promote a more intelligent public health policy." He went on to say, "Although the public may resent the costs of research, it behooves us to recall that the development of effective vaccines and antibiotics did not happen overnight" (Altman, 1982). His injunction in 1982 is as relevant today as it was then.

More recently, Altman commented on the West Nile virus. In his column, "The Doctor's World," in *The New York Times*, he cautioned, "When you hear hoofbeats, don't think of zebras. To doctors, the axiom is a call to focus on common ailments and not waste time on the exotic. But on those rare occasions when they do detect a zebra, doctors say they need to take extra steps in their investigation to make sure they have identified the right one. A case in point is the encephalitis outbreak that is blamed for at least three deaths in New York City" (Altman, 1999).

The transmission of microbes from animals resulting in outbreaks of disease among humans has occurred throughout history. In recent years there has been speculation, in some cases supported by evolutionary genetic studies, that tuberculosis and plague were introduced into human populations in ancient times from domestic animal reservoirs (Achtman *et al.*, 1999).

An important recent finding relates to the origin of HIV and AIDS. Gao *et al.* (1999) provide persuasive evidence that HIV-1 came to humans from the chimpanzee *Pan troglodytes,* a species that harbors SIVcpz, a simian immuno-deficiency virus related to HIV. Previous reports showed that the HIV in West Africa was derived from Sooty mangabey monkeys carrying SIVsm. Other evidence indicates that human T lymphotropic virus type 1 (HTLV-1) originated from related simian viruses, including the STLV-1 in chimpanzees. Weiss and Wrangham (1999) comment on the work of Gao *et al.*, noting the role of human behavior in transmission of retroviruses from primates to humans, as follows: "Hunters dismember chimpanzees with primitive butchery, and so expose themselves to the risk of zoonotically transmitted disease." The hunters "are paid by timber companies to provision logging camps, while the kills of free-lance hunters are traded as far as the cities," where the meat is served as a delicacy in certain restaurants.

It is likely that consumption of chimpanzee meat has been practiced for centuries and that, from time to time, frequent mutant viruses arose that overcame the species barrier to infect humans who processed the raw meat. The portal of entry of the virus could have been the mouth or abrasions of the skin. If this is what happened, it is likely that sporadic cases of HIV/AIDS occurred during the past several centuries, if not longer, and were confined to small rural villages and dwellings in the outback of central Africa, at a time when migration and travel were difficult and rare. The personal reminiscences of non-Africans who resided in central Africa during the 1930s provide vivid descriptions of the arduous effort needed to travel slowly by foot between small, isolated village compounds (Hahn, 1995). It is also likely that there were few secondary cases, or as population biologists would express it, that the reproductive rate was $R_0 < 1$.

Commenting in 1992 on the AIDS epidemic, Anderson and May applied a mathematical model to the emergence of the epidemic based on the initial sluggish interactions between rural villages. As they note, such models reveal "a pattern where the numbers of cases of HIV infection (and thence of AIDS) increase faster as time goes on, in compound-interest fashion" (Anderson and May, 1992, p. 58). Taking all these matters into consideration, it might have taken 100 years or more for AIDS to escalate from the putative source to the outbreak of the 1980s. What we do know is that the epidemic gained momentum and emerged from hibernation with the influence of changing cultural patterns, urbanization, migration of peoples, and long-distance land travel.

In the twenty-first century, the HIV pandemic will clearly proceed from bad to worse before it waxes and wanes. Indeed, the future looks bleak. As Fauci notes in his article, "The AIDS Epidemic—Considerations for the 21st Century," "unless methods of prevention, with or without a vaccine, are successful, the worst of the global pandemic will occur in the twenty-first century" (Fauci, 1999, p. 1049).

The theme from these examples is now commonplace. The evolution of microbes and the vagaries of human behavior conspire to fan the flames of regional diseases that can be transmitted easily worldwide by human travel and behavior (e.g., sexual practices, in the case of HIV infection). Our ability to identify and contain emerging and reemerging microbes is challenged daily and will continue to be challenged in the future. Many national and international efforts to combat emerging infectious diseases are supported by private and public agencies and by international organizations such as the World Health Organization, as summarized in a recent issue of the journal *Emerging Infectious Diseases* (1998).

One international health problem that deserves special attention is malaria, which continues to be one of the most serious infectious diseases in developing countries because of the emergence of mosquitoes resistant to DDT and of malaria parasites resistant to chloroquin. In 1997 an alliance of organizations and individuals concerned about malaria formed the Multilateral Initiative on Malaria (MIM). The purpose of MIM is to maximize the effect of scientific research on malaria in Africa and to utilize the "lessons learned" to combat malaria elsewhere in the world. Indeed, for malaria as well as for many infectious diseases, Sir Ronald Ross's "happenings" (1911), as he termed microbial outbreaks, continue to surprise us.

The Wellcome Trust coordinated MIM for the first 18 months. During this time, new programs have been initiated to delay the spread of resistance to antimalarial drugs, to accelerate the development of novel drugs and vaccines, and to organize a repository of standardized research reagents. New efforts undertaken in Africa include the training of researchers and health professionals and the funding of multicenter research studies.

In May 1999 the Wellcome Trust transferred coordination of MIM to the John E. Fogarty International Center (FIC) of the U.S. National Institutes of Health. On the transfer of the MIM Secretariat to the FIC, Dr. Gerald Keusch, the FIC director, noted the following: "Continuing the momentum on the MIM will depend on the strong partnerships already in place with African scientists, on cooperative efforts of existing sponsoring agencies, and on the energetic participation of new partners in this important global initiative. MIM is a new model for international research cooperation and we will work hard to ensure its continued success" (The Wellcome Trust, 1999).

This new model and other international efforts to cope with emerging infectious diseases are founded on an understanding of epidemics gained over the past century. In my first introduction to this book, I noted the contributions of William Farr and Florence Nightingale, who pioneered public health efforts to prevent the spread of infectious diseases in the nineteenth century. I also noted that William Farr is generally credited with being the first person to develop a numeric description of an epidemic. He developed his own mathematical formula to explain the relationships of factors that enhance the occurrence of cholera and other infections.

The analytical study of epidemics has now become a highly advanced science, as noted by Roy Anderson in the second chapter of this book. Anderson states that "transmission, persistence, and control are the conventional areas for the application of mathematical techniques in the study of infectious diseases. The real challenge for analytical approaches in the coming years lies in the field of molecular epidemiology, pathogen evolution, and pathogen interaction with the host immune responses.

Finally, a cautionary note is needed concerning biological warfare, the deliberate use of infectious disease agents for warfare or terrorism, a subject that has become of ever greater concern than it was when the chapters for the first edition of this book were written. I have elected not to add a chapter on this subject, but rather to refer the reader to a recent publication entitled *Biological Weapons, Limiting the Threat*, edited by Joshua Lederberg (Lederberg, 1999). In his introduction, Lederberg sets forth the issue in a succinct paragraph:

> The transcendence of biological warfare (BW)—over medicine and public health, private criminal acts, terrorism, interstate warfare, and international law directed at the elimination of BW—makes this one of the most intricate topics of discourse, poses very difficult security problems, and opens some novel challenges in the ethical domain. . . . That same transcendence confounds efforts to organize governmental and intergovernmental measures of control: health authorities will need to negotiate with the military, with law enforcement, and with environmental managers. And all will have to cope with how to enhance security without imposing intolerable stresses on personal liberties and on freedom of travel and of commerce. (Lederberg, 1999, p. 4)

Lederberg speaks particularly to those who wrote and will read this book concerning the public health and scientific response to this threat:

> If despite deterrence, law, and moral suasion, the means of attack cannot be forsworn, the obligation remains to be prepared to blunt them. Physicians and local health services, along with police and firefighter first-responders, are in the front lines to deal with health emergencies. This same apparatus is needed to deal with natural disease outbreaks: recall Legionella, Influenza A-II5N1, and *Escherichia coli* 0157:H7 of recent vintage. The local responders also need to be trained in exercises entailing support from the Public Health Service and, if need be, military personnel. While BW attacks may be widely dispersed, they are amenable to medical intervention far more than trauma from explosives or chemicals, provided diagnosis is timely and resources can be mobilized. In many cases, there may be little or no advance warning. Vigilance in understanding the fate of victims near the dose-epicenter might provide an alert for the much larger cohorts likely to receive smaller doses and exhibit longer incubation times—a window of opportunity for treatment. (Lederberg, 1999, pp. 7–8)

As this edition of *Emerging Infections* goes to press at the beginning of the twenty-first century, it is worth recalling the work of William Farr and his contributions to vital statistics, social hygiene, and the laws of epidemics. Writing in 1840, Farr informed us that "epidemics appear to be generated at intervals in unhealthy places, spread, go through a regular course, and decline; but of the cause of their evolutions no more is known than of the periodical paroxysms of ague. The body, in its diseases as well as in its functions, observes a principle of periodicity; its elements pass through prescribed cycles of changes, and the diseases of nations are subject to similar variations" (Farr, 1840, p. 95).

Commenting on the work of Jacob Henle of Berlin, who had written about the germ theory of disease, Farr added: "Each epidemic disease has its specific animal contagion, its specific genera of infusoria. Henle has proved the existence of this cause, and the truth of the theory in every way but one; he has never seen the epidemic infusoria. . . . The infusorial hypothesis does not satisfactorily explain the cause of epidemics; it accounts for them by the creation of animalcules, but does not shew [sic] why the animalcules are created at distant times in swarms. The phenomena of swarms of insects, of blight, and of infusorial generation, may suggest investigation; but in the present state of pathology, they cannot supply its place" (Farr, 1840, p. 95). Farr continued: "If the latent cause of epidemics cannot be discovered, *the mode in which it operates may be investigated*. The laws of its action may be determined by observation, as well as the circumstances in which epidemics arise, or by which they may be controlled" (Farr, 1840, p. 95, italics added).

Writing specifically about an epidemic and local outbreaks of smallpox in England, Farr noted that these events are governed by certain general laws, which he investigated based on his own study of total quarterly deaths from smallpox during the 1838–39 epidemic. He compared these total deaths with a "regular series of numbers . . . calculated upon the hypothesis that the fall of the mortality took place at a uniformly accelerated rate." That is, the epidemic rises until it reaches a point of maximum intensity, "where it remains stationary, like a projectile at the summit of the curve which it is destined to describe." The epidemic then declines "until the disease attains the minimum intensity, and remains stationary" (Farr, 1840, pp. 96–97).

Farr was not initially convinced that the germ theory of disease was correct, but he came to accept this view as the evidence became irrefutable. Yet we should remember how much was achieved with sanitation and social reform, before the germ theory of disease existed. Proof of the theory was followed by application of the principles of bacteriology to public health; chlorination of water and pasteurization of milk both had a dramatic effect on morbidity and mortality from infectious diseases. Nothing comparable to these advances was achieved again until the advent of antibiotics and childhood vaccines. Now, however, as evident from the AIDS epidemic, these achievements of the twentieth century are no match for the evolution of microbes when coupled with changes in human behavior (e.g., lifestyles, demography, agricultural practices). Research to develop new and creative ways to confront the emergence of infectious diseases must remain a high priority in the twenty-first century.

REFERENCES

Achtman, M., Zurth, K., Morelli, G., Torrea, G., Guiyoule, A., and Carniel, E. (1999). *Yersinia pestis,* the cause of plague, is a recently emerged clone of *Yersinia pseudotuberculosis. Proc. Natl. Acad. Sci. U.S.A.*

Altman, L. K. (1982). The Doctor's World—Infections still a big threat. *The New York Times,* July 20, p. C2.

Altman, L. K. (1999). The Doctor's World—Encephalitis outbreak teaches an old lesson. *The New York Times,* September 28, p. F8.

Anderson, R. M., and May, R. M. (1992). Understanding the AIDS pandemic. *Sci. Am.* **266**(5), 58–66.

Centers for Disease Control and Prevention. (1999). Outbreak of Hendra-like virus—Malaysia and Singapore, 1998–1999. *Morbidity & Mortality Weekly Report* **48**(13), 265–269.

Duke, L. (1999a). New York outbreak gets new diagnosis—West Nile fever has made its U.S. debut. *The Washington Post,* September 28, p. A3.

Duke, L. (1999b). A watch is posted for dead crows—West Nile virus may be flying south from New York, health officials warn. *The Washington Post,* September 29, p. A14.

Emerging Infectious Diseases. (1998). **Special issue, 4(3).**

Farr, W. (1840). "Letter, Second Annual Report of the Registrar-General on the Progress of Epidemics—Epidemic of Small Pox," App., 91–98. General Register Office, London.

Fauci, A. S. (1999). "The AIDS epidemic—considerations for the 21st century." *N. Engl. J. Med.* **341**(14), 1046–1050.

Gao, F., Bailes, E., Robertson, D. I., Chen, Y., Rodenburg, C. M., Michael, S. F., Cummins, L. B., Arthur, I. O., Peeters, M., Shaw, G. M., Sharp, P. M., and Hahn, B. H. (1999). Origin of HIV-1 in the chimpanzee *Pan troglodytes troglodytes. Nature (London)* **397**, 436–444.

Hahn, E. (1995). Personal history—I say this. *The New Yorker,* July 31, pp. 35, 38–39.

Krause, R. M. (1981). *"The Restless Tide: The Persistent Challenge of the Microbial World."* The National Foundation for Infectious Diseases, Washington, D.C.

Lederberg, J. (1999). "Biological Weapons: Limiting the Threat." The MIT Press, Cambridge.

Ross, R. (1911). "The Prevention of Malaria (with addendum on the theory of Happenings)." Murray, London.

Weiss, R. A., and Wrangham, R. W. (1999). The origin of HIV-1: From *Pan* to pandemic. *Nature (London)* **397**, 385–386.

The Wellcome Trust. (1999). International fight against malaria enters a new phase (press release). 26 May.

INTRODUCTION TO EMERGING INFECTIOUS DISEASES; STEMMING THE TIDE

RICHARD M. KRAUSE

Fogarty International Center
National Institutes of Health
Bethesda, Maryland

I feel myself a traveler on a journey that is far longer than the history of nations and philosophies, longer even than the history of our species.

　　　　　　　　　　　　　　　　Loren Eiseley

If disease is an expression of individual life under unfavorable conditions, then epidemics must be indicators of mass disturbances in mass life.

　　　　　　　　　　　　　　　Rudolf L. K. Virchow

There are similarities between the diseases of animals or man, and the diseases of beer and wine—if fermentations were diseases, one could speak of epidemics of fermentation.

　　　　　　　　　　　　　　　　　L. Pasteur

The fear that new plagues are in the making is not unjustified. In most parts of the world we are unprepared for any new pestilence. We have not enough water, not enough food, not enough shelter, and no peace.

　　　　　　　　　　　　　　　I. J. P. Loefler

Principiis obsta; sero medicina paratur
Cum mala per longas convaluere moras.

　　　　　　　　　　　　　　　　　Ovid
　　　　　　　　　　　　　43 B.C.–A.D. 17

Prevent (disease) at the start; medicine (physicians) must act Before disease becomes persistent through long delays.

　　　　　　　　　　　　　Free translation
　　　　　　　　　　　　　　　　R.M.K.

I. EMERGING INFECTIONS

I began this introduction the first week of January 1997 by browsing in the newspapers for recent accounts of infectious disease outbreaks. The journalists had been busy. I read about brucellosis in Wyoming bison, rabies in raccoons in the Eastern United States that is spreading to the Midwest and posing a threat to humans and domestic animals. And arguments have arisen concerning the use of cattle feed in the United States that is derived in part from the carcasses of livestock. The issue here concerns the possibility that the agent of "mad cow disease" might be transmitted to cattle, and then in time into the U.S. food chain. Appropriate steps are under consideration to prevent this occurrence.

The notice concerning brucellosis in bison is of particular relevance to a discussion on the factors that lead to the emergence of infectious diseases. Some years ago bison were on the verge of extinction. Protective laws were enacted and were successful in reversing the decline in the population. Since then, herds in northwest Wyoming have grown and, as a consequence of population pressure, are expanding their range into the grazing regions of Montana where beef cattle roam. It is feared that brucellosis will be transmitted to the cattle. If this happens brucellosis could once again enter the food chain as it did before this disease was eliminated from domestic cattle in the United States through a rigorous system of culling infected animals.

When I was a house officer at the Barnes Hospital in St. Louis in 1952, if a worker on the slaughter lines of a meat packing plant came into the emergency room with a high fever, the provisional diagnosis was assumed to be brucellosis until proved otherwise. The threat of this disease was high in these workers, because brucellosis had not yet been eliminated from livestock through vigorous control measures, including culling *all* of the infected animals. Some regions of the country were slow to adopt the use of a serologic blood test developed in the 1930s that was used to identify infected animals.

And yet here we are in 1997 with the threat that brucellosis might reemerge in people by exposure to the meat products of infected livestock if steps are not taken now to prevent this occurrence. The first batch of bison have been slaughtered, but a consensus between conservationists and ranchers has not yet been reached on this course of action. This possible threat of brucellosis raises an important issue that will be repeated again and again in this volume. Microbes pursue every possible avenue to escape from the barriers that are erected to contain them, and we must be forever on our guard. They will seek out undercurrents of opportunity and reemerge.

The final example in the current popular press concerns the reemergence of pertussis, or whooping cough. It has been generally accepted by the public and the medical profession that pertussis is primarily a childhood disease which has been effectively prevented by immunization with the pertussis (combined with diphtheria and tetanus toxoids) vaccine. And yet there are now numerous reports on the occurrence of pertussis in young adults. For example, 26% of 130 college students with a severe lingering cough treated at the UCLA health clinic were diagnosed as having pertussis as confirmed by bacterial culture.

The reason pertussis occurs in adults is that the immunity from the vaccine received in childhood declines after 10 to 15 years. Historically, pertussis was thought to be primarily a childhood disease, because of its frequency and severity in that age group. Certainly, prior to 1940 when the vaccine was introduced, pertussis was among the most serious illnesses of childhood and a frequent cause of mortality. After the introduction of the vaccine the number of cases in the United States dropped from more than 200,000 annually to approximately 1000 in 1976. The disease has made a comeback since then, and in 1993 more than 6500 cases were reported, with more than 4400 children being hospitalized during that year.

There are several reasons for this reemergence of pertussis. One the one hand, a sizable percentage of the children born each year do not receive the pertussis vaccine. So a significant number of preschool children are not immune, and they are a potential reservoir for pertussis. Second, with the disease clearly established in the adult population, there are frequent opportunities for the microbe to be transmitted to children. Because the pertussis vaccine was so effective in decreasing the incidence of the disease, it is not surprising that the public and public health physicians believed that the problem had been solved. In fact, the microbe had not been eliminated. It had been driven underground, so to speak, where it sought out undercurrents of opportunity—in this case the young adult and adult population. It is now time to reexamine immunization policy in regard to pertussis, and for physicians to recognize that pertussis can be an adult disease (Mortimer, 1994).

A quick review of the scientific literature that crossed my desk the first week of January 1997 was rewarded with numerous accounts concerning infectious diseases, including the emergence of familiar ones, as well as those that have been rare or uncommon in the past. Loefler, a surgeon in Nairobi Hospital, Kenya, published a paper in *Lancet* entitled *"Microbes, Chemotherapy, Evolution, and Folly,"* and for this he received the Wakely Prize (Loefler, 1996). The *Lancet* established this award for the individual who submitted the most meritorious essay of no more than 2000 words, on any clinical topic of international health importance. In commenting on the winning essay for this year, Richard Horton, Editor of *Lancet,* noted common themes that emerged in the 70 Wakely Prize entrants included tuberculosis, acquired immunodeficiency syndrome (AIDS), and health problems now afflicting poorer nations. Little wonder then that the award went to Loefler who has spent a lifetime in surgical practice in Africa. In graceful prose, he touches on the major themes that are embraced by the authors of this book concerning the factors that influence the emergence of infectious diseases. Loefler wrote, "The fear that new plagues are in the making is not unjustified. In most parts of the world we are unprepared for any new pestilence. We have not enough water, not enough food, not enough shelter, and no peace."

When I became Director of the National Institute of Allergy and Infectious Diseases (NIAID) in 1975 the "end of infectious disease" was a popular idea, and this was reflected in the National Institutes of Health (NIH) budget. Following the declaration of war on cancer and heart disease, the NIH budget doubled between 1970 and 1975. During that period the NIAID budget increased by 20%. So the political reality reflected the scientific judgment.

During that era not all scientists were so sanguine about the benign future of infectious diseases. The late Alfred Evans wrote about possible emerging infections as did Edwin Kilbourne. During the first few years that I was Director of NIAID, I developed a strategy to reverse the downward trend in research support. In reviewing the rationale for infectious disease research, the conclusions were not unexpected, but they posed a warning. Hidden behind the success with antibiotics for bacterial infections was an empty closet of antiviral drugs, and there was much yet to be learned before an antiviral drug strategy could be developed. At the same time the Tropical Disease Research program on the great neglect of diseases of humankind under the World Health Organization (WHO) sponsorship focused once again on the unbelievable toll in human life from parasitic diseases, with malaria alone, the King of Diseases, accounting for at least 1 million deaths of infants and children each year in Africa.

In my Congressional testimony in support of the NIAID budget and in a series of lectures for scientific and general audiences, I warned about the challenge of emerging infectious diseases. These lectures became a series of essays published in a book in 1981 entitled *The Restless Tide: The Persistent Challenge of the Microbial World* (Krause, 1981). It was my purpose to inform the general public and the U.S. Government that the threat of epidemics was real and that it would persist. I warned that we had become complacent about such threats, because of the success of antibiotics for the treatment of common pyogenic infections, and the success of vaccines for the prevention of common childhood virus infections such as polio, rubella, and measles. But this general optimism, I cautioned, overlooked the alarms off stage—the rising tide of antibiotic resistance among microbes, the genetic drift of microbes in the evolutionary stream, and the microbial emergence in unexpected places and unexpected times as a consequence of modifications of lifestyle, medical practices, commerce, agriculture, war, and mass transportation.

My book was at best a small factor that led to public and scientific recognition since the late 1980s that infectious diseases are and will remain a persistent challenge. Rather, it was microbial mischief that attracted attention. Microbes conspired with the changing circumstances of our times and fomented a succession of unexpected events; an epidemic of genital herpes, Legionnaires' disease, toxic shock syndrome, Lyme disease, and chloroquin-resistant malaria and DDT-resistant mosquitoes, to name a few. Unlike the yearly budgets from 1970 to 1975, from 1976 to 1981 the NIH budget increased by 30% and the NIAID budget increased 100%. And then came AIDS. All of these events were widely reported in the popular press, which gradually changed the perception about the demise of infectious diseases.

By 1975 the rising tide of antibiotic resistance was well underway. And yet our first experience with bacterial resistance to drugs occurred much earlier, during World War II. Sulfonamide-resistant group A streptococci first appeared at Great Lakes Naval Training Center when this drug was being used on a wide scale to prevent and treat streptococcal pharyngitis. Because recruits who harbored the resistant organisms were constantly deployed out of the Training Center to all parts of the globe, within weeks the sulfa-resistant streptococci were transmitted worldwide to military personnel and civilians. This occurred just before penicillin was widely available, and in the absence of a vaccine, so there was no second line of defense. Fortunately penicillin was waiting in the wings, and soon thereafter production bottlenecks were overcome; however, it was a close call. Had penicillin not become available at just that time there would have been epidemics of streptococcal pharyngitis for which we had no treatment.

The next major threat that caught us entirely unprepared was the appearance of sulfonamide-resistant meningococci in the early 1960s. Before that event, an outbreak of meningitis was nipped in the bud by administering sulfonamide prophylactically to everyone who had close contact with the index case. For example, with the occurrence of one or several cases of meningitis, the children in a city's schools or the recruits on an Army base were all given sulfonamide. The effect of this mass use of sulfonamide was sure and dramatic. An epidemic was prevented. Then suddenly in the early 1960s epidemics of meningococcal meningitis occurred throughout the country and elsewhere in the world, despite the use of sulfonamide prophylaxis. Hard hit were Army recruit training camps. These strains of meningococci turned out to be sulfonamide resistant, and mass prophylaxis was not effective. There was no second line of defense, no second drug for mass prophylaxis programs, and no vaccine. Indeed, we had become so overconfident that only a few U.S. bacteriologists had remained committed to research on meningococci. Fortunately the story does not end with failure.

Between 1965 and 1969 the first successful vaccine against meningococcal meningitis was developed by a group of military physician–scientists at the Walter Reed Army Institute of Research. A novel chemical approach was employed to extract the capsular polysaccharide in a native state and use it in a vaccine formulation (Goldschneider *et al.,* 1971). In recognition of this achievement, the leader of this group, Emil C. Gotschlich, received the Albert Lasker Clinical Research Award in 1979. The work of Gotschlich paved the way for the subsequent success of John Robbins and collaborators (Schneerson *et al.,* 1980) in developing a vaccine for *Hemophilus influenzae* meningitis. This disease of infants and children has now been all but eliminated in the United States and much of western Europe following widespread use of the vaccine.

Embodied in this tale are all the elements of past and future success in confronting emerging infectious diseases: surveillance for early detection; fundamental research in epidemiology, microbiology, immunology, molecular biology, and pathogenesis; and implementation of a strategy to combat the outbreak. All of these matters are discussed in the IOM/NAS report on *Emerging Infections: Microbial Threats to Health in the United States* (Lederberg *et al.,* 1992). The report provides a useful working definition of emerging infectious diseases. They are "clinically distinct conditions whose incidence in humans has increased regionally or worldwide." Further, "Emergence may be due to the introduction of a new agent, to the recognition of an existing disease that has gone undetected. . . . Emergence, or more specifically re-emergence may also be used to describe the re-appearance of a known disease after a decline in incidence." The report has a lengthy, useful table giving examples of emerging virus and bacterial diseases, and those due to protozoa, helminths, and fungi; it notes many examples of emergence for each type of infection. Also noted are the factors that have been important in emergence or reemergence. The report concludes with recommendations for future action to minimize the influence of future emerging infectious diseases.

II. THE DYNAMICS OF MICROBIAL EMERGENCE

A. "Microbial Traffic"

Microbes know no country. There are no barriers to prevent their migration across international boundaries. As recounted by McNeil (1976), this "microbial traffic" is as ancient as civilization. Microbes traveled with people and armies on the Roman roads from the Levant to the English Channel, and on the silk highway from China to Europe, and vice versa. And today, as stowaways, they travel with one million people who fly each day of the year as they circumnavigate the globe at a velocity approaching the speed of sound.

From A.D. 165 to 180 measles was spread along the caravan routes, and from A.D. 251 to 266 smallpox traveled these same trails. One-third of the population died. Such a major catastrophe did not recur until the bubonic plague spread from Asia to Europe in the thirteenth and fourteenth centuries. This occurred when the horsemen of the Mongol armies raced across the steppes of Asia, transmitting the disease from the point of origin in northern Burma. They carried fleas infected with the plague bacillus. From there the plague moved farther westward via the caravan routes to Europe and elsewhere. After 1492, the oceans became highways that further extended the dispersal of disease agents. Through shipment of cargo and transport of people, it became possible for plagues such as smallpox and measles to circle the globe within a year. The oceans remained the predominant route of transmission until the present era of mass air travel. Today, airborne travelers incubating infections can reach any point on the globe in 24 hr. Each day one million passengers board daily domestic and international flights. As a consequence, worldwide exposure to a highly infectious virus, such as influenza, occurs in a matter of weeks.

Much has been written about the introduction of infectious diseases, such as measles, tuberculosis, and smallpox, into the Western hemisphere after the expeditions of Columbus and others who followed soon thereafter. The historian and microbiologist Oswald A. Bushnell (1993) has recounted the arrival of Captain Cook in the Hawaiian islands in 1778, and the diseases introduced by his sailors.

Epidemic contagious diseases, including sexually transmitted diseases (STD), had died out among the isolated Hawaiians following their migration from the Marquesa Islands and Tahiti 2000 years ago. And then epidemics including those of venereal disease followed the arrival of Captain Cook in Hawaii in 1778. Cook, I should say, tried his best to prevent sailors with "the venereal," as he called it, from mingling with the natives. Sailors were not allowed ashore until the ship's surgeon had examined them for outward signs of disease. Most likely venereal infection was latent and undetectable in many of the sailors, but was still transmissible to contacts among the natives. Within a year venereal diseases spread rapidly throughout the Hawaiian Islands. There was no natural immunity in this isolated population, and yet the matter is more complicated than that. The Hawaiian people had not developed cultural taboos about sexuality to contain venereal diseases, because they had no historical experience with them. In *The Gifts of Civilization: Germs and Genocide in Hawai'i,*

Bushnell describes the open sexuality of the Hawaiian people. This is the major reason syphilis and gonorrhea spread like wildfire after Captain Cook's arrival.

> Sexual experience was not in itself an evil thing to Hawaiians. Like all sensible Polynesians, they were rather relaxed about pre-marital and extramarital sexual activities. Commoners at least put no price upon virginity, thought chastity a most unnatural state and seldom evolved a jealous lover . . . Before 1778, their license to experiment and to enjoy the pleasures of sex brought no complications of any kind. Prostitutes were not needed and an illegitimate child did not exist. "A child born too soon" simply joined his mother's family, as welcome a member of the group, as were his siblings . . . [Bushnell, 1993]

The presence of foreigners did not change the attitude of the Hawaiians towards sex. A girl who once or twice lay with a transit mariner was not scorned. It is very possible that with time the Hawaiian people would import sexual restraints or have made "folk adaptations," to use McNeill's phrase, to contain the STD epidemic. But there was no time for this, because tuberculosis, measles, typhoid fever, and other epidemic diseases took their toll. Within a generation, the Hawaiian population was reduced from 300,000 to 30,000. The elders who survived and could look back lamented: *"Ua hala, !a!ole ho!i hou mai."* "They have gone, never to return."

Advances in population genetics and in immunogenetics have led to interesting theories concerning the reasons for the severity of the infections introduced by foreigners to Hawaiians and native Americans. Mortality was very high due to tuberculosis, measles, and much more. One theory holds that the native populations may have been so inbred that they were genetically predisposed to high morbidity and mortality from contagious diseases. The first Hawaiians arrived from the Marquesas Islands and Tahiti 2500 years ago, and it is likely all descendants were derived from a very small gene pool, from perhaps only 100 men and women. This had two consequences.

First, the contagious pathogens they brought with them died out because there were too few susceptible subjects to sustain transmission. A recent example illustrating this point concerns the demise of acute respiratory infections on an isolated island after all contact with the outside world was cut off. Prior to air travel this occurred in 1931 in Spitzbergen after the last boat arrived the end of September. The island was then ice-bound and isolated for the whole winter season. Thereafter, for 6 to 8 weeks, the incidence of acute respiratory disease in this small circumscribed population decreased so that by spring there were very few cases. This decline occurred because the population became immune to a finite number of infectious agents. [From the work of Mogabgab *et al.* (1975), it is known that immunity can develop to a specific common cold virus even though at least 100 serologically different viruses cause this illness. This was observed when a very limited number of serotypes was circulating in a restricted military population.]

The following spring after the arrival of the first boat in Spitzbergen following the breakup of the winter ice, there was a sudden outbreak of acute respiratory disease among the islanders, because "new" pathogens were introduced to which the population was not immune. Frequent cases of respiratory disease persisted throughout the summer as people passed frequently to and fro between the island and the mainland. And then the following fall, the cycle repeats itself. If such a demise in acute respiratory disease could occur in Spitzbergen during 8 months of isolation, it is likely that all of the *contagious* pathogens brought to Hawaii over 2000 years ago were eclipsed in a similar fashion during the following centuries.

The second consequence concerning the common ancestry of the Hawaiian population at the time of Captain Cook's discovery concerns the susceptibility to infectious diseases of an inbred population. Of necessity, the initial small colony of Hawaiians was enlarged by inbreeding. If these assumptions are accepted, then in the course of centuries, by the process known as genetic drift, potential weaknesses or strengths in the ancestral gene pool would have been enhanced among interrelated descendants. Genetic drift, the so-called Sewall Wright effect, refers to the change in a population genotype. It results from the chance sampling errors in gene distribution that arise from the interbreeding of small populations or breeding units. When the gene has reached a frequency of 1 it is said to be fixed. When it reaches 0, it is lost. This distortion of the original gene pool can lead to gene loss or gene fixation. In a small population, an advantageous gene may never get established and a deleterious gene may remain fixed.

As Francis Black (1993) has pointed out, in a lead article in *Science* entitled "Why did they die?," the new world people and the Polynesians have much less heterogeneity in the highly polymorphic loci that control the immune system—the class I and the class II major histocompatibility complex (MHC) antigens—than do the Europeans. He postulates that the measles virus became more virulent with passage from person to person in the inbred population, whereas in an outbred population passage from person to person does not enhance virulence. Kilbourne (1996) has referred to this process as "measles in the new world and the cost of consanguinity."

The Hawaiian experience illustrates the consequences when new infectious diseases are introduced into a society that is not prepared biologically, historically, or culturally to cope with it. The parallel to the human immunodeficiency virus (HIV) epidemic is too obvious to merit further comment.

B. Other Factors Leading to Emergence/Reemergence of Microbial Diseases

In addition to all the avenues of microbial traffic noted above, other factors include changes in human behavior such as medical care, personal hygiene, and sexual practices; ecological disruptions associated with natural disasters, economic development, new patterns of land use, and agriculture; the migration of people from the rural countryside to urban centers; the disruption of public services and health care following natural disasters, civil strife, and war; the shifting relationships among the population dynamics of microbes, vectors, and animal and human hosts; and ecological and climatic changes that foster the multiplication of microbes, vectors, and animal reservoirs. The rhythms of climate and weather influence the occurrence of cholera and malaria, and an outbreak of coccidiomycosis in southern California occurred following the last major earthquake in 1994. This outbreak of pneumonitis occurred because the spores of this soil organism became air-borne in the plumes of dust arising from the turbulance of the shock waves. Refer to Morse (1993, 1995) and Krause (1992) for additional details.

Each of these factors that influence the emergence/reemergence of microbes cannot be discussed here. Each is a chapter or a book in its own right. The authors of the following chapters have noted the particular circumstances that have led to the emergence/reemergence of the diseases under discussion. One demographic statistic is compelling and has now and will have an even greater influence on microbial emergence in the future. In 1900, only 5% of the world's population lived in cities of over 100,000 people. By 2025, 61% of humanity, or more than 5 billion people, will be living in cities, as noted in Lutz (1994). What microbes will fester in these cauldrons, only to boil over as tuberculosis did in the industrial slums of Victorian cities? Louria (1996) has reviewed the societal variables that influence the emergence of microbial diseases.

But we need not wait until 2025 for unexpected diseases to emerge in our cities. Recall the outbreak of cryptosporidia gastroenteritis in Milwaukee in 1994. Over 400,000 people were afflicted. The source: a polluted watershed caused by the runoff of manure from nearby cattle holding pens, and a waterworks filtration system not equipped to kill cryptosporidia. Water purification methods designed to eliminate *Escherichia coli* and typhoid are not sufficiently stringent to kill cryptosporidia. These matters are discussed in this volume by Mahmoud (1997).

In retrospect we should have been warned about possible danger from cryptosporidia by the unexpected occurrence in 1982 of cryptosporidia diarrhea in AIDS patients. We should have suspected that there was a weak point in either our food chain or water supply. The early AIDS patients with cryptosporidia diarrhea were like the canaries employed in coal mines to detect poisonous gas. Those early AIDS patients sounded the alarm, but we did not take the warning, and the Milwaukee outbreak was the result. The take-home lesson: we must be ever alert to the new routes that microbes discover to enter the water supply or food chain.

The historian William McNeil (1976) has observed that throughout history folkways have been adopted that allowed people to coexist with infectious diseases. They have adopted a pattern of life which does not eliminate the disease, but contains it to a tolerable extent. I learned an example of this adaptation when I was in Japan during the height of the *E. coli* outbreak in the summer of 1996 that caused hundreds of cases of severe diarrhea and dozens of cases of severe hemolytic–uremic syndrome. I recall the remarks of Dr. Hiroo Imura, a physician and President of Kyoto University; when he was a boy his mother imposed a ban on eating raw fish (sushi) or raw meat from May until October. Many contemporary Japanese and visiting Americans have disregarded this warning. And, as recorded elsewhere in this volume, toxic *E. coli* have entered the food chain and the water supply in the United States. A few years ago *E. coli* gastroenteritis in the northwestern states was traced to hamburger that had not been thoroughly cooked, and most recently to nonpasteurized apple juice. Will we never learn?

C. Microbial Factors that Influence Emergence/Reemergence

Thus far, I have focused on various environmental factors that enhance microbial emergence, and these derive from what people do or what Nature does. And yet, microbes are not idle bystanders, waiting for new opportunities offered by human mobility, ignorance, or neglect. New viruses may occur as a consequence of genetic mechanisms such as point mutations and genetic reassortment. These and other genetic mechanisms that contribute to microbial emergence will be noted in the following chapters. And yet there are restraints on genetic changes that stem from the need to preserve the structural integrity of the microbe and its relation to a specific ecological niche. This accommodation has evolved over time and will not be abandoned lightly. But it can and does happen. The emergence of new microbes or old editions in new garments stems from genetic evolution. Microbes (bacteria, viruses, parasites, fungi) possess remarkable genetic versatility that enables them (under favorable circumstances) *to develop new pathogenic vigor; to escape population immunity by acquiring new antigens; and to develop antibiotic or drug resistance.* Moreover, vectors are equally facile in circumventing barriers raised to contain them. In the case of bacteria, this genetic versatility is enhanced by "gene traffic," the movement of genes via the "transforming principle," plasmids, and bacteriophage, *within* and *between* microbial species. The rising tide of antibiotic resistance is in large measure the consequence of such gene traffic. It is likely that this gene traffic has played a role in the emergence of streptococcal toxic shock syndrome, as discussed in this volume by Musser and Krause (1997).

One example of a sudden epidemic outbreak of an infectious disease that is the consequence of microbial genetic events concerns an outbreak of diphtheria in Manchester, England (Pappenheimer and Murphy, 1983). The resident non-toxin-producing diphtheria microbe (which did not cause disease) was converted to a toxin-producing, virulent form. The origin of this conversion was traced to a child who had become infected with a pathogenic toxin-producing diphtheria bacillus in Africa prior to returning by plane to Manchester. This African organism contained the phage that carried the diphtheria toxin gene. Because multiple infections allowed toxigenic diphtheria bacilli to commingle in the Manchester community with nontoxic diphtheria bacilli, the toxin gene was spread to the latter by phages that possessed the toxin gene.

D. The Diversity of Microbial Epidemics

Microbial epidemics display various patterns of emergence and decline. Such diverse behavior derives from the previously described ecological factors that influence emergence, the infectious properties and genetic adaptability of microbes, and the status of herd immunity. This is well illustrated by the striking contrast between the one major epidemic of polio in the twentieth century and the repetitious rapid fire epidemics of influenza during the last several centuries.

Endemic polio has been around for a long time, for centuries, if not millennia. A stone relief of an Egyptian with a paralyzed leg supported by a crutch is often used to document the occurrence of polio in ancient Egypt. Unlike the periodic genetic shift and drift of the influenza virus, we know that the three serotypes of poliovirus have been pretty stable for the past 40 years. We know this because the vaccine containing the three serotypes has been so successful in preventing paralytic polio.

During the twentieth century polio emerged from an endemic disease to an epidemic one, and the possible factors that led to this emergence deserve comment. One of the first reports of a rare polio outbreak in the United States was in Rutland, Vermont, in 1894, which was published by Caverly (1894) in the first issue of the *Yale Medical Journal*. It is worth quoting from the first two paragraphs of his paper.

> The disease prevailed chiefly in the city of Rutland up to about the middle of July, when other towns about this city began to report cases. . . . From my own observation and conversation with other physicians, and the general feeling of uneasiness that was perceptible among the people in regard to the *new disease* [emphasis added] that was affecting the children, I determined during the last of July to undertake a systematic investigation of the outbreak. . . . This investigation . . . soon convinced me that this region had been affected by an epidemic of a nervous disease *very rarely observed* [emphasis added].

The rest is history. The polio epidemic in the United States gained momentum and peaked in the 1940s and early 1950s, as recounted by Robbins (1994). What brought on the polio epidemic in the twentieth century? Many factors surely, but most of our own doing. Civilization, our way of life and so on, was changing in this century—a population increase with a demographic shift from rural to urban centers—and those changes fostered the polio epidemic. Ironically, perhaps, the greatest culprit was sanitation improvements. This led to the postponement of the first exposure to the virus until the second and third decades, whereas throughout the prior millennia infection in early life resulted in a less paralytic form of the disease. Such early exposure to the virus resulted in lifelong immunity against subsequent infection that was more likely to be paralytic.

Why did the polio epidemic build to a crescendo in the 1940s and early 1950s? One possibility it seems to me is the cumulative effects of World Wars I and II on disruption of the social order. Never in history had there been such worldwide disruption of civil life, with large armies on the move. In World War II, the United States had 15 million men and women under arms who were sent to and returned from the four corners of the globe. The combined armies of the Allies and Axis powers exceeded 50 million. Millions of refugees, fleeing the ravages of war, migrated hither and yonder. Malnutrition and starvation were widespread. Such chaos and turbulence enhance the emergence of infectious diseases. There is an old adage in the folklore of the Mediterranean people—"Malaria flees before the plow"—to which I have added Krause's corollary: "And it returns on the winds of war." And so it did. During World War I, malaria occurred north of the Arctic Circle for the first time in European history. Just so, I believe the polio epidemic rose to a crescendo in the 1940s and 1950s as a consequence of the chaotic events generated by World Wars I and II. By 1955 the epidemic was waning, even before the introduction of the Salk and Sabin vaccines.

Through the use of mass vaccination programs, polio has been eliminated in the western hemisphere, and the omens are good that this will be achieved in Asia in the next few years. Polio may join smallpox and become the second microbial dinosaur.

Contrast the decades long buildup of the polio epidemic in this century to the unpredictable but periodic reoccurrence of influenza epidemics, as noted in this volume by Webster (1997). Major influenza pandemics have occurred periodically for centuries. In each case this has been due to a "new" virus, a new genetic recombinant to which we were not immune. We live in the shadow of the 1918 influenza pandemic. Perhaps a total of 100 million people died. The population was scared half to death. I remember my mother recalling the measures she took to protect my two older brothers during that time. She used hot boiling water to decontaminate anything that might carry the invisible germ, including the spot on the front sidewalk where a passerby would expectorate sputum. And the demographics of that epidemic make clear how frightful it was. There was a high death rate in those 20 to 40 years of age. In contrast, the typical flu epidemic is most severe for the very young and the very old. So the 1918 epidemic was due to a strain of flu which was unlike any preceding virus that occurred during the prior 30 to 40 years. Thus, 20 to 40 year olds lacked immunity to the 1918 flu.

Influenza is one of the best examples of a *new* virus that can cause an epidemic. It is bred from a novel recombination of the segmented genomes of two parents. And yet how fast the new virus circumnavigates the globe depends on us—how many of us travel, how fast, how far, and how wide. No single mass immunization campaign will curtail the appearance of successive epidemics of influenza, as is the case for polio. For the foreseeable future the best we can do is keep the vaccine up to date with components that match as closely as possible the new virus when it emerges. And the detection of a new virus depends on worldwide surveillance systems. All of this was brought into focus in 1975 when the influenza community was alarmed over the occurrence of swine flu at Fort Dix, New Jersey. One Army recruit died, and at least 500 to 1000 were infected with a virus that was *serologically similar* to the putative virus that cause the 1918 epidemic. The swine flu virus reacted serologically with sera of people who had lived through the 1918 flu pandemic, and therefore had antibodies to that virus. Was the 1918 flu virus or a close relative making a long awaited return?

After much consultation and discussion at the highest levels of the U.S. Government, in which I was involved (having become Director of NIAID only 6 weeks before), and after extensive consultation with influenza and infectious disease experts, the Public Health Service launched a crash program to manufacture sufficient swine flu vaccine to immunize over 50 million people. This was achieved during voluntary mass immunization in October, 9 months after the alarm was sounded. The epidemic did *not* occur, however; it turned out to be a false alarm, and the public and much of the scientific community accused us of overreaction. As someone noted, it was the first time we have been blamed for an epidemic that did not occur.

It is worth noting that there is an upside to the swine flu affair. In some ways it was one big fire drill. We proved it was possible to organize a mass influenza immunization program from start to finish: identify the virus, grow up stocks, prepare and field test the vaccine, and immunize a large segment of the population, all within 9 months. We learned a great deal from that drill, and I am sure we can do better next time. The day will come when we will again retrace this race against time.

I have called the uncertainty that surrounds any response to a microbial outbreak the Fog of Epidemics, analogous to the Fog of War of which the historians speak.

The Fog of War: Uncertainty [1]
    Where is the enemy?
    What is his strength?
    What counterattack?

The Fog of Epidemics: Uncertainty
    Where is the microbe?
    How many; how virulent; how communicable?
    What counterattack?

Perceived Miscalculations
    1975   Swine flu outbreak
            Response too rapid
    1981   HIV/AIDS occurrence
            Response too slow

In the case of swine flu, we may have acted too soon. And in the case of AIDS, we may have acted too slowly. Read the book authored by Neustadt and Fineberg (1978) entitled *The Swine Flu Affair: Decision-making on a Slippery Disease* for a full account of our perceived folly in regard to swine flu. For an account of the perception that, as Director of NIAID, I dithered and dithered over the onset of HIV, read what Shilts (1987) says about me in *And the Band Played On*. I shall return to that matter later.

I relate these personal reminiscences because many who read this book will be on the firing line when future epidemics threaten, and they will either erupt or fizzle out. You will be in a fog, and you will need to exercise the best judgment you can on the basis of an assessment of the surveillance information. Roy Anderson and others (1996) are currently on the firing line in the United Kingdom as they advise on policies to contain bovine spongiform encephalopathy (BSE) in cattle and the possible transmission to humans. No matter what you do, you will be criticized. But as President Truman said, "If you can't stand the heat, stay out of the kitchen."

To repeat, the contrast between the repetitive rapid fire epidemics of influenza and the one major epidemic of polio in the twentieth century, which rose to a crescendo over a period of several decades, reveals the diversity of epidemics due to different microbes. Contributing to this diversity, on the one hand, are the genetic variables of the microbe and, on the other hand, the various environmental and societal factors mentioned earlier that either favor or impede transmission. Because of this diversity and complexity, no single strategy can be employed to intervene or contain epidemics. But on one matter there is common agreement. There must be a worldwide network of surveillance to detect, at the earliest possible moment, the emergence of an infectious disease, *and* we must have the scientific knowledge to diagnose, treat, and prevent it.

In recent times, the description of and predictions about epidemics have been expressed in the elegance of mathematical formulations, reviewed in this

---

[1] One of the cited examples by war historians is General Lee's uncertainty concerning the whereabouts of the Union Army before the battle of Gettysburg. Jeb Stuart, who led Lee's cavalry, was on a scouting mission of his own riding around the Union Army, and during this time he and Lee were not in communication. The cavalry was a major source of military intelligence in those days. Lacking intelligence briefings from Stuart, Lee was in a fog.

volume by Anderson (1997). These calculations have added new power to the containment strategies employed by public health agencies to counter an epidemic, including measures as diverse as vaccination programs for measles and rubella, targeting the use of condoms by risk groups to control the spread of gonorrhea and HIV, and a strategy to contain the epidemic of BSE in cattle in the United Kingdom.

We should not forget that these powerful mathematical models on the dynamics of epidemics have their origins in the nineteenth century. They arose from the growing interest in vital statistics in the 1830s, a time when purely political discussion gave way to statistical analysis in the search for solutions to public health issues. Foremost among those social Victorians with a statistical flair was the physician William Farr. His contributions and his collaboration with Florence Nightingale to reform sanitation and much more has been recorded by Eyler (1979) and others.[2] Farr was attracted to medical statistics as a young physician. The purpose for statistics, a word derived from the German word *Staat* (meaning state), was to bring together facts that would illustrate the conditions and prospects of society—the state of the state, or health of the state, if you will. Farr was an idealistic physician who used statistics to advance the general welfare. For these reasons, Victorian statistics became a science for social reform, particularly in the hands of Florence Nightingale, who almost single-handedly reformed the War Office after the medical disaster of the Crimean War.

Like the mathematical modelers of today, Farr developed a mathematical formula of his own to explain the relationships of factors that enhance the occurrence of cholera and other infections. Farr was best known for his study of epidemics. There is general agreement that he was the first person to develop a numeric description of an epidemic. He did this by extending the actuary's technique of discovering the law of mortality in a life table, and this he applied to smallpox epidemics. To quote: "It appears probable, however, that the smallpox increases at an accelerated rate and then at a retarded rate; that it declines first at a slightly accelerated rate, then at a rapidly accelerated, and lastly at a retarded rate until the disease attains the minimum intensity and remains stationary." A plot of the number of deaths at each quarter during the epidemic forms a normal curve. From the historical record, Eyler (1979) notes that Farr did not perceive this and did not plot the data on an $x-y$ axis. And yet using the law derived from the observations of actual number of cases, he was able to predict the course of an epidemic. For example, by knowing the rate at which smallpox had increased in the first two quarters of a prior epidemic, Farr was able to predict the rise of the current epidemic of smallpox.

	Number of smallpox deaths for time period			
	1	2	3	4
Registered deaths	60	104	170	253
Farr's calculation of deaths	60	99	163	267

[2] In this discussion about Farr and Nightingale, I have drawn also on the accounts of these events in *Eminent Victorians* by Lytton Strachey, *Creative Maladies* by George Pickering, and *The Reason Why* by Cecil Woodam Smith.

His conclusion: Recommend universal smallpox vaccination. For Farr, success in predicting the course of the epidemic was not an end in itself. He was very critical of the Poor Laws for failure to authorize vaccination programs. Farr was pragmatic. He was well grounded in mathematical theory, but for him mathematics was a means to an end. I feel some kinship with Farr. I am a physician with a baccalaureate degree in mathematics.

E. Reflections on the HIV/AIDS Epidemic

As Director of NIAID during the first 3 years of the epidemic, what did I do, and why did I do it? Remember that in 1982 AIDS was referred to as the 4 H disease (for Haitians, heroin addicts, hemophiliacs, and homosexuals), a pejorative designation, and one we should not be proud of. In 1983, Thomas Quinn, Clifford Lane—both NIAID Public Health Service officers—and I went to Haiti to learn what we could about AIDS in the Haitian people. Why Haiti? Haiti, it seemed to me, was the key. We learned two clinical facts; first we learned that 40% of the patients were women. I said then that AIDS would become a heterosexual venereal disease. And so it has. Second, we learned that 30% of the patients had tuberculosis as the AIDS-defining opportunistic infection before that was recognized here in the United States. Again, I recall musing about what this would mean as the AIDS epidemic gained momentum in Africa and Asia, where tuberculosis was so prevalent.

We also learned about historical events that influenced the spread of HIV out of Africa. We learned that thousands of Haitians had gone to Zaire in the 1960s to take managerial jobs left vacant by the departing Belgians. In the 1970s, because of political events, the Haitians were required to leave Zaire. They migrated back to Port-au-Prince, and to New York, to Paris, to Montreal, cities in which AIDS was first reported. For some time after they left Zaire, the virus was quietly multiplying in their lymphoid tissue. In this sense they represented a modern day Trojan horse.

From that Haitian experience two decisions were made. First, we established Project SIDA in Kinshasa, Zaire, to learn what events led to the widespread occurrence of AIDS in that country. Funds, personnel, and resources were committed to this effort, and Dr. Anthony Fauci, who succeeded me as Director of NIAID, continued that commitment. Quinn and Fauci (1997) in this volume summarize what was learned during 7 years at Project SIDA, and by others in Uganda and Kenya, about the factors in Africa that fueled the AIDS epidemic. I would mention six: (1) migration from rural areas, where the infection was endemic, to urban centers; (2) separation of men from families; (3) the rise of prostitution; (4) other STDs as a risk factor in transmission; (5) the disruption of health services; and (6) the perinatal transmission from HIV-positive mothers to the newborn. These demographic and social factors unleashed the virus, which had previously been confined to the outback of Central Africa, and where in all probability it had jumped from primates to humans several centuries ago.

The high rate of perinatal transmission of HIV from pregnant women to the newborn offered an opportunity to disrupt transmission by use of drugs and immune serum, and there are reports on success with the use of zidovudine

(AZT). Other studies are underway in Uganda, championed by Frederick C. Robbins, to prevent neonatal transmission with HIV-IgG. The hyperimmune HIV-IgG has been collected from the plasma of HIV patients who have a high concentration of neutralizing antibody. If HIV infection is prevented in the newborn infants, then vaccine development should focus on the HIV antigen(s) that stimulates the neutralizing antibodies in the plasma of HIV-infected patients. Relevant to this effort is the earlier success in the prevention of the transmission of hepatitis B from infected mothers to newborn infants with immunoglobulin, as reported by Beasley et al. (1983). The immunoglobulin was given to the infants in the neonatal period, and this was very effective in preventing transmission. Such success implies that a vaccine, to be effective, must also stimulate neutralizing antibodies. This has been achieved with the hepatitis B vaccine.

It is worth recalling in the history of vaccine development that success in passive immunity has been the key element in devising a successful strategy to develop a vaccine for active immunization. In each case success with passive immunity led the way; for tetanus, diphtheria, rubella, poliomyelitis, meningococcal and H. influenzae meningitis, pneumococcal pneumonia, and hepatitis A and B. *It turns out that the microbial antigen that stimulates convalescent humoral immunity following an infection is identical to the vaccine antigen that induces a protective humoral immune response.*

To be successful a vaccine must stimulate humoral immunity in which the specific neutralizing antibody concentration reaches a level of 4×10^{12} antibody molecules per milliliter of serum. Such a high number of antibodies insures an instantaneous rapid fire bombardment of the infectious microbe at the point of invasion, the portal of entry, or in the subsequent dissemination into the bloodstream. The antibody–microbe interaction leads to a rapid cascade of events that includes phagocytosis of the microbe. For bacteria, the pathway of *in vivo* microbial killing in phagocytic cells has been well worked out by W. Barry Wood, Christian de Duve, Zanvil Cohn, and James Hirsch, and more recently in biochemical terms by Seymor Klebanoff and others, as reviewed by Densen and others (1995). The *in vivo* antibody-initiated processes that result in virus degradation in white blood cells and macrophages are less clear. A stimulation of T-cell immunity is also a requirement for a successful vaccine. But T-cell immunity alone cannot initiate the processes of microbial neutralization at the time and place of the inoculum, and the clearance of the microbe from the blood stream. This requires specific neutralizing antibodies.

It has been estimated that a patient infected with HIV produces 1×10^9 virus particles per day, as reported by Perelson et al. (1996). If all the virus particles are infectious, within 1000 days sufficient virus will have been produced to infect all of the lymphocytes in the adult human, approximately 2×10^{12} cells. Clearly these calculations are not precise, and there is a wide margin for error in these estimates, in large part because not all of the virus particles are infectious, and throughout the course of the disease young lymphocytes are generated to replace those that are destroyed. Both pharmacological and immune mechanisms are needed to decrease or terminate virus production. It is equally important, however, that the 1×10^9 virus particles produced each day be neutralized and cleared from the blood by circulating and fixed macrophages so that they do not infect other lymphocytes. As noted above, at least 4×10^{12}

molecules of antibody per milliliter of blood are needed to neutralize and clear the viremia. In the late stages of HIV infection antibody level falls, often below detectable levels, and for this reason both pharmocological and immunological methods are being employed to maintain an effective antibody concentration.

One important fact should be kept in mind in developing strategies for the treatment and prevention of HIV with either active or passive immunity. *The serum antibody concentrations required for treatment are 10 to 100 times higher than those required for prevention.* There is a large body of literature substantiating this fact, dating back to the successful treatment of pneumococcal pneumonia with antiserum, as reported by Avery *et al.* (1917) and reviewed by Casadevall (1996) and Robbins *et al.* (1995).

The second decision I made in regard to AIDS was conditioned by my experience with rheumatic fever for many years at Rockefeller University Hospital. Very much on my mind were the long-term studies on the *natural history* of rheumatic fever by Jones and Bland (1952). For 20 years they observed the progression to severe rheumatic heart disease and death. Many patients, however, improved over time or remained stable. Some lost all stigmata of heart disease. With so little known at the time about the clinical course of AIDS, it was clear to me that we needed long-term studies on the natural history of HIV and HIV infections.

In 1983, we started such an effort that came to be known as the MACS Study (the Multi-Center AIDS Cohort Study). The plan was straightforward; the execution, time-consuming and costly: enroll 1000 HIV-positive and -negative gay men and observe what happens to them in the years, even decades ahead, providing such medical care and advice on advoidance of risk behavior that was acceptable practice at the time. This was the only way we could learn about the natural history of the disease. When and how did it start? What was the duration? Were there different clinical patterns? And so on.

One of the important discoveries that emerged from the MACS Study was the recognition of rapid progressors and long-term survivors. Perhaps 10% or more of HIV-positive patients have lived 10 to 15 years and have not yet developed AIDS. The infection was kept under control, in part, because there was sufficient neutralizing antibody in the blood to clear the viremia. In the rapid progressors, the virus multiplies rapidly in the lymphoid tissue, and the immune system decays. So the search is on, through pharmacology and immunology, to halt virus replication and postpone, perhaps forever, the onset of AIDS in HIV-positive patients, to duplicate with pharmacology and immunology what Mother Nature has achieved on her own in the long-term survivors.

Now what did I neglect to do? Reflection on these matters might be useful for those who will confront the next emerging infectious disease. I neglected to go to the autopsy room and see the gross and microscopic pathology on patients who died with AIDS. How stupid that was. As a physician, I surely knew better. I even overlooked a January 1985 letter to the Editor of *Lancet* by P. Racz, an old time pathologist, a Hungarian refugee working in Hamburg (1985). He reported HIV-like virus particles in the cells of lymphoid tissue as seen by electron microscopy. So as early as 1983–1984 there were pathologists who knew that the HIV infection was a *productive infection with virus multiplication* in the lymphoid tissue and with death of lymphocytes and dendritic cells. HIV was not just a latent infection with its genome folded into the human genome. HIV was a chronic, productive infection. This has become an active area of research pursued by Anthony Fauci, Ashley Haase, David Ho, and others, but I should have tumbled to that recognition earlier and mobilized research efforts to combat an active infection.

III. STRATEGIES FOR THE FUTURE

A series of thoughtful recommendations were included in the IOM/NAS Report on Emerging Infectious Diseases (Lederberg *et al.*, 1992), and the U.S. Centers for Disease Control and Prevention (CDC) and NIAID have developed initiatives that buttress these recommendations. In developing strategies to confront the emerging infectious diseases, there are lessons to be learned from the historical record of past achievements. Recall again the work of William Farr and Florence Nightingale which I have already noted.

A word more needs to be said about Florence Nightingale, a forceful and dynamic figure for change—not reflected in the portrait of "the lady with the lamp"—and her subsequent collaboration with William Farr which changed forever sanitation, hospital practices, and nursing care. For Florence, all of this began with her experience in health care during the catastrophe of the Crimean War, where medical care for the sick and wounded was atrocious.

It was to this chaotic scene that Florence Nightingale, at the invitation of the British Government, took 38 nurses. Once in Scutari, however, she was ignored by hostile and suspicious British doctors. But Miss Nightingale persisted. She had anticipated the sorry state of affairs of Scutari and had made numerous purchases en route. Her nurses were put to work sorting these supplies and putting them away. Gradually and firmly she gained ground on both the hostility of the doctors and the surrounding chaos of the hospital.

Supplies were pathetically low. Only Miss Nightingale seemed to know what was needed and how to go about getting it. She wrote a friend: "I am kind of general dealer in socks, shirts, knives, and forks, wooden spoons . . . cabbages and carrots . . . operating tables, towels and soap . . . scissors, bed pans, and stump pillows." She ordered the lavatories and the wards cleaned, and she rented a house in which she had patients' clothes washed by soldiers' wives. "The strongest will be wanted at the washtub," said Florence. (In Strachey's view, "it was upon the washtub, and all that the washtub stood for, that she expended her greatest energies.")[3]

---

[3] Certainly Farr and Nightingale deserve much credit for the changes that took place in social medicine in the nineteenth century, but, in all fairness, there was a rising tide of interest in sanitation, hygiene, and public health practices to improve health. Semmelweiss and Oliver Wendell Holmes come to mind in regard to the prevention of childbed fever. Less well known are the earlier practices of Captain Cook to prevent diseases, including scurvy, among the sailors under his command of His Majesty's Ship *Resolution* during a 3-year voyage.

Cook described his formula for success in disease prevention (and in today's jargon, promotion of "wellness") in the *Philosophical Transactions of the Royal Society* (1776): "As many gentlemen have expressed some surprise at the uncommon good state of health that the crew of *The Resolution* under my command experienced during her late voyage; I take the liberty to communicate to you the methods that were taken to obtain that end." After giving all due credit to the importance of sauerkraut and lemons for the prevention of scurvy, he goes on to review the sanitary practices

Her reports and recommendations transformed the War Office, the British Army, the Indian Army, sanitation, hospitals, nursing, the Poor Laws, and the scientific disciplines of epidemiology and statistics. In here efforts to benefit the "sick poor"—whether they were British soldiers or mothers in lying-in hospitals—Florence Nightingale despaired at the intransigence of those in power. Again and again she had to work through and around those who were blocking her efforts to reform the War Office. She had great difficulty, particularly with Mr. Gladstone, who became Prime Minister in 1868. She wrote to a friend, "One must be as miserably behind the scenes as I am to know how miserably our affairs go on. . . . What would Jesus have done," she asked plaintively, "if he had had to work through Pontius Pilate?"

Early on, Miss Nightingale learned to master the arcane ways of government. By the time she finished her report, she realized that a report, by itself, would solve nothing. As she wrote, "Reports are *not self-executive* [italics added]." So she had subcommissions created to *act* on the reports. Chief among these subcommissions was the one that founded a statistical department for the Army (epidemiology, in other words). When the work of these subcommissions was accepted by the Prime Minister, Miss Nightingale had won the battle. Let

---

which he imposed with great attention to detail and execution. Where possible he obtained fresh vegetables, "to make nourishing and wholesome messes and was the means of making the people eat a greater quantity of greens than they would have done otherwise." His plan? "The crew were at three watches—by this means they were not so much exposed to the weather as if they had been at watch and watch: and they had generally dry clothes to shift themselves when they happened to get wet. Proper methods were employed to keep their persons, hammocks, bedding, clothes, etc., constantly clean and dry. Equal pains were taken to keep the ship clean and dry between decks—cleanliness as well in the ship as amongst the people. Too great attention cannot be paid; the least neglect occasions of putrid, offensive smell below—these smells will be attended with bad consequences."

The fat that boiled out of the salt beef and pork he never used, but discarded. "I never failed to take in water wherever it was to be secured, even when we did not seem to want it, because I look upon fresh water from the shore to be much more wholesome than that which has been kept for some time on board. For this essential article we were never at an allowance, but had always abundance for every necessary purpose. I am convinced that with plenty of fresh water and close attention to cleanliness, the ship's company will seldom be afflicted with the scurvy." He closes by noting that these were the methods instituted over 3 years and 18 days in all climates from 52° north to 71° south, with the loss of only four men; two were drowned, one killed by a fall, and one died from a complicated and lingering illness without the signs of scurvy.

It is striking to me that there is no mention of death or serious illness due to infectious diseases. It would be of interest to learn, through research into the historical record, if Cook's ship was unique in the British navy at that time in its attention to a strict sanitary code and good nutrition. It seems likely that the sailors of his ship would have had less severe diarrhea than was common in the navy at large. The voyage of George Anson (1780) stands in stark contrast to that of Captain Cook: 1300 seamen out of 1955 died due to or with scurvy, and it is likely death was hastened by various infections.

For his astonishing achievement in maintaining health, and particularly for the scientific merits of his exploits in the far reaches of the world, Captain Cook was elected a Fellow of the Royal Society in 1776. Among the 28 of those who signed the certificate of election were John Hunter and Henry Cavendish. In that same year he was awarded the Copley Medal for his paper, "Giving an Account of the Method He Had Taken to Preserve the Health of the Crew of HM Ship *The Resolution* During Her Late Voyage Around the World." So he was Captain James Cook, R.N., FRS. Bushnell, who has reviewed the history of that era, says of Cook, "He was generous, enlightened, observant, far in advance of contemporary naval officers in his regard for the health of his seamen." It is ironic that the infamous Captain Bligh served under him as a younger man.

this lesson not be lost on all those who eagerly press for one report after another. Reports are not self-executive. We today should heed her admonition from the very beginning that a report is being planned. Early thought should be given to the implementation of the report's recommendations.

What's to be done? It is likely that epidemics will continue to occur in the future as they have in the past. Changes in human social behavior and customs will continue to provide opportunities for microbes to produce unexpected epidemics. We must be aware of the possible consequences of altering our behavior individually and as a society, discussed more fully by Plotkin and Kimball (1997). Furthermore, science cannot halt the future occurrence of "new" microbes. These emerge from the evolutionary stream as a consequence of genetic events and selective pressures that favor the "new" over the "old." It is Nature's way. Therefore, public health requires the renewal and expansion of research on the epidemiology and biology of microbes, vectors, and intermediate hosts, and awareness to the possibility that new epidemics can and will emerge in unexpected places. Although clinicians in New York and California detected the first cases of AIDS, it was the nationwide surveillance system of the CDC that identified the epidemic nature of the disease and its mode of transmission. Such cooperation between astute, observant clinicians and epidemiologists was also true for toxic shock syndrome, Legionnaires' disease, Lyme disease, and the new strains of cholera.

It is necessary to expand surveillance efforts both in the United States as well as in other regions of the world. Surveillance alone, however, cannot detect the unexpected emergence of future plagues or prepare the defense against them. The accomplishments thus far in the fight against AIDS are based on decades of biomedical research. Research is needed into the survival of microbes, vectors, and intermediate hosts, and their adaptation to new habitats. It is likely we will have much to learn from population and evolution biologists, as discussed in detail by Ewald (1996). In addition, it is necessary to understand the genetic makeup of microbes and their ability to cause disease, including mechanisms of pathogenesis as well as the immunological processes that are mobilized by the body against microbial invasion and infection (National Institute of Allergy and Infectious Diseases, 1992).

In summary, a complex matrix of social, economic, political, and ecological factors play a major role in the emergence of epidemics due to microbes with which the human population has coexisted for centuries, if not millennia. Throughout history epidemics have followed on the heels of migration of people along the highways of commerce. But beyond these factors that contribute to the emergence of new infectious diseases we must also recognize changes in microbial agents, human populations, insect vectors, and the ecological relationships among them. This is true for influenza, for polio, for AIDS, and for the other infections described in this volume. A new microbe as the cause of an epidemic is far less common, but there are examples such as influenza that emerge as a consequence of genetic events that enhance virulence, pathogenicity, or escape from herd immunity. These various factors must be considered in the development of strategies to cope with future epidemics.

Finally, microbes and vectors swim in the evolutionary stream, and they swim much faster than we do. Bacteria reproduce every 30 minutes. For them, a millennium is compressed into a fortnight. They are fleet afoot, and the pace of our research must keep up with them, or they will overtake us. Microbes were here on earth 2 billion years before humans arrived, learning every trick for survival, and it is likely that they will be here 2 billion years after we depart.

To confront the challenges of emerging microbes:

Let thy Studies be as free as thy Thoughts and Con-
templations, but fly not only upon the wings of
Imagination; Joyn Sense unto Reason, and Experi-
ment unto Speculation, and so give life unto Embryon
Truths, and Verities yet in Their Chaos.

Christian Morals *(1716)*
Sir Thomas Browne

REFERENCES

Anderson, R. M. (1997). Analytical Theory of Epidemics. *In* "Emerging Infections" (R. M. Krause, ed.), pp. 23–49 (this volume). Academic Press, New York.

Anderson, R. M., Donnelly, C. A., Ferguson, N. M., Woolhouse, M. E. J., Watt, C. J., Udy, H. J., MaWhinney, S., Dunstan, S. P., Southwood, T. R. E., Wilesmith, J. W., Ryan, J. B. M., Hoinville, L. J., Hillerton, J. E., Austin, A. R., and Wells, G. A. H. (1996). Transmission dynamics and epidemiology of BSE in British cattle. *Nature (London)* **382,** 779–788.

Avery, O. T., Chickering, H. T., Cole, R., and Dochez, A. R. (1917). Acute lobar pneumonia; prevention and serum treatment. Monograph 7, pp. 1–109. Rockefeller Institute for Medical Research, New York.

Beasley, R. P., Hwang, L.-Y., Stevens, C. E., Lin, C.-C., Hsieh, F.-J., Wang, K.-Y., Sun, T.-S., and Szmuness, W. (1983). Efficacy of hepatitis B immune globulin for prevention of perinatal transmission of hepatitis B virus carrier state: Final report of randomized double-blind, placebo-controlled trial. *Hepatology* **3,** 135–141.

Black, F. L. (1993). Why did they die? *Science* **258,** 1739–1740.

Bushnell, O. A. (1993). "The Gifts of Civilization: Germs and Genocide in Hawai'i." Univ. of Hawaii Press, Honolulu, p. 188.

Casadevall, A. (1996). Crisis in infectious disease: Time for a new paradigm? *Clin. Infect. Dis.* **23,** 790–794.

Caverly, C. S. (1894). Preliminary report of an epidemic of paralytic disease occurring in Vermont, in the summer of 1894. *Yale Med. J.* **1,** 1.

Cook, J. (1776). The method taken for preserving the health of the crew of His Majesty's Ship *The Resolution* during her late voyage round the world. *Philos. Trans. R. Soc.* **66,** 402–406.

Densen, P., Clark, R. A., and Nauseef, W. M. (1995). Granulolytic phagocytosis. *In* "Principles and Practices of Infectious Diseases" (G. L. Mandell, J. E. Bennet, and R. Dolm, eds.), pp. 78–101. Churchill Livingston, New York.

Ewald, P. W. (1996). Guarding against the most dangerous emerging pathogens: Insights from Evolutionary Biology. *Emerging Infectious Diseases* **2,** 245–257.

Eyler, J. M. (1979). "Victorian Social Medicine, the Ideas and Methods of William Farr." Johns Hopkins Univ. Press, Baltimore.

Goldschneider, I., Lepow, M. L., and Gotschlich, E. C. (1971). Immunogenicity of the group A and group C meningococcal polysaccharides in children. *J. Infect. Dis.* **125,** 509–519.

Jones, T. D. and Bland, E. F. (1952). The natural history of rheumatic fever. *In* "Rheumatic Fever" (L. Thomas, ed.), pp. 5–16. Univ. of Minnesota Press, London.

Kilbourne, E. (1996). The emergence of "emerging diseases": A lesson in holistic epidemiology. *Mount Sinai J. Med.* **63,** 159–166.

Krause, R. M. (1981). "The Restless Tide: The Persistent Challenge of the Microbial World." The National Foundation for Infectious Diseases, Washington, D.C.

Krause, R. M. (1992). The origin of plagues: Old and new. *Science* **257**, 1073–1078.

Lederberg, J., Shope, R., and Oaks, S. C. (1992). Report: Institute of Medicine. "Emerging Infections: Microbial Threats to Health in the United States." National Academy Press, Washington, D.C.

Loefler, I. J. (1996). Microbes, chemotherapy, evolution, and folly. *Lancet* **348** (9043), 1703–1704.

Louria, D. B. (1996). Emerging and re-emerging infections: The societal variables. *International Journal of Infectious Diseases* **1**, 59–62.

Lutz, W. (1994). The Future of World Population. Population Bulletin No. 49. Population Reference Bureau, Inc., Washington, D.C.

McNeil, W. (1976). "Plagues and Peoples." Doubleday, New York.

Mahmoud, A. F. (1997). New intestinal parasitic protozoa. *In* "Emerging Infections" (R. M. Krause, ed.), pp. 431–446 (this volume). Academic Press, New York.

Mogabgab, W. J., Holmes, B. J., and Pollock, B. (1975). Antigenic relationships of common rhinovirus types from disabling upper respiratory illnesses. International Symposium on Immunity to Infections of the Respiratory System in Man and Animals, London 1974. Dev. Biol. Stand. **28**, 400–411.

Morse, S. S. (1993). "Emerging Viruses." Oxford Univ. Press, New York.

Morse, S. S. (1995). Factors in the emergence of infectious disease. *Emerging Infectious Diseases* **1**, 7–15.

Mortimer, S. A. (1994). Pertussis vaccine. *In* "Vaccines" (S. A. Plotkin and E. A. Moritmer, eds.), pp. 91–135. Saunders, Philadelphia.

Musser, J. M., and Krause, R. M. (1997). The revival of group A streptococcal diseases, with a commentary on staphylococcal toxic shock syndrome. In "Emerging Infections" (R. M. Krause, ed.), pp. 185–218 (this volume). Academic Press, New York.

National Institute of Allergy and Infectious Diseases (1992). Report of the task force on microbiology and infectious diseases. National Institutes of Health Publication No. 92-3320. Bethesda, Maryland.

Neustadt, R. E., and Fineberg, H. V. (1978). "The Swine Flu Affair: Decision-making on a Slippery Disease." U.S. Department of Health, Education and Welfare, Washington, D.C.

Pappenheimer, Jr., A. M., and Murphy, N. R. (1983). Studies on the molecular epidemiology of diptheria. *Lancet* 2(8356)923-926.

Perelson, A. S., Neumann, A. U., Markowitz, M., Leonard, J. M., and Ho, D. D. (1996). HIV-1 dynamics in vivo: Virion clearance rate, infected cell life-span, and viral generation time. *Science* **271** (5255), 1582–1586.

Plotkin, B. J., and Kimball, A. M. (1997). Designing an international policy and legal framework for the control of emerging infectious diseases: First steps. *Emerging Infectious Diseases* **3** (1), 1–9.

Quinn, T. C., and Fauci, A. S. (1997). The AIDS epidemic: Demographic aspects, population biology, and virus evolution. *In* "Emerging Infections" (R. M. Krause, ed.), pp. 327–364 (this volume). Academic Press, New York.

Robbins, F. C. (1994). Polio-Historical. *In* "Vaccines" (S. A. Plotkin and E. A. Mortimer, eds.), pp. 137–154. Saunders, Philadelphia.

Robbins, J. B., Schneerson, R., and Szu, S. (1995). Perspective hypothesis: Serum IgG antibody is sufficient to confer protection against infectious diseases by inactivating the inoculum. *J. Inf. Dis.* **171**, 1387–1398.

Schneerson, R., Barrera, O., Sutton, A., and Robbins, J. B. (1980). Preparation, characterization, and immunogenicity of *H. influenzae*-type b polysaccharide–protein conjugates. *J. Exp. Med.* **152**, 361–376.

Shilts, R. (1987). "And the Band Played On: Politics, People, and the AIDS Epidemic." St. Martin's Press, New York.

Tenner-Racz, K., Racz, P., Dietrich, M., and Kern, P. (1985). Altered follicular dendritic cells and virus-like particles in AIDS and AIDS-related lymphadenopathy. *Lancet* **1** (8420), 105–106.

Webster, R. G. (1997). Influenza: An emerging microbial pathogen. *In* "Emerging Infections" (R. M. Krause, ed.), pp. 275–300 (this volume). Academic Press, New York.

2
ANALYTICAL THEORY OF EPIDEMICS

ROY M. ANDERSON
The Wellcome Trust Centre for the Epidemiology of Infectious Disease
University of Oxford
Oxford United Kingdom

I. INTRODUCTION

The aim of this chapter is to show how mathematical models of the within-host dynamics of pathogen population growth and of the transmission of infectious agents within human communities can aid in the interpretation of observed epidemiological pattern, in the collection of data to improve understanding, and in the design of programs for the treatment and control of infection and disease. A central theme is to further understanding of the interplay between the variables that determine the typical course of infection within an individual and the variables that control the pattern of infection and disease within communities of people. This requires an understanding of the basic similarities and differences between different infections in terms of the number of population variables (and consequent equations) needed for a sensible characterization of the system, the typical relations among the various rate parameters (such as birth, death, recovery, and transmission rates), and the role of the many heterogeneities both genetic and behavioral that influence transmission success or pathogen population growth.

Model construction, whether mathematical, verbal, or diagrammatic, aims to reduce complex biological processes into a more simple, idealized, and easily understandable sequence of events. Consequently, the use of a precise quantitative language in model formulation, as provided by the framework of mathematics, is a common exercise in scientific study. The biologist's or epidemiologist's distrust of mathematics is easy to understand as it often centers not so much on an unfamiliarity with the language, but more on the simplicity of the assumptions embedded in models in the face of known biological complexity. In mathematical terms, simple assumptions are more likely to permit precise analytical investigation, but this argument rarely convinces the experimentalist. A stronger argument rests on the underlying philosophy of starting simply and slowly adding complexity in a manner akin to that adopted by the bench scientist, where one or a few variables are allowed to vary while others are held constant in the experimental design. This approach has many advantages because even simple models of nonlinear biological phenomenon may exhibit very complicated patterns of dynamical behaviour, including chaotic patterns and multiple stable states each with its own zone of attraction. Because most problems in the study of infection and immunity exhibit nonlinear relationships between key variables, it seems sensible to fully understand the simplest of descriptions before tackling the interplay between the many variables and parameters that characterize the interaction between host and infectious agent (Anderson and May, 1991). Carefully building complexity on a simple framework can greatly facilitate an understanding of the major factors that influence or control a particular biological process or pattern. Sensibly used, mathematical models are no more and no less than tools for thinking about things in a precise way.

The origins of modern mathematical epidemiology owe much to the work of Hamer, Ross, Soper, Kermack, and McKendrick who, in different ways, began to formulate specified theories about the transmission of infectious disease in simple but precise mathematical statements, and to investigate the properties of the resulting models (Ross, 1911; Kermack and McKendrick, 1927; Soper, 1929). Their work generated one of the key concepts in mathematical epidemiology via the hypothesis that the course of an epidemic depends on the rate of contact between susceptible and infectious individuals. This is the so-called mass action principle in which the net rate of spread of infection is assumed to be proportional to the density of susceptible people multiplied by the number of infectious individuals. This simple principle gives rise to the threshold theorem according to which the introduction of a few infectious people into a community of susceptibles will not give rise to an epidemic outbreak unless the density or number of susceptibles is above a critical level.

Since these early beginnings the growth in the literature has been very rapid, particularly in the 1980s and 1990s, and reviews have been published by Anderson and May (1991) and Dietz and Schenzle (1985). The simple template constructed by Ross and others in the early part of the twentieth century has been modified and adapted to take account of many different facets of the interaction between infectious agent and host. These include the wide range of observed parasite life cycles encompassing horizontal and vertical transmission plus the involvement of vectors or intermediate hosts (Anderson and May, 1979; May and Anderson, 1979), the many heterogeneities that influence the transmission process (May and Anderson, 1987), the demography of the host population (Anderson et al., 1988), the population genetics of the pathogen (Gillespie, 1975; Beck, 1984), the evolution of virulence (Nowak and May, 1994), the influence of spatial processes and seasonality in transmission (Grenfell et al., 1995), antigenic variation within pathogen populations (Agur et al., 1989; Antia et al., 1996; Haraguchi and Sasaki, 1997), the within-host dynamics of infectious agents (Hetzel and Anderson, 1996; Nowak et al., 1991, 1995;

Nowak and Bangham, 1996; McLean, 1992, and competition between pathogen strains circulating in the same host population (Gupta *et al.*, 1994a, 1996).

The chapter is organised as follows. The first section considers simple models of host invasion by a new pathogen with the aim of identifying what population characteristics determine whether infection will result post exposure and the typical course of infection in the host. The second section considers the simplest of models of disease spread within a population and describes the main determinants of the reproductive success of an infectious agent. This is followed by a brief section on the influence of the many heterogeneities that impact on the transmission process, including seasonality and spatial factors. The fourth section examines the issue of competition between different strains of the same infectious agent in an attempt to delineate what influences the likelihood of a new genetic variant becoming establishing within the host population. The final section addresses the important issue of drug resistance in pathogens with the aim of furthering understanding of the factors that determine persistence and spread. The chapter ends with a brief discussion of how theory can help us understand more about the emergence and spread of new infectious diseases.

II. WITHIN-HOST DYNAMICS

Before turning to the question of transmission between hosts, we briefly examine what lessons can be learnt from models of the infection process itself. To address this issue we need to consider the events immediately post the exposure of the host to an inoculum of the infectious agent. In other words we must examine the within-host dynamics of pathogen population growth and decay. To provide specificity, the example considered is that of an agent which infects a specific host cell type in order to replicate, such as viruses and many bacteria and protozoa. Three variables are considered, namely, uninfected host cells, X, infected cells, Y, and free viral particles, V (or bacterial or protozoan life cycle stages, e.g., the merozoites of *Plasmodium falciparum*; Fig. 1). For illustrative purposes, staying with the viral example, infected cells are assumed to be produced via contact between uninfected cells and free virus particles at a net rate βXV and die at a net rate αY, where β is the per capita rate of cell invasion and α is the virus-induced host cell death rate. Free virus is produced by infected cells at a net rate rY, where r is the replication rate, and dies at a rate dV. Uninfected cells are produced at a constant rate Λ within the host (e.g., immune system cells) and have a life expectancy of $1/\mu$ where μX is their net death rate. These assumptions give rise to the following system of differential equations for viral population growth in the absence of specific or nonspecific immunological responses targeted at the pathogen (see Hetzel and Anderson, 1996; Nowak and Bangham, 1996):

$$
\begin{aligned}
dX/dt &= \Lambda - \mu x - \beta XV \\
dY/dt &= \beta XV - \mu Y \\
dV/dt &= rY - dV.
\end{aligned}
\tag{1}
$$

Analysis of this very simple system reveals that the establishing inoculum of the virus will only grow and induce a replicating infection provided the rate at

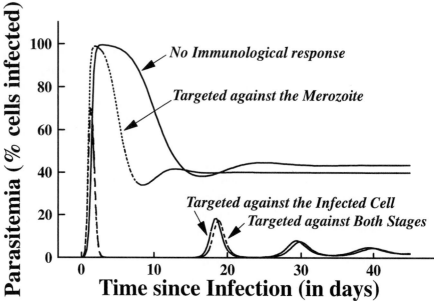

FIGURE 1   Predictions of the within-host dynamics of the merozoite stage of *Plasmodium falciparum* and infection of the host red blood cell population (see Hetzel and Anderson, 1996). The trajectories record temporal changes in the percentage of infected red blood cells post infection under four different assumptions concerning the action of the host's immunological response (no response, response targeted against merozoites, response targeted against infected host cells, and response targeted against both stages). Note the epidemic curve of infection even in the absence of an immune response.

which infected cells generate new infected cells in the very early stages post exposure (when most cells are uninfected) is greater than unity in value. The rate is called the within-host basic reproductive number, R_0, and is given by the quantity

$$R_0 = \beta r \Lambda / (\alpha \mu d).$$ (2)

This expression makes clear that a number of different population attributes of the virus determine its likelihood of growing within the host. Explicitly these are the rate of cell invasion, β, the replication rate of the virus within its host cell, r, the rate at which uninfected cells are produced, and the death rates α, μ, and d of, respectively, infected cells, uninfected cells, and free virus.

Considering new pathogen variants, or emerging infectious agents, produced by mutation or recombination, their genotypes must produce phenotypes with the properties necessary to ensure $R_0 > 1$ if they are to overcome the first hurdle of establishment within the host population, namely, the production of a fulminating infection in a single host. The values of the different parameters will vary from one pathogen genotype to the next, and, indeed, for a fixed genotype they may vary from host to host due to genetic variability in the host population influencing parameters such as Λ, μ, α, and β. As such the average phenotype of the new pathogen must have properties leading to $R_0 > 1$ for the average host in the population. The use of the word average is somewhat misleading since even if $R_0 > 1$ in individuals that constitute some small fraction of the host population, pathogen establishment and persistence may result depending on other factors that will be examined in the following section on transmission.

The model defined by Eq. (1) is of course far too simple in structure to mimic reality. However, it is a very useful starting template on which to graft complexity, particularly that induced by immunological attack. Even the simple structure, however, produces a result showing how a conglomeration of specific parameters determine the likelihood of establishment.

Infectious agents illicit immunological responses which act to limit reproduction and raise pathogen death rates. Despite rapid advances in our understanding of the human immune system far too little experimental immunology is directed at the quantification of how different arms of the immune system influence pathogen birth and death rates. Of equal importance in terms of the within-host dynamics of pathogens is the functional relationships between pathogen abundance and specific immunological responses, such as T-cell proliferation rates, and the life expectancy of different immune system cells (Nowak *et al.*, 1995; Mitchie *et al.*, 1992; Hetzel and Anderson, 1996). In the absence of precise quantitative data for any one specific host–parasite interaction, we can formulate models in general terms to illustrate both the importance of quantification and the role of immunological responses in determining the pattern of infection.

For simplicity one arm of the immune response is considered, namely, the action of cytotoxic T lymphocytes (CTL) targeted at infected cells since this serves to illustrate a number of general principles. We need to define a new variable Z for the abundance of CTL specific to infected cells and to modify one of the equations in our simpler model:

$$dY/dl = Y(\beta XV - \alpha - \gamma Z) \qquad (3)$$
$$dZ/dt = Z(cY - b).$$

Here, γ defines the rate at which CTL kill infected cells, c defines the rate at which CTL proliferate via contact with infected cells, and $1/b$ is the life expectancy of the CTL. Simple inspection of Eq. (3) for Z reveals that the pathogen will only stimulate a CTL response provided the density of infected cells, Y, exceeds a critical value:

$$Y > b/c. \qquad (4)$$

If there is an active CTL response then it will act to significantly decrease the virus load and increase the abundance of uninfected cells.

More generally, the rate at which CTL (or indeed any other cells in the immune system) proliferate in response to the viral load will be nonlinear such that the rate attains a maximum value at high burdens. In this case a variety of outcomes are possible. First, $R_0 < 1$, the pathogen will fail to establish. If $R_0 > 1$, the pathogen may outrun the CTL response and continue to grow exponentially until a very high fraction of the host cells are infected, resulting in the death of the host. Alternatively, the CTL response may regulate the viral population to an equilibrium level approached via damped oscillations. Finally, the pathogen may attain an equilibrium at a very low level below that required to stimulate a CTL response.

If the biological model is broadened to include pathogens not replicating within cells, and different arms of the immune response, these outcomes divide into four basic types, namely, failure to establish, unconstrained pathogen growth and subsequent host death, regulated pathogen growth to a stable equilibrium reflecting persistent infection at low or high levels of pathogen abundance, and pathogen growth regulation and elimination by the immune system reflecting host recovery from a fulminating infection. Which type pertains depends on the quantitative details of the values of the many population parameters that characterize even these simple mirrors of pathogen–immune system interactions. This area is one of urgent need for further development given its significance to a quantitative understanding of drug action within the host and the potential influence of immunotherapy or vaccines.

In the context of emergent infections, however, the key concept is the definition of a consortium of parameters that define whether a fulminating infection could result. Added to this is the concept that the likelihood of transmission on to a new host will depend on the abundance of the pathogen in the host over the entire duration of the infection. In other words infectiousness of a given host will depend on the integral of $V(t)$ in Eq. (1) and (3) over the entire course of the infection. Failure to establish provides the trivial result that the value of this integral is zero. In the nontrivial cases, particularly for persistent infections, the weighting to be applied to viral or, more generally, pathogen burden at any one time may depend on the nature of pathogen evolution or parasite development within the host over the incubation period of the disease. The selective pressures acting on the agent within the host will differ from those that influence transmission between hosts. One facet of parasite evolution within the host of great relevance to the likelihood of a new pathogen emerging and persisting in the host population is that of antigenic variation. Many infectious agents adopt the strategy of generating antigenic variation by mutation, recombination, or differential gene expression to facilitate persistence within the host in the face of severe immunological attack. The specificity of the mammalian immune system acts as a strong selective pressure which gives advantage to new "escape" variants that express novel or variant antigens as yet unseen by the immune system prior to their emergence. This acts to prolong the typical duration of infection which, in turn, acts to enhance the likelihood of transmission on to a new host.

In the generation of new variants, antibodies or cell-mediated attack directed to a specific antigen may simply have decreased efficacy to a variant of this antigen or of an epitope of that antigen. This is termed cross-reactivity. The human immunodeficiency virus (HIV-1) is a good example of the advantage conferred by antigenic variation in terms of inducing long-term, persistent infection. For example, changes in amino acid sequences of the envelope (gp120) gene of a quasi species of HIV in an infected patient accumulate rapidly over time via substitutions (Wolfs et al., 1990). The fast rate of amino acid substitution in the variable V3 region of the envelope gene is an example of adaptive evolution driven by the specificity of the human immune response. Phylogenetic analyses of viral samples taken over the course of the incubation period of acquired immunodeficiency syndrome (AIDS) reveal that nonsynonymous substitution in the putative antigen-determining sites accumulate at a rate several times greater than synonymous substitutions in other regions of the same gene. This is clear evidence that selection is promoting the generation of antigenic variation.

Mathematical studies of this process using a novel framework which represents cross-reactivity between variants as a continuous distribution reveal a series of important insights into how the degree of cross-reactivity between variants influences the persistence of the virus within its host. If the width of cross-reactivity is narrower than a critical value, antigenic variants gradually evolve as a moving wave and the total pathogen population persists at an equilibrium density. If the width of cross-reactivity is greater than the critical value, however, the moving wave loses stability and the distribution of antigenic variants fluctuates over time. In certain cases pathogen density may fluctuate widely post infection as a series of intermittent peaks caused by distantly separated antigenic types (Haraguchi and Sasaki, 1997; Nowak *et al.*, 1995).

The pattern of antigenic variation and selection by the immune system taking place within the host reflects to a degree the larger scale processes taking place at the host population level. If an agent is highly transmissible such that during an epidemic most people are exposed, and recovery from infection induces long lasting immunity to reinfection, the generation of new antigenic variants is a key strategy to avoid the effects of herd immunity. The influenza viruses provide a nice example of this larger scale process where shift and drift in a key surface antigen generate new variants (probably via the reassortment of genes from avian and human strains in mixed infections) which evade the herd immunity created by their recent ancestors. Cross-reactive responses also play a key role since drift induces small changes where the new variant antigens are recognized, but at a reduced precision, by the immune responses generated by memory of the previous closely related variant. The issues of herd immunity and the coexistence of related pathogen strains are addressed in latter sections of this chapter.

III. TRANSMISSION AND HERD IMMUNITY

Mathematical methods have been widely used in the study of the growth and decay of epidemics of infectious disease agents. More than most areas of mathematical biology, this area of theory has a strong tradition of confronting prediction with observation, and of the use of epidemiological data for parameter estimation. As such the insights gained from such application are of great relevance to our understanding of emerging infectious agents and their potential for spread and persistence.

Much can be learned from the simplest of models which describes the spread of a directly transmitted viral or bacterial pathogen in a closed population of susceptible people. Traditionally, the host population is divided into susceptibles, X, infecteds who are infectious, Y, and immunes who have recovered from infection, Z. Additional components can be added such as infecteds not yet infective, newborns with maternal derived immunity, and persistent carriers of infections. Such refinements are straightforward and their effects well understood (see Anderson and May, 1991), but the simplest of models suffices to illustrate the key principles and concepts. Under the assumption that net transmission is proportional to the density of susceptibles times the density of infected individuals, a basic model for the simple epidemic following the introduction of a few infecteds into a susceptible population of constant size N (where $N = X + Y + Z$) is of the form

$$dX/dt = \mu N - \mu X - \beta XY$$
$$dY/dt = \beta XY - (\mu + \sigma)Y \qquad (5)$$
$$dZ/dt = \sigma Y - \mu Z.$$

Here μ denotes the per capita death rate of the host, β is the transmission probability, and $1/(\sigma + \mu)$ is the average duration of infection where σ is the recovery rate. Immunity post recovery from infection is assumed to be lifelong.

After invasion of the susceptible host population (of size N), an epidemic will result provided the case reproductive number, R_0, is greater than unity in value. R_0 defines the average number of secondary cases of infection generated by one primary case in a susceptible population and is given by

$$R_0 = \beta X/(\mu + \sigma). \qquad (6)$$

Given the constraint that $R_0 > 1$ for an epidemic to occur, Eq. (6) can be expressed in terms of a critical density of susceptibility, X_T, necessary for the persistence of the invader in the host population where

$$X_T = (\mu + \sigma)/\beta \qquad (7)$$

Equations (6) and (7) reveal that for invasion and subsequent persistence, transmission success as denoted by the parameter β must be sufficiently large relative to the average duration of infection in a population of defined density. This simple system of equations exhibits damped oscillations to a stable point at which the equilibrium numbers of susceptibles, infecteds, and immunes are

$$X^* = (\mu + \sigma)/\beta$$
$$Y^* = [\mu/(\mu + \sigma)][N - X^*] \qquad (8)$$
$$Z^* = [\sigma/(\mu + \sigma)][N - X^*]$$

If the duration of infection is short [i.e., $1/(\mu + \sigma)$ is small] relative to host life expectancy $(1/\mu)$, the fraction infected in the population at equilibrium may be very small. Most of the population will be immune if R_0 is moderate to large. An illustration of this point is presented in Fig. 2 which records the fraction immune (=seropositive) as a function of the age of an individual in a modified version of Eq. (5) in which an extra class is added to denote the presence of maternally derived antibodies that confer protection to infection in newborn infants for an average period of 6 months. The parameter values in this example are set to mimic the transmission of the measles virus prior to wide scale immunization.

The temporal dynamics of a successful invasion by a new pathogen into a virgin population can be established from Eq. (5) using numerical methods. An illustration of this is presented in Fig. 3 for an HIV-1 epidemic in a population of male homosexuals. For this particular example Eq. (5) is somewhat simpler given that current evidence argues against the existence of acquired immunity. Infected individuals remain so for life with a greatly enhanced mortality rate in

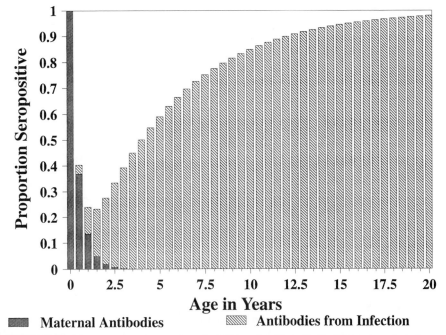

FIGURE 2 Diagrammatic representation of a cross-sectional (stratified by host age) serological profile with parameters set to mimic transmission of the measles virus prior to the introduction of immunization in cities in developed countries (see Anderson and May, 1991). The graph records the fraction with
maternally derived antibodies (decaying with a half-life of 6 months) and the fraction with antibodies
specific to measles virus antigens post infection, with an average age at infection of 5 years.

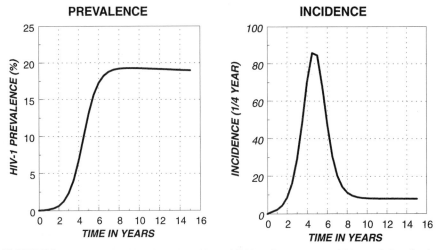

FIGURE 3 Simple epidemic curve simulated for an HIV-1 epidemic in a susceptible population of male
homosexuals, recording temporal changes in the incidence of infection and the percentage seropositive.

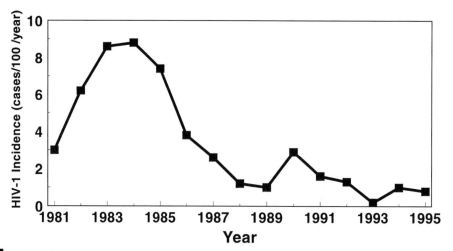

FIGURE 4   Observed longitudinal trend in annual incidence of HIV-1 infection in a male homosexual cohort in Amsterdam from 1981 to 1995 (Coutinho et al., 1996).

the infected class. Figure 3 reveals a number of important features of the invasion by a new pathogen of a susceptible population. The epidemic is bell shaped in form where incidence rises to a peak and then decays to a stable endemic state. Observed patterns mimic this trend, as illustrated in Fig. 4 where longitudinal changes in incidence are recorded for a gay community in Amsterdam. A somewhat different example is plotted in Fig. 5 which records reported deaths

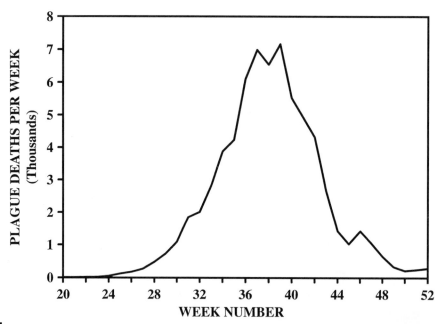

FIGURE 5   The 1665 plague epidemic in London as recorded by reported plague deaths per week in the diary of Daniel Defoe.

in the plague epidemic in London in 1665. In the case of HIV-1 and indeed other epidemics in virgin populations, the interpretation of longitudinal changes in incidence is made complicated by the impact of the epidemic on the behaviours that influence the rate of transmission. Temporal changes in this rate can induce a more rapid decline in incidence than would be observed in their absence, but teasing out the contribution of changes in behaviour from that of the natural bell-shaped form of the incidence curve can be difficult in practice unless quantitative information is available on behavioral changes over time.

The reason why the incidence curve is bell shaped in a simple epidemic relates to temporal changes in the reproductive success of the pathogen (i.e., the magnitude of the effective reproductive number, R). On invasion the vast majority of hosts are susceptible, and hence provided $R_0 > 1$ the epidemic expands as illustrated in Fig. 6. However, as the epidemic progresses, more and more of the contacts made by an infected host are either immune or already infected. As such the effective reproductive number declines, and eventually at equilibrium it settles to the value of unity where each infected person generates an average of one secondary infection.

A wide variety of factors influence the likelihood that a new pathogen will successfully invade a defined host community. Heterogeneities in various processes can influence the likelihood, and these are briefly discussed in the following section. However, before turning to these a number of modifications to the

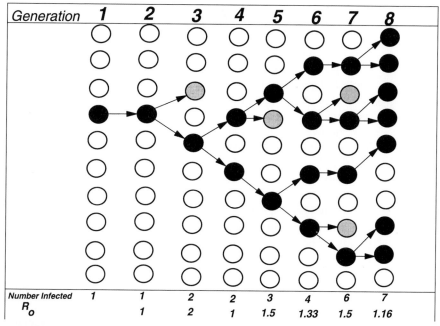

FIGURE 6 Diagrammatic representation of chains of transmission of an infectious agent starting with one infectious person in a population of 10 people. The value of the case reproductive number R_0 per generation cycle is recorded at the bottom of the graph. Since $R_0 > 1$ on average across eight time periods of the generation of secondary cases, the chain is an expanding one over the recorded time span. The black circles denote individuals who successfully transmit on the infection and the grey circles denote "dead ends".

simple model defined in Eq. (5) illustrate how different biological processes affect the likelihood of invasion and the patterns of change in incidence of infection post invasion.

Host genetic background plays an important role in determining the typical course of infection. Variability in this background often results in some fraction of the host population harboring persistent infection which may or may not be cleared in the long term. An example is hepatitis B virus where a fraction of those infected become carriers of infection for many years and are able to transmit infection on to others. Pathogens can exploit such niches within the host population such that even if the magnitude of R_0 is too low to allow persistence in the majority of the population the presence of a small number of carriers allows persistence in the longer term. The model defined in Eq. (5) can be modified to take account of a small fraction f of carriers in whom the typical duration of infection is long, with an average duration of $1/(\mu + \sigma_2)$, by comparison with that in the majority of the host population, $1/(\mu + \sigma_1)$. The basic reproductive number of the modified model is given by

$$R_0 = \frac{(1 - f)\beta_1 X}{(\mu + \sigma_1)} = \frac{f\beta_2 X}{(\mu + \sigma_2)} \tag{9}$$

where β_1 and β_2 are the transmission probabilities for noncarriers and carriers, respectively. Equation (9) illustrates clearly how the magnitude of R_0 can be maintained above unity by transmission induced by the carriers alone. As such, for new and emerging infectious agents persistence may depend on genetic heterogeneity within the host population and more specifically on a small fraction of people (0.1 to 0.2% in the case of hepatitis B) who are very susceptible to infection and who remain infectious for prolonged periods. This is certainly the case for hepatitis B virus and may also be true for other hepatitis viruses and to a lesser degree for HIV-1 and HIV-2. Other factors can also enhance the likelihood of establishment and persistence such as vertical transmission from mother to infant (Anderson and May, 1991).

Following an initial epidemic in a virgin population a variety of outcomes may arise. If host population size is small stochastic effects are more than likely to drive the infection to extinction, particularly if the typical direction of infectiousness is short and immunity post recovery long. Even in large populations extinction may be frequent for viral and bacterial infections with very short infectious periods, such as measles, pertussis, and the influenza viruses. An illustration of this phenomenon is provided in Fig. 7 which records longitudinal trends in the incidence of measles virus in Iceland prior to wide scale immunization. New epidemic posextinction events were stimulated by the immigration of a few infectives at infrequent intervals (Cliff *et al.*, 1981). Interestingly, analyses of the frequency distribution of the magnitudes of these epidemics and the intervals between them reveal power laws that suggest more regularity in the processes during these epidemics than is apparent from a visual inspection of the observed trends (Rhodes and Anderson, 1996a,b).

The model of a simple epidemic defined in Eq. (5) suggests damped oscillations to a stable point. In reality, the propensity to oscillate on a longer term basis, perhaps with a regular interepidemic period, such as the 2-year interval

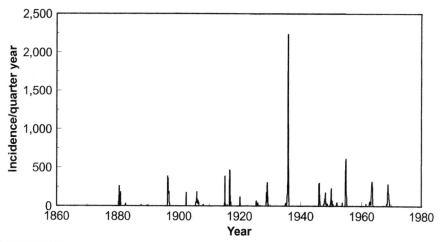

FIGURE 7   Time series of the reported incidence of measles in the Faroe Islands (monthly returns) from 1912 to 1969 (Rhodes and Anderson, 1996a,b).

for measles in developed countries prior to mass vaccination, is enhanced by both stochastic effects, spatial heterogeneity, and seasonality in the transmission rate (Bolker and Grenfell, 1993; Grenfell *et al.*, 1995; Ferguson *et al.*, 1997) (Fig. 8). Simple models based on the division of the host population into susceptible, incubator, infectious, and immune individuals, with a seasonally forced transmission rate, are well able to capture the biannual cycles in measles incidence when viewed at a countrywide level. On finer spatial scales a wealth of dynamical complexity emerges (Ferguson *et al.*, 1997; Keeling and Grenfell, 1997).

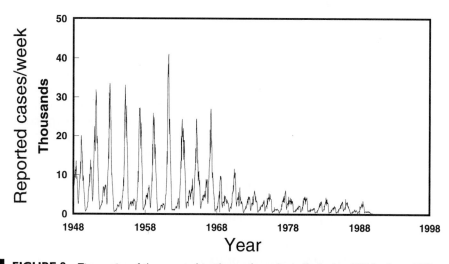

FIGURE 8   Time series of the reported incidence of measles in England and Wales from 1948 to 1994. Mass vaccination was introduced in 1968. Note the 2-year cycles prior to immunization and the seasonal cycles each year.

IV. HETEROGENEITIES IN THE TRANSMISSION PROCESS

The likelihood that an infectious agent is able to persist within a defined host community depends on a wide variety of factors relating to both the typical course of infection in an individual and transmission between individuals. The latter process is subject to many heterogeneities relating to chance effects, variation in host behavior, and genetic variation in both host and pathogen populations. In this section brief comment is made on the heterogeneities that influence the transmission process and, concomitantly, the persistence of the infectious agent within the host population.

Sexually transmitted diseases provide a good model to illustrate the influence of variation in human behavior because sexual activity is very variable both within and between different human communities (Anderson, 1994a,b). One of the key variables determining the likelihood of transmission in a defined community is the rate of new sexual partner acquisition per unit of time. For a sexually transmitted disease (STD) that does not induce immunity in those who recover from infection, the basic reproductive rate may be defined as

$$R_0 = \beta c D. \tag{10}$$

Here β is the average transmission probability defined per sexual partner, c is the average rate of sexual partner acquisition, and D is the average duration of infectiousness of an infected person. The term average hides much complexity because all these parameters are subject to variation due to either host genetic background or the type of pathogen strain infecting the host. For example, as illustrated in Fig. 9, reported rates of sexual partner acquisition are typically very variable, where the variance, σ^2, greatly exceeds the mean, m, in value. Under the assumption that sex partner choice occurs at random and in proportion to the number of individuals in each sexual activity class (defined on the basis of the rate of partner acquisition), our definition of R_0 is modified as follows:

$$R_0 = \beta(m + \sigma^2/m)D. \tag{11}$$

This simple expression suggests that high variability can act to enhance the likelihood of disease persistence ($R_0 > 1$), where those who form many partnerships act as a core group maintaining chains of transmission. More generally, however, a key component of the role of such behavioral heterogeneity depends on how individuals in one sexual activity group choose partners from all other groups. In other words, "who mixes with whom" determines the net rate of transmission within the host population. If we defined p_{ij} as the fraction of partnerships formed by individuals in sexual activity class i (defined in terms of the rate of partner acquisition) with those in group j, then the case reproductive number of secondary cases of infection generated in group i by one infectious person in group j is given by

$$R_{(0)ij} = \beta c_j p_{ij} D \tag{12}$$

where c_j is the mean rate of sexual partner acquisition in group j. Equation (12) makes clear that the infection can persist in the population if only certain of the $R_{(0)ij}$ values exceed unity. Thus a few groups could maintain the infection in the total population, whereas in their absence the infection would die out.

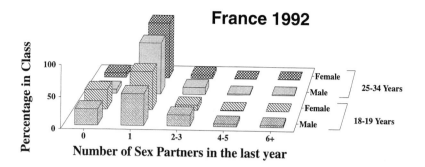

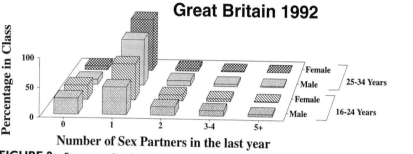

FIGURE 9 Frequency distributions of reported numbers of sexual partners per annum stratified by age group and sex, for two surveys conducted in France (ACSF, 1992) and Great Britain (Johnson et al., 1992). The distributions are heterogeneous in form, where the magnitude of the variance exceeds that of the mean (Anderson and Medley, 1988).

To determine the likelihood of persistence within the total population it is therefore necessary to determine the elements of the who mixes with whom matrix. To date very little research has been directed toward this issue, but a few studies suggest that mixing is assortative where those with high rates of partner change tend, on average, to choose partners from within their own sexual activity class (Garnett and Anderson, 1993; Garnett et al., 1996; Ghani et al., 1997). With all other parameters held constant, the pattern of mixing has a major influence on temporal changes in the prevalence or incidence of infection. This is illustrated in Fig. 10 which records simulated epidemics of HIV-1 in a male homosexual population under the assumption that mixing is either weakly or moderately assortative (like with like), or random. When mixing is assortative the epidemic takes off rapidly, but a small fraction of the population acquires infection since transmission events are largely restricted to the high sexual activity class. Waves of infection may be discernible in the overall epidemic as infection spreads from the high, to the medium, to the low activity classes. Conversely, if mixing is disassortative the epidemic takes off slowly, but eventually a high fraction of the population becomes infected. The general point illustrated by this example is that the likelihood of an emergent infection persisting depends both on heterogeneity in the behaviors that determine transmission and on the degree to which hosts with different behaviors mix in the population.

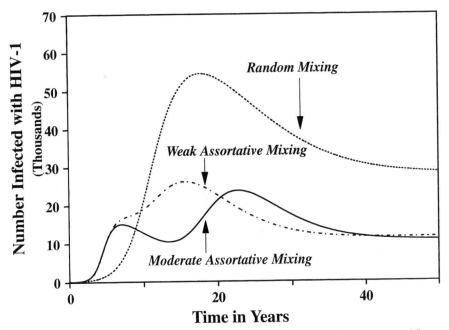

FIGURE 10   Simulated epidemics of HIV-1 in a male homosexual population under three different assumptions concerning the pattern of mixing between different sexual activity classes defined in terms of the rate of sexual partner acquisition per unit of time. The three cases are random mixing, weak assortative mixing, and moderate assortative mixing (Anderson, 1994a,b). In all three cases parameter values were fixed except for the pattern of mixing. Surveys of mixing patterns suggest moderate assortative mixing (Garnett and Anderson, 1993; Ghani et al., 1997).

A further important source of heterogeneity relates to the spatial distribution of hosts in a defined habitat. For many of the most important directly transmitted viral and bacterial infections, close contact between susceptible and infected individuals is required for transmission. As such the spatial distributions of individuals, and their movements between spatial locations, are important determinants of disease spread and persistence. A multipatch version of a susceptible–incubator–infectious–immune (SEIR) model is easy to formulate with the net rate or force of infection in patch i defined as

$$\lambda_i = \Sigma_j \, \beta_{ij} Y_j \tag{13}$$

where β_{ij} denotes the transmission probability from patch j to susceptibility in patch or spatial location i and Y_j is the density of infectious individuals in patch j. The term β_{ij} describes mixing between patches where the diagonal elements represent within-patch mixing.

Numerical studies of stochastic versions of such models reveal much about the importance of spatial heterogeneity and between-patch mixing on the likely persistence of an infectious agent in the space occupied by the total population (a summation over all patches) (Bartlett, 1957; Murray and Cliff, 1975; Keeling and Grenfell, 1997; Ferguson et al., 1997). The details of how populations in different patches interact with each other (between-patch mixing) has a significant influence on observed pattern. At one extreme all patches could be coupled, which may represent major cities linked by road, rail, and air transport systems. Alternatively, interactions may be restricted to nearest neighbors such as neighboring schools or boroughs. Simulation studies of such model systems (with the added refinement of seasonality in transmission) reveal much complexity in terms of local and metapopulation dynamics. Provided coupling between patches is sufficiently strong, local extinction in defined patches may be frequent such that temporal and spatial patterns reveal a boom and bust picture in any one patch but with asynchronous behavior across different patches.

For simulations structured to mimic the transmission dynamics of measles in developed countries, a critical total population density of around 2 million people within the coupled patch network is required to prevent extinction of the disease as illustrated diagrammatically in Fig. 11 (Bolker and Grenfell, 1993; Ferguson *et al.*, 1997). The more coupled the patches are (i.e., the greater the degree of between-patch mixing), the lower the critical community size required for the long-term persistence of measles.

More broadly, these types of analyses reveal the importance of chance effects, seasonality in transmission, and spatial structure of the population on the likely persistence of infection in any single spatial location and in the total community. Even if simple deterministic models suggest that the characteristics of an infectious agent are such that its case reproductive number, R_0, is likely to exceed unity in value, fully stochastic spatial models are required to assess the likelihood of long-term persistence and the possible patterns of incidence at various spatial scales. In the past such studies have required very large computers to simulate large populations within a finely stratified spatial arena. More recently the increased power of desktop computing facilities has enabled rapid progress in studies of the spatial spread and persistence of infectious agents (Keeling and Grenfell, 1997).

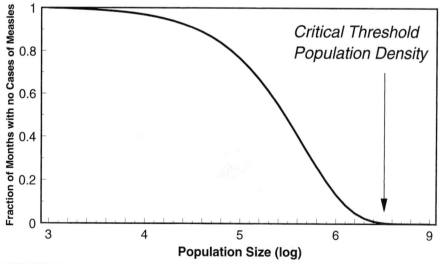

FIGURE 11   Diagrammatic representation of the relationship between "fade out" of measles (defined in terms of the percentage of months in which no cases are reported) and community population size (log scale) predicted by stochastic models of the transmission dynamics of the virus (Ferguson *et al.*, 1997).

V. STRAIN STRUCTURE IN PATHOGEN POPULATIONS

Molecular epidemiological studies are beginning to reveal much detail of the genetic constitution of pathogen populations. Relating this detail, which is often based on sequence information from defined regions of the pathogen genome, to phenotypic properties is more problematic. Certain phenotypic properties are measurable, such as the immunological details of the host response to variant pathogen antigens or clinical features that might reflect the pathogenicity or virulence of the invading strain. Generally, the key property of relevance to the transmission dynamics of a particular strain in a host population in which many strains of the infectious agent are circulating is the uniqueness of the antigens exposed to the host's immunological attack. The emergence of a new strain, via mutation or recombination, with antigenic properties novel to the host population gives the new variant a distinct advantage in terms of overcoming the herd immunity generated by the circulation and transmission of its ancestors. Typically, new variants will share some antigens with their ancestors but will often have novel surface antigens to enable them to partially evade the prevailing herd immunity created by their relatives. Thus, when considering the spread and persistence of new variants concomitant with the continued transmission of their ancestors, it is necessary to consider the respective roles of strain-specific and cross-reactive immunological responses.

A simple model serves to illustrate some key principles. The degree of cross-protection afforded by immunity to one strain against infection by another strain is defined by a single parameter c where if $c = 1$ there is no cross-immunity and if $c = 0$ there is total cross-protection (i.e., immunity to one confers immunity to all). In the case of two strains circulating in the same population, there are four possible outcomes, namely, both strains fail to persist, both persist (coexistence), and one persists and the other is competitively displaced by the herd immunity generated by its competitor (and vice versa). The occurrence of each of these outcomes depends on the magnitude of the degree of cross-protection (assumed to act on both strains equally) and the basic reproductive numbers of the two strains, R_{01} and R_{02}. Figure 12 plots the parameter domains that lead to each outcome for the simple two-strain case (see Gupta et al., 1994a). Four cases are plotted for $c = 0$, $c = 0.2$, $c = 0.5$, and $c = 1.0$. The two extremes depict competitive exclusion ($c = 0$), where the strain with the biggest R_0 value displaces the other strain, and coexistence ($c = 1$) in the absence of cross-protection. The other two cases reveal that coexistence can occur provided the magnitude of the R_0 values for the two strains are not too dissimilar. If they are then the strain with the greatest reproductive (=transmission) success will competitively displace the other. More generally, the occurrence of cross-immunity facilitates the persistence of a series of antigenically distinct strains in the same host population.

In practice the situation is somewhat more complex than that outlined above, owing to a number of biological factors. In our simple example we only consider the population biology of the two strains in the absence of any genetic mechanisms. For many important pathogens of humans, such as *Plasmodium falciparum* and HIV-1, exchange of genetic material between strains is frequent due to various mechanisms including sexual processes (*P. falciparum*), coinfec-

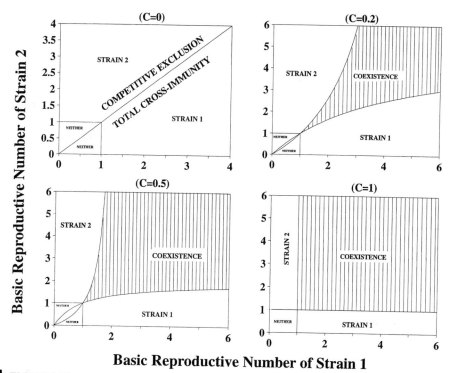

FIGURE 12 Influence of the degree of cross-immunity ($c = 0$, total cross-immunity; $c = 1$, no cross-immunity) on patterns of coexistence (at equilibrium) of two strains of a pathogen. (From Gupta et al., 1994a.) The graph records domains of coexistence and persistence of either strain in the two-dimensional space of values of the respective basic reproduction numbers (R_{0i} of the two strains; see text).

tion of the same host cell (HIV-1), or transformation in bacteria. However, in some cases strain structure appears to be maintained in pathogen populations over long periods despite evidence for frequent exchange of genetic material. This appears at first sight to be paradoxical since genetic recombination should result in disintegration of associations between loci in the genome of the pathogen.

Before examining this problem in more detail it is important to focus on what exactly is meant by a strain. In epidemiological terms it may be defined simply in terms of the loci that affect its transmission rather than by its entire genetic constitution. Transmission will be influenced by a variety of pathogen factors including antigenic constitution, ability to replicate within the host, and ability to persist in the host in the face of immunological attack. The host immune response will have the most profound effect, and important loci will therefore encode a set of immunogens. A subset of these will be dominant in the sense that they elicit the immune response that is most effective in eliminating the parasite from the host. Thus, in the case of two immunologically dominant loci, each with two possible alleles or variants, four possible types of strains will occur, namely, *ay, ax, bx,* and *by,* where *a* and *b* are alleles at one locus, and *x* and *y* are alleles at the second locus.

Work on the dynamics of transmission and persistence of the two loci–two allele case reveals much complexity in outcome (Gupta *et al.,* 1996). In the simplest of cases recovery from infection with a given strain may confer lifelong strain-specific immunity. However, what happens when this immune host is challenged by a strain which has different alleles at one or both loci? If we assume that infection is possible but the invading strain is at a disadvantage such that the ability of the host to transmit these strains with shared alleles (with the strain to which the host is immune) is reduced by a factor γ (the degree of cross-protection), then a variety of outcomes are possible. If cross-protection is complete ($\gamma = 1$), then one set of discordant strains (i.e., ay, bx or ax, by) will always dominate, but the other discordant pair will persist due to recombinant events. If cross-protection is absent then the system will permit all strains to coexist at prevalences dependent on the magnitudes of their respective R_0 values. The move from this "symmetric" equilibrium to the "asymmetric" case, where one discordant pair dominates, occurs when the degree of cross-protection exceeds a critical value, γ_c, where

$$\gamma_c = \left(1 - \frac{1}{R_{0j}}\right)\bigg/\left(1 - \frac{1}{R_{0i}R_{0q}}\right). \tag{14}$$

Here R_{0i} and R_{0q} are the basic reproductive numbers of the dominant pair of strains, and R_{0j} is the lower value of the basic reproductive numbers of the subordinate pair. When γ exceeds this critical value, the symmetric equilibrium (all equal if the R_0 value for all four strains are the same) bifurcates to give upper and lower asymmetric branches with one pair of discordant strains dominant (Gupta et al., 1996).

The asymmetric case is very stable to invasion by new strains. For example, if a new recombinant (say, by) is introduced into the host population in which a strain (say, ax) is already endemic, it fails to persist even when its case reproductive number is much greater in value than that of the resident strain provided the average duration of infectiousness is short. This suggests that distinct subsets of all possible strains are likely to persist over long periods. An empirical test of this theory based on analysis of the two principal antigenic regions of Neisseria meningitidis (regions VR1 and VR2) revealed that the strains present in a defined population had nonoverlapping epitope combinations over the period 1989–1991 despite the occurrence of reassortment by horizontal genetic transfer. In other words, despite the presence of frequent recombination events only a few nonoverlapping combinations of epitopes persist at high frequency (Gupta et al., 1996).

The main point revealed by these analyses is that immune selection at the population level can cause populations of infectious pathogens to be organized into a stable collection of independently transmitted strains with nonoverlapping repertoires of dominant polymorphic determinants, despite the effects of recombination. Furthermore, the prevailing strain structure may be difficult to invade by a new recombinant unless its fitness is greatly in excess of those of the resident strains.

Numerical studies of models of these two or more loci systems reveal further complexity including oscillating behavior, chaotic patterns, and for many loci multiple bifurcation points. The oscillating and chaotic behavior patterns are of particular interest because they suggest that many antigenically variable viruses, bacteria, and protozoa that can be classified into strains (i.e., serotypes) may exhibit complex changes over time in the dominance patterns of the different strains. Such fluctuations may simply result from dynamic processes in the absence of any changes in the host population or its environment. This is an interesting area for further study and one of great importance given much current research on the development of vaccines for pathogen populations that exhibit strain structure (e.g., the pneumococci and *P. falciparum*). In further work in this area it will be important to add further biological complexity including mutation, many alleles, and differing degrees of cross-protection induced by the alleles at different loci. The general aim must be to create a new theoretical template that melds population genetics with transmission dynamics.

VI. EVOLUTION OF DRUG RESISTANCE

Starting with the widespread use of antibiotics in the 1940s there has been a steady rise in the number of previously susceptible species of pathogenic and commensal microorganisms for which resistant strains have been recorded. In the broad field of new and emerging infectious agents, drug-resistant viruses, bacteria, and protozoa constitute one of the most urgent problems. The modern era of antimicrobial chemotherapy has been likened to an "arms race" between the pharmaceutical industry and bacterial evolution. It is a race in which the microbes are increasingly gaining the upper hand (Neu, 1992).

The problem of the emergence and spread of drug-resistant pathogens can be viewed at a series of levels including the within-host population dynamics of the infectious agent and its response to chemotherapy, the evolution and spread of resistance in hospital or managed care settings, and the spread and persistence in communities of people. The majority of research to date has focused on what happens within the treated patient, but increasingly epidemiological study and surveillance are providing records of longitudinal trends in the frequency of resistance in hospitals, communities, and countries. One example of community-based surveillance is recorded in Fig. 13 which shows the evolution of resistance to penicillin in middle-ear isolates of *Moraxella catarrhalis* in children in Finland over the period 1978 to 1993. Over this time the frequency of resistance in the isolates rose from close to 0 to over 90% (Nissenen *et al.*, 1995).

The study of the population dynamics and evolution of resistance has accelerated due to increased interest in the question of how best to use combinations of drugs to suppress viremia and bacteria in patients who will develop serious disease in the absence of chemotherapy. A typical example at present is the treatment of patients with HIV-1 by combinations of antivirals targeted at different functions within the virus or its host cell. Trials involving different combinations of drugs aimed at patients at different stages of the HIV-1 incubation period are complicated in design, difficult to organize in practice, and raise important ethical issues (Cohen, 1997). Most importantly, the key issue for combination therapy is the length of time over which viremia will be suppressed to low levels. Given a mean incubation period for HIV-1 of around 10 years, choices in how best to deliver combinations cannot wait until long-

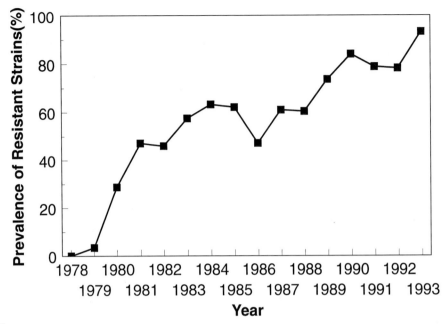

FIGURE 13   Development of β-lactamase resistance to penicillin in middle-ear isolates of *Moraxella catarrhalis* in children in Finland from 1978 to 1993. (From Nissenen *et al.*, 1995.)

term trials are completed. In these circumstances, analytical approaches to the study of the within-host population dynamics and evolution of the virus have an important role to play.

Mathematical models that meld components of pharmacodynamics, population dynamics, and evolution are ideally required. If it is assumed that resistance emerges via mutation events then simple models suggest that using three drugs together as early as possible post infection will produce the best long-term results by suppressing viremia to very low levels and hence minimize the net rate of evolution of resistance to all three drugs (Perelson *et al.*, 1996). Indeed, some have suggested that the decay characteristic of the virus post triple therapy suggests that HIV-1 could be eliminated from plasma and lymphocytes in circulation in 2–3 years (Perelson *et al.*, 1996). However, the models on which these predictions are based do not take into account the recombination events which are known to be frequent in HIV-1-infected patients (Robertson *et al.*, 1995) and the longer term persistence of the virus in cells in lymph nodes within the patient. Further work is required in the development of more sophisticated models that meld population genetics with population dynamic processes where the selective pressure at any one time is prescribed by the pharmacokinetics of the combinations of drugs within given treatment regimes (Bonhoeffer and Nowak, 1997). Such approaches have begun to be developed for single therapy situations, particularly in the context of antibiotics and bacterial diseases (Austin *et al.*, 1997; Lipsitch and Levin, 1997).

A more traditional approach can be adopted for the study of the emergence and persistence of resistant organisms within defined communities. The appropriate analytical framework must employ components of conventional epidemiology, such as rates of transmission, and elements of population genetics, such as the relative fitness of susceptible and resistant strains and the biological mechanisms that lead to the evolution and spread of resistance genes. In all such study a key component is the measurement of selective pressure defined by the volume of drug used per unit of time per patient. In practice there is much heterogeneity in drug use patterns between patients, where some may complete a full course of treatment while others cease treatment once clinical symptoms abate. Such heterogeneity is thought to be a key factor in the evolution of resistance within patients who fail to comply with a full course of treatment.

For many important bacterial populations, the infectious agent often persists within the host population as a commensal and only causes disease in a few patients due to a variety of causes including immunosuppression and the evolution of a virulent strain of the bacteria. Epidemiological models can be constructed to mimic these situations based on the description of transmission, drug use patterns, and the genetics of resistance (Massad *et al.*, 1993; Austin *et al.*, 1997; Levin *et al.*, 1997). Such models delineate the critical level of community-based drug treatment required to eliminate both the commensal and pathogen populations in terms of the case reproductive number of both organisms and the frequency of treatment of individuals within the community. Once resistance evolves the situation becomes much more complex. Assuming that resistance carries a fitness cost in the absence of treatment, a variety of outcomes are possible given that the case reproductive numbers exceed unity in value in the absence of treatment. These include coexistence of both sensitive and resistant strains, where the relative abundance of the two strains is determined by the magnitude of the fitness cost. Certain drug use patterns can lead to total replacement of the sensitive strains by resistant varieties.

The major role of such models is to highlight the effect of different antibiotic prescription policies on the frequency of resistance. First, they show that with a constant pattern of drug use the frequency of resistance will change over time from close to zero post its evolution or establishment, via a sigmoid pattern, to an equilibrium frequency whose value is determined by many factors including the volume of drug use, the fitness cost, and the mechanism of the spread of resistance genes (Fig. 14). These patterns of change in gene frequency under a constant selection pressure are well understood by population geneticists. They are much less widely understood in medical microbiology, however, where many interpret the rise in the frequency of resistance as reflecting a rising volume of drug use. Growth in the volume of drug use will change the pattern of sigmoid growth (as shown in Fig. 13), but the slow rise followed by a phase of rapid change before a slow approach to a steady state will occur with a constant selective pressure.

Analytical studies also reveal what happens after the cessation of antibiotic use (or a switch to an antibiotic with a different mode of action). In general, such changes will lead to a gradual reduction in the frequency of resistance, but the rate of decline will typically be slower than the rate of its emergence. The theory of the evolution of drug-resistant infectious agents is in its infancy, but the field has great potential to help to inform policy makers about options to better manage (on a community-wide basis) this important problem.

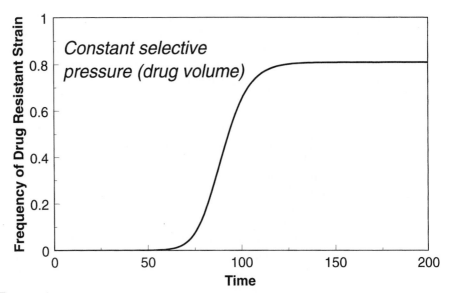

FIGURE 14 Predicted temporal change in the fraction of a bacterial population resistant to a defined drug post the evolution or arrival of the resistant strain, under a constant selective pressure which equates to a constant level of drug use in the host population. This example is purely illustrative of the predictions of a class of models of the population genetics of the evolution of resistance.

VII. CONCLUSIONS

Transmission, persistence, and control are the conventional areas for the application of mathematical techniques in the study of infectious diseases. The real challenge for analytical approaches in the coming years lies in the fields of molecular epidemiology, pathogen evolution, and pathogen interaction with the host immune responses. In all these areas information is accumulating rapidly, but the focus of this explosion of information is largely descriptive at present. Unraveling how the detail fits into a broader landscape that encompasses population-level phenomena and evolutionary processes, whether within the host or in communities of hosts, will require more focus on the dynamics of the interaction between host and parasite and more attention on the quantification of the many rate parameters that control parasite population growth and evolution (Anderson, 1994a,b).

In some areas, a growing volume of epidemiological research is addressing the issue of how to quantify rates of transmission (Babad et al., 1995; Gregson et al., 1996). In others, experimental studies are beginning to provide precise information on the rate processes that control parasite population growth and decay within individuals (Hetzel and Anderson, 1996; Nowak and Bangham, 1996; Perelson et al., 1996). Advances have been stimulated by the availability of more precise molecular and immunological techniques for the quantification of pathogen abundance within patients and the use of mathematical tools to estimate rates of change from observed trends in parasite population growth and decay. Examples of this approach are limited in number, however, and

geneity in the parasite population on the transmission dynamics of malaria. *Proceedings of the Royal Society of London Series B* **256**, 231–238.

Gupta, S., Maiden, M., Feavers., I., Nee, S., May, R. M., and Anderson, R. M. (1996). The maintenance of strain structure in populations of recombining infectious agents. *Nature Medicine* **2**, 437–442.

Haraguchi, Y., and Sasaki, A. (1997). Evolutionary patterns of intra-host pathogen antigenic drift: Effect of cross-reactivity in immune response. *Philos. Trans. R. Soc. Ser. B* **352**, 11–20.

Hetzel, C., and Anderson, R. M. (1996). The within-host cellular dynamics of bloodstage malaria: Theoretical and experimental studies. *Parasitology* **113**, 25–38.

Johnson, A. M., Wadsworth, J., Wellings, K., Bradshaw, S., and Field, J. (1992). Sexual lifestyles and HIV risk. *Nature (London)* **360**, 410–412.

Keeling, M. J., and Grenfell, B. T. (1997). Disease extinction and community size: Modeling the persistence of measles. *Science* **275**, 65–67.

Kermack, W. O., and McKendrick, A. G. (1927). A contribution to the mathematical theory of epidemics. *Proc. R. Soc. London* **115**, 700–721.

Levin, B. R., Lipsitch, M., Perrot, V., Schrag, S., and Antia, R. (1997). The population genetics of antibiotic resistance. *Clin. Infect. Dis.* **24**(Suppl. 1), 309–16.

Lipsitch, M., and Levin, B. R. (1997). The population dynamics of antimicrobial chemotherapy. *Antimicrob. Agents Chemother.* in press.

Massad, E., Lundberg, S., and Yang, H. M. (1993). Modeling and simulating the evolution of resistance against antibiotics. *International Journal of Biomedical Computing* **33**, 65–81.

May, R. M., and Anderson, R. M. (1979). Population biology of infectious diseases: Part II. *Nature (London)* **280**, 455–461.

May, R. M., and Anderson, R. M. (1987). Transmission dynamics of HIV infection. *Nature (London)* **326 (or 342)**, 137–142.

McLean, A. R. (1992). T memory cells in a model of T cell memory. *In* "Theoretical and Experimental Insights into Immunology" (A. S. Perelson and G. Weisbuch, eds.), Vol. H66, pp. 149–162. Springer-Verlag, Berlin.

Mitchie, C. A., McLean, A., Alcock, C., and Beverly, P. C. L. (1992). Lifespan of human lymphocyte subsets defined by CD45 isoforms. *Nature (London)* **360**, 264–265.

Murray, G. D., and Cliff, A. D. (1975). A stochastic model for measles epidemics in a multi-region setting. *Inst. Br. Geogr.* **2**, 158–174.

Nissenen, A. P., Gronvoos, P., Huovinen, E., Herra, M., Katila, T., Klaublea, S., Konliainen, O., Oinonen, S., and Makela, H. (1995). Development of β-lactamase mediated resistance to penicillin in middle-ear isolates of *Moraxella catarrhalis* in Finnish children 1978–1993. *Clin. Infect. Dis.* **21**, 1193–1196.

Nowak, M. A., and Bangham, C. R. M. (1996). Population dynamics of immune responses to persistent viruses. *Science* **272**, 74–79.

Nowak, M. A., and May, R. M. (1994). Superinfection and the evolution of parasite virulence. *Proc. R. Soc. London Ser. B* **255**, 81–89.

Nowak, M., Anderson, R. M., McLean, A. R., T., Wolfs., Goudsmit, J., and May, R. M. (1991). Antigenic diversity thresholds and the development of AIDS. *Science* **254**, 936–969.

Nowak, M. A., May, R. M., Phillips, R. E., Rowland-Jones, S., Lalloo, D. G., McAdam, S., Klenerman, P., Koppe, B., Sigmund, K., Bangham, C. R. M., and McMichael, A. J. (1995). Antigenic oscillations and shifting immunodominance in HIV-1 infections. *Nature (London)* **375**, 606–611.

Perelson, A. S., Neumann, A. U., Markowitz, M., Leonard, J. M., and Ho, D. D. (1996). HIV-1 dynamics in vivo: Virion clearance rate, infected cell lifespan and viral generation time. *Science* **271**, 1582–1586.

Rhodes, C. J., and Anderson, R. M. (1996a). Power laws in the incidence of measles virus infection in small isolated island populations. *Nature (London)* **381**, 600–602.

Rhodes, C. J., and Anderson, R. M. (1996b). A scaling analysis of measles epidemics in a small population. *Philos. Trans. R. Soc. Ser. B* **351**, 1679–1688.

Robertson, D. L., Sharp, P. M., McCutchan, F. E., and Hahn, B. H. (1995). Recombination in HIV-1. *Nature (London)* **374**, 124–126.

Ross, R. (1911). "The Prevention of Malaria (with addendum on the theory of Happenings)." Murray, London.

Soper, M. A. (1929). The interpretation of periodicity in disease prevalence. *J. R. Statistical Soc. Ser. A* **92**, 34–61.

Wolfs, T. F. W., de Jong, J.-J., van den Berg, H., Tunagel, J. M. G. H., Krone, W. J. A., and Goudsmit, J. (1990). Evolution of sequences encoding the principal neutralization epitope of human immunodeficiency virus 1 is host dependent, rapid and continuous. *Proc. Natl. Acad. Sci. U.S.A.* **88**, 9921–9826.

3

PERSISTING PROBLEMS IN TUBERCULOSIS

JOHN D. MCKINNEY, WILLIAM R. JACOBS, JR., AND BARRY R. BLOOM
Howard Hughes Medical Institute
Department of Microbiology and Immunology
The Albert Einstein College of Medicine
Bronx, New York

I. LESSONS FROM THE PAST

There is a dread disease which so prepares its victim, as it were, for death; which so refines it of its grosser aspect, and throws around familiar looks, unearthly indications of the coming change—a dread disease, in which the struggle between soul and body is so gradual, quiet, and solemn, and the result so sure, that day by day, and grain by grain, the mortal part wastes and withers away, so that the spirit grows light and sanguine with its lightening load, and, feeling immortality at hand, deems it but a new term of mortal life; a disease in which death takes the glow and hue of life, and life the gaunt and grisly form of death; a disease which medicine never cured, wealth warded off, or poverty could boast exemption from; which sometimes moves in giant strides, and sometimes at a tardy pace; but, slow or quick, is ever sure and certain.

Charles Dickens, *Nicholas Nickleby,* 1870

Even at that early date Dr. Janeway's great skill in physical diagnosis was recognized, and he had a class at Bellevue for physical diagnosis to which I belonged. He received me cordially and began the examination at once. When this was concluded he said nothing. So I ventured, "Well, Dr. Janeway, you can find nothing the matter?" He looked grave and said, "Yes, the upper two-thirds of the left lung is involved in an active tuberculous process." . . . I think I know something of the feelings of the man at the bar who is told he is to be hanged on a given date, for in those days pulmonary consumption was considered as absolutely fatal. I pulled myself together, put as good a face on the matter as I could, and escaped from the office after thanking the doctor for his examination. When I got outside, as I stood on Dr. Janeway's stoop, I felt stunned. It seemed to me that the world had suddenly grown dark. The sun was shining, it is true, and the street was filled with the rush and noise of traffic, but to me the world had lost every vestige of brightness. I had consumption—that most fatal of diseases!

Edward Livingston Trudeau, *An Autobiography,* 1916, p. 71

A. Origins of Tuberculosis

The evolutionary origins of *Mycobacterium tuberculosis,* the bacterium that causes human tuberculosis, are uncertain. Tuberculosis may have originated in prehistoric humans as a zoonotic infection transmitted from tuberculous animals. A likely candidate for the most recent ancestor of *M. tuberculosis* is the closely related species *Mycobacterium bovis,* the cause of bovine tuberculosis. Alternatively, these two species may have diverged independently from a common precursor. On an evolutionary timescale, this divergence was a recent event, as indicated by the high degree of DNA sequence homology (>98%) between the genes encoding the 16SS rRNA in the two species (Smida *et al.,* 1988). Today, *M. bovis* is endemic in several domestic and wild mammals— including bovines, badgers, and Australian possums—and is still a significant cause of zoonotic disease in humans. For *M. tuberculosis,* on the other hand, humans are the only natural reservoir. Genetic polymorphisms between modern-day isolates of *M. tuberculosis* and *M. bovis* suggest that these species probably diverged at least 15,000 to 20,000 years ago (Kapur *et al.,* 1994a). The discovery of pathological signs of tuberculous disease in ancient human remains recovered from widely dispersed sites also supports the antiquity of *M. tuberculosis* as a human pathogen. (Bates and Stead, 1993; Daniel *et al.,* 1994; Haas and Haas, 1996). In North Africa, mummified human remains, exhumed from Egyptian tombs dating from 2000 to 4000 B.C., displayed lesions characteristic of tuberculosis. These presumptively tuberculous lesions included the skeletal abnormalities associated with tuberculosis of the long bones and of the spine (Pott's disease) as well as fibrotic pulmonary lesions typical of advanced and partially healed lung disease. In the Americas skeletal deformities suggesting Pott's disease were found in 2000-year-old human remains, and acid-fast bacilli (presumably mycobacteria) were recovered from the lungs of a South American mummy dating from A.D. 700.

The question whether tuberculosis has caused widespread human disease for millennia remains contentious. Skeptics have argued that other infections may cause lesions similar to those of tuberculosis, and that acid-fast environmental mycobacteria could have contaminated archaeological specimens. More recently, however, DNA specific for pathogenic tubercle bacilli was detected in the tracheobronchial lymph node of a 1000-year-old Peruvian mummy with tuberculous lung pathology, establishing that pulmonary tuberculosis was endemic in the Americas prior to the first contact of indigenous Americans with European populations in the late fifteenth century (Salo *et al.,* 1994). The DNA sequences identified in this study were derived from the insertion element *IS6110,* which is common to all member species of the *M. tuberculosis* complex (*M. tuberculosis, M. bovis, M. microti,* and *M. africanum*). Therefore, the study does not rule out the possibility that the Peruvian mummy harbored *M. bovis,* a zoonotic pathogen, rather than *M. tuberculosis,* whose only natural host is *Homo sapiens* (Stead *et al.,* 1995). If the bacilli detected in the Peruvian mummy were indeed *M. tuberculosis,* it would follow that infected humans introduced the contagion to the New World sometime between the last ice age (15,000–35,000 years ago), when the first wave of human migrants crossed the Bering land brige, and the completion of deglaciation (~6000 years ago), when rising sea levels submerged the Bering passage (Cavalli-Sforza and Cavalli-Sforza, 1995).

The possibility that *M. tuberculosis* originated within prehistoric human populations is supported by computer modeling studies of tuberculosis transmission dynamics, which suggest that the disease may become endemic in human communities of ≥200 contiguous individuals (McGrath, 1988). Stable human communities of this size have likely existed in some parts of the world since the origin of agriculture, approximately 10,000 years ago (McNeill, 1976). In the twentieth century, Francis Black documented the persistence of tuberculosis in demographically isolated tribal communities inhabiting the South American Amazon basin (Black *et al.,* 1974; Black, 1975). The persistence of *M. tuberculosis* in small, isolated human populations contrasts strikingly with the behavior of many other familiar pathogens, such as the measles and smallpox viruses, which persist only within human communities of ≥250,000 contiguous individuals (Black, 1966, 1975; Black *et al.,* 1974; Keeling and Grenfell, 1997). These pathogens cause acute disease soon after entering the host, and they engender a sterilizing immune response that provides lifelong protection against reinfection (Slifka and Ahmed, 1996). Consequently, smallpox and measles require uninterrupted access to fresh susceptibles for their propagation and "burn out" after infecting every individual in a small, isolated community. The immune response to *M. tuberculosis,* on the other hand, may protect against disease but does not eradicate the pathogen, and infected individuals typically develop lifelong chronic disease or latent infection. The inadequacy of acquired host immunity to tuberculosis is also indicated by the observation that previously infected individuals can be reinfected following reexposure to virulent tubercle bacilli. Individuals with chronic tuberculosis may shed tubercle bacilli in their pulmonary secretions for many years, even decades, thereby serving as long-standing reservoirs of contagion. Individuals with latent tuberculosis are not infectious, but they may reactivate later in life and resume transmitting to susceptible contacts. As many as half of all active cases of tuberculosis are thought to arise from the reactivation of latent infection. This unusual tenacity of *M. tuberculosis* accounts for the antiquity of tuberculosis as a human disease and contines to pose a formidable obstacle to present-day efforts to eradicate tuberculosis from infected individuals and communities.

The gradual urbanization of human populations in the last several millennia, with attendant increases in community size and crowding, created optimal conditions for the transmission of many infectious diseases, including tuberculosis (McNeill, 1976, 1992; Cohen, 1995). Until the advent of modern sewage systems and other public health measures in the latter half of the nineteenth century, towns and cities were places of demographic decay and could not sustain themselves biologically without a steady influx of immigrants from the healthier countryside. In Europe and North America, the prevalence of tuberculosis probably peaked during the eighteenth or early nineteenth centuries and has been in slow decline since (Fig. 1A). At its zenith, tuberculosis may have accounted for as many as one in four deaths in major cities in Europe and the United States (Dubos and Dubos, 1952). The concentration of morbidity and mortality in young adults of reproductive age, among whom tuberculosis was the leading cause of death well into the twentieth century, presumably exerted

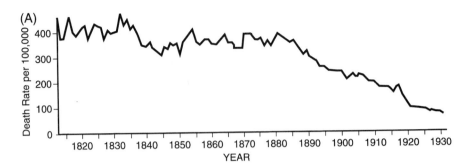

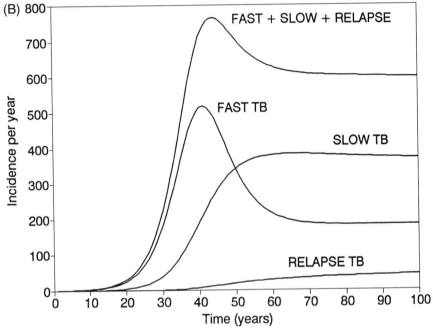

FIGURE 1 (A) Pulmonary tuberculosis death rate in New York, Philadelphia, and Boston. Tuberculosis epidemics occur over very long time spans. The urban death rate (deaths per 100,000 population per annum) from pulmonary tuberculosis in the United States remained at a constant high level throughout the first half of the nineteenth century, entering a period of steady decline only after 1880. (Redrawn from Drolet, 1946.) (B) Numerical simulation of a tuberculosis epidemic using the detailed transmission model of Blower *et al.* (1995). The simulation illustrates the relative contributions of three types of tuberculosis (TB) to the disease incidence rate: fast (progressive primary tuberculosis), slow (reactivation tuberculosis), and relapse tuberculosis. The decline of the overall incidence rates is due to the intrinsic transmission dynamics of the epidemic, and it occurs in the absence of any changes in the parameters. The epidemic was initiated by entering one infectious case of tuberculosis at time zero in a susceptible population of 200,000. A fourth source of disease, exogenous reinfection, which is not included in the model, would presumably increase the peak level and duration of the epidemic. (Reprinted from Blower *et al.*, 1995. *Nature Medicine* 1, 815–821. With permission.)

considerable selective pressure for genetic traits conferring resistance. The evolution of resistant genotypes among European races may have contributed to the steady decline of tuberculosis morbidity and mortality since the mid-1800s.

During the nineteenth and early twentieth centuries, as tuberculosis entered previously unaffected human populations, high rates of infection, morbidity, and mortality were observed. Previously "naive" races, particularly sub-Saharan Africans, were also reported to develop more fulminating disease than did Europeans, in whom tuberculosis tended to pursue an indolent, chronic course (Rich, 1944; Dubos and Dubos, 1952). These observations lent further support to the idea that genotypic resistance to tuberculosis gradually evolved among European races during centuries of exposure. Evidence of this sort must be interpreted with caution, however, as social and economic differences between racial and ethnic groups may also influence susceptibility to infectious disease, quite apart from genetic factors.

B. Epidemiological Considerations

Although the geographic origins of the disease remain obscure, as infected Europeans traveled to increasingly remote areas, tuberculosis gradually spread to all corners of the globe, invading previously unexposed nations and races. Today, as much as one-third of the world's population may be infected with *M. tuberculosis* (Ad Hoc Committee, 1996). Ironically, the medical practices of the nineteenth century probably contributed to the global dissemination of tuberculosis, as the climates of exotic, far-away countries were considered salubrious for tuberculous sufferers, who were prepared to travel great distances in search of a "cure" (Rothman, 1994). In much of the developing world, tuberculosis may still be considered an emerging infection, as the contagion spreads to previously unaffected regions and case rates continue to rise among the newly afflicted. In the remote interior of sub-Saharan Africa, tuberculosis was virtually unknown well into the twentieth century, but it has since become a major cause of morbidity and mortality, with approximately 1 million new cases and 0.5 million deaths occurring in 1990 (De Cock *et al.*, 1992; Raviglione *et al.*, 1992, 1995; Ad Hoc Committee, 1996). Tuberculosis was apparently rare in India and China until the end of the nineteenth century, and it did not reach the island populations of Papua New Guinea until the middle of the twentieth century; today, however, southeast Asia is the focus of the global tuberculosis problem. According to the World Health Organization, of the 50.5 million deaths occurring in 1990 (worldwide, of all causes), approximately 2 million were due to tuberculosis, with 98% of these deaths occurring in the demographically developing nations (Ad Hoc Committee, 1996). Clearly, the "White Plague" continues to cause more human suffering and death than any other infectious disease, despite the misperception in industrialized nations that tuberculosis has been conquered.

Compared to epidemics of more rapidly progressing infectious diseases, which typically rise, peak, and decline over brief intervals of time, epidemics of tuberculosis wax and wane very slowly (Drolet, 1946; Grigg, 1958; Blower *et al.*, 1995) (Fig. 1A). On the basis of morbidity and mortality data for Europe and the United States, Grigg (1958) proposed that a tuberculosis epidemic will peak within 50–100 years of its inception, and that morbidity and mortality will decline slowly thereafter as the genotypically more resistant individuals survive and reproduce. This pattern was emphasized by a modern epidemiological model of tuberculosis that evaluated both progressive primary ("fast") tuber-culosis and reactivation ("slow") tuberculosis: via computer simulation, the model predicted that the natural history of a tuberculosis epidemic will encompass from one to several centuries (Fig. 1B), rising from very low numbers to epidemic proportions and then slowly declining to an intermediate endemic level (Blower *et al.*, 1995). This study suggested that the rise and fall of a tuber-culosis epidemic are largely due to the intrinsic transmission dynamics of tuber-culosis, independent of efforts to control the disease. In light of this concept, it is noteworthy that tuberculosis incidence in Europe and North America began to decline before the infectious etiology of the disease had been established (in 1882) and public health measures instituted to prevent transmission.

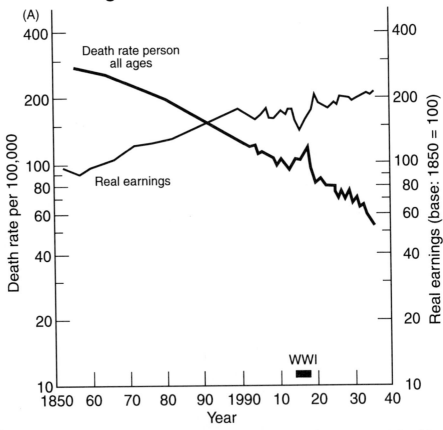

FIGURE 2   (A) Data for changes in working-class real earnings linked for successive periods and data for pulmonary tuberculosis mortality (per 100,000 population) for both sexes and all ages from Registrar General's Reports. (Adapted from Hart and Wright, 1939). (B) The influence of war on the incidence of pulmonary tuberculosis. The steady decline in the tuberculosis death rate in England and Wales from 1904–1945 was temporarily interrupted between 1914–1918 (World War I) and 1939–1945 (World War II). (Reprinted from Dubos, et al. 1949 J Exp. Med. 86, 159–174. With permission.)

The decline of tuberculosis in the industrialized nations from the mid-1800s onward was probably accelerated by improvements in public health and hygiene, nutrition, and living conditions in cities, illustrating the critical role that general well-being plays in maintaining the delicate balance between pathogen and host (Fig. 2A). On three occasions in the twentieth century, the downward trend in tuberculosis incidence in the industrialized nations was temporarily reversed. The first two occasions were the first and second world wars, during which the incidence of tuberculosis rose steeply but transiently in virtually all involved countries (Fig. 2B). Presumably, host resistance was weakened and transmission was promoted by the familiar stresses of war—physical fatigue, psychological trauma, crowding in shelters and barracks, inadequate nutrition, etc. The deleterious effects of stress and malnutrition on resistance to tuberculosis were particularly well documented in a study of Danish concentration camp prisoners during World War II, among whom tuberculosis was responsible for considerable morbidity and mortality (Helweg-Larsen *et al.*, 1952). In the worst of the "KZ-camps," pulmonary tuberculosis may have afflicted as much as 20–40% of the inmate population and may have accounted for as much as 40% of all deaths. Following the war, the downward trend in tuberculosis incidence resumed and continued steadily for the next 40 years, with a discernable downward inflection in the 1950s attributable to the introduction of antituberculous chemotherapy. In the United States, from 1953, when national surveillance was initiated, the annual incidence of tuberculosis declined at an average rate of 5.3% per annum. An ominous reversal of this trend was seen first in large urban centers such as New York City, which in 1979 reported its first increase in tuberculosis incidence in decades, followed by escalating tu-

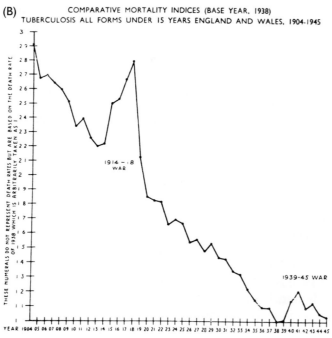

(B) COMPARATIVE MORTALITY INDICES (BASE YEAR, 1938)
TUBERCULOSIS ALL FORMS UNDER 15 YEARS ENGLAND AND WALES, 1904-1945

FIGURE 2 *Continued*

berculosis case rates each year thereafter until 1992 (Hamburg, 1995). From 1985 through 1992, the annual incidence of tuberculosis in the United States as a whole rose by about 20%, with similar increases occurring in several Western European nations (Raviglione *et al.*, 1993).

Why has the global tuberculosis epidemic worsened in recent years? In industrialized nations such as the United States, at least four factors have contributed to the reemergence of tuberculosis: (i) the decay of social programs in inner cities, leading to increased homelessness and substance abuse; (ii) immigration of infected individuals from high-prevalence countries; (iii) the emerging epidemic of the acquired immunodeficiency syndrome (AIDS); and (iv) withdrawal of support for tuberculosis control programs. Although much notoriety has, quite appropriately, accompanied the resurgence of tuberculosis in the industrialized nations, the impact of the growing burden of disease has been felt most keenly in the developing nations, where demographic, socioeconomic, political, and epidemiologic factors have all contributed to the problem. In the developing world, the crowding of humanity within increasingly large and unsanitary cities has created optimal conditions for transmission of infectious diseases. In the last two centuries, urbanization has proceeded even more rapidly than has overall population growth: in 1800, the world was populated by approximately 0.9 billion humans, about 2% of whom were urban dwellers; by 1995, the world's population had grown to 5.7 billion, with 45% of persons living in cities (Cohen, 1995). Because of the changing age structure of human populations in developing countries, the number of young adults (the age group most susceptible to tuberculosis) has increased disproportionately (Cohen, 1995). Countries with the fastest rates of population growth are among the world's poorest. In these demographically evolving regions, where exposure to *M. tuberculosis* is intense and lifelong, widespread poverty and malnutrition increase the risk of disease progression for millions of infected individuals.

The plight of sub-Saharan Africa is instructive. For the 660 million inhabitants of this region, the annual risk of infection ranges from 1.5 to 2.5%. By adulthood, more than half of all sub-Saharan African adults have been infected with *M. tuberculosis*, resulting in a prevalence of 2 to 3 million active cases and an annual toll of approximately 0.5 million tuberculous deaths (Murray *et al.*, 1990; De Cock *et al.*, 1992; Raviglione *et al.*, 1992, 1995; Ad Hoc Committee, 1996). The concentration of tuberculous disease and mortality in the age range 25 to 55 years, the most economically fruitful years of life, causes substantial losses in productivity and contributes to the impoverishment of afflicted regions (Murray *et al.*, 1990; Murray, 1994a; Raviglione *et al.*, 1995). According to the United Nations Development Programme, 27 of the world's 40 poorest countries are in sub-Saharan Africa (World Bank, 1993). In many sub-Saharan African nations (19 of 47), the median age at death is below 10 years. Malnutrition—caused by inadequate food intake and infectious diseases (particularly diarrheal diseases)—is an ever-present and growing problem, accounting for nearly a third of the total disease burden in 1990 (Ad Hoc Committee, 1996). Between 1970 and 1991, per capita food production in sub-Saharan Africa declined substantially; during the 1980s, the per capita gross national product declined by more than 1% per year, as population growth continued unabated (World Bank, 1993). Medical expenditures are just US$1–4 per person per year in many countries in this region. Much hardship is attributable to internecine political instability and warfare: since 1957, when Ghana became the first black African nation to achieve independence, 28 of sub-Saharan Africa's 47 nations have suffered at least one violent change of government (Thomas, 1996). In 1990, war ranked as the eighth leading cause of mortality in this region, accounting for approximately 0.3 million deaths (Ad Hoc Committee, 1996).

Unfortunately, these problems are not peculiar to sub-Saharan Africa. In 1994, for example, an estimated 49 million individuals worldwide were displaced from their homes, about half of whom were thought to be infected with *M. tuberculosis* (World Health Organization, 1996). Medical care for these displaced persons has been woefully inadequate. In 1990, the combined foreign aid programs of all wealthy nations provided just $16 million for global tuberculosis control assistance to developing nations (World Health Organization, 1996). In contrast, the United States alone spent an estimated $700 million in 1990 to combat 26,000 domestic cases of tuberculosis; nearly one-third of these cases occurred among the foreign-born, suggesting that wealthy nations could significantly reduce their own burden of tuberculosis by investing in tuberculosis control abroad. The American Lung Association has estimated that improvements in global tuberculosis control could increase economic output from developing nations by as much as $24 billion annually. In both economic and humanitarian terms, tuberculosis control measures have been identified as one of the most cost-effective interventions for investments in global health (Ad Hoc Committee, 1996).

In developing and industrialized nations alike, epidemiologic factors have exacerbated the global tuberculosis problem. Perhaps the most important contributor has been the emerging epidemic since the early 1980s of infection with the human immunodeficiency virus (HIV), the etiologic agent of AIDS (Barnes *et al.*, 1991; Lucas and Nelson, 1994). HIV infection accelerates the clinical course of tuberculosis by crippling and destroying the CD4$^+$ T cells and macrophages that mediate protective immunity. HIV coinfection is the greatest single risk factor known for progression from tuberculous infection to active tuberculous disease. For immunocompetent individuals infected with *M. tuberculosis,* the lifetime risk of developing clinical disease is just 10% (Comstock, 1982; Styblo, 1991). In contrast, individuals coinfected with *M. tuberculosis* and HIV have a 5–15% *annual* risk of disease progression (Selwyn *et al.*, 1989). HIV-infected individuals are more likely to develop progressive tuberculosis following primary infection (or reinfection) and are more likely to reactive a previously acquired latent infection. Disease progression is also accelerated in HIV-infected individuals, in whom tuberculosis often pursues a fulminant course resembling disease in infants. Worse still, the deadly partnership between HIV and *M. tuberculosis* may be mutually reinforcing, as active tuberculosis seems to enhance viral replication and accelerate progression of AIDS (Pape *et al.*, 1993; Whalen *et al.*, 1995; Goletti *et al.*, 1996). Tuberculosis may stimulate HIV replication by triggering the release of cytokines such as tumor necrosis factor α (TNFα), which induces viral transcription, and by activating the CD4$^+$ T cells that support viral replication (Fauci, 1996).

Unfortunately, HIV infection rates are highest among the most sexually active age group (15–45 years), the same age group at greatest risk for developing progressive tuberculosis. As of 1994, an estimated 5.6 million individuals were coinfected with *M. tuberculosis* and HIV worldwide (Raviglione *et al.*, 1995). In sub-Saharan Africa, where coinfection rates are highest, HIV seroprevalence rates among tuberculosis cases range from 20 to 70% (De Cock *et al.*, 1992; Raviglione *et al.*, 1992; Lucas and Nelson, 1994). In a postmortem analysis of HIV-infected individuals in Africa, tuberculosis was the attributable cause of death in one-third of cases and was considered a contributing cause in even more (Lucas and Nelson, 1994). Ominously, in Asia, currently home to two-thirds of tuberculosis cases, the HIV epidemic is spreading more rapidly than anywhere else in the world, portending increased rates of HIV-related tuberculosis in the near future. Among the estimated 1 million HIV-infected and 10–19 million *M. tuberculosis*-infected individuals in the United States, approximately 80,000 persons are thought to be coinfected (Raviglione *et al.*, 1995). A 1991 HIV seroprevalence survey of U.S. tuberculosis clinics revealed a median coinfection rate of approximately 8% among active tuberculosis cases (range 0 to 61%) (Cantwell *et al.*, 1994). Coinfection rates are highest in urban centers: of the 2,995 reported tuberculosis cases in New York City in 1994, 34% were HIV-positive, 22% were HIV-negative, and the HIV status of the remaining 44% was unknown (Hamburg, 1995). One estimate suggests that 50–60% of excess tuberculosis cases in the United States from 1985 through 1992 were attributable to HIV infection (Cantwell *et al.*, 1994).

Another epidemiological factor contributing to the failure of tuberculosis control measures is the evolution of strains of *M. tuberculosis* resistant to multiple antimycobacterial drugs. Multidrug-resistant tuberculosis (MDR-TB) is defined as resistance to at least isoniazid and rifampicin, the two most effective first-line drugs. The worldwide emergence and spread of MDR-TB is one of the saddest chapters in the history of disease control, underscoring the conclusion that scientific progress alone cannot provide enduring solutions to the problems of infectious disease, if substantive long-term commitment to public health is lacking. Antimycobacterial chemotherapy seemed full of promise following the clinical introduction of effective and inexpensive antituberculous drugs in the 1940s and early 1950s: streptomycin in 1945, *p*-aminosalicylic acid in 1946, and, most importantly, isoniazid in 1952. The "disease which medicine never cured" became curable, and the prognosis for millions of tuberculous patients, even those suffering from advanced pulmonary disease, was suddenly and radically altered. Many believed that vigorous application of the new drugs would reverse the course of the global tuberculosis epidemic and might even lead to its complete eradication.

Ironically, the early and spectacular successes of modern chemotherapy may have laid the groundwork for its ultimate failure. Tuberculous patients receiving chemotherapy often experience significant clinical improvement and disappearance of tubercle bacilli from their sputum ("sputum conversion") within a few weeks. Consequently, following the introduction of effective chemotherapy, there was a shift from traditional inpatient treatment of tuberculosis to outpatient treatment, with patients administering their own medications at home without supervision. Tuberculosis, however, is a notoriously stubborn adversary, requiring 6 to 12 months of multidrug therapy. In the absence of close medical supervision, patients whose symptoms have abated tend to stop taking their medications, or to take them sporadically, or to take only one drug at a time. In some poor urban settings, even within the United States, up to 90% of tuberculosis patients receiving chemotherapy fail to complete treatment after discharge from the hospital (Brudney and Dobkin, 1991). The emergence of drug-resistant strains of *M. tuberculosis* under such conditions is hardly surprising; indeed, it became clear within a few years of its inception that streptomycin monotherapy was ineffective due to the rapid emergence of streptomycin-resistant mutants (Yegian and Vanderlinde, 1950). The algorithm describing the emergence of drug resistance is simple: mycobacterial persistence necessitates prolonged multidrug therapy, which leads to patient noncompliance, which selects for drug-resistant mutant strains of *M. tuberculosis*. Failure to appreciate the gravity of treatment failure and to apply vigorous measures to ensure patient compliance has permitted the emergence of strains of the tubercle bacillus resistant to all first-line antimycobacterial drugs.

Although the global prevalence of MDR-TB is not known precisely, the World Health Organization has estimated that approximately 50 million persons already are infected with MDR strains of *M. tuberculosis* (Ad Hoc Committee, 1996). In the United States, approximately 14% of all clinical isolates in the early 1990s were resistant to one or more drugs; appallingly, 4% were resistant to five or more drugs (Bloch *et al.*, 1994). Drug-resistant tuberculosis was most common in New York City, where 32% of isolates were resistant to one or more drugs, and 19% were MDR (Frieden *et al.*, 1993). Epidemiologic data for the Bronx borough of New York City indicated that nearly 40% of tuberculosis cases in 1990–1992 were due to recent transmission; of these isolates, 49% were resistant to one or more drugs, and 24% were MDR (Alland *et al.*, 1994). Treatment of MDR-TB patients with less effective second-line drugs requires extended periods of treatment at enormously increased expense. In the United States, the cost of treating one patient with MDR-TB can approach $250,000, about 10–15 times the cost of treating a case of fully drug-sensitive tuberculosis (Arno *et al.*, 1993). Even when treated aggressively, patients harboring MDR-TB suffer vastly increased rates of treatment failure, relapse, and death: 40–60% for immunocompetent patients, and >80% for immunodeficient patients (Goble *et al.*, 1993; Iseman, 1993). Transmission of MDR-TB to healthy contacts of MDR-TB cases, as indicated by tuberculin skin-test conversion,[1] has been documented in several outbreaks occurring in recent years. Although most of these infected individuals have not yet developed disease, they remain at risk for development of reactivation MDR-TB later in life—certainly a worrisome prospect for the affected individuals, and potentially a disaster for public health. In countries where access to antituberculous drugs is unrestricted and treatment is unsupervised, the potential for a major emergence of MDR-TB is real and frightening. A mere half-century after the introduction of effective chemotherapy, the spectre of incurable tuberculosis haunts us once again.

The emergence and global spread of MDR-TB are all the more tragic because they could have been prevented by the committed implementation of established control strategies. Studies have shown that aggressive case identifica-

---

[1] Tuberculin is an autolysate of the tubercle bacillus, containing bacillary proteins, lipids, and carbohydrates, that will elicit a local delayed-type hypersensitivity response when injected intradermally into an infected individual. The purified protein derivative (PPD) of tuberculin is most commonly used for skin testing.

tion and treatment are the most cost-effective measures available for combating tuberculosis, and cure rates >90% can be achieved by the simple strategy of directly observed therapy (DOT), in which the administration of each dose of medications is supervised by an attending health care or social worker. Even in very poor countries such as Tanzania, Malawi, and Mozambique, cure rates of 85–95% were achieved, and the emergence of drug resistance was suppressed, by implementation of DOT and other strategies to assure patient compliance (Murray *et al.*, 1994). Due largely to the misperception than unsupervised outpatient chemotherapy is much less expensive, DOT has not been used widely: of the more than 7 million new tuberculosis cases occurring annually worldwide, about 5 million receive some drug treatment, but only 0.5 million receive DOT (World Health Organization, 1996). Although it is true that the personnel costs for administration of DOT are greater than for unsupervised outpatient chemotherapy, studies suggest that these expenses are offset by savings that accrue from reductions in retreatment costs and prevention of drug resistance.

Without increased investment in intervention strategies, the global tuberculosis situation is expected to worsen by the beginning of the next millennium. In the decade 1990–2000, an estimated 88 million new cases of tuberculosis are predicted, 8 million of which will be attributable to HIV infection (Raviglione *et al.*, 1995). In the same period, 30 million tuberculous deaths are predicted, 2.9 million of them HIV-related. The number of new cases occurring each year is predicted to rise from 7.5 million in 1990 to 10.2 million in the year 2000. Demographic factors will account for nearly 80% of this increase, with epidemiological factors (predominantly HIV coinfection) accounting for the remaining 20%. The greatest increases will occur in sub-Saharan Africa, where the number of new cases is expected to double by 2000, and in southeast Asia; together, these regions will account for more than 60% of all tuberculous deaths in the 1990s.

Is the problem of tuberculosis being addressed adequately? Worldwide, tuberculosis currently ranks as the seventh leading source of disease burden, measured in terms of disability-adjusted life years (DALYs[2]) lost; by current projections, no improvement is expected in the next several decades (Table I). The global target for tuberculosis control recommended in 1992 by the World Health Organization would achieve 70% detection of new cases and an 85% cure rate by the year 2000 at a cost of US$200 million per annum (World Health Organization, 1992). If implemented, this program is predicted to reduce the prevalence of tuberculosis, as well as the incidence of new cases, by as much as 50% in less than a decade; the cost per life saved would be just US$30–85, making the proposed program one of the most cost-effective health interventions available today. Even a temporary improvement in the cure rate, particularly for infectious (sputum-smear positive) cases, would immediately decrease transmission, leading to sustained reductions in tuberculosis prevalence. In economic terms, any investment in tuberculosis control that improves cure rates, even temporarily, will generate sustained long-term yields. Given the magnitude

---

[2] DALYs are calculated as the sum of years of life lost due to mortality plus years of productive life lost due to incapacitating illness. Murray (1994b) provides a detailed exposition of the DALY concept.

■ **TABLE I**   Leading Causes of Disability-Adjusted LIfe Years Lost Worldwide in 1990 with Their Projected Contribution to the Global Burden of Disease in 2020[a]

	1990		2020	
Cause of DALYs	Rank	% of total DALYs	Rank	% of total DALYs
Lower respiratory infections	1	8.19	6	3.07
Diarrheal diseases	2	7.22	9	2.67
Perinatal conditions	3	6.69	11	2.50
Unipolar major depression	4	3.68	2	5.66
Ischemic heart disease	5	3.39	1	5.93
Cerebrovascular disease	6	2.79	4	4.42
Tuberculosis	7	2.79	7	3.06
Measles	8	2.65	25	1.11
Road-traffic accidents	9	2.49	3	5.13
Congenital anomalies	10	2.39	13	2.23
Malaria	11	2.30	24	1.12
Chronic obstructive pulmonary disease	12	2.11	5	4.15
Falls	13	1.93	19	1.53
Iron-deficiency anemia	14	1.78	39	0.54
Protein-energy malnutrition	15	1.52	37	0.56

[a] Data from Ad Hoc Committee (1996).

of the problem, and the immediate improvement that could be achieved by the committed application of existing control measures, global tuberculosis management is a sound investment. At the same time, sustained investment in basic and applied research is crucial if we are ever to progress from control to elimination of tuberculosis as a public health problem. Research and development (R & D) have an excellent record for generating gains in the quality and cost-effectiveness of health care yet receive proportionately little investment of funds. In 1992, for example, the world's nations collectively spent US$1,600 billion on health care, yet health-related R & D received just 3.4% (US$56 billion) of all health-related expenditures (Ad Hoc Committee, 1996). Sadly, although not surprisingly, R & D addressing the health problems of the developing nations is particularly neglected: in 1992, the combined problems of pneumonia, diarrheal diseases, and tuberculosis received a mere US$133 million dollars (0.2% of global health-related R & D spending), while accounting for one-fifth of the global disease burden (Ad Hoc Committee, 1996). Tuberculosis alone accounts for approximately 3% of the global disease burden each year yet receives less than 0.1% of health-related R & D expenditures. Correcting these "monumental mismatches" will require greater understanding of the problems that beset the developing world, and recognition by the wealthy nations that many of the problems of the world's less fortunate peoples are our problems as well.

II. PATHOGENESIS

> Although consumption has often been labeled "the great white plague," in reality it bore
> little resemblance to the epidemics that had earlier ravaged Europe. Consumption did not
> suddenly appear, devastate a population, and then abruptly recede. Its course was at once
> less precipitous and more tenacious, taking a grim toll year after year. Its sufferers did
> not usually succumb within a matter of days. Acute attacks alternated with remissions;
> the process of wasting and dying could take a few years or span several decades. In effect,
> consumption was a chronic disease, not a plague.
>
> Shiela M. Rothman, *Living in the Shadow of Death: Tuberculosis and the Social
> Experience of Illness in American History*, 1994, p. 13

A. Progressive Primary Tuberculosis

For centuries, the cause of pulmonary tuberculosis, known from ancient times
as "phthisis" (wasting) or "consumption," was shrouded in mystery. Observing
that the disease often ran in families, pursuing an indolent, chronic course quite
unlike that of the familiar acute infectious diseases, many believed that consumption arose from a heritable constitutional disorder, and this bias persisted
among medical professionals well into the nineteenth century. The idea that
consumption might arise from airborne contagion due to a microorganism was
first explicitly articulated by an obscure English physician, Benjamin Marten,
in *A New Theory of Consumption, more especially of a Phthisis or Consumption of the Lungs,* published in 1722 in London. With remarkable perspicacity,
Marten hypothesized that

> The Original and Essential Cause, then, which some content themselves to call a vicious
> Disposition of the Juices, others a Salt Acrimony, others a strange Ferment, others a Ma-
> lignant Humour, may possibly be some certain Species of *Animalculae* or wonderfully
> minute living creatures that, by their peculiar Shape or disagreeable Parts are inimicable
> to our Nature; but, however, capable of subsisting in our Juices and Vessels. . . . [T]he
> minute animals or their seed . . . are for the most part either conveyed from Parents to
> their Offspring hereditarily or communicated immediately from Distempered Persons to
> sound ones who are very conversant with them. . . . It may, therefore, be very likely that
> by habitual lying in the same Bed with a consumptive Patient, constantly eating and
> drinking with him or by very frequently conversing so nearly as to draw in part of the
> Breath he emits from the Lungs, a Consumption may be caught by a sound Person. . . .
> I imagine that slightly conversing with consumptive Patients is seldom or never sufficient
> to catch the Disease, there being but few if any of those minute Creatures . . . communi-
> cated in slender conversation. [Dubos and Dubos, 1952, pp. 94–95.]

In 1865, Jean-Antoine Villemin lectured before the French Academy of Science on his findings that normal rabbits or guinea pigs would develop tuberculosis if injected with sputum or caseous tissue from tuberculous patients, and
that the contagion could be transmitted by inoculation from animal to animal
in unending series. Although these experiments fairly established an infectious
etiology for tuberculosis, there were ignored or discounted by the pundits of the
day. This skepticism arose partly from the deep-rooted prejudice that tuberculosis was a constitutional disorder, and partly from the fact that the "germ
theory" of disease had not yet taken root. In subsequent years, numerous investigators, including the pre-eminent Louis Pasteur, tried and failed to identify a
microorganism as causative agent of tuberculosis. In retrospect, their failure

was probably due to the extremely slow growth rate of the tubercle bacillus, which divides just once every 24 hr; overgrowth of contaminating microorganisms would have frustrated any attempt to culture the bacillus from tuberculous lesions by the methods then available (i.e., inoculation of liquid broth with infected tissue).

The German physician and self-taught microbiologist Robert Koch succeeded where others failed in part because of his serendipitous "discovery" of solid culture medium for the isolation of pure bacterial populations. The story of this discovery provides a striking example of Pasteur's dictum, "Chance favors the prepared mind." According to legend, Koch one day noticed that the smooth surface of a cut potato left lying about was covered with small, discrete nodules of various hues which, on microscopic inspection, were found each to consist of a homogeneous population of a single microbe. It occurred to him that the simple expedient of using solid medium for the cultivation of microbes (sliced potatoes were soon replaced by coagulated blood serum) would permit the isolation of pure cultures of a particular species from an initially heterogeneous population. The invention of solid medium, and new techniques for staining and visualizing the tubercle bacillus within infected tissues, permitted Koch to make his greatest discovery. Even with these tools in hand, a lesser scientist may well have failed; the discovery of the tubercle bacillus owed even more, perhaps, to Koch's tenacity and personal courage than to his technical prowess:

> The stroke of "genius" that saved Koch's experiments from failure was his practice, at exactly the right time and place, of the homely quality of patience. Until Koch set out to isolate and cultivate the tubercle bacillus, he had never come across a germ that would not appear visible to the naked eye within two or three days after the implantation of material within the culture tube. With such an experience of five years or more behind him, it is thinkable that practically every man, Koch included, would have arrived at the general idea that bacteria were relatively rapidly growing forms of life which from single germs would heap up into noticeable colonies within a very few days of their localization at a favorable site, and, further, that a failure to do so would denote their absence in any material under scrutiny.
>
> But, though Koch did not know it at the time, here again, as in staining capacities and in favorable media, the tubercle bacillus is a "different" kind of germ. Those forms of it that cause disturbance in man and other mammals will take nearer two or three weeks than as many days after removal from tissues to first come to view in cultures. Accordingly, at the expected time after planting his first tubes of blood-serum, Koch saw nothing significant astir in his tubes. Even at the end of the first week these tubes remained as unchanged as the day he "inoculated" them. One can well imagine the anxious and despairing heart of the eager investigator who first faced this issue. . . . It is here, in the extraordinarily long interval between planting and "sprouting", that Koch's "genius" as an investigator underwent its supreme test.
>
> Given the same circumstances—the same pioneering quest, the same limited knowledge of possibilities, the same imperfection of initial media—and the view is almost compelling that only the rarest of investigators would have maintained the prolonged and heartbreaking vigil for the earliest "showing through" of minutest particles that belonged, as yet, only to the shadowy realm of "working hypothesis". [Krause, 1932.]

On March 24, 1882, in an historic address to the Berlin Physiological Society, Koch revealed his identification of *Mycobacterium tuberculosis* as the etiological agent of tuberculosis. Koch was barred from presenting his revelation to the *Berliner Pathologische Gesellschaft* because Rudolph Virchow, the reigning authority in pathology, vehemently opposed the theory that tuberculosis was a communicable, rather than an organic, disease (Krause, 1932). Virchow himself attended Koch's lecture and, in the stunned silence that followed the overwhelming evidence marshalled by the young physician, left the lecture hall without uttering a word. Koch delivered his proofs like hammer blows: "To prove that tuberculosis is a parasitic disease, that it is caused by the invasion of bacilli and that it is conditioned primarily by the growth and multiplication of the bacilli, it was necessary to isolate the bacilli from the body; to grow them in pure culture until they were freed from any disease-product of the animal organism which might adhere to them; and, by administering the isolated bacilli to animals, to reproduce the same morbid condition. . . ." (Koch, 1882). These beautifully simple and rigorous criteria for the identification of the tubercle bacillus as the causative agent of tuberculosis, known since as "Koch's Postulate," established the classic standard for proof of etiology in infectious diseases.

Koch was the first to infect experimental animals with *M. tuberculosis* by the aerogenous route, which is now accepted as the principal route for acquisition of pulmonary tuberculosis. Controversial for some time, the aerogenous route of transmission was firmly established by the classic studies of Richard Riley and William Wells in the 1950s, demonstrating that airborne particles generated by tuberculous patients could infect guinea pigs inspiring air shunted from the patients' rooms (Riley, 1957; Riley *et al.*, 1959, 1962; Sultan *et al.*, 1960). These infectious particles, generated in tremendous numbers when the tuberculous individual coughs, sneezes, or merely speaks, are very small (a droplet 5 μm in diameter may contain from 1 to 10 tubercle bacilli) and can remain suspended in the air for long periods. Conversion of the exposed individual from negative to positive tuberculin skin-test reactivity, between 2 and 6 weeks after exposure, signals the establishment of infection. As with many other infectious diseases, it is critical to distinguish between tuberculous infection and tuberculous disease. Approximately 90% of infected individuals will never develop clinical disease (Comstock, 1982; Styblo, 1991; Enarson and Murray, 1996). Of the 10% of infected persons who do develop disease, half will develop progressive primary tuberculosis within the first 2 to 3 years following infection, while the other half will control the infection initially, only to develop tuberculosis later in life due to the endogenous reactivation of latent tubercle bacilli persisting within residual lesions. Given the remarkable proclivity of the tubercle bacillus to persist for the lifetime of the infected host, it is not clear whether the immunity that develops in the 90% of infected individuals who never develop clinical disease actually results in sterilization or merely effective lifelong suppression of latent tubercle bacilli. The long-term maintenance of tuberculin skin-test reactivity suggests that tubercle bacilli may persist in many infected individuals, although the issue of whether maintenance of skin-test reactivity actually requires the continued presence of viable bacteria remains unsettled.

Tuberculous disease can arise from three sources of contagion: primary progression of infection arising from inhalation of small numbers of tubercle bacilli, endogenous reactivation of the latent residuum from an earlier infection, and exogenous reinfection by the aerogenic route following resolution of a previous infection. As discussed later, the pathology of primary progressive tuberculosis is distinct from that of postprimary tuberculosis arising from endogenous reactivation or exogenous reinfection. In order to understand the events leading to the development of progressive primary or postprimary tuberculosis, it is crucial to understand the pathogenetic events occurring in the early stages of infection. Much of our knowledge of the pathogenesis of tuberculosis derives from histopathological studies of tuberculous human cadavers conducted largely in the first half of the twentieth century, and from experimental aerogenic infection of laboratory animals, in particular the classic work of Max Lurie with inbred rabbits and of Donald Smith with guinea pigs. The stages of untreated tuberculous infection and disease will be described only briefly here; a more detailed account is provided by Dannenberg and Rook (1994).

In humans, pulmonary tuberculosis usually initiates with the implantation of a single infectious particle in the alveolar spaces of the lower lungs, as indicated by the presence of just one primary lesion in the majority of tuberculous individuals. Most primary lesions occur within the terminal ramifications of the respiratory tree, just under the pleura; presumably, inspired particles implanting at higher levels of the tree are removed by the mucociliary elevator. Only particles of diameter $\leq 5-10$ μm, containing from 1 to 10 tubercle bacilli, are small enough to reach the alveoli (Sonkin, 1951), where they are likely phagocytosed by resident alveolar macrophages. In some cases, depending on the innate resistance of the host and the virulence of the infecting strain of *M. tuberculosis,* the engulfed tubercle bacilli may be destroyed. Bacilli that escape destruction replicate within the infected macrophage until the bacillary burden becomes cytotoxic. The mechanisms by which tubercle bacilli kill macrophages remain a fundamental but poorly understood problem. In part, cytotoxicity may be due to the exquisite sensitivity of infected cells to TNFα, which is secreted copiously by macrophages exposed to tubercle bacilli (Filley and Rook, 1991; Filley *et al.,* 1992). Components of the mycobacterial cell wall, in particular lipoarabinomannan (LAM), trigger TNFα secretion (Moreno *et al.,* 1989). Induction of apoptosis by an unknown mechanism may also play a role in the destruction of infected host cells (Molloy *et al.,* 1994). Death of infected cells by apoptosis is associated with reduced bacillary viability, suggesting that this mechanism of tissue damage may represent a protective response of the host. Bacilli released from dying macrophages are ingested by neighboring alveolar macrophages and by monocytes and neutrophils emigrating from the bloodstream. These phagocytes are attracted to the site of infection by chemotactic bacillary products and by inflammatory chemokines, such as monocyte chemotactic protein-1 (MCP-1) and interleukin-8, secreted by infected macrophages (Kasahara *et al.,* 1994). Accumulation of blood monocyte-derived macrophages at the focus of infection generates a compact granulomatous structure, the characteristic nodule or "tubercle" first described by Franciscus Delaboe Sylvius in his *Opera Medica* of 1679. Infiltrating neutrophils may also contribute to the early stages of tubercle formation by ingesting and killing tubercle bacilli (Jones *et al.,* 1990) and by releasing monocyte chemotactic factors; as the nascent lesion develops, however, monocytes replace neutrophils as the predominant cell type.

In an immunologically naive host, unrestricted exponential growth of tubercle bacilli within nonactivated alveolar macrophages and blood monocyte-derived macrophages continues for several weeks following infection. As bacillary growth progresses, the tubercle gradually expands outward and penetrates the lung parenchyma. Small numbers of tubercle bacilli escape from the primary parenchymal lesion and enter the lymph nodes draining the affected region of the lung. Replication of tubercle bacilli within the draining lymph nodes frequently causes pronounced lymphadenopathy, which is indicative of primary tuberculosis, as significant nodal involvement seldom occurs in postprimary tuberculosis, except in cases of severe immunosuppression. The primary parenchymal lesion and associated lymph node(s) are referred to as the "primary complex" or "Ghon complex" (Ghon, 1923). From the draining nodes, tubercle bacilli are conducted via the thoracic duct to the bloodstream, whence they are disseminated to the extrapulmonary organs, as well as to previously unaffected regions of the lungs. Within these metastatic foci of infection, and within the primary lesion, exponential bacillary growth continues until the development of a specific T-cell response between 3 and 6 weeks post infection. Thereafter, the rate of bacillary growth declines abruptly as inhibition or destruction of tubercle bacilli occurs within macrophages activated by T-cell derived cytokines, such as interferon-γ. The emergence of an *M. tuberculosis*-specific T-cell population is signaled by the development of a delayed-type hypersensitivity (DTH) response to intradermally injected tuberculin PPD.

Sensitization of the host to the products of the tubercle bacillus, and direct cytotoxicity of those products, causes the centers of expanding tuberculous lesions to undergo caseating necrosis, a "tissue-damaging response" that may benefit the host by slowing bacillary growth (Dannenberg and Rook, 1994). The necrotic center of a tuberculous lesion is surrounded by a sheath of monocytes and T cells, which effectively suppress the intracellular growth of tubercle bacilli once strong cell-mediated immunity (CMI) is established. However, the emergence of effective CMI usually requires several weeks, during which the necrotic center of the lesion expands. In a susceptible host, effective CMI may never develop, and intracellular proliferation of tubercle bacilli may continue within nonactivated macrophages at the periphery of the lesion. In the absence of effective CMI, the unchecked intracellular replication of tubercle bacilli causes progressive destruction of lung tissue and, ultimately, the death of the host.

The mechanisms responsible for destruction of host tissues within caseous necrotic foci are ill-defined but probably reflect injury caused by excessive host immune responses. Direct injury of cells by toxic bacillary products, such as "cord factor" (trehalose dimycolate, a component of the mycobacterial cell wall), may also contribute to tissue damage. Immunopathological mechanisms may include the local overproduction of TNFα and other toxic cytokines, reactive oxygen and reactive nitrogen intermediates released by activated macrophages, and cytotoxic T lymphocytes directed against infected host cells. The "tissue-damaging response" may slow disease progression at early stages of infection, prior to the emergence of effective CMI, by lysing parasitized nonactivated macrophages that cannot suppress intracellular bacillary growth. Although the tubercle bacillus is not an obligate intracellular pathogen, extracellular growth of bacilli within the acellular necrotic centers of caseous lesions is inhibited by the environment of the solid caseous tissue. Factors contributing to the suppression of extracellular bacilli may include toxic fatty acids released from moribund host cells, reduced pH, and anoxia. As long as the caseous tissue remains solid, bacillary growth is suppressed. Unfortunately, solid caseum may undergo liquefaction, creating a rich culture medium for tubercle bacilli, which may then grow to enormous numbers. Liquefaction is associated with a strong host response to bacillary antigens and is likely an example of immunopathology caused by an inappropriate or excessive host CMI response.

Unrestricted dissemination of tubercle bacilli from the lungs may lead to the establishment and progression of numerous metastatic lesions in multiple tissues. This form of acute disseminated disease is known as "miliary" tuberculosis, due to the resemblance of the metastatic lesions to millet seeds (Baker & Glassroth, 1996). As a complication of progressive primary tuberculosis, miliary tuberculosis is most common among individuals with limited CMI, including children, the elderly, and immunocompromised individuals (such as those with AIDS). Miliary disease may also occur as a late-stage complication of adult-type pulmonary tuberculosis, following the erosion of a tuberculous lesion into a major blood or lymph vessel. Miliary tuberculosis, involving the dissemination of large numbers of bacilli, must be distinguished from the early hematogenous dissemination described previously, which occurs in most if not all cases of infection and which seeds only very small numbers of bacilli to the extrapulmonary organs. Although uncommon, miliary tuberculosis is rapidly progressive and invariably fatal if not treated promptly.

Erosion of a tuberculous lesion into an airway leads to the bronchogenic spread of bacilli within the lung, the establishment of metastatic pulmonary lesions, and expulsion of bacilli from the lungs when the infected person coughs or expectorates. The appearance of tubercle bacilli in the sputum signals the onset of infectivity. Identification and treatment of "sputum positive" cases currently form the most effective means of combating tuberculosis, as it is they who are ultimately responsible for transmission and perpetuation of the contagion. The expulsion of liquified caseous material from a lesion excavates a cavity. During the prechemotherapy era, roughly 80% of patients with tuberculous pulmonary cavities died within the first year of diagnosis (Barnes and Barnes, 1928). In the absence of treatment, healing proceeds by deposition of a fibrotic capsule around the lesion and progressive calcification of necrotic tissue. Culture of residual lesions at autopsy suggests that complete calcification correlates with sterilization of the lesion. However, tuberculous cavities seldom heal completely without chemotherapy, and, like smouldering caseous necrotic lesions, may continue to harbor viable tubercle bacilli for the lifetime of the host.

Established tuberculous lesions may oscillate between periods of activity and dormancy; chronic or sporadic disease is common, and chronically ill individuals may continuously shed tubercle bacilli for years or decades. One case study, remarkable for the detail with which the course of disease was followed, will serve to illustrate this point; the report describes a patient who suffered from chronic cavitary tuberculosis from 1927 until her death in 1967:

> A white woman had active pulmonary tuberculosis of 41 years' duration. She had bilateral bronchiectasis, cavitation of the right upper lobe, a collapsed left lung, pleural effusion, and an induced left pneumothorax of 36 years' duration [Fig. 3]. She had received no antituberculous drugs for 28 years from the onset of her illness, and these drugs, when

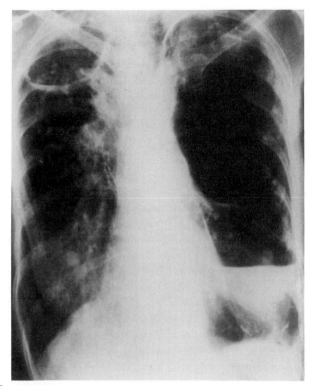

FIGURE 3 Final chest X-ray before death of a patient with chronic pulmonary tuberculosis of 41 years' duration (1927–1967). The patient had a huge cavity in the right apex, pleural effusion, bilateral bronchiectasis, and an induced left pneumothorax of 36 years' duration with collapsed left lung. (Reprinted from Edwards *et al.,* 1970, *American Review of Respiratory Diseases* 102, 448–455. With permission.).

> finally prescribed, were taken inadequately. Pleural effusion present for 28 years required aspiration. *Mycobacterium tuberculosis* recovered by culture of sputum was resistant to all antituberculous drugs except streptomycin. [Edwards *et al.,* 1970.]

During most of her illness, the patient was ambulatory and sputum positive, and presumably was responsible for considerable transmission of multidrug-resistant tuberculosis to susceptible persons in her community. The ability of *M. tuberculosis* to establish such a long-term relationship with the infected host, in which the host survives but continues to shed bacilli, doubtless contributes to the success of this tenacious pathogen in establishing and maintaining itself within human populations.

B. Postprimary Tuberculosis: Latency and Endogenous Reactivation

Even when host resistance is sufficient to suppress active disease, tubercle bacilli frequently persist as a latent infection. The latently infected individual may show no signs of clinical illness for many years but remains at risk for reactivation of the latent focus later in life. In the words of George Comstock, one of the pre-eminent epidemiologists of tuberculosis: "Following infection, the incubation period of tuberculosis ranges from a few weeks to a lifetime" (Comstock et al., 1974a) Reactivation tuberculosis often occurs in individuals whose primary infection was unsuspected, and the "healed" residuum of the primary lesion of remote infection is discovered only after the latent infection reactivates to cause clinical disease. Early investigators demonstrated that viable tubercle bacilli may persist for many years in the tissues of individuals with no clinical signs or symptoms of tuberculosis, and that these latent bacilli retain the ability to cause disease when injected into experimental animals (Wang, 1916, and references therein). Autopsy studies of individuals with "healed" pulmonary lesions demonstrated that the Ghon complex is usually sterilized (particularly when calcified), whereas the metastatic lesions established by the early hematogenous dissemination of tubercle bacilli frequently continue to harbor viable organisms (Opie and Aronson, 1927; Robertson, 1933; Feldman and Baggenstoss, 1938, 1939). Consequently, reactivation of "healed" primary lesions occurs only rarely (Pagel, 1935; Reichle and Gallavan, 1937), and most cases of reactivation tuberculosis arise from residual metastatic lesions located in the apical regions of the lung (see below). It is not clear why the host response is able to eradicate the large numbers of bacilli in the primary complex but is unable to sterilize secondary lesions, which typically contain fewer bacilli. In general, little is known of the host and pathogen factors that contribute to the establishment and maintenance of latent infection, or of the events transpiring during reactivation from latency. These issues are not merely academic, as approximately half of all active cases of tuberculosis are thought to arise from endogenous reactivation of latent infection. Better understanding of latency might lead to improved interventions to prevent the establishment and reactivation of latent infection, or to eradicate latent bacilli.

Reactivation tuberculosis typically results from immunosuppression of the latently infected host. Even a rather mild impairment of host immunity can cause reactivation. Individuals with HIV infection are at increased risk for reactivation of latent tuberculosis from the earliest stages of progression to AIDS, when CD4$^+$ T cell counts are still relatively high and other opportunistic infections have not yet appeared. In developing nations, where coinfection with M. tuberculosis and HIV is common, clinical tuberculosis is a sentinel disease, the most frequent first indicator of underlying HIV infection. Other immunosuppressive conditions that may increase the risk of reactivation tuberculosis include end-stage renal disease, diabetes, corticosteroid therapy, cancer (particularly Hodgkin's disease and malignant lymphomas), silicosis, gastrectomy, and intravenous drug abuse.

The gradual decline in CMI that accompanies old age may cause reactivation of latent tuberculous infection, and it is responsible for the majority of cases of tuberculosis occurring in industrialized countries, where the prevalence of tuberculosis is low. If latent tuberculous infection is acquired in childhood, the hormonal changes that occur during puberty may decrease resistance, leading to reactivation and accelerated progression of disease. Indeed, in developing countries today, young adults experience the highest rates of tuberculous disease and death, as was true also in the industrialized nations well into the twentieth century (Fig. 4). Age-related changes in susceptibility to tuberculosis were convincingly documented by Wade Hampton Frost (1939), who determined mortality rates for the same-age cohort followed from 1880 to 1930, a time when

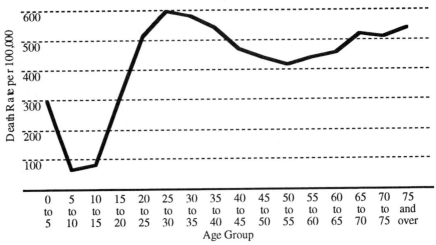

FIGURE 4   Pulmonary tuberculosis death rate (deaths per 100,000 population per annum) in the United States in 1900, by age group (United States Census Bureau statistics). (Redrawn from Rich, et al. 1944, "The Pathogenesis of Tuberculosis." Charles C. Thomas, Springfield, IL. p. 181. With permission.)

most individuals in the United States were infected during childhood (Opie, 1922; Drolet, 1934). In Frost's cohort, mortality rates were very high in early childhood (age 0 to 4 years) but declined steeply thereafter, reaching a nadir during late childhood (age 5 to 9 years); mortality rates rose again at puberty, peaked during early adulthood (age 20 to 30 years), and then declined gradually. Similarly, Drolet (1938) showed that case fatality rates (ratio of deaths to cases) during the 2-year period 1933–1934 were high (~46%) in the age group 0–4 years, lowest (~15%) in the age group 5–14 years, and high in the age groups 15–24 years (~39%) and 25–44 years (~50%) (Table II). More recently, in a prospective study of 82,269 PPD-positive (i.e., *Mycobacterium*-infected) Puerto Rican children, Comstock *et al.* (1974a) found that "infected children who pass unscathed through the risks of childhood face another period of high risk in adolescence." Stead and Lofgren (1983) further suggested that for adults over 60 years of age, the risk of reactivating a latent tuberculous infection may rise rather dramatically. The physiological mechanisms underlying the increased susceptibility to tuberculosis during infancy, young adulthood, and old age are not well understood and have received little attention. Rich's classic treatise on tuberculosis (1944) is still one of the richest sources discussing this important and neglected issue.

C. Experimental Studies of Latency

Although an intact host response is clearly essential to prevent reactivation of latent infection, the immune mechanisms required to establish and maintain latency have not been defined. The increased reactivation rate of HIV-infected individuals harboring latent *M. tuberculosis* suggests that CD4+ T-cell function is essential to prevent reactivation, although precisely which subclass(es) of CD4+ T cells is important is not known. Reactivation of latent *Leishmania* infection in HIV-infected individuals has also been documented, and, in a mouse

TABLE II Tuberculosis Case Fatality Rates (% Fatal Cases and Deaths/Cases), by Age Group, in the Period 1933–1934[a]

Locale	Population	0–4 years		5–14 years		15–24 years		25–44 years	
		%	Deaths/cases	%	Deaths/cases	%	Deaths/cases	%	Deaths/cases
New York State	5,880,649	47%	150/317	9%	110/1,244	32%	841/2,618	51%	2,722/5,376
New York City	7,226,164	56%	342/616	25%	202/813	34%	1,400/4,058	40%	3,812/9,552
New Jersey	4,231,000	72%	117/163	13%	96/726	41%	747/1,826	52%	1,906/3,651
Chicago	3,490,700	36%	141/390	12%	135/1,088	33%	776/2,336	38%	1,756/4,680
England and Wales	40,467,000	46%	3,609/7,891	15%	2,959/19,797	42%	13,340/31,967	54%	24,810/45,674
London	4,230,200	44%	287/658	16%	247/1,556	34%	1,537/4,486	44%	2,804/6,368

[a] Data from Drolet (1938).

model of *Leishmania* latency, ablation of CD4$^+$ T cells was shown to cause reactivation (Aebischer, 1994). Stenger *et al.* (1996) demonstrated that chronic production of nitric oxide at the site of *Leishmania* persistence is essential for maintenance of latency, as treatment of latently infected mice with inhibitors of nitric oxide synthase triggered parasite replication and disease progression. Although nitric oxide was shown to play a critical role in protection against *M. tuberculosis* in the mouse model (Denis, 1991a; Flesch and Kaufmann, 1991; Chan *et al.*, 1992), a role for nitric oxide in the maintenance of mycobacterial latency has not yet been demonstrated.

The neglect of mycobacterial latency as a topic for experimental investigation is puzzling, given the scientific and medical significance of the phenomenon. The most extensive experimental studies of tuberculosis latency were conducted by Walsh McDermott and colleagues at Cornell University in the 1950s and 1960s (McCune and Tompsett, 1956; McCune *et al.*, 1956, 1966a,b). The Cornell group found that 3 months' chemotherapy of tuberculous mice with the antimycobacterial drugs isoniazid and pyrazinamide consistently achieved "sterilization." Extensive efforts to detect persisting tubercle bacilli in the treated mice, using assays sensitive enough to reveal even a few residual organisms, were unsuccessful. Nonetheless, during the postchemotherapy observation period, a proportion of the treated animals displayed spontaneous bacteriological relapse beginning a few months after withdrawal of antimicrobials. Immunosuppression with corticosteroids provoked reactivation in all of the treated animals. The tubercle bacilli reappearing in the treated animals were fully sensitive to both isoniazid and pyrazinamide, demonstrating that their persistence was due to phenotypic drug tolerance rather than to genotypic drug resistance. McDermott concluded that isoniazid and pyrazinamide were highly effective against the actively replicating tubercle bacilli responsible for clinical disease but were not effective against latent "persisters." Later studies described the same phenomenon when infected mice were treated with rifampicin substituted for pyrazinamide (Grosset, 1978).

The failure of antimicrobials to eradicate latent bacilli from tuberculous patients has also been documented in clinical studies (Beck and Yegian, 1952; Falk *et al.*, 1954; Hobby *et al.*, 1954; Auerbach *et al.*, 1955; Vandiviere *et al.*, 1956; Wayne and Salkin, 1956; Canetti, 1965; Kopanoff *et al.*, 1988; Girling, 1989). The lengthy and complex chemotherapeutic regimens required to treat tuberculosis reflect the difficulty of eradicating slowly-growing and latent organisms. The likelihood of patient noncompliance rises steadily with increasing duration of treatment. In a study conducted in Botswana, default rates during the twentieth week of chemotherapy were approximately five times higher than during the first six weeks of treatment (Murray, 1994). Patient noncompliance results in high rates of treatment failure and the emergence of bacterial drug resistance. Clearly, improved understanding of the factors that influence the establishment and maintenance of mycobacterial persisters, and the development of drugs targeting this recalcitrant subpopulation of bacilli, should be high priorities for tuberculosis research.

Little is known about the metabolic state of tubercle bacilli during latency. It is clear that there is no significant expansion of the latent bacillary population over time; however, two very different models could account for the mainte-

(B)

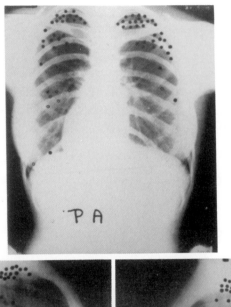

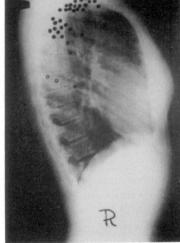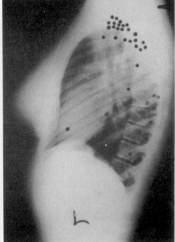

FIGURE 6 *Continued*

The metastatic lesions established by hematogenous dissemination of bacilli occur anywhere in the lung, yet tubercles arising in the apical or subapical regions progress more rapidly than do tubercles occurring in the lower lung. Since tubercle bacilli require oxygen for growth, the higher oxygen tension in the upper lung could plausibly account for the more rapid progression of the tuberculous process there (Rasmussen, 1957; Riley, 1960). It is possible, however, that the susceptibility of the lung apex may be due to some other physiological peculiarity of inadequate perfusion (e.g., a deficiency of immune effector cell numbers or function), or sluggish lymph drainage (Goodwin and Des Prez, 1983). The greater susceptibility of the upper lung fields is particularly graphic in cases of miliary tuberculosis, when the lung is riddled with numerous small tubercles: a gradient of tubercle size is often apparent, with tubercles largest at

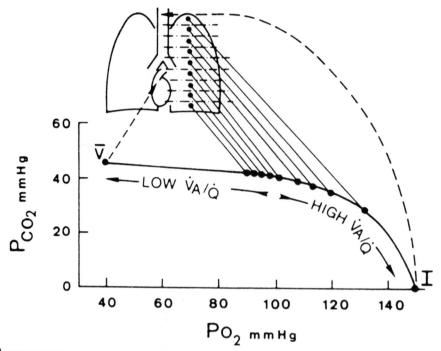

FIGURE 7 Regional differences in ventilation (V_A) and perfusion (Q) in the lung of the upright human. In the upper lung, the V_A/Q ratio is high, resulting in low CO_2 tension and high O_2 tension. In the lower lung, the $V_A Q$ ratio is low, resulting in high CO_2 tension and low O_2 tension. (Reprinted from West, 1995 "Respiratory Physiology the Essentials." 5th Edition. Williams & Wilkins, Baltimore, Maryland. With permission.).

the apex and diminishing in size toward the base (Fig. 8A). Suggestively, this gradient of tubercle size is not apparent in cases of miliary tuberculosis among recumbent infants (Fig. 8B). Cavitary lesions, arguably the most severe form of pulmonary damage in tuberculosis, show a marked proclivity for the upper lung fields (Fig. 9). Among 152 consecutive autopsies of persons dying with tuberculosis, Ewart (1882) observed that tuberculous cavities occurred approximately 25 times more frequently in the upper regions of the lungs than in the bases of the lower pulmonary lobes. These findings were repeatedly confirmed by later investigators (Weigel, 1924; Sweany *et al.*, 1931; Rich, 1944; Medlar, 1948). The disparity in resistance of the upper and lower lung fields is so extreme that progression and cavitation of lesions in the apex may occur while lesions in the base are undergoing regression and healing, as recognized by Arnold Rich more than half a century ago:

> That inhaled tubercle bacilli induce progressive infection in the apical portions and fail to do so in the lower portions appears to be due to some as yet undetermined influence that favors their multiplication at the apex. The resistance of the lower portions of the lungs of white adults to tuberculous infection is often amazing. It is not so unusual to find a cavity in one upper lobe, with only a few small lesions in the lower parts of the lungs, though bacilli had been freely discharged into the bronchi in large numbers over a considerable period of time, providing abundant opportunities for aspiration. In such cases the opposite apex, however, not infrequently becomes diseased, whether by the haematogenous or aerogenous route. Infection of the portions of the lungs below a cavity, as a

(A)

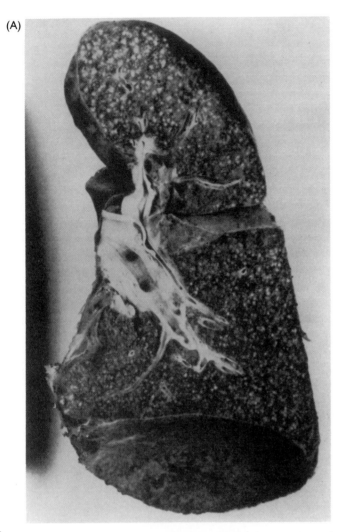

FIGURE 8 (A) Miliary tuberculosis in an 8-year-old child. Tubercles are largest at the lung apex and diminish in size toward the base. (B) Miliary tuberculosis in a 5-month-old infant. Tubercules at the lung apex are not larger or more numerous than elsewhere. (Reprinted from Rich, *et al.* 1944, "The Pathogenesis of Tuberculosis." Charles C. Thomas, Springfield, IL. p. 775. With permission.)

result of aspiration of bacilli, is, of course, common, but in spite of the fact that the opportunity for clumps of bacilli to find their way into the alveoli is unparalleled under these circumstances, the sites of infection are commonly only small, discrete ones that tend to become encapsulated and arrested. When the apex of the opposite lung becomes infected, however, the tendency toward progressive disease is much greater. [Rich, 1944, pp. 770–771.]

That the susceptibility of the lung apex is attributable to the upright posture was demonstrated by Edgar Medlar in a classic experiment (Medlar and Sasano, 1936). Medlar observed that *M. tuberculosis*-infected rabbits, whose four-footed stance situates the base (caudal end) of the lung above the apex (rostral end), developed progressive tuberculous disease and cavitation in the

(B)

FIGURE 8   *Continued*

most elevated (dorso-caudal) lung fields (Fig. 10A). In contrast, when infected rabbits were supported in an upright posture, the progressive tuberculous process shifted from the base (caudal end) to the apex (rostral end) of the lung (Fig. 10B). Medlar likewise observed that the primary lesions of bovine tuberculosis in cattle occurred more or less randomly throughout the lungs, as do primary lesions in humans, whereas cavities were clustered in the dorso-caudal lung region—again, in the most elevated region of the lung (Medlar, 1940).

   The lesions of postprimary tuberculosis, whether of exogenous reinfection or endogenous reactivation, occur almost exclusively in the apical and subapical lung fields (Fig. 6B). This observation suggests that acquired host resistance can effectively protect the lower lung but cannot extend the same protection to the upper lung. The entire question of protective immunity, then, which is an enormously vexing problem in tuberculosis, may possibly hinge on a single crucial

Loan Receipt
Liverpool John Moores University
Library and Student Support

Borrower Name: Leonard,Sam
Borrower ID: **********8117

Epidemiology /
31111012594543
Due Date: 02/02/2012 23:59

Infection prevention and control :
31111012783336
Due Date: 02/02/2012 23:59

Emerging infections /
31111008809343
Due Date: 02/02/2012 23:59

Total Items: 3
26/01/2012 12:43

Please keep your receipt in case of
dispute.

Loan Receipt
Liverpool John Moores University
Library and Student Support

Borrower ID: 21111146663117

Loan Date: 10/02/2010

Loan Time: 11:49 am

Epidemiology /
31111012594543

Due Date: 17/02/2010 23:59

Please keep your receipt
in case of dispute

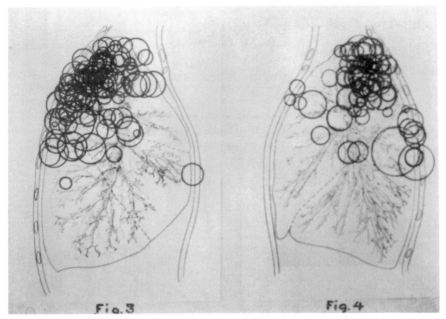

FIGURE 9 Location of 268 cavities in 204 patients with cavitary pulmonary tuberculosis. Cavities are located almost exclusively in the upper lobes. (Reprinted from Sweany *et al.*, 1931 *American Review of Tuberculosis* 24, 558–582. With permission.)

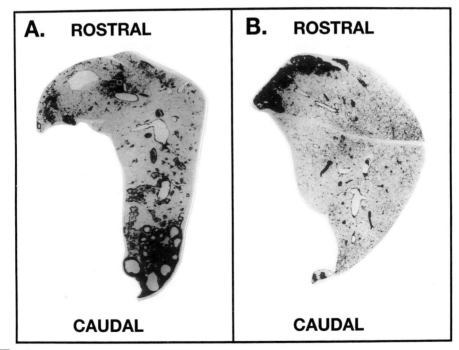

FIGURE 10 Progressive pulmonary tuberculosis in the rabbit develops in the most elevated region of the lung. (A) Infected rabbits standing normally (body horizontal), on all four feet, develop progressive tuberculosis in the dorso-caudal region of the lung. (B) When infected rabbits are foced to stand upright, the progressive tuberculous process shifts to the rostral end of the lung. (Reprinted from Medlar and Sasano, 1936 *American Review of Tuberculosis* 34, 456–476. With permission.)

issue: why acquired resistance does not protect the apex of the lung (Balasubramanian *et al.,* 1994). In recent years, this area of tuberculosis research has received astonishingly little attention, considering its manifest significance, as articulated by Rich:

> Actually, one should, perhaps, first place the question: What is it that renders the infra-apical portions of the lungs of adults so highly resistant? rather than the usual question: What is it that renders the apices susceptible? . . . The importance of the problem becomes evident when it is realized that, were it not for the failure of the apices to share equally in the increased resistance of the infra-apical portions, there would be almost no pulmonary tuberculosis in white adults. [Rich, 1944, p. 778.]

Apical localization of tuberculosis in individuals with acquired immunity may also shed light on the protective efficacy, and shortcomings, of vaccination with the bacille Calmette–Guerin (BCG), a live attenuated strain of *M. bovis.* In a prospective study of children living in tuberculous households, Kereszturi *et al.* (1935) observed that unvaccinated children who developed tuberculosis displayed radiographic profiles typical of primary disease, with profound adenopathy of the regional lymph nodes, whereas previously vaccinated children who developed tuberculosis showed little or no nodal involvement. Similarly, Dahlstrom (1952), in a comparison of 36,235 BCG-vaccinated and 25,239 nonvaccinated Swedish defense force soldiers, found that vaccinated soldiers who developed tuberculosis were less likely to display the features characteristic of primary tuberculosis. These clinical observations suggest that BCG vaccination provides protection at the level of the local lymph nodes, restricting the extension of the tuberculous process within and beyond the regional nodes, as does the resistance conferred by preexisting natural infection. Unfortunately, these studies did not report whether the tuberculous process was confined to the upper regions of the lungs in vaccinated individuals. To the best of our knowledge, the question has not been posed whether BCG vaccination mimics natural infection by protecting the lower but not the upper lung fields from reinfection, or whether BCG is unable to protect either region of the lung. Conceivably, this issue could still be addressed by examination of extant records from BCG field trials, such as the British MRC trial (MRC, 1956, 1959, 1963, 1972) and the Swedish defense force trial (Dahlstrom and Difs, 1951; Dahlstrom, 1952), conducted among individuals who were tuberculin skin-test negative (uninfected) when vaccinated.

III. THE TUBERCLE BACILLUS

> Consumption, that great destroyer of human health and human life, takes the first rank as an agent of death. Any facts regarding a disease that destroys one-seventh to one-fourth of all that die, cannot but be interesting.
>
> Lemuel Shattuck, 1849

More than a century has passed since Koch's identification of the tubercle bacillus. The intervening years have brought significant advances in our understanding of the epidemiology, pathogenesis, and immunology of tuberculosis. Arguably, the biology of the tubercle bacillus and its interaction with the infected host remain the areas of our greatest "wealth of ignorance." For many years, progress in understanding the genetic basis of virulence was stalled due to the lack of basic tools for the genetic manipulation of *M. tuberculosis*. The first demonstration of stable transformation of tubercle bacilli by foreign DNA was reported by Snapper *et al.* in 1988. Since then, steady progress has been made in the development of new tools, including (i) novel "phasmids" that replicate as plasmids in *Escherichia coli* and as phages in mycobacteria; (ii) shuttle plasmids that can be maintained as episomes in mycobacteria and *E. coli;* (iii) site-specific integrating plasmids based on the mycobacteriophage L5 attachment–integrase system; (iv) an efficient plasmid transformation mutant of the fast-growing saprophyte *Mycobacterium smegmatis;* and (v) promoter-based vectors for expression of foreign genes in mycobacteria (Hatfull and Jacobs, 1994; Falkinham and Crawford, 1994; McAdam *et al.,* 1994; Colston and Davis, 1994; Burlein *et al.,* 1994).

Future efforts to identify mycobacterial virulence determinants will require the development of new methods for the generation and phenotypic analysis of mutants. Despite considerable effort, construction of defined mutants by homologous recombination between exogenously introduced DNA and the mycobacterial chromosome has been frustrated by high frequencies of illegitimate recombination (Colston and Davis, 1994). To date, only four examples of allelic exchange have been described in slow-growing mycobacteria: three in the vaccine strain *M. Bovis* BCG (Reyrat *et al.,* 1995; Azad *et al.,* 1996; Baulard *et al.,* 1996) and one in virulent *M. tuberculosis* (Balasubramanian *et al.,* 1996). Several methods for promoting homologous exchange, such as the use of long linear recominbation substrates or counterselectable markers to eliminate illegitimate recombinants, have been proposed (McFadden, 1996). These efforts would benefit from improved understanding of the recombination mechanisms of slow-growing mycobacteria, a subject that has received remarkably little attention. The possibility of generating random mutant libraries by transposon mutagenesis has been suggested by the identification of several transposable elements that will "hop" randomly into the mycobacterial chromosome from exogenously introduced DNA (McAdam *et al.,* 1994). The utility of these transposons has been limited by the low transformation frequencies obtained with slow-growing mycobacteria. However, William Jacobs and colleagues (personal communication, 1997) have developed a novel system for efficient transposon delivery, based on temperature-sensitive (*ts*) replication-defective mutants of mycobacteriophages. With these tools in hand, the years ahead hold exciting prospects for genetic analysis of the biology and pathogenicity of the tubercle bacillus.

Tubercle bacilli are small, nonmotile rods, sometimes slightly curved, measuring between 1 and 4 μm in length and 0.3 to 0.6 μm in diameter. Although they stain poorly by the usual Gram stain procedure, owing to the impermeability of their thick, waxy cell walls, tubercle bacilli are considered nonetheless to belong to the gram-positive class of bacteria, as they possess only a single, cytoplasmic membrane. *Mycobacterium tuberculosis* is not a fastidious organism, and it will grow (albeit slowly!) on simple media composed of salts, trace elements, and a source of carbon and nitrogen (Wayne, 1994). Metabolically, tubercle bacilli appear to be unexceptional, possessing a complete Embden–Meyerhof pathway and Krebs cycle for generation of energy by oxidation of simple or complex carbon substrates to carbon dioxide and water (Ratledge, 1982; Wheeler and Ratledge, 1994). Three aspects of mycobacterial physiology deserve special mention, as they profoundly influence the pathogenesis of tuberculous disease.

First, as discussed previously, the tubercle bacillus is an obligate aerobe, incapable of growing by anaerobic fermentation alone, although at least some of the enzymes required for fermentative growth (e.g., lactate dehydrogenase) have been detected in bacillary extracts. The requirement for oxygen is thought to be at least partly responsible for the tissue tropism of the tubercle bacillus, which grows most luxuriantly within oxygen-replete tissues, such as the lungs, and more slowly within oxygen-poor tissues, such as the liver (Corper *et al.*, 1927). Oxygen availability is probably limited as well as by the intracellular growth environment, as tubercle bacilli replicating within macrophages are more sensitive to reduced ambient oxygen tensions than are bacilli replicating extracellularly (Meylan *et al.*, 1992).

Second, the tubercle bacillus grows very slowly, with a doubling time under "optimal" conditions of approximately 20–24 hr. The indolent replication of tubercle bacilli obviously contributes to the slowly progressing, chronic nature of tuberculous disease. Elucidation of the molecular basis for slow growth may not be a simple problem, as the pathogenic mycobacteria have adapted to slow growth at many levels (Wheeler and Ratledge, 1994). The time required to replicate the *M. tuberculosis* genome (10 to 11 hr) is much longer than the time required to replicate the *E. coli* genome (0.9 to 1.0 hr), despite the similarity in size of the two genomes (about 4 megabases). Similarly, the time required to transcribe a ribosomal RNA (rRNA) locus is about 0.12 hr in *M. tuberculosis* as compared to just 0.013 hr in *E. coli;* production of rRNA is also limited in *M. tuberculosis* by the presence in the genome of just one rRNA operon, as compared to seven in *E. coli. A priori,* the slow population doubling time of *M. tuberculosis* could be explained by either of two growth behaviors: there might be a broad range of interdivision times for individual cells in the population, with some cells dividing rapidly and others dividing seldom or never; alternatively, there might be a narrow range of interdivision times, with each cell in the population replicating at the same slow rate. This issue has not been addressed definitively, but the latter model (indolent replication of each cell in the population) is supported by time-lapse slide-culture studies, which revealed little variation in interdivision times for individual bacilli (Kahn, 1929; Espersen, 1949; Engbaek, 1952; Kudoh, 1956). Although the teleologic and mechanistic reasons for slow growth are still unclear, it is tempting to speculate that slow growth has evolved as a mechanism for prolonging host survival, thereby ensuring transmission and perpetuation of the tubercle bacillus. Slow growth may also contribute to the unusual resistance of mycobacteria to the microbicidal mechanisms of macrophages.

Third, the mycobacteria possess an unusually complex and lipid-rich cell envelope, which may account for many of their unusual characteristics: their resistance to chemical and physical stresses; their ability to withstand the killing mechanisms of macrophages and to replicate intracellularly; their impermeability to many antimicrobials; and their potent adjuvant activity, which may contribute to the immunopathological manifestations of tuberculous disease.

One of Robert Koch's seminal contributions to bacteriology was the development of techniques for staining tubercle bacilli, which are refractory to staining by conventional methods. Koch serendipitously discovered that heating the bacilli was critical for dye fixation, and that, once stained, tubercle bacilli resisted decolorization by acid treatment (hence the term acid-fast bacilli). From these observations, Koch postulated, correctly, that mycobacteria must possess extraordinary surface properties.

Biochemical characterization of the cell wall of tubercle bacilli has made impressive strides, thanks largely to the superb biochemical studies of Patrick Brennan and colleagues (McNeil and Brennan, 1991; Brennan and Draper, 1994; Besra and Chatterjee, 1994). The model of mycobacterial cell wall structure that has emerged from this important work is depicted in Fig. 11. The cytoplasmic membrane of mycobacteria is encapsulated by a rigid layer of peptidoglycan, consisting of parallel chains of alternating N-acetylglucosamine and muramic acid units cross-linked via tetrapeptide side chains attached to the muramic acid residues. The peptidoglycan of mycobacteria is unusual in two respects: first, the muramic acid residues are N-acylated with glycolic acid rather than the more usual acetic acid; second, some of the cross-links between tetrapeptide side chains occur between two residues of *meso*-diaminopimelic acid (DAP) in place of the more usual *meso*-DAP-D-alanine linkages (Draper, 1982). Via an unusual disaccharide phosphate linker region, the peptidoglycan backbone is attached to the arabinogalactan, a branched-chain polysaccharide consisting of a proximal galactose chain linked to a distal arabinose chain. Arabinose residues within the arabinogalactan backbone are linked to terminal pentaarabinose units, some of which are esterified with mycolic acids, which are long-chain 1-alkyl 2-hydroxy fatty acids peculiar to the mycobacteria and related genera (*Corynebacterium* and *Nocardia*).

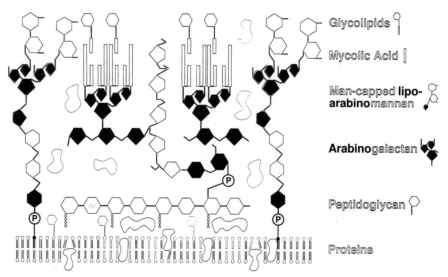

FIGURE 11　Schematic representation of the cell wall of *M. tuberculosis*. (Courtesy of Patrick Brennan and Delphi Chatterjee.)

The covalently linked peptidoglycan–arabinogalactan–mycolate structure forms the "cell wall skeleton." The lipids, glycolipids, and proteins that are non-covalently associated with this skeleton vary considerably from species to species within the genus *Mycobacterium*. Among the most prominent of the cell wall-associated glycolipids in *M. tuberculosis* is lipoarabinomannan (LAM). LAM is anchored in the plasma membrane by a phosphatidylinositol–diacylglycerol domain at the proximal end, linked to a branched-chain arabinomannan polysaccharide. LAM displays a number of biological activities relevant to tuberculosis pathogenesis, including inhibition of T-cell activation and of macrophage activation by interferon-γ, scavenging of reactive oxygen intermediates (ROI), inhibition of protein kinase C activity, and induction of macrophages to secrete several cytokines, including TNFα. In virulent strains of *M. tuberculosis,* mannose capping of the nonreducing termini of the arabinomannan has been shown to reduce the cytokine-inducing activity of LAM, which may promote survival within host macrophages. With the advent of mycobacterial genetics and molecular biology, identification and characterization of genes encoding enzymes involved in biosynthesis of the mycobacterial cell wall have become major goals. These enzymes will surely offer attractive targets for development of novel therapeutic drugs, as many of the unusual chemical structures found in the mycobacterial cell wall are not found in human tissues.

A number of mycobacterial genes have been identified that may contribute to pathogenesis (Collins, 1996). To date, however, genetic evidence establishing a role in mycobacterial virulence has been provided for just two genes: *rpoV,* encoding the major housekeeping sigma factor for RNA polymerase, and *katG,* encoding a bifunctional catalase/peroxidase. The *rpoV* gene was identified by complementation of an avirulent mutant of *M. bovis* for growth in guinea pigs (Collins *et al.,* 1995). Sequencing of the mutant allele revealed a single missense mutation in a sigma factor domain known to be involved in recognition of the −35 promoter region. Growth of the *rpoV* mutant appeared normal *in vitro,* suggesting that misexpression of a relatively small set of genes may account for the *in vivo* growth defect. A role for catalase in mycobacterial virulence was suspected for some time, as catalase-negative clinical isolates of *M. tuberculosis* often displayed attenuated virulence for guinea pigs (Zhang and Young, 1993). (As discussed later, catalase-negative strains frequently arise in patients receiving isoniazid chemotherapy, as loss of *katG* causes isoniazid resistance; approximately 50% of isoniazid-resistant clinical isolates carry mutations in *katG*.) Presumably, catalase is necessary to detoxify reactive oxygen species encountered by tubercle bacilli growing in the infected host. The observation that some catalase-negative clinical isolates retained full virulence cast doubt, however, on the role of *katG*. Unequivocal proof of the importance of *katG* was provided by the demonstration that an attenuated catalase-negative mutant of *M. bovis* could be restored to full virulence by provision of wild-type *katG* on a plasmid (Wilson *et al.,* 1995a). Apparently, fully virulent catalase-negative clinical isolates of *M. tuberculosis* have acquired secondary gain-of-function mutations in *ahpC,* encoding an alkylhydroperoxidase enzyme, which may compensate for loss of *katG* (Sherman *et al.,* 1996; Wilson and Collins, 1996).

The term parasite is derived from the greek παρασιτοσ, meaning "one who eats at the table of another." That tubercle bacilli must obtain nutrients from the infected host in order to grow and divide is obvious, but the mechanisms involved in nutrient acquisition have not been identified. The issue is fundamental, as tuberculous disease arises only following the survival and massive expansion of an initially minute inoculum of tubercle bacilli. Bacilli replicating intracellularly presumably derive nutrients from the host cell cytoplasm, although, as discussed later, it is not yet clear whether phagocytosed bacilli replicate within a membrane-bound vacuole or within the cytoplasm following escape from the phagosome. At late stages of tuberculous infection, within tissue that has undergone caseation and liquefaction, tubercle bacilli replicate extracellularly, but the nutritional milieu provided by liquefied host tissue has not been defined. It will be important to determine what compounds serve as sources of carbon, nitrogen, energy, etc. for tubercle bacilli growing *in vivo*, both intra- and extracellularly. Such studies are likely to offer surprises. For example, studies of virulent tubercle bacilli isolated directly from the tissues of infected animals suggested that fatty acids may serve as a major source of carbon and energy *in vivo* (Segal and Bloch, 1956; Segal, 1984; Wheeler and Ratledge, 1988). In future, the construction of mutants defective for uptake and utilization of specific nutrients should permit the definition of the *in vivo* growth environment, which may, in turn, suggest new possibilities for antimycobacterial drug development.

One of the most exciting recent developments in the field of microbial pathogenesis is the elucidation of the complete nucleotide sequences of the genomes of several important human pathogens, including *Haemophilus influenzae, Mycoplasma genitaleum, Streptococcus pneumoniae,* and Group A *Streptococcus.* A joint World Health Organization/World Bank committee examining "Investments in Health Research and Development" identified sequencing the genomes of major human pathogens as one of the best buys for investment in R & D relating to "continually changing microbial threats," including tuberculosis (Ad Hoc Committee, 1996). The information gleaned from the sequencing of microbial genomes is expected to accelerate the identification of protective antigens for vaccine development and of new targets for antimicrobial drug development. At the time of this writing in early 1997, the Sanger Centre in Cambridge, England, has undertaken the determination of the complete nucleotide sequence of the *M. tuberculosis* genome, with a predicted time to completion of 2 years at a projected cost of as little as US$1.5 million.

That common access to the complete *M. tuberculosis* genome sequence will revolutionize mycobacterial research is widely taken for granted; the question that naturally arises, however, is, what will we do with this information? The identification of protective antigens will require the definition of immunologic correlates of protection, and the development of improved animal models for testing the protective efficacy of novel vaccination protocols. Experience with experimental BCG vaccination suggests that this will prove no easy task: in a study designed to rank the protective efficacies of 13 commonly employed vaccine strains of BCG, there was no concordance between laboratories employing different assay systems (Chamberlayne, 1972; Fine, 1989; Bloom and Fine, 1994). The identification of potential targets for new antituberculous drugs will require the analysis of mycobacterial functions for their role in bacillary survival, growth, and pathogenicity *in vivo*. Predictably, one major approach will

be to generate targeted disruptions of genes that show homology to known virulence determinants of other pathogens, and then to assess the resulting mutant
strains in model systems. On the other hand, for the bacterial genomes sequenced thus far, 30–40% of the predicted open reading frames appear to be
unique to each organism, showing no homology to sequences in the databases.
Some of these novel genes may encode attractive targets for the development of
species-specific drugs. For the genetic analysis of mycobacterial pathogenicity,
there is an urgent need for improved techniques for introducing defined mutations into the *M. tuberculosis* chromosome; ideally, it should be possible to generate not only gene disruptions, but also more subtle mutant alleles encoding
altered proteins. Many outstanding questions cry out for attention. What mycobacterial functions are important for survival and growth within macrophages? What products of the tubercle bacillus are responsible for cytotoxicity,
and for the destruction of host tissues? What bacillary components are important for engendering a protective (or damaging) host response? How does the
tubercle bacillus survive prolonged periods of latency, and what functions are
mobilized during reactivation from latency?

Infectious disease is a relationship: ultimately, the frontier of research in
tuberculosis lies at the interface between pathogen and host. As advances in
mycobacterial genetics open the door to analysis of the virulence determinants
of the tubercle bacillus, improved model systems will be essential to assess the
contribution of specific mycobacterial genes to different aspects of the disease
process. At best, however, model systems are problematic; at worst, they are
irrelevant. Perhaps the greatest challenge for future research will be the extension of the information generated from laboratory studies in model systems to
the practical problems of human tuberculosis in the clinic.

IV. HOST RESISTANCE

> The slightest reflection makes it obvious that native resistance is an exceedingly impor-
> tant biological phenomenon. By reason of the presence or absence of species resistance,
> many different types of micro-organisms are rendered highly dangerous, or perfectly
> harmless, for a given species; differences in the degree of racial resistance determine
> marked differences in the pathology and clinical course of an infection in different races
> of the same species; and differences in individual native resistance are responsible for
> many of the marked differences in the reaction of different individuals of a given race to
> first infection with the same dose of a given micro-organism. In the pathogenesis of tu-
> berculosis, species, racial and individual native resistance all play a highly important role,
> and it is necessary, therefore, to attempt to gain as clear an understanding as possible
> concerning the nature, the mechanism and the effects of native resistance in this disease.
> Arnold R. Rich, *The Pathogenesis of Tuberculosis*, 1944, p. 119

A. Genetic Determinants of Resistance and Susceptibility

In 1929–1930 in the northern German town of Lubeck, 251 newborns were
inoculated with what was presumed to be the newly introduced BCG vaccine
strain of tubercle bacillus. Administration was *per os* in three separate doses
within the first 10 days following birth. Of the 251 vaccinees, 72 died of tuber-

culosis within the first year of life, while another 175 who were still alive after 4 years' observation showed arrested macroscopic lesions, predominantly of the mesenteric lymph nodes. Apparently, only 4 infants escaped infection, or suppressed the infection so efficiently that no pathology became evident. This disaster cast a pall over the use of the BCG vaccine and incited a flurry of invective against its inventors, Albert Calmette and Camille Guerin. In many countries, skepticism about the safety of BCG vaccination remains to this day, even though Calmette, in experiments that were remarkable given the paucity of diagnostic tools available at the time, definitively established that the bacilli administered to the unfortunate infants was not BCG, but a contaminating laboratory strain of virulent *M. tuberculosis*. Reasons for persistence of the belief that BCG could revert to virulence were related, rather sardonically, by Philip D'Arcy Hart:

> Calmette died in 1933, it is said a disappointed man, since, though the court verdict declared that a separate virulent culture had been given at Lubeck (this having been demonstrated by laboratory tests) and exonerated the B.C.G. strain, the idea that the latter could show virulence survived. One of its chief advocates was an eminent bacteriologist, whose laboratory I chanced to visit in 1935 in a sanatorium in New York State, where at that time many of the staff were themselves suffering from tuberculosis. As he showed me a culture plate originally sown with B.C.G. and now carrying virulent mycobacteria he was overcome by a fit of coughing. [Hart, 1967.]

Although the appalling morbidity and mortality suffered by the victims of the Lubeck tragedy understandably riveted the world's attention, even more remarkable, perhaps, is the fact that fully 71% of the infected infants, possessing no previously acquired resistance whatsoever, survived massive infection with highly virulent *M. tuberculosis*. From this perspective, the Lubeck disaster provides a striking example of the native resistance of the human species to the tubercle bacillus.

The resistance of *Homo sapiens* to tuberculosis appears all the more remarkable when contrasted to the exquisite susceptibility of several other mammalian species; for example, the rhesus monkey and the guinea pig uniformly succumb to infection with even minute numbers (1–20 colony-forming units) of tubercle bacilli. Rodents and rabbits, on the other hand, are relatively resistant. The underlying reasons for these interspecies differences in native resistance are not understood, but they could reflect either immune or nonimmune mechanisms (e.g., lung anatomy and physiology). Racial differences in native resistance to tuberculosis have also been described for a number of species, including rabbits (Lurie, 1964), guinea pigs (Wright and Lewis, 1921), mice (Buschman and Skamene, 1988), and humans (Rich, 1944; Buschman and Skamene, 1988), but, once again, the mechanisms underlying these differences are poorly understood. Clearly, the genetic component of resistance or susceptibility to tuberculous infection remains an underexplored and important area for future research.

In the mouse, marked differences in resistance to tuberculosis have been described for different inbred strains, but the underlying mechanisms accounting for these differences have not been defined (Fig. 12) (Buschman and Skamene, 1988). Resistance to infection with the attenuated *M. bovis* BCG strain of tubercle bacillus is influenced by a single autosomal dominant locus, the *Bcg* gene, which also influences susceptibility to infection with the intracellular pathogens *Leishmania major* and *Salmonella typhimurium* (Buschman and Ska-

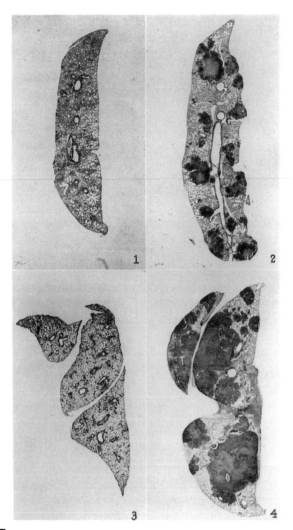

FIGURE 12 Tuberculous pathology in genetically resistant and susceptible strains of mice. Mice were infected intravenously with approximately 1.5×10^6 (panels 1 and 2) or 5×10^8 (panels 3 and 4) colony-forming units of virulent *M. tuberculosis*. At 3 weeks post infection, lungs from genetically resistant Swiss Albino mice (panels 1 and 3) show small granulomas without necrosis, whereas lungs from genetically susceptible dba mice (panels 2 and 4) show large granulomas with accumulation of epithelioid cells and extensive necrosis. (Reprinted from Pierce *et al.*, 1947 *Journal of Experimental Medicine* 86, 159–174. With permission.)

mene, 1988). Resistance is expressed as a greater inhibition of early growth of BCG in mice carrying the allele determining resistance (Bcg^r) as compared to mice carrying the allele determining susceptibility (Bcg^s). The Bcg phenotype is expressed at the level of the macrophage: macrophages of Bcg^r mice suppress the intracellular growth of BCG more effectively than do Bcg^s macrophages. The nucleotide sequence of the cloned Bcg gene predicts a transmembrane protein (designated Nramp, for natural-resistance-associated macrophages protein) with limited homology to yeast nitrate transporters, suggesting the intriguing possibility that Nramp may be involved in the generation of mycobactericidal reactive nitrogen intermediates (Vidal *et al.*, 1993). The significance of the *Bcg* gene remains unclear. Its effect is suppressed by high-dose challenge with BCG, or by local rather than systemic challenge with *Leishmania*. Evidence was presented that *Bcg* allelism does not affect the response to infection with virulent *M. tuberculosis* (Medina and North, 1996), suggesting that bacilli of sufficient virulence can overcome host resistance mechanisms. Whether targeted disruption of the *Bcg* gene in transgenic mice would affect resistance remains to be determined. In humans, it has not yet been established whether allelism for the *Bcg* homolog, located within a syntenic region of human chromosome 1, affects resistance to tuberculosis and other intracellular pathogens (Blackwell, 1996).

In humans, the role of heredity in resistance to tuberculosis has been hotly debated for decades, owing to the difficulty of dissociating genetic determinants of resistance from social, economic, and environmental factors that can also profoundly influence the interaction of pathogen and host. Nonetheless, it would be imprudent to disregard the selective force imposed by a pathogen that has accounted for such a large proportion of human deaths over so many years. Individual differences in native resistance were suggested by studies of tuberculosis in human twins, which demonstrated a greater concordance of risk for tuberculosis among monozygotic than among dizygotic twins (Kallmann and Reisner, 1943; Simonds, 1963; Comstock, 1978); however, questions have been raised about the interpretation of twin studies in tuberculosis (Fine, 1981). Native resistance to tuberculosis may be determined, in part, by the ability of the individual to present mycobacterial antigens to T cells, as several studies have revealed disproportionate representation of certain human leukocyte antigen (HLA) haplotypes among tuberculosis cases (Buschman and Skamene, 1988; Ivanyi and Thole, 1994). For example, the HLA-DR15 (DR2) locus has been associated with increased risk of tuberculous disease, and particularly with severe disseminated forms of tuberculosis. Racial differences in native resistance were suggested by the high rates of morbidity and mortality that occurred when tuberculosis was introduced into previously naive human populations, such as North American Indians, Alaskan Eskimos, and sub-Saharan Africans, among others.

The extant medical literature from the late nineteenth and early twentieth centuries, when tuberculosis was still the leading cause of death among young adults in the United States, consistently reported a greater burden of tuberculosis among American blacks as compared to whites. Unequal partitioning of disease burden along racial lines was evident when socially and economically similar groups were compared. Clinicians also recognized differences in the pathology of tuberculosis in black and white adults, as summarized by Arnold Rich:

> Anyone who has an opportunity to compare continually the pathological aspects of tuberculosis in the negro and in the white cannot fail to be struck by the fact that it is no great rarity to encounter in the adult city-dwelling negro the so-called "infantile" or "childhood" type of progressive pulmonary tuberculosis, which is well known to be extremely rare in the white adult. This type of tuberculosis, which is typical of the white or negro infant or young child, . . . is characterized by rather rapidly progressive caseation with little associated fibrosis; by marked enlargement and caseation of the regional lymph

nodes, which not infrequently erode into an adjacent bronchus and thereby discharge bacilli into the corresponding portion of the lung; and by a tendency to widespread generalization. Furthermore, the site of origin of the progressive pulmonary lesion is not confined to the apical or sub-apical portion of the lung, but may be situated in the lower portions and even at the very base. [Rich, 1944, p. 133.]

Similarly, Pinner and Kasper (1932) suggested that socioeconomic factors might explain the higher incidence of tuberculosis among blacks as compared to whites but could not explain the striking differences in the "pathologico-anatomical" characteristics of disease in the two races (Opie, 1930; Everett, 1933). More recently, William Stead has championed the idea that genetic differences in susceptibility may play an important role in human disease, and that tuberculosis may have exerted a significant selective force on the evolution of the human genome in populations exposed over many generations. Comparing rates of tuberculous infection among more than 53,000 residents of racially integrated nursing homes in Arkansas, Stead *et al.* (1990) found that blacks were twice as likely as whites to become infected (as indicated by tuberculin skin-test conversion) following similar exposure to infectious cases. Black individuals who developed tuberculosis were also more likely to become sputum-smear positive, suggesting that disease was more severe in these patients. On the other hand, during an outbreak of tuberculosis in a racially integrated elementary school, no difference was seen in the rates of infection among black versus white children, although infected black children were significantly more likely to develop disease, as indicated by chest radiographic abnormalities (Hoge *et al.*, 1994).

The most compelling evidence for genetically determined differences in susceptibility to tuberculosis was provided by the classic work of Max Lurie (1964). Conceptually, differences in native resistance can be distinguished at two levels: as differences in resistance manifest early in the course of infection, prior to the development of a specific T-cell response, and as differences in the ability to acquire specific immunity. Both levels of innate resistance were beautifully demonstrated by Lurie, who, in the 1930s, bred strains of rabbits that were genetically resistant or susceptible to tuberculosis. It is one of the tragedies of tuberculosis research that, after Lurie's death, the inbred rabbit colonies he developed so painstakingly were not preserved for future study.

Lurie, who himself suffered from tuberculosis, as did so many of the pioneers in tuberculosis research, devoted his long professional life entirely to the study of native and acquired resistance to tuberculosis. Having developed methods for the aerogenous infection of rabbits with small numbers of tubercle bacilli (i.e., approximating the conditions of natural infection in humans), Lurie showed that natively resistant rabbits were less susceptible to the establishment of infection with virulent *M. tuberculosis*. The number of inhaled bacilli required to establish a single pulmonary lesion in resistant rabbits ($\sim$1000 bacilli) was approximately 15 times the number required to generate a single lesion in susceptible rabbits ($\sim$70 bacilli) (Fig. 13). This difference between rabbit strains was not seen when animals were infected with *M. bovis*, which was known to be more virulent for rabbits than *M. tuberculosis* (Cobbett, 1932). In both strains of rabbits, each inhaled infectious particle of *M. bovis* generated a visible lesion, indicating that native host resistance to initial infection could be overcome by bacilli of sufficient virulence. Relative susceptibility to establishment of

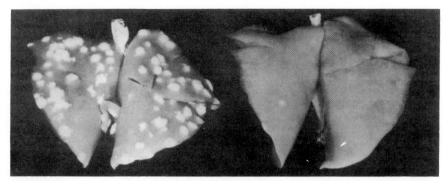

FIGURE 13 Lungs of a susceptible rabbit (left) and of a resistant rabbit (right) 33 days after the inhalation of ~10,000 colony-forming units of *M. tuberculosis*. In the susceptible rabbit, 36 inhaled bacilli were sufficient to produce 1 tubercle; in the resistant rabbit, 932 inhaled bacilli were required to generate 1 tubercle. (Reprinted from Lurie, 1964, "Resistance to Tuberculosis: Experimental Studies in Native and Acquired Defensive Mechanisms." Harvard University Press, Cambridge, MA, p. 196. With permission.)

infection with *M. tuberculosis* was not determined by differential implantation of bacilli in resistant and susceptible rabbits, since the number of bacilli reaching the lungs was shown to be the same in both rabbit strains, but was due to a marked difference in the subsequent fate of implanted bacilli. Natively resistant rabbits inhibited the early survival and growth of inspired tubercle bacilli more effectively than did natively susceptible rabbits. Consequently, the "lag phase" for growth of tubercle bacilli in the lungs of resistant rabbits was significantly longer than the "lag phase" in susceptible rabbits (Fig. 14).

The greater suppression of tubercle bacilli in the lungs of resistant rabbits was inferred to reflect a greater capacity of their alveolar macrophages to inhibit

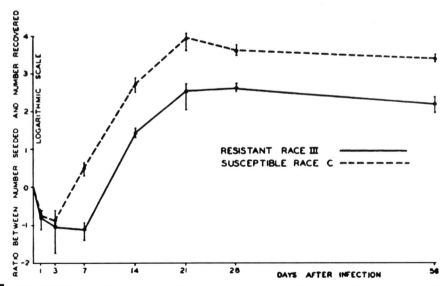

FIGURE 14 Population changes of *M. tuberculosis* in the lungs of natively resistant and susceptible rabbits after quantitative airborne infection.

or destroy bacilli following phagocytosis. This conclusion was supported by the observation that macrophages derived from susceptible rabbits permitted more luxuriant intracellular growth of tubercle bacilli *in vitro* than did macrophages derived from resistant animals. Growth of inspired tubercle bacilli began after just 3 days' lag in susceptible rabbits but did not begin until after 7 days' lag in resistant rabbits. Consequently, by the time the bacilli began multiplying in the lungs of resistant rabbits, the census of bacteria in susceptible rabbits had already undergone a 30-fold expansion. Following the lag phase, however, surviving tubercle bacilli grew with equal rapidity in the lungs of resistant and susceptible animals until around 3 weeks post infection (Fig. 14). At this time, the emergence of a specific host response, signaled by the onset of tuberculin skin-test reactivity, was accompanied by the rather abrupt curtailment of bacillary growth. The bacillary census then stabilized in both strains of rabbits, showing little fluctuation over the ensuing weeks, although remaining ~30-fold higher in the susceptible animals (Fig. 14).

B. Immunopathology of Tuberculosis

Among animal models of tuberculosis, pathology in the rabbit is most faithful to the pathology of human disease. In Lurie's studies, from 3 weeks post infection onward, bacillary growth was effectively inhibited in both resistant and susceptible rabbits. Nonetheless, relative native resistance was manifest at this time as differences in the progression of disease. Tubercles in the lungs of resistant rabbits were smaller, contained fewer bacilli, and showed a more rapid and complete breakdown of caseous material than did tubercles in the lungs of susceptible animals. Lymphocytic infiltration, maturation of epithelioid cells and giant cells, vascularization, and fibroplasia all proceeded more rapidly and extensively within and around the tubercles of resistant rabbits. In susceptible animals, the intraalveolar pneumonic process was more pronounced and interstitial inflammation was less apparent. Although dissemination of tubercle bacilli from the primary pulmonary lesion(s) to the draining lymph nodes was actually accelerated in resistant animals, growth of the bacilli that reached the nodes, and within the metastatic lesions generated by hematogenous dissemination, was markedly less. Beginning at about 2 months post infection, and following the full development of tuberculin hypersensitivity, the caseous centers of tubercles in the lungs of resistant rabbits began to soften, and cavities were excavated by the discharge of liquefied, bacilli-laden material into the airways. Liquefaction was accelerated by prior immunization and was not seen in susceptible animals, suggesting a link to a strong host immune response. This conclusion was supported by later studies showing that liquefaction and cavity formation could be prevented by desensitization of rabbits to the products of the tubercle bacillus (Yamamura *et al.*, 1974).

The occurrence of liquefaction only in previously sensitized *resistant* rabbits nicely illustrates a central paradox of the acquired immune response to *M. tuberculosis*. It is clear that an intact host response is essential for protection, as individuals with depressed immunity—including elderly persons, AIDS patients, and patients receiving immunosuppressive chemotherapy—are at increased risk for development of disease from exogenous (re)infection or from reactivation of latent infection. However, the immune response to bacillary antigens is also thought to contribute to the destruction of host tissue and the pathogenesis of tuberculous disease. In seminal experiments conducted during the late nineteenth century, Koch first discovered this dual nature of the host response, ultimately to his chagrin. Koch observed that intradermal injection of a crude glycerin extract of the tubercle bacillus ("Old Tuberculin") into a tuberculous guinea pig caused necrosis and sloughing of the tissue both at the site of injection and at the site of previous infection. This rapid necrotic response, known since as the Koch phenomenon, was not seen when Old Tuberculin was injected into nontuberculous animals. Encouraged by this demonstration of seemingly beneficial immune modulation, Koch announced in 1890, before the Tenth International Congress of Medicine in Berlin, that he had discovered a substance capable of ameliorating tuberculosis. Koch's pronouncement was greeted with accolades from the medical and popular presses, such as the following from an article in the December 1890 issue of the English *Review of Reviews*:

> Europe witnessed a strange but not unprecedented spectacle last month. In the Middle Ages the discovery of a new wonder-working shrine, or the establishment of the repute of the grave of a saint as a fount of miracles, often led to the same rush which has taken place last month to Berlin. . . . The consumptive patients of the Continent have been stampeding for dear life to the capital of Germany. The dying have hurried thither, sometimes to expire in the railway train, but buoyed up for a time by a new potent hope. . . . The news that the German scientist had discovered a cure for consumption must have sounded as the news of the advent of Jesus of Nazareth in a Judaean village. The whole community was moved to meet him. His fame went throughout the region around about, and telegrams in the newspapers announced that all the sleeping cars had been engaged for months to come to convey the consumptives of the Riviera to the inclement latitude of Berlin. [Dubos and Dubos, 1952, pp. 104–105.]

Alas, jubilation gave way to disillusionment and then to outrage, as clinicians everywhere found that Koch's "cure" did not stand up to its early reputation. At best, the administration of Old Tuberculin to tuberculous patients was found to provide little or no lasting benefit; in many cases, the injections actually exacerbated tissue damage and accelerated the course of disease, or precipitated a lethal "tuberculo-shock" (Lancet, 1890). The limitations and dangers of Old Tuberculin were apprehended at an early date by the famous author and physician Sir Arthur Conan Doyle, who wrote:

> Koch has never claimed that his fluid kills the tubercle bacillus. On the contrary, it has no effect upon it, but destroys the low form of tissue in the meshes of which the bacilli lie. Should this tissue slough in the case of lupus [tuberculosis of the skin], or be expelled in the sputum in the case of phthisis [pulmonary tuberculosis], and should it contain in its meshes all the bacilli, then it would be possible to hope for a complete cure. When one considers, however, the number and the minute size of these deadly organisms, and the evidence that the lymphatics as well as the organs are affected by them, it is evident that it will only be in very exceptional cases that the bacilli are all expelled. Your remedy does not treat the real seat of the evil. It continually removes the traces of the enemy, but it still leaves him deep in the invaded country.

Worse still, as Doyle recognized, the administration of tuberculin frequently

> . . . stirs into activity all those tubercular centres which have become dormant. In one case which I have seen, the injection, given for the cure of a tubercular joint, caused an

> ulcer of the eye, which had been healed for twenty years, to suddenly break out again,
> thus demonstrating that the original ulcer came from a tubercular cause. It may also be
> remarked that the fever after the injection is in some cases so very high . . . that it is hardly
> safe to use it in the case of a debilitated patient. [Dubos and Dubos, 1952, pp. 105–106.]

Eventually, the use of Old Tuberculin for the treatment of tuberculosis was abandoned. However, the lessons painfully learned during the "tuberculotherapy era" were not without enduring value. The extreme and detrimental reactions observed following injections of Old Tuberculin gave rise to the important concept that the Koch phenomenon might account, in part, for the progressive destruction of tissue that occurs in the course of tuberculous disease. In modern parlance, the Koch phenomenon reflects an excessive cellular immune response of the tuberculous host to inflammatory products of the tubercle bacillus.

C. Delayed-Type Hypersensitivity, Immunologic Protection, and Tissue Damage

Since Koch's time, much effort has been directed toward elucidating the immune responses responsible for acquired resistance, on the one hand, and for hypersensitivity and tissue damage, on the other. Several lines of evidence suggested the possibility of dissociating the protective and damaging aspects of the host response, beginning with early clinical experience of tuberculotherapy itself. As mentioned before, the initial enthusiasm for Koch's tuberculotherapy was quickly tempered by the discovery that injection of large quantities of Old Tuberculin into tuberculous patients could stimulate excessive and harmful responses, sometimes resulting in the patients' death. Clinicians therefore resorted to repeated injections of smaller quantities, which was considered less dangerous, and discovered that this treatment often led to the loss of tuberculin hypersensitivity. By injecting graded doses of tuberculin, it was found that patients eventually could be rendered nonresponsive to enormous doses without any apparent worsening of their clinical condition, suggesting that desensitization did not entail the loss of acquired resistance (Lister, 1890). Similar results were obtained by desensitization of tuberculous guinea pigs (Rich, 1944). In long-term experiments, it was shown that guinea pigs vaccinated with attenuated strains of tubercle bacilli gradually lost their hypersensitivity, but without any accompanying loss of acquired resistance (Rich, 1944).

The early findings have since been interpreted as suggesting a dissociation between delayed-type hypersensitivity (DTH), revealed by the tuberculin skin test, and cell-mediated immunity (CMI), responsible for acquired resistance. However, desensitization is a temporary and often a local phenomenon, and it probably does not affect the memory T cells that are responsible for acquired immunity. A more substantial argument derives from field trials of BCG vaccination in humans that have failed to demonstrate a correlation between establishment of hypersensitivity and protection. In the randomized placebo-controlled British MRC vaccine trial in Wales and England, initiated in 1950, vaccinees were inoculated with either BCG or the vole bacillus, *M. microti*. Both vaccines provided protection approaching 80%, even though the vole bacillus generated hypersensitivity in less than 30% of vaccinees whereas BCG established hypersensitivity in >85% of recipients. The directors of the MRC trial summed up their collective experiences with the statement, "It is concluded that with *highly effective* tuberculosis vaccines, the degree of protection conferred *on the individual* is independent of the degree of tuberculin skin sensitivity induced in that individual by the vaccination" (Hart *et al.*, 1967).

One of the critical questions in tuberculosis is whether immunopathology is the unavoidable price paid for protection. It is by no means clear that protection and tissue damage are mediated by distinct mechanisms, as tissue necrosis could result from excessive or mislocalized protective responses. Improved understanding of the effector mechanisms responsible for protection and damage will be essential for the rational development of improved vaccines and immunotherapies for the prevention and treatment of tuberculosis, as discussed in several recent reviews (Chan and Kaufmann, 1994; Barnes *et al.*, 1994; Rook and Bloom, 1994; Cooper and Flynn, 1995).

D. The Role of the Macrophage

At the cellular level, the central role of the macrophage in tuberculosis was recognized soon after Koch's identification of the tubercle bacillus. Histopathological studies of infected tissue from humans and experimentally infected animals revealed that tubercle bacilli growing *in vivo* were located within large mononuclear cells (macrophages), as recognized first by Koch (1882). Lurie found that the accumulation of bacilli within macrophages *in vivo* was much reduced in resistant rabbits as compared to susceptible rabbits (Lurie, 1964). Later work suggested that macrophages from resistant rabbits were also less sensitive to the cytopathic effects of tubercle bacilli (Hsu, 1965a,b). Lurie further demonstrated that tubercle bacilli grew readily within macrophages from normal rabbits but were inhibited within macrophages from tuberculous or BCG-vaccinated animals (Lurie, 1964). However, the efficiency of inhibition varied between experiments, and Lurie's results could be not reproduced in tissue culture experiments using highly purified macrophages, suggesting the involvement of additional, unrecognized factors. Rich (1944) had noted the abundance of lymphocytes accumulating at the periphery of tuberculous lesions and postulated their involvement in antituberculous immunity, but he had not been able to ascribe any specific function to them. In a classic experiment, Chase (1945) demonstrated that cutaneous hypersensitivity to tuberculin could be transferred to naive guinea pigs, not by serum, but by cells derived from the spleens, lymph nodes, or peritoneal exudates of sensitized animals, thus establishing the foundation of cellular immunology. Evidence that lymphocytes could respond to specific antigens by triggering nonspecific macrophage antimicrobial functions was provided by Mackaness (Mackaness and Blanden, 1967). The demonstration that lymphocytes (now known to be T cells) would respond to specific antigens by secreting diffusible factors (later termed lymphokines and cytokines) was provided independently by Bloom and Bennett (1966) and David (1966). Later work established interferon-γ (IFNγ) and TNFα as key cytokines responsible for activation of macrophage antimycobacterial functions (Rook *et al.*, 1986a; Flesch and Kaufmann, 1987).

Despite the ability of M. *tuberculosis* to parasitize a wide variety of host cell types *in vitro*, bacilli growing within the tissues of the infected host are found almost exclusively within cells of the mononuclear phagocyte lineage. It is not clear why this pathogen has evolved to parasitize a professional phagocytic cell capable of mounting potent antimicrobial responses. The interaction of *M. tuberculosis* and macrophages is one of the most important and vexing problems in tuberculosis research, as the mechanism of uptake and the intracellular fate of tubercle bacilli *in vivo* remain unclear. The adherence of mycobacteria to macrophages may be mediated by several macrophage cell-surface molecules, including complement receptors, mannose receptors, the lipopolysaccharide (LPS) receptor (CD14), Fc receptors, and possibly the scavenger receptor. Phagocytosis of virulent tubercle bacilli into macrophages via mannose receptors or the complement receptors CR1 and CR3 does not trigger the release of reactive oxygen intermediates (ROI) (Schlesinger *et al.*, 1990; Schlesinger, 1993). In a classic experiment, Armstrong and Hart (1975) showed that preopsonization of tubercle bacilli with specific antibody promoted uptake by Fc receptors, which does elicit ROI, yet did not affect the ability of the bacilli to survive and replicate intracellularly.

When phagocytosed by macrophages, bacteria typically enter a specialized endosome or phagosome that undergoes progressive acidification (to pH 5.0– 5.5) followed by fusion with lysosomes. Within the mature phagolysosome, degradation of ingested bacteria, particularly after killing by macrophage microbicidal mechanisms, is accomplished by a variety of acid hydrolases. In contrast, tubercle bacilli reside within a phagosome that fails to acidify fully (phagosomal pH remains between 6.0 and 6.5), apparently because of the absence of the vacuolar proton pump required for vacuolar acidification (Sturgill-Koszycki *et al.*, 1994). The mechanism by which *M. tuberculosis* excludes the vacuolar proton pump remains unclear, but it does not seem to reflect a general block in delivery of membrane proteins to the infected phagosome (Xu *et al.*, 1994; Clemens and Horwitz, 1995).

There exists a long-standing controversy whether phagosomes containing *M. tuberculosis* actually fuse with lysosomes. By transmission electron microscopy, Armstrong and Hart (1971, 1975) demonstrated that the majority of bacilli-containing vacuoles failed to fuse with prelabeled secondary lysosomes. Importantly, preopsonization of bacilli with specific antibody promoted phagosome–lysosome fusion, yet did not affect bacillary survival and intracellular growth. It was later reported that sulfatides present on the mycobacterial surface could block phagosome–lysosome fusion (Goren *et al.*, 1976), although the significance of this mechanism *in vivo* remains unclear. In contrast, Myrvik *et al.* (1984) demonstrated significant fusion of bacilli-containing vacuoles with lysosomes. More recently, McDonough *et al.* (1993) reported that bacilli-containing vacuoles fused with lysosomes at early stages of infection, but that, as infection progressed, replicating bacilli were found within unusual vacuoles with tightly apposed membranes. At 2–3 days post infection, a subpopulation of intracellular tubercle bacilli was observed to lack detectable enclosing vacuolar membranes, suggesting that some bacilli might have escaped from the vacuole into the cytoplasm. A potential mechanism for lysis of vacuolar membranes was suggested by the identification of mycobacterial genes encoding phospholipases (King *et al.*, 1993; Leao *et al.*, 1995; Johansen *et al.*, 1996). Whether tubercle bacilli do escape from the vacuole remains controversial, however, as studies using freeze-substitution electron microscopy, which may permit better visualization of membranes, indicated that tubercle bacilli reside exclusively within membrane-bound vacuoles and are not found free in the cytoplasm (Xu et al., 1994; Clemens and Horwitz, 1995; Clemens, 1996). Whatever the final resolution of this controversy may be, it seems clear that escape from the vacuole is not essential for intracellular replication, since there is general agreement that the attenuated BCG bacillus replicates solely within the confines of a membrane-bound compartment.

Elucidation of the intracellular fate of phagocytosed tubercle bacilli is important for several reasons. First, the growth environment within the vacuole may be more restrictive than the environment within the nutrient-rich cytoplasm. It is not clear to what extent the nutrients required for bacterial growth can penetrate the vacuole; the failure of a leucine auxotroph of BCG to replicate within infected mice or cultured human macrophages suggests that the nutritional milieu within vacuoles may be sparse (McAdam et al., 1995; Bange et al., 1996). Second, accumulated evidence that major histocompatibility complex (MHC) Class I-restricted T cells are important for protection indicates that antigenic peptides must somehow gain access to the cytoplasm and endoplasmic reticulum (ER). Relevant to this issue is the demonstration that M. tuberculosis can promote the presentation of soluble antigens by MHC Class I (Mazzaccaro et al., 1996). Soluble antigens such as ovalbumin are endocytosed by macrophages and presented to T cells in the context of MHC Class II, and they are generally not presented by MHC Class I, which instead presents antigens derived from the macrophage cytoplasmic compartment. However, when soluble ovalbumin was added to macrophages infected with M. tuberculosis, the antigen was efficiently presented to MHC Class I-restricted T cells. This process was dependent on TAP (the transporter of antigenic peptides), a membrane-spanning transporter that translocates peptides from the cytoplasm into the lumen of the ER. Within the ER, peptides are loaded onto MHC Class I, and the complex is then translocated to the cell surface. The mechanism by which M. tuberculosis promotes presentation of soluble antigens by MHC Class I is not known, but it presumably reflects either lysis of the enclosing vacuolar membrane or permeabilization of the vacuolar membrane so as to permit leakage of antigen into the cytoplasm. It is tempting to speculate that M. tuberculosis has evolved such a mechanism to acquire needed nutrients from the cytoplasm, and perhaps to allow killing of infected lung macrophages by toxic bacillary products, thereby producing the tissue necrosis and liquefaction that promote transmission.

E. Mycobactericidal Mechanisms

Murine macrophages activated by IFNγ produce ROI, but are not capable of suppressing the growth of intracellular tubercle bacilli. Murine macrophages can be activated to kill phagocytosed tubercle bacilli by treatment with both IFNγ and TNFα. TNFα is secreted by infected macrophages and accumulates within tuberculous lesions (Barnes et al., 1990; Filley and Rook, 1991; Filley et al., 1992). TNFα release is induced by mycobacterial cell wall components, such as muramyl dipeptide and LAM (Moreno et al., 1989). Chronic production of TNFα is thought to be responsible for many of the symptoms of advanced tuberculosis, such as fever, wasting, fatigue, and night sweats; thus, TNFα is thought to play a dual role, mediating both protection and immuno-pathology. The importance of IFNγ and TNFα for resistance to *M. tuberculosis* was underscored by infection of transgenic "knockout" mouse strains defective for production of IFNγ or the p55 TNFα receptor: both strains were exquisitely susceptible and succumbed rapidly to infection (Cooper *et al.*, 1993; Flynn *et al.*, 1993, 1995a). In humans, an inherited defect in the gene encoding the IFNγ receptor has been associated with increased susceptibility to severe my-cobacterial infections, including fatal disseminated BCG-osis (Jouanguy *et al.*, 1996; Newport *et al.*, 1996). In contrast to murine macrophages, human mac-rophages activated with IFNγ and TNFα are not capable of killing tubercle bacilli, although activation with 1,25-dihydroxyvitamin D_3 has been shown to slow the rate of intracellular bacillary replication (Rook *et al.*, 1986b; Crowle *et al.*, 1987; Crowle and Elkins, 1990; Denis, 1991b).

The macrophage antimycobacterial effector mechanisms triggered by IFNγ and TNFα have been a topic of intensive investigation. Early investigators sug-gested that the production of ROI by activated murine macrophages might con-tribute to the killing of intracellular tubercle bacilli. The role of ROI in anti-mycobacterial defense remains problematic, however, as tubercle bacilli are remarkably resistant to killing by oxygen radicals and hydrogen peroxide *in vitro* (Chan *et al.*, 1992) and may deploy several mechanisms to evade ROI *in vivo*: (i) mycobacteria can enter macrophages by engaging cell-surface mannose receptors or the complement receptors CR1 and CR3, ligation of which does not trigger the macrophage oxidative burst; (ii) sulfatides secreted by tubercle bacilli can downregulate ROI production; (iii) LAM inhibits protein kinase C activity and transcriptional induction by IFNγ, both of which are important for ROI generation; (iv) LAM is itself an efficient scavenger of ROI. One additional point deserves mention. In response to sublethal challenge with ROI, bacteria such as *E. coli* adapt by upregulating genes involved in defense against ROI (e.g., genes encoding DNA repair enzymes). Transcriptional regulation of ROI defense genes is under the control of the OxyR transcription factor. *A priori*, one might expect tubercle bacilli, which replicate within macrophages capable of generating ROI, to possess a functional OxyR regulon. However, cloning and sequencing of the *M. tuberculosis* homolog of *oxyR* revealed that the gene is clearly inactive due to the accumulation of multiple mutations (Deretic *et al.*, 1995). Whether restoration of *oxyR* function to *M. tuberculosis* would affect virulence remains to be determined.

More recent work has established an essential role for reactive nitrogen intermediates (RNI), such as nitric oxide, in antituberculous immunity. Treat-ment of murine macrophages with IFNγ and TNFα was shown to promote synthesis of the inducible isoform of nitric oxide synthase (iNOS), produc-tion of RNI, and killing of ingested tubercle bacilli (Denis, 1991a; Flesch and Kaufmann, 1991; Chan *et al.*, 1992). This is currently the only mechanism known for killing of tubercle bacilli by macrophages. The significance of RNI for inhibition of bacillary growth *in vivo* was established by treatment of infected mice with specific chemical inhibitors of iNOS, aminoguanidine or N-monomethylarginine (NMMA), both of which promoted unrestricted bacil-lary growth culminating in the death of the animals (Chan *et al.*, 1995). As yet,

impaired in their ability to activate mycobacterium-specific CD4$^+$ T cells. Prostaglandins released by infected macrophages may mediate downregulation of B7, as indomethacin treatment can restore B7 expression to normal (preinfection) levels (Saha *et al.*, 1994). Downregulation of MHC Class II may be autoregulated, at least in part, by the release of transforming growth factor β (TGFβ) from infected macrophages (Hirsch *et al.*, 1994, 1996). In tuberculous patients, TGFβ production has been demonstrated in blood monocytes and in giant cells and epithelioid cells within pulmonary lesions (Toossi *et al.*, 1995a). Both LAM (Dahl *et al.*, 1996) and PPD (Toossi *et al.*, 1995b) have been shown to induce TGFβ secretion by human monocytes. TGFβ displays several potent immunosuppressive activities that may be relevant to the pathogenesis of tuberculosis. T-cell responses to antigen, including IL-2-dependent proliferation and secretion of Th1 cytokines, are inhibited by TGFβ, as are the effector functions of cytotoxic T cells and NK cells. TGFβ also interferes with macrophage effector functions, including generation of reactive oxygen and nitrogen intermediates in response to IFNγ and TNFα (Tsunawaki *et al.*, 1988; Ding *et al.*, 1990). Replication of *M. tuberculosis* within human monocytes is enhanced in the presence of TGFβ, suggesting that induction of this immunosuppressive cytokine may be a strategy by which the tubercle bacillus manipulates the host response to its own advantage (Hirsch *et al.*, 1994). On the other hand, production of TGFβ may benefit the infected host by limiting the extent of tissue damage caused by excessive CMI responses, and may help contain infection by promoting fibrosis and healing of tuberculous lesions. These potentially beneficial effects of TGFβ and other immunomodulatory cytokines, such as IL-10, may be particularly useful for the resolution of tuberculous lesions in patients receiving antimycobacterial chemotherapy.

The complexity of interactions occurring among the T-cell subsets and cytokines engaged by *M. tuberculosis* infection emphasizes the difficulty of defining which responses are necessary and sufficient for protection. Further elucidation of antimycobacterial host mechanisms should contribute to the development of surrogate markers for protective immunity, which will be necessary to permit meaningful evaluation of new vaccine candidates.

V. VACCINATION

> We have in B. C. G. vaccination an example of the difficulties in taking knowledge developed in the technically advanced countries and simply transferring it to the less advanced, when it is to be applied there on a mass scale. We may have to rethink, reshape, even freshly to investigate basic concepts.
>
> Philip D'Arcy Hart, 1967

A. Live Attenuated Vaccines

The existence of acquired resistance to tuberculosis was suggested as early as 1886 by Antonin Marfan, who oberved that pulmonary tuberculosis occurred only rarely in individuals who had overcome lupus vulgaris (tuberculosis of the skin) or tuberculous adenitis (Rich, 1944). In 1890, Koch described experiments demonstrating acquired resistance in experimentally infected guinea pigs. Koch noted that naive guinea pigs inoculated subcutaneously with live tubercle bacilli showed little reaction initially at the site of injection. Local proliferation of the bacilli, however, generated a hard nodule which underwent ulceration and necrosis after 2 to 3 weeks; meanwhile, bacilli disseminated to the draining lymph nodes, and from there to the extrapulmonary organs. When tubercle bacilli were inoculated into a tuberculous animal, however, the sequence of events was quite different: induration and ulceration occurred at the inoculation site within 2 to 3 days, the local growth of the injected bacilli was strongly inhibited, and dissemination of bacilli from the inoculation site was reduced or prevented. The same accelerated necrotic response (Koch phenomenon) was seen when killed bacilli or a cell-free bacillary derivative (Old Tuberculin) was injected into tuberculous animals. These experiments provided the first description of the delayed-type hypersensitivity (DTH) response to *M. tuberculosis* antigens that is still used clinically for the diagnosis of tuberculous infection.

Koch's observation that tuberculin elicited a powerful DTH response in sensitized animals spurred efforts to develop an effective vaccine based on cell-free derivatives of tubercle bacilli. All attempts to use nonliving immunogens were unsuccessful. Louis Pasteur's contemporaneous work with viral pathogens had established the feasibility of using live attenuated strains to induce a specific protective immune response. Efforts to develop an attenuated vaccine strain of the tubercle bacillus were initiated at the Pasteur Institute of Lille by Albert Calmette and Camille Guerin. These investigators built on Pasteur's observation that repeated *in vitro* passage of a virulent organism could cause attenuation (Geison, 1995). In 1902, Edmond Nocard isolated a virulent strain of *Mycobacterium bovis,* a close relative of the human tubercle bacillus, from the udder of a tuberculous cow. Calmette and Guerin demonstrated in 1908 that this isolate (called by them "souche lait Norcard" or "milk strain of Nocard") was highly virulent in guinea pigs; that same year, they began to passage the bacillus in potato medium containing beef bile to promote dispersed growth. After about 230 serial passages, completed between 1908 and 1919, Calmette and Guerin succeeded in deriving a stably attenuated strain of tubercle bacillus, subsequently designated "bacille Calmette–Guerin" or BCG in their honor. Exhaustive experimentation demonstrated that BCG never reverted to virulence in animal models ranging from monkeys to mice. Vaccination trials in cattle established that BCG was safe and provided significant protection against subsequent challenge with virulent tubercle bacilli.

In 1921, Benjamin Weill-Halle and Raymond Turpin administered BCG for the first time to a human being, a child born to a tuberculous mother who was deemed at high risk for developing tuberculosis; the child survived, and it remained free of tuberculous disease thereafter. Following that first success, BCG quickly spread throughout Europe and to other continents, and today it remains the most widely used vaccine in the world. Originally, BCG was administered orally, but the failure of many vaccinees to convert to tuberculin skin-test positivity, which was believed to correlate with protection, prompted a switch to subcutaneous administration. This method, however, frequently gave rise to lymphadenitis and cold abscesses, sometimes requiring surgical treatment. In

1928, Arvid Wallgren and Johannes Heimbeck introduced intradermal inoculation, which they found to cause tuberculin conversion in most vaccinees. Intradermal inoculation, however, generated a localized, slowly healing skin ulcer, and in a few cases was associated with ulceration of the draining lymph nodes. These undesirable side effects were not seen with the "multiple puncture technique," introduced by S. R. Rosenthal in 1939. To date, BCG has been administered to more than 3 billion people worldwide, making it the most widely used of all vaccines.

BCG possesses many attributes of an ideal vaccine: (i) BCG is remarkably safe, with a very low rate of serious adverse side effects (the mortality rate among vaccinees is just 60 per billion); (ii) BCG is inexpensive, costing just U.S.$0.06 per immunization; (iii) BCG can be given at or any time after birth; (iv) a single inoculation with BCG can engender long-lasting immunity; (v) the BCG vaccine does not require refrigeration and is accessible in the poorest countries where the "cold chain" cannot reach; and (vi) BCG vaccination usually generates a scar, which is useful for epidemiological surveillance. Importantly, in several field trials, BCG vaccination of human children has been shown to provide high-level protection against the most severe forms of childhood tuberculosis, including tuberculous meningitis and disseminated miliary tuberculosis. Unfortunately, these many advantages are overshadowed by the most important consideration of all: despite its widespread use for decades, the efficacy of BCG vaccination in protecting adults against pulmonary tuberculosis remains a subject of controversy. In large case-controlled and randomized placebo-controlled trials, the protection conferred by BCG vaccination has ranged from nil to nearly 80% in different regions of the world (Table III). Most disturbingly, in the largest field trial yet conducted, carried out in the Chingleput district of India and involving over 350,000 individuals, BCG vaccination displayed no protective efficacy in any adult age group (Tuberculosis Prevention Trial, Madras, 1980).

Several explanations have been proposed for the tremendous trial-to-trial variability in the efficacy of BCG vaccination (Bloom and Fine, 1994): (i) methodological flaws in the design of vaccine trials introduced biases that caused the efficacy of BCG to be underestimated in some trials; (ii) heterogeneity among the vaccine strains and handling methods used in different regions caused the potency of the administered vaccines to vary between trials (Brewer and Colditz, 1995); (iii) genetic differences between the populations enrolled in different trials affected the efficacy of vaccination; (iv) populations in different trial areas were exposed to *M. tuberculosis* strains of differing virulence; (v) BCG vaccination protected in trial regions where primary infection and endogenous reactivation were the primary source of contagion but failed to protect in regions where most disease was due to exogenous reinfection; and (vi) a high prevalence of infection by environmental mycobacteria in some regions masked the effect of BCG vaccination by conferring partial immunity prior to vaccination (Wilson *et al.*, 1995). All of the aforementioned considerations may have contributed to the observed variability in protection. However, the "environmental mycobacteria" hypothesis currently provides the most plausible explanation for the geographic pattern of vaccine efficacy, which increases with

TABLE III Summary of Major Randomized Controlled BCG Vaccination Trials[a,c]

Trial or study (reference)	Age	Follow-up (years)	Number of subjects studied		Number of tuberculosis cases		Protective efficacy	Atypical exposure
			Not vaccinated	Vaccinated	Control	Vaccinated		
Haiti (Vandiviere et al., 1973)	<20 years	3	629	2,545	15	25	80%	?
Canada, Qu'appelle Cree Indians (Ferguson and Simes, 1949)	<3 months	6.5	303	306	29	6	80%	Low
British MRC (Hart and Sutherland, 1977)	14 years	20	12,867	13,598	248	62	77%	Low[b]
North American Indians (Aronson et al., 1958)	0–19 years	20	1,451	1,541	372	108	75%	Low[b]
Chicago, high risk infants (Rosenthal et al., 1961)	<3 months	23	1,665	1,716	65	17	75%	Low
South Africa, miners (Coetzee and Berjak, 1968)	30 years	3	17,135	20,623	74	48	37%	?
Puerto Rico (Comstock et al., 1974b)	1–18 years	6.3	27,338	50,634	141	186	31%	High[b]
India, Madanapalle (Frimodt-Moller et al., 1973)	All ages	14	5,808	5,069	46	28	31%	High
Georgia and Alabama (Comstock et al., 1976)	>5 years	14	17,854	16,913	32	26	14%	High
Georgia (Comstock and Webster, 1969)	6–17 years	20	2,341	2,398	3	5	None	High
India, Chingleput	>1 year	12.5	79,398	272,455	93	192	None	High
skin test <7 mm induration (Tuberculosis Prevention Trial, 1980)			30,000	60,000			None	

[a] From Bloom and Fine (1994), with corrections suggested by Philip D'Arcy Hart.

[b] Indicates presumed, not reported, level of exposure.

[c] Protective efficacy is defined as follows: [(incidence rate in unvaccinated − incidence rate in vaccinees)/(incidence rate in unvaccinated)] × 100.

increasing distance from the equator, where infection with environmental mycobacteria is most common (Wilson et al., 1995b).

Even in the most successful vaccine field trials, it was clear that the protection
engendered by BCG vaccination was far from 100%. The limited protective efficacy of BCG vaccination has also been demonstrated in vaccination studies in
animal models. In guinea pigs, BCG vaccination prior to aerogenous challenge
with virulent tubercle bacilli was shown to slow the rate of bacillary growth, to
retard and reduce the dissemination of bacilli from the lungs, and to prolong
survival, yet it did not prevent death in a large proportion of animals.[3] (Smith
et al., 1988). That BCG should provide only limited protection against virulent
tubercle bacilli is hardly surprising, given the natural history of tuberculous disease. As discussed previously, the acquired resistance elicited by infection with
virulent tubercle bacilli often fails to eradicate this tenacious pathogen, and it
provides only limited protection against reinfection. Vaccination with attenuated strains of tubercle bacilli, such as BCG, may confer still less effective protection, since these strains are capable of only limited growth and persistence
in vivo. That in vivo bacillary growth is necessary for eliciting protective immune responses was established by experiments in which early treatment of infected animals with isoniazid to inhibit bacillary growth was shown to prevent
the development of effective acquired resistance (Dubos and Schaefer, 1956;
McDermott et al., 1956; Schmidt, 1956; Peizer et al., 1957). Similarly, vaccination with non-living (Smith et al., 1988) or non-replicating (Kanai, 1966)
bacilli was less effective than vaccination with bacilli capable of in vivo growth.
The vexing question arises, then, how to design a vaccine that is safe yet provides better protection than that afforded by natural infection with fully virulent
tubercle bacilli?

One potentially fruitful approach would be to manipulate the host response
at the time of BCG vaccination so as to promote optimal protective responses.
In the case of tuberculosis and other diseases caused by intracellular pathogens,
such as leishmaniasis, manipulation of the host response should be directed toward promoting effective CMI. In a mouse model of Leishmania infection,
Heinzel et al. (1993) showed that injection of susceptible animals with IL-12 led
to the development of strong CMI and resolution of infection. Furthermore,
when injected along with leishmanial antigens, IL-12 acted as a potent adjuvant,
eliciting robust antigen-specific CMI responses and strong protection against
subsequent challenge with virulent Leishmania (Afonso et al., 1994). Along
similar lines, injection of IL-12 into tuberculous mice was shown to reduce bacterial replication, ameliorate lung pathology, and increase survival time; unfortunately, IL-12 administration did not enhance the protective efficacy of BCG
immunization (Flynn et al., 1995b; Cooper et al., 1995). Recent efforts have
been directed toward improving the immunizing capacity of BCG itself, using
recombinant DNA technology and newly developed techniques for the genetic
manipulation of mycobacteria. Recombinant BCG (rBCG) strains engineered to
secrete various mouse cytokines, particularly IFNγ, IL-2, or granulocyte–
macrophage colony-stimulating factor (GM-CSF), evoked stronger mycobacterium antigen-specific T-cell responses than did nonrecombinant BCG (Murray
et al., 1996). Although these results are encouraging, the ultimate question is

---

[3]Radiographic and autopsy studies have demonstrated similar effects in BCG-vaccinated humans (Lindgren, 1961, 1965; Sutherland and Lindgren, 1979).

whether vaccination with cytokine-producing rBCG will engender better protection *in vivo* against challenge with virulent *M. tuberculosis*.

In addition to improving the protective efficacy of BCG vaccination against tuberculosis, efforts have been made to increase the versatility of BCG by using it as a multivalent vaccine vehicle. Vaccination of mice with rBCG strains encoding antigens from various pathogens, whether viral (HIV), bacterial (*Borrelia burgdorferi*), or protozoan (*Leishmania*), elicited long-lasting humoral and CMI responses to the cognate foreign antigens (Aldovini and Young, 1991; Stover *et al.*, 1991, 1993; Connell *et al.*, 1993; Langermann *et al.*, 1994; Honda *et al.*, 1995). Given that BCG is currently the most widely used vaccine in the world, administered to as many as 100 million children each year, rBCG vaccines may eventually provide wide coverage for several common childhood infectious diseases in addition to tuberculosis. Because BCG can be given to infants, rBCG vaccines may prove particularly useful for vaccination against important childhood diseases such as measles, for which there is currently no vaccine that can be administered at birth. The live attenuated measles vaccine in use today cannot be given for 9–12 months postpartum due to interference from inherited maternal antibodies, which prevent the attenuated vaccine strain from "taking" but do not protect against virulent measles virus. In developing countries, where infection is common in the first year of life, an effective vaccine for neonates would contribute significantly to the reduction of mortality from measles, which claimed more than 1 million young lives in 1995.

Although BCG vaccination only rarely provokes adverse responses, concerns have been raised about the safety of administering a live attenuated vaccine to immune-compromised individuals, including those with AIDS. These concerns are not gratuitous, as several cases of disseminated BCG-osis and death have been reported in HIV-infected individuals (von Reyn *et al.*, 1987; Braun and Cauthen, 1992; Weltman and Rose, 1993) and in individuals genetically defective in the IFNγ gene (Jouanguy *et al.*, 1996; Newport *et al.*, 1996). As the AIDS epidemic spreads throughout nations where BCG vaccination is widely used, concern over the safety of BCG is likely to increase. One approach to this problem would be to attenuate BCG still further by introducing additional mutations (e.g., in genes required for amino acid biosynthesis). Several auxotrophic strains of BCG have been shown to confer protection against tuberculosis equal to that conferred by wild-type (prototrophic) BCG when administered to immune-competent mice (Guleria *et al.*, 1996). Importantly, the auxotrophic BCG strains were nonpathogenic even for severe combined immunodeficiency (SCID) mice, which succumbed rapidly to infection with wild-type BCG. A major effort is underway to identify and mutate *M. tuberculosis* genes that are essential for pathogenicity in hopes of creating a vaccine strain that is safe yet more potent than BCG.

B. Subunit Vaccines

The possibility of developing an effective subunit vaccine has been reassessed by several groups, with emphasis placed on the secreted antigens present in culture filtrates of *M. tuberculosis*. Obviously, a cell-free subunit vaccine would bypass the safety issues raised by a live vaccine. The focus on actively secreted rather than somatic antigens was prompted by early observations that living tubercle bacilli evoked protective immune responses in animals, whereas bacilli killed with heat or chemicals were ineffective. Also, time-course experiments suggested that the secreted antigens were among those first recognized by the host immune system, making them logical targets for immunization to inhibit the early progression of tuberculous infection (Chan and Kaufmann, 1994). In the mouse (Andersen, 1994; Roberts *et al.*, 1995), immunization with secreted antigens in adjuvant was shown to confer short-term protection comparable to that provided by BCG vaccination. In the guinea pig, immunization with a mixture of secreted proteins was shown to provide limited short-term protection, resulting in an approximately 10-fold reduction in the bacillary load (Pal and Horwitz, 1992; Horwitz *et al.*, 1995). Although these results are encouraging, it is not yet clear whether a subunit vaccine can be designed to generate long-lasting protection comparable to that provided by BCG; indeed, this issue is difficult even to address in short-lived animal models of tuberculosis.

C. DNA Vaccines

Following the latest developments in vaccinology, two groups have demonstrated the feasibility of developing "naked DNA" vaccines against tuberculosis. Mice were vaccinated by intramuscular injection of plasmid DNAs encoding either the secreted antigen 85 (Ag85) of *M. tuberculosis* (Huygen *et al.*, 1996) or the highly conserved 65-kDa heat shock protein (hsp65) or 36-kDa proline-rich antigen of *M. leprae* (Tascon *et al.*, 1996). Vaccinated mice mounted strong cellular and humoral responses to the encoded antigens, including generation of IFNγ and cytotoxic T cells, and showed enhanced resistance to challenge with virulent *M. tuberculosis*. The advantages of naked DNA vaccines include their low cost, ease of administration, and stability at ambient temperatures. However, the ability of naked DNA vaccines to elicit long-lasting protective responses remains unproved. The safety of naked DNA vaccines has also been questioned, since integration of injected DNA into human chromosomes has the potential to cause deleterious mutations or to activate oncogenes.

D. Evaluating the Protective Efficacy of New Vaccine Candidates

Although the developments cited above inspire hope that improved vaccines against tuberculosis may be developed in the near future, perplexing problems remain to be addressed outside of the laboratory. In particular, it is not clear how the protective efficacy of any new vaccine candidate will be tested in humans at risk for tuberculosis, which is clearly the most important issue of all. The incidence of tuberculosis in the industrialized nations is now so low that significant results could be obtained only if coverage were very high, entailing considerable expense. Furthermore, the real question is not whether a new vaccine is capable of providing *any* protection, but whether it provides better protection than BCG; addressing this issue would require even larger study populations, at still greater expense. Given the disillusionment created by the failure of several large and expensive BCG vaccination trials, it will likely prove dificult to generate the necessary enthusiasm, cooperation, and funding for similar long-term vaccine field trials. New conceptual approaches to the problem of vaccine evaluation are therefore urgently needed.

One approach might be to vaccinate recently infected healthy individuals (PPD converters) and to monitor progression to disease within a defined period. This approach is potentially problematic, as natural infection may bias the immune system in ways that cannot be overcome by vaccination; also, vaccination may have no effect on the reactivation of preexisting latent infection. Such studies would benefit from the development of rational animal models for evaluating the protective efficacy of novel vaccination protocols against reactivation tuberculosis. A second approach would entail the identification of immunologic correlates of protection, and the testing of new vaccine candidates for their ability to induce and maintain such correlates in small Phase I and II clinical trials. Although tuberculin skin-test reactivity has been interpreted as indicating "successful" vaccination, current evidence suggests that it does not reliably predict protection. Correlates of additional protective mechanisms (cytotoxic T cells, macrophage RNI production, etc.) need to be developed. With optimal induction of immunologic correlates of protection, it might be possible to assess large-scale field studies more efficiently by sampling small numbers of individuals within a population and projecting the outcome in a larger population using computer modeling techniques.

VI. CHEMOTHERAPY[4]

> Lulled by a slightly falling death rate, we are apt to forget that in England and Wales tuberculosis still carries off some 400 people a week. An epidemic which killed on such a scale would arouse intense anxiety, and the sense of resignation which prevents stronger action against this endemic disease is quite irrational. . . . Is tuberculosis really receiving the share of our medical resources which it deserves as a disabling and destructive disease of young people? Far from it—this infection, which could be virtually eradicated from this country within a few years if we took enough trouble, is allowed to reproduce itself continuously because, for historical reasons, it has been assigned a relatively low medical priority.
>
> *The Lancet,* 1949

> To make a long story short, *plenty of work is waiting for microbial geneticists in the field of tuberculosis.* I have never understood why the only infection in which resistance acquired *in vivo* is so frequent has not attracted their attention. Is it because everything takes place at such a slow pace in tuberculosis? This certainly increases the amount of undesirable side changes occurring in the experimental system. But, on the other hand, with such a slow-motion picture of the processes, what an opportunity for closer observation!
>
> Georges Canetti, 1965

When Robert Koch delivered his epoch-making lecture to the Berlin Physiological Society in 1882, sitting in the audience was a young man named Paul Ehrlich, who later wrote: "The evening stands in my memory as my greatest scientific experience." Ehrlich of course, would later rise to preeminence in the

---

[4] For a detailed history of antituberculous chemotherapy, the reader is referred to Frank Ryan's engrossing chronicle *The Forgotten Plague: How the Battle Against Tuberculosis Was Won—and Lost* (Little, Brown and Co., 1993).

field of medical microbiology. Among his many seminal contributions was the invention of histopathology using selective stains to visualize microscopic details of diseased tissues. From these studies emerged the revolutionary concept of treating infectious diseases with "magic bullets," compounds that would selectively kill the offending microbe without harming the host, based on the principle *"Corpora non agunt nisi fixata"* (Ehrlich, 1913). Observing that certain dyes stained pathogenic microorganisms without altering the surrounding host tissue, Ehrlich hypothesized that bipartite antimicrobials could be devised by attaching toxic "side chains" to his microbe-specific dyes (Ehrlich, 1913). Ehrlich's idea was elegant: the dye moiety would combine specifically with the microbial invader, thereby restricting the delivery of the toxin moiety to its proper target.[5] In a mouse model of sleeping sickness (trypanosomiasis), Ehrlich administered compound after compound and finally found one (trypan red) that would cure infected mice, although it had no activity against trypanosomiasis in humans. Undeterred, Ehrlich tested other dyes, constructing hundreds of subtle variants, and, after 605 failures, finally devised "Compound 606" (dioxydiamidoarsenobenzol), which he named "Salvarsan." On the entirely mistaken notion that trypanosomes, the protozoa that cause sleeping sickness, were closely related to *Treponema pallidum,* the spirochete that causes syphilis, Ehrlich tested Salvarsan in syphilitic rabbits and found that it possessed potent curative properties. Following this first success in the laboratory, the effectiveness of Salvarsan in the treatment of human syphilis was established in clinical trials.

It would be difficult to exaggerate the impact that this ground-breaking discovery has had on the course of medical history. For centuries, syphilis was regarded with especial dread as consigning its victims to protracted and disfiguring physical degeneration, ending finally, after years of lingering physical misery and social ostracism, in insanity and death (Quetel, 1992). The advent of Salvarsan therapy abruptly changed the prognosis for these unfortunates: with a few doses of Ehrlich's miracle drug, most patients were rid of their "incurable" disease. The invention of Salvarsan ushered in the modern science of antimicrobial chemotherapy, which has radically altered the relationship of humanity and its parasites—although not always to our benefit.

Building on Ehrlich's pioneering work, many investigators searched for new compounds with antimicrobial activity and, in particular, for a compound capable of curing the greatest killer of all, tuberculosis. In December of 1932, Gerhard Domagk and Josef Klarer, working for the Bayer Company in Germany, showed that the new sulfonamide drug Prontosil was capable of curing mice suffering from a lethal streptococcal infection. In 1933, Prontosil was administered for the first time to a human patient, a young woman dying of virulent streptococcal infection; despite her unfavorable prognosis, the patient recovered rapidly and completely. Eventually, Prontosil found wider use in the

[5] It is interesting, and rather humbling, to note that Ehrlich's adversaries, the disease-causing microbes, "invented" this strategy long before he did. The "AB-type toxins" of *Corynebacterium diphtheriae, Vibrio cholera, Bordetella pertussis, Shigella dysenteriae,* and *Clostridium tetani* (among others) are all bipartite in precisely the sense envisaged by Ehrlich: the nontoxic "B" subunit binds to its cognate receptor on the surface of the target cell, delivering the associated toxic "A" subunit with exquisite specificity.

treatment of infections caused by other gram-positive bacteria, including the pneumococcus and staphylococcus, but the question remained: Would Prontosil cure tuberculosis? In 1938, Arnold Rich and Richard Follis, of the Johns Hopkins Medical School in Baltimore, reported that administration of sulfanilamide (a derivative of Prontosil) to tuberculous guinea pigs ameliorated, but did not cure, their disease (Rich and Follis, 1938). Unfortunately, the extremely high doses of sulfanilamide required to achieve any benefit were far too toxic for use in human patients. Nonetheless, these experiments offered the first glimmer of hope that chemotherapy might eventually provide a potent weapon in the war against the formidable tubercle bacillus. In the early 1940s, William Feldman and H. Corwin Hinshaw, working at the Mayo Clinic in Minnesota, conducted the first clinical trials of Promin and Promizole, two less toxic and more potent derivatives of sulfanilamide (Hinshaw et al., 1943, 1944). Although Feldman and Hinshaw's studies demonstrated only limited efficacy for these drugs, and toxicity continued to pose a problem for long-term treatment, these trials provided the first evidence that chemotherapy could benefit human patients suffering from pulmonary tuberculosis.

A radically different approach to the development of antimicrobials was suggested by Alexander Fleming's serendipitous discovery of the first antibiotic in the course of his researches at St. Mary's Hospital in London. In 1928, Fleming observed the killing of virulent staphylococcus bacteria on a petri dish accidentally contaminated with *Penicillium notatum,* the fungus that produces penicillin (Fleming, 1929). Although penicillin was never shown to be effective against the tubercle bacillus, Fleming's observations triggered a chain of events leading to the discovery of streptomycin, the first effective antibiotic for the treatment of human tuberculosis (Comroe, 1978). In 1939, Fleming attended the Third International Congress of Microbiology in New York City, where he spoke with Selman Waksman, a soil microbiologist at Rutgers University in New Jersey. Fleming's description of his experiments with *P. notatum* galvanized Waksman. Several years previously, a student of Waksman's had observed that avian tubercle bacilli were capable of multiplying in sterile soil but were inhibited and even killed in septic soil, particularly if manure were added (Rhines, 1935). Although Waksman had failed to follow up those early observations, Fleming's revelation in 1939 convinced him that he should revive his search for "pay dirt." Waksman was further encouraged by his former student, Reneé Dubos, also in attendance at that fateful conference, who had described the isolation of a new antibiotic, tyrothricin, from the soil-dwelling bacterium *Bacillus brevis.* In 1943, Waksman's student Albert Schatz, having already screened hundreds of microbial species isolated from various environmental sources, discovered that the actinomycete *Streptomyces griseus* produced a substance capable of killing virulent staphylococcal bacteria. Schatz quickly showed that this antimicrobial substance, dubbed "streptomycin," was also active against tubercle bacilli *in vitro* (Schatz et al., 1944), and provided enough purified material to Feldman and Hinshaw for their first experimental trials of streptomycin in tuberculous guinea pigs, begun in April of 1944 (Feldman, 1954). On November 20, 1944, streptomycin was administered for the first time to a human patient, a young woman suffering from advanced pulmonary tuberculosis; following 5 months of treatment, her case was considered cured, and, 10 years afterward, she was reported to remain in good health (Pfuetze *et al.*, 1955). Merck & Co. began producing streptomycin in December of 1944, permitting Feldman and Hinshaw to begin the first large-scale clinical trials in 1945 (Feldman, 1954; Hinshaw, 1954).

At the same time that Albert Schatz was beginning his experiments with soil microbes, Jorgan Lehmann, working across the Atlantic at Sahlgren's Hospital in Gothenburg, Sweden, invented the first effective antituberculous drug by a process of pure deductive reasoning (Comroe, 1978). In a letter sent to the Ferrosan pharmaceutical company on March 3, 1943, Lehmann predicted that the compound *para*-aminosalicylic acid (PAS) should possess antimycobacterial activity. Lehmann's prediction arose from earlier observations that a significant increase in the rate of bacillary respiration occurred when salicylic acid was added to cultures of tubercle bacilli. Lehmann reasoned that salicylic acid must be metabolized by the tubercle bacillus, and he speculated that a structural analog might poison bacillary metabolism by competitive inhibition. Noting the similarity in the chemical structures of salicylic acid and the sulfonamide drugs, Lehmann predicted that attaching an amino group to the para position of the salicylic acid ring (i.e., at the same position where an amino group occurs in sulfanilamide) would endow the otherwise harmless molecule with the ability to poison mycobacterial metabolism. Remarkably, Lehmann's predictions proved correct. In December of 1943, Lehmann found that PAS synthesized for him by Ferrosan dramatically inhibited growth of tubercle bacilli *in vitro*, and he demonstrated forthwith the efficacy of PAS in the treatment of experimental tuberculosis in guinea pigs. In October of 1944, PAS was administered for the first time, successfully, to a human patient suffering from advanced pulmonary tuberculosis.

Despite the early successes of streptomycin and PAS in treating pulmonary tuberculosis, the serious shortcomings of these two antimicrobials became painfully clear soon after their introduction (Yegian and Vanderlinde, 1950). Although a few patients with advanced pulmonary disease were successfully cured by treatment with streptomycin or PAS alone, the majority of treated patients, despite having shown clinical improvement initially, later relapsed and died. The bacteria recovered from relapse cases were almost invariably resistant to the drug that the patient had received (Crofton and Mitchison, 1948). A major step forward was taken in 1948, when the British Medical Research Council initiated a clinical trial to compare the efficacy of streptomycin or PAS alone with the two drugs administered in combination. The results were dramatic: administration of the two drugs together effectively suppressed the emergence of drug resistance and prevented bacteriological relapse (MRC, 1949), results that were confirmed by workers in the United States (Karlson *et al.*, 1949). One of the first beneficiaries of combination chemotherapy was none other than William Feldman, who had contracted a life-threatening pulmonary infection with *M. tuberculosis* H37Rv, the virulent laboratory strain he had used in his ground-breaking experiments with guinea pigs. Of course, the idea of using multidrug therapy to prevent the emergence of resistant organisms was hardly novel; as early as 1913, in an address to the Seventeenth International Congress of Medicine, Paul Ehrlich had articulated the concept in his characteristically colorful style:

> Now, it is a frequent practice of many uncivilised peoples, in order to be certain of killing their enemies, that they not only rub over their arrow with one kind of poison, but with two or three totally different kinds of poison. And so it also appeared advisable to imitate this procedure against the parasites . . . and to poison our synthetically poisoned arrows not singly but doubly. . . . A further advantage of combined therapy is, that under the influence of two different medicines the danger of rendering the parasites immune . . . which naturally would be a very great obstacle in connexion with further treatment, is apparently greatly minimised. . . . For all these reasons I think that combined therapy will in the future conquer an ever-increasing field of action. [Ehrlich, 1913.]

The biological basis for the efficacy of combination chemotherapy is simple: in a previously unselected bacterial population, spontaneous mutants resistant to either streptomycin or PAS arise at a frequency of about 2×10^{-8}; that is, about 1 in 50 million bacteria are singly resistant. Bacillary loads far exceeding this number commonly occur in advanced cases of pulmonary tuberculosis, particularly within liquified caseum and tuberculous cavities. In retrospect, it is hardly surprising that monotherapy so frequently led to the emergence of drug-resistant strains. The chance that a single bacterial cell will evolve resistance to two drugs simultaneously is simply the product of the individual drug resistance frequencies, that is $(2 \times 10^{-8}) \times (2 \times 10^{-8}) = 4 \times 10^{-16}$; thus, in a previously unselected bacterial population, spontaneous mutation to resistance to both streptomycin and PAS will occur in only 1 in 2.5×10^{15} bacteria. To appreciate the enormity of this number, it may be helpful to consider that the production of this quantity of bacteria in the laboratory would require about one thousand liters of saturated broth culture. Clearly, not even patients with far-advanced disease harbor such huge numbers of bacilli.

The development of combination chemotherapy radically changed the prognosis for tuberculous patients receiving antimicrobials, although drug resistance, particularly resistance to streptomycin, continued to pose serious problems. The next breakthrough came in 1952, when three pharmaceutical companies—Bayer in Germany, and Squibb and Hoffman La Roche in the United States—independently announced their discovery that isoniazid, a drug first synthesized in 1912, possessed greater antituberculous activity than either streptomycin or PAS (McDermott, 1969). That same eventful year saw the first clinical trial of pyrazinamide, another potent new antituberculous drug produced by the Lederle pharmaceutical company in the United States. Isoniazid and pyrazinamide displayed stunning activity in early clinical trials: for the first time, cure rates for tuberculous patients receiving combination chemotherapy approached 100%, even among patients with advanced disease. The superiority of the new drugs was exhaustively demonstrated in a series of experimental studies conducted by Walsh McDermott and colleagues at the Cornell Medical Center in New York City. Using the murine model of tuberculosis, the Cornell group demonstrated that 3 months' administration of isoniazid plus pyrazinamide would routinely sterilize infected animals; extensive efforts to detect even a few residual tubercle bacilli were fruitless (McCune and Tompsett, 1956; McCune et al., 1956, 1966a,b). Compared with streptomycin and PAS, which could suppress but not cure tuberculous infection in mice, the potency of isoniazid and pyrazinamide was striking. Finally, in 1963, following in the tradition of Waksman and Schatz, the pharmaceutical company Lepetit identified rifamycin B, a new class of antibiotic produced by the actinomycete *Amycolatopsis mediterranei,* and in 1966 the CIBA pharmaceutical company introduced rifampicin, a more potent derivative of rifamycin B. Modern "short-course" chemotherapy based on rifampicin reduced treatment time from 12–24 months (Bobrowitz and Robins, 1967) to just 6 months (Fox and Mitchison, 1975; American Thoracic Society, 1980).

Currently, the United States Centers for Disease Control and Prevention recommends a short-course regimen consisting of a 2-month intial phase of rifampicin, isoniazid, pyrazinamide, and streptomycin or ethambutol (introduced in 1966 by Lederle) administered daily, followed by a 4-month continuation phase of rifampicin plus isoniazid administered twice weekly. This regimen, if completed, will cure ~90–95% of patients with fully drug-sensitive tuberculosis (Perriens *et al.,* 1995).

A. Drug Tolerance and Latency

The tubercle bacillus is a daunting adversary. Beginning a few months after cessation of chemotherapy, McDermott's group observed bacteriological relapse in some of the mice that had been "cured" (i.e., rendered culture-negative) with isoniazid and pyrazinamide. Remarkably, they found that 100% of the treated animals relapsed if immunosuppressed with corticosteroids. The bacilli that had survived 3 months' chemotherapy with isoniazid and pyrazinamide remained fully sensitive to both drugs when tested *in vitro*. McDermott interpreted these results to mean that a small population of latent bacilli in the infected mice were "indifferent" to drugs that were highly effective against actively replicating bacteria. Presumably, the reactivation and outgrowth of these drug-tolerant "persisters" after the withdrawal of chemotherapy were responsible for the bacteriological relapses observed. In agreement with these findings, clinical evidence suggests that even prolonged chemotherapy may not eradicate latent tubercle bacilli in humans (Beck and Yegian, 1952; Falk *et al.,* 1954; Hobby *et al.,* 1954; Auerbach *et al.,* 1955; Vandiviere *et al.,* 1956; Wayne and Salkin, 1956; Canetti, 1965; Girling, 1989). In one recent study, 91 (15%) of 601 recurrent cases of tuberculosis occurred in individuals who had previously completed an "adequate" course of chemotherapy, yet the bacilli isolated from these individuals were, in most cases, fully drug-sensitive (Kopanoff *et al.,* 1988).

The drug tolerance of latent tubercle bacilli is probably due to their persistence in the stationary phase of the growth cycle, as nonreplicating bacilli can survive prolonged exposure to antimicrobials that would rapidly kill replicating bacteria. Clinical and experimental evidence suggests that mycobacterial "persisters" may arise in response to oxygen limitation within tuberculous lesions. Unfortunately, nothing is known of the mechanisms that permit mycobacteria to persist under these conditions, although a few investigators are beginning to tackle this problem (de Maio *et al.,* 1996; Wayne and Hayes, 1996; Yuan *et al.,* 1996). Wayne and Sramek (1994) have demonstrated that oxygen-starved tubercle bacilli are rapidly killed by the drug metronidazole, which is used clinically for the treatment of infections caused by anaerobic microbes; it is not yet known, however, whether metronidazole has any activity against latent tubercle bacilli *in vivo*.

The failure of the Cornell group to eradicate tubercle bacilli from infected animals despite 3 months' combination chemotherapy underscores the major problem that clinicians experience in treating human tuberculosis, namely, the complex and lengthy regimens required to achieve reliable cure (Girling, 1989). Even with the best modern multidrug regimens, and with infection caused by fully drug-sensitive bacilli, patients must receive prolonged treatment to ensure against bacteriological relapse. Worse still, the treated patient remains at risk for development of reactivation tuberculosis later in life, due to the persistence of latent bacilli. In this respect, chemotherapy is radically different for tuberculosis than for the familiar acute bacterial infections, which generally require no more than a few weeks' treatment. As the duration of chemotherapy increases, patient compliance deteriorates, resulting in treatment failure and the emergence of strains of *M. tuberculosis* with acquired drug resistance. These newly drug-resistant strains are readily transmitted to susceptible contacts, resulting in tuberculous infection with primary drug resistance. In stepwise fashion, strains of *M. tuberculosis* have evolved that are resistant to all first-time antituberculous drugs, such as the infamous "Strain W" now circulating in New York City and elsewhere (Heym *et al.*, 1994).

B. Mechanisms of Drug Resistance

Following the development of methods for the genetic manipulation of pathogenic mycobacteria, rapid progress has been made in elucidating the genetic basis of resistance to the major antimycobacterial drugs. The mechanisms of resistance to rifampicin, streptomycin, and fluoroquinolones in *M. tuberculosis* are similar to resistance mechanisms in other bacteria. Missense mutations in *rpoB*, encoding the β subunit of the bacterial DNA-dependent RNA polymerase, confer resistance to rifampicin (Telenti *et al.*, 1993; Heym *et al.*, 1994; Kapur *et al.*, 1994b; Williams *et al.*, 1994). Mutations that alter either the ribosomal protein S12 (*rpsL*) or 16S ribosomal RNA (*rrs*) confer resistance to streptomycin (Finken *et al.*, 1993; Honore and Cole, 1994). Fluoroquinolone resistance is associated with missense mutations in *gyrA*, encoding one of the subunits of the bacterial DNA gyrase (Takiff *et al.*, 1994). Resistance to pyrazinamide is conferred by loss of a functional *pncA* gene, encoding a pyrazinamidase required for covalent modification and activation of pyrazinamide (Scorpio and Zhang, 1996). (Interestingly, wild strains of *M. bovis* lack this pyrazinamidase activity and are correspondingly resistant to pyrazinamide.) Mutations conferring resistance to ethambutol have been identified in a novel gene, *embB*, which may be involved in biosynthesis of the arabinogalactan component of the mycobacterial cell wall (Telenti *et al.*, 1997).

Resistance to isoniazid appears to be complex. Early studies of isoniazid-resistant clinical isolates of *M. tuberculosis* revealed that many were deficient in catalase activity (Middlebrook, 1954; Cohn *et al.*, 1954a,b; Middlebrook *et al.*, 1954); more recently, Zhang *et al.* (1992) demonstrated that such isolates carry mutations in the *katG* gene, encoding a bifunctional catalase/peroxidase enzyme. It has been suggested that isoniazid is a prodrug that requires covalent modification and activation by KatG, although the putative activated form of isoniazid has not been identified (Zhang and Young, 1993). A potential target for isoniazid was suggested by genetic studies in the fast-growing saprophyte *M. smegmatis,* in which missense mutations in the *inhA* gene, or overexpression of wild-type *inhA*, were shown to confer resistance to both isoniazid and the second-line drug ethionamide (Banerjee *et al.,* 1994). Similar results were subsequently reported for virulent *M. bovis* (Wilson *et al.,* 1995a). On the basis of sequence similarity and *in vitro* enzymatic activity, *inhA* was proposed to encode an enoyl–acyl carrier protein (ACP) reductase involved in the biosynthesis of cell wall-associated mycolic acids. These findings provided a molecular foundation for the decades-old hypothesis that isoniazid kills mycobacteria by interfering with cell wall biosynthesis, based on the early observation that tubercle bacilli treated with isoniazid lost their acid-fast staining properties. The rational design of new drugs effective against isoniazid-resistant strains of *M. tuberculosis* should benefit from further genetic and biochemical studies, and from the solution of the crystal structure of InhA (Dessen *et al.,* 1995).

It has been suggested, however, that although InhA is the target of isoniazid in *M. smegmatis,* it may not be the target in *M. tuberculosis,* and isoniazid may interfere with a different step in mycolic acid biosynthesis in the latter species (Barry and Mdluli, 1996). The argument is based on the observation that treatment of *M. tuberculosis* with isoniazid leads to the accumulation of long-chain (C_{24}–C_{26}) saturated fatty acids rather than the short-chain (C_{16}–C_{19}) fatty acids whose synthesis is thought to be dependent on InhA. The accumulation of long-chain saturated fatty acids in isoniazid-treated tubercle bacilli suggests that isoniazid may inhibit an as-yet unidentified C_{24} desaturase enzyme. On the other hand, some isoniazid/ethionamide-resistant clinical isolates have been shown to carry mutations in the *inhA* structural gene or promoter analogous to mutations conferring isoniazid/ethionamide resistance in *M. smegmatis* (Heym *et al.,* 1994; Musser *et al.,* 1996; Ristow *et al.,* 1996). Because of the difficulty of genetic manipulation in *M. tuberculosis,* it has not yet been possible to establish unequivocally whether these mutations are responsible for isoniazid resistance. Approximately 20% of isoniazid-resistant clinical isolates carry mutations in *inhA* and 50% carry mutations in *katG*; the remaining 30% are apparently wild type for *inhA* and *katG*, suggesting that additional mechanisms of resistance likely exist (J. Musser, personal communication, 1997).

C. Tuberculosis Control and Chemotherapy: Cost-Effectiveness Analysis

Despite the availability of drug combinations effective against tuberculosis, public health officials have been pessimistic that chemotherapy could make a significant impact on the disease burden, for two reasons. First, the most rapidly effective drug combinations are expensive, discouraging their use in poor countries where disease is most prevalent. Second, the lengthy regimens of chemotherapy required to cure tuberculosis result in poor patient compliance and high rates of treatment failure. These two problems are closely linked. Patients undergoing chemotherapy for tuberculosis commonly experience rapid improvement of their clinical signs and symptoms, and, with the apparent return of health, may take their medications haphazardly or discontinue treatment prematurely, placing themselves at risk for relapse. Consequently, patient noncompliance remains the major obstacle to successful treatment, accounting for more cases of treatment failure, relapse, and acquired drug resistance than any other cause (Addington, 1979; Kopanoff *et al.,* 1988; Murray *et al.,* 1990).

The most important factor affecting patient compliance is the duration of chemotherapy: as the length of treatment is extended, patient compliance steadily deteriorates (Murray *et al.,* 1990; Murray, 1994a). One approach to improving patient compliance was the development of "short-course" chemotherapy employing more rapidly effective drugs for shorter times (Fox and Mitchison, 1975; Girling, 1989). In the industrialized nations, the introduction in the early 1970s of 6-month short-course chemotherapy regimens based on isoniazid, rifampicin, and pyrazinamide led to significant improvements in patient compliance and cure rates. Developing nations, however, continued to rely on lengthier (12-month) but less expensive "standard" chemotherapy regimens consisting of isoniazid, streptomycin, and thiacetazone. The work of Karel Styblo, of the International Union Against Tuberculosis and Lung Diseases, demonstrated that cure rates in poor countries such as Tanzania, Malawi, and Mozambique could be increased dramatically by the application of short-course chemotherapy under direct supervision for at least the first 2 months of treatment (Table IV). From Styblo's studies, which were conducted not as small-scale pilot projects but as whole-country control programs, emerged cogent data demonstrating the greater overall cost-effectiveness of short-course chemotherapy, due largely to improved patient cure rates, reduced transmission, and prevention of bacterial drug resistance. Short-course chemotherapy reduced transmission not only by improving long-term cure rates, but also by promoting more rapid sputum conversion: after 2 months of treatment, 85–95% of patients receiving short-course chemotherapy will have converted to sputum-negative (noninfectious) status, as compared to just 50% of patients receiving standard 12-month chemotherapy (Murray *et al.,* 1990). Currently, short-course chemotherapy for tuberculosis is recognized as one of the most cost-effective public health interventions for adults to gain healthy years of life, costing less than US$250 per death averted (Barnum, 1986; Murray *et al.,* 1990; Murray, 1994a).

Despite the advantages offered by short-course chemotherapy, 6 months is still a long time to expect unsupervised patients to continue taking their medications faithfully. Directly observed therapy (DOT) offers a cost-effective strategy to combat the problem of noncompliance. The DOT concept is simple: the administration of each dose of antimicrobials to a patient on DOT is supervised by an attending health care professional or social worker. Some cities have implemented community-based DOT programs, whereby outreach workers travel to the patients' homes, schools, or workplaces in order to ensure that individuals who would not attend a chest clinic nonetheless receive treatment. The example of Baltimore, Maryland, is instructive (Chaulk *et al.,* 1995). In 1978, Baltimore implemented a clinic-based DOT program, and the city "upgraded" to community-based DOT in 1981. Between 1965 and 1978, Baltimore consistently ranked among the top three cities in the United States with respect to tuberculosis incidence rates; by 1992, Baltimore was ranked 28th. While many major U.S. cities experienced an upsurge in tuberculosis incidence

TABLE IV Fate of 100 Cases of Smear-Positive Tuberculosis Receiving No Treatment, 12-Month Standard Chemotherapy, or 6-Month Short-Course Chemotherapy[a]

Year	No treatment			Standard chemotherapy (12 months)			Short-course chemotherapy (6 months)		
	Cured	Excreting bacilli	Dead	Cured	Excreting bacilli	Dead	Cured	Excreting bacilli	Dead
0	0.0	100.0	0.0	0.0	100.0	0.0	0.0	100.0	0.0
1	18.5	61.3	20.1	61.1	23.2	15.8	79.6	11.6	8.8
1.5	27.8	42.0	30.2	61.1	23.2	15.8	81.3	7.9	10.7
2	27.8	38.6	33.6	61.1	21.3	17.7	81.3	7.3	11.4
3	27.8	31.7	40.5	61.1	17.5	21.4	81.3	6.0	12.7
4	30.3	24.9	44.9	62.4	13.7	23.8	81.8	4.7	13.5
5	32.5	18.0	49.2	63.7	9.9	26.2	82.2	3.4	14.3

[a] From Murray et al. (1970). Figures give number per 100 cases.

between 1985 and 1992, case rates in Baltimore actually declined by 29.5%, by far the steepest rate of decline among any of the 20 cities with the highest tuberculosis incidence rates in 1981. Compared to individuals receiving unsupervised chemotherapy, Baltimore patients receiving DOT showed significantly higher rates of sputum conversion (indicating transition to noninfectious status) as well as lower rates of relapse and acquired bacterial drug resistance. Similarly, in a study conducted in Tarrant County, Texas, implementation of DOT in 1986 led to substantial declines in the rates of treatment failure, relapse, and bacterial drug resistance (Weis *et al.*, 1994). Importantly, implementation of DOT was shown to reduce the rates not only of acquired drug resistance, but also of primary drug resistance, indicating that transmission of drug-resistant strains was interrupted. New York City, too, has benefited from DOT: from 1978 through 1992, the incidence of tuberculosis in the city climbed steadily, but it has since declined by about 30%, largely in response to efforts to prevent nosocomial transmission and by the implementation of an aggressive community-based DOT program (Frieden *et al.*, 1995; Hamburg, 1995) (Fig. 15). Importantly, these programs led to an even greater decline (about 70%) in the incidence of multi-drug resistant tuberculosis.

Despite its proven efficacy, DOT is currently administered to only about 10% of all tuberculosis cases treated in the United States, due to the perception that DOT must be more expensive than unsupervised therapy. Wherever analyzed carefully, however, DOT has been found to be more cost-effective than unsupervised therapy, as improved cure rates eliminate the need for costly retreatment of noncompliant patients (Barnum, 1986; Murray *et al.*, 1990; Iseman *et al.*, 1993). In the long term, DOT can be expected to generate even greater savings by preventing the evolution of bacterial drug resistance. Considering that the cost to treat a single patient with MDR-TB can approach US$250,000, a strong argument can be made for administering DOT to every tuberculosis patient.

Nationwide institution of new DOT programs will require a substantial investment of financial and human resources. In New York City, the downturn

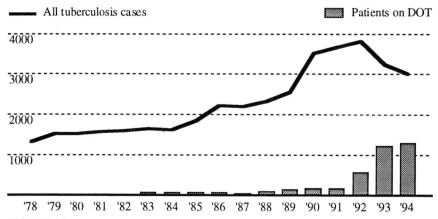

FIGURE 15   Tuberculosis cases and cases receiving directly observed therapy (DOT) in New York City from 1978 to 1994. (Redrawn from Hamburg, 1995.)

in tuberculosis incidence after 1992 was fueled by increased funding for tuberculosis control: between 1988 and 1994, the budget of the New York City Department of Health's Bureau of Tuberculosis Control rose from US$4 million to more than US$40 million, with an accompanying increase in staff from 144 to more than 600 (Arno et al., 1993; Frieden et al., 1995; Hamburg, 1995). These funds were used to trace and treat contacts of tuberculosis cases, to improve clinical laboratory procedures for diagnosis and drug susceptibility testing, to educate health care workers and the public, to renovate and expand inpatient tuberculosis care facilities, and to implement a community-based DOT program. To a considerable extent, the reemergence of tuberculosis in the 1980s was attributable to the decay of the public health infrastructure in the United States following the virtual elimination of funding for tuberculosis control after 1972 (Bloom and Murray, 1992). Basic health measures required for effective tuberculosis control had been increasingly neglected for some years, but they were hardly new: most of them were advocated at the turn of the twentieth century by Hermann Biggs, then director of the New York City Department of Health. As Biggs, the father of the public health movement in the United States, liked to say, "Health is purchasable." If we neglect to invest in tuberculosis control now, we can expect to pay dearly later.

Given the increasing rates of MDR-TB in many countries, a priority for future research must be the development of new antimicrobials to replace those whose utility is now compromised by bacterial resistance. Of particular benefit would be novel drugs capable of accelerating the cure rate for tuberculosis by attacking the bacillary persister population; the shorter treatment regimens afforded by such drugs would reduce the direct and indirect costs of chemotherapy and substantially improve patient compliance. Improvements in patient compliance would, in turn, further reduce the cost of chemotherapy by lowering the rates of treatment failure and relapse, by interrupting transmission, and by suppressing the evolution of bacterial drug resistance. None of the currently available antimicrobials is rapidly effective against mycobacterial persisters. Development of new drugs capable of attacking this small but obstinate subpopulation of bacilli will require a better understanding of the biology of persistence, which has historically received little attention.

Unfortunately, drug companies in the industrialized nations at present see little incentive to invest in the development of new antituberculous drugs, as their domestic markets are too small to offer substantial profits. In the developing world, although tuberculosis is a public health problem of the first order, the necessary funds and infrastructure for new drug development are lacking. Because only about 1 in 100 compounds that look promising in the laboratory proceed all of the way from testing to licensing, the cost to bring a new drug to market is estimated at US$361–500 million (Office of Technology Assessment, 1993; Di Masi, 1995). The effective life of a patent is about 9 years; consequently, responsible corporate planners are unable to invest in drugs that will not rapidly produce returns sufficient to cover development costs and generate reasonable profits.

If the pharmaceutical companies of wealthy nations are to be persuaded to invest in R & D for new antituberculous drugs, special mechanisms and financial incentives must be provided by the international community. A number of innnovations have been suggested which, if implemented, could spur the development of drugs and vaccines most relevant to the needs of nonaffluent individuals in industrialized nations and poor populations in developing countries: (i) protection of intellectual property rights against piracy; (ii) extended patent protection for specified products; (iii) tax incentives and abatements; (iv) provision of market information for developing nations; (v) loans and rotating funds to assure markets; (vi) streamlining and harmonization of regulatory standards; and (vii) cooperation of public and private sectors (World Bank, 1993; Ad Hoc Committee, 1996; Institute of Medicine, 1997). Currently, tuberculosis imposes an enormous financial burden on the world's developing nations, contributing substantially to their continued impoverishment. As trafficking of peoples and goods between nations increases, the world community is becoming ever more global and interdependent; in the future, the health problems of the developing nations can be expected to have an ever greater impact on our own well-being. Clearly, infectious diseases do not respect national borders. Investment by wealthy nations in the development and implementation of improved global tuberculosis control measures is justified not only by economic and humanitarian considerations, but also, let us not forget, by enlightened self-interest.

VII. REFLECTIONS ON THE PAST AND ON FUTURE PRIORITIES

> But most important, the ancient foe of man, known as consumption, the great white plague, tuberculosis, or by whatever other name, is on the way to being reduced to a minor ailment of man. The future appears bright indeed, and the complete eradication of the disease is in sight.
> Selman A. Waksman, *The Conquest of Tuberculosis*, 1964, p. 217

> It is a profound and necessary truth that the deep things in science are found not because they are useful; they are found because it was possible to find them.
> J. Robert Oppenheimer, *The Making of the Atomic Bomb*, 1986

As one surveys our "wealth of ignorance" of the basic biology of the tubercle bacillus, of the pathogenesis of tuberculous disease, and of host responses contributing to protection and pathology, a broad portfolio of research is warranted. Priorities for operational research are discussed by Murray (1994a) and by De Cock *et al.* (1996). Here, we adumbrate five basic research problems that we believe have not received adequate attention from the scientific community. With the impending availability of the complete nucleotide sequence of the *M. tuberculosis* genome, the research we would emphasize increasingly focuses on fundamental biological questions relating to the interaction between pathogen and host. Increased investment in addressing these problems would, we believe, yield both important knowledge and new strategies to combat the disease.

1. Mycobacterial Persistence, Latency, and Reactivation. The long-term persistence of viable tubercle bacilli in the infected host creates an enormous and stable reservoir of infecteds, which is responsible for the very slow dynamics of tuberculosis epidemics and the long-term endemicity of the contagion.

Mycobacterial persistence also necessitates lengthy regimens of chemotherapy, generating increased treatment costs, treatment failure, and the evolution of bacterial drug resistance. The campaign against smallpox launched by the World Health Organization in 1967, based on the aggressive identification of cases and vaccination of contacts, led to the global eradication of this pathogen in just 10 years' time. The eradication of smallpox in such a short time interval was possible only because the disease had no reservoir outside of humans, an effective vaccine was available, and the variola virus does not cause persistent infection. Without a vaccine capable of preventing establishment and reactivation of latent tuberculous infection, or drugs capable of eliminating latent tubercle bacilli, the eradication of *M. tuberculosis* will be very difficult, if achievable at all. The development of new tools to combat latent tuberculosis will require better understanding of several fundamental scientific questions. What is the role of the host response in the induction and maintenance of latency? What are the events occurring during the reactivation from latency? What mycobacterial genes are essential for latency? Would the products of such "latency genes" represent new targets for drug development? Can immunotherapies be devised that would promote the eradication of latent bacilli?

2. **Continuing Need for New and Improved Drugs.** New drugs are urgently needed now, and for the forseeable future, to replace those compromised by bacterial resistance. In particular, novel drugs capable of killing drug-tolerant "persisters" could profoundly improve treatment by accelerating the course of chemotherapy and eliminating the risk of long-term relapse. Identification of new drug targets should benefit from the information provided by the *M. tuberculosis* Genome Project, and from the development of tools for the genetic manipulation of pathogenic mycobacteria. The development of better screens for drugs effective against tubercle bacilli, and particularly latent bacilli, will require improved understanding of mycobacterial physiology and the development of meaningful *in vitro* models for studying the different stages of the disease process.

3. **Mechanisms of Pathogenesis.** A fundamental question in the pathogenesis of tuberculosis is the cause of pathology: to what extent is pathology generated by the pathogen itself, and to what extent does it derive from the host response against the pathogen? Are the protective immune responses engaged in localizing, inhibiting, and killing tubercle bacilli distinct and dissociable from those causing tissue damage? Is pathology the unavoidable price that the host pays for protection, or can immune responses be manipulated to protect without causing tissue damage? What is the physiological and molecular basis for the predilection of tuberculosis for long tissue, and particularly for the lung apex? What is the ecology (nutritional and growth environment) of tubercle bacilli growing in the infected host within different tissues and at different stages of infection? What mycobacterial functions are important for survival and growth within macrophages? Does *M. tuberculosis* infect somatic cell types other than macrophages within the host? How does *M. tuberculosis* kill the cells it infects? Which bacillary components contribute directly or indirectly (by inducing the host response) to tissue damage? Identification of the virulence determinants of the tubercle bacillus is just beginning, and many outstanding questions cry out for attention. Progress in addressing these problems will be

contingent on the development of more effective methods (e.g., allelic exchange, transposon mutagenesis) for the genetic manipulation of *M. tuberculosis*.

4. Mechanisms of Protection and Vaccine Development. The fulminant course of tuberculosis in immunodeficient individuals emphasizes the critical role of the immune system in protecting against tuberculosis. What are the necessary and sufficient conditions for protection against disease? Why is the host response to *M. tuberculosis* not eradicative? Would it be possible to induce host responses that would prevent the establishment and reactivation of latent infection? Vaccination with BCG appears to provide some protection against disease but not infection; could new approaches to immunization induce host responses capable of preventing infection, as well? Answers to these questions would greatly facilitate the development of more effective vaccines, which could ultimately provide the most cost-effective means to control tuberculosis. In order to evaluate new vaccine candidates in an economically feasible way, it will be crucial to identify immunologic correlates of protection that can be rapidly and inexpensively evaluated in large study populations. Perhaps the greatest future challenge for tuberculosis vaccine research will be the design of vaccine field trials, which must be economically and logistically feasible, scientifically informative, and amenable to rapid evaluation.

5. Genetics of Host Resistance and Susceptibility. The case fatality rate of untreated tuberculosis ranges from 40 to 60% (Murray *et al.*, 1990; Styblo, 1991). Given the global prevalence of tuberculous infection, it is inconceivable that this disease has not served as a significant selective force on the evolution of the human genome. We know that tuberculosis is more severe in some strains of mice, rabbits, and guinea pigs, and in some human ethnic populations. Yet with the possible exception of the human major histocompatibility complex, which affects the immune responses to myriad antigens, little is known about genes controlling resistance and susceptibility to tuberculosis in any species. With the enormous progress being made in the Human Genome Project, and the availability of genetic linkage maps for mouse and man, it should be possible to learn much about the genetic basis for susceptibility and resistance to tuberculosis. Such genetic studies promise valuable insights into evolution of the human genome and may suggest unanticipated new approaches for interventions in this and other microbial infections.

If there is an overarching lesson to be gleaned from the rich history of tuberculosis research, it is the need for humility by the scientific community. The discovery of the tubercle bacillus by Koch in 1882 was truly one of the great triumphs of scientific rigor and personal courage. But let us not forget that in 1890 Koch proclaimed *the* cure for tuberculosis, that in 1921 Calmette and Guerin introduced *the* vaccine against tuberculosis, that in 1946 Schatz and Waksman isolated *the* antibiotic that killed the tubercle bacillus—yet, tuberculosis remains today the leading cause of death in the world from a single infectious disease.

The most critical time in the control of an infectious disease is when case rates are dropping and public support and interest wane. In the United States and Europe, the incidence of tuberculosis is currently declining, but in many developing countries in Asia and Africa, as the full impact of the AIDS epidemic hits, case rates are rising and resistance to antimycobacterial drugs in increasing. One widely held view in public health, dating back to the nineteenth century, is that we already "have the tools" to eradicate tuberculosis—it is only a matter of their implementation! It is our view that more knowledge of the basic scientific issues discussed here, and the translation of that knowledge into new tools and control strategies, will be vital to the success of future efforts to combat the persistent threat of tuberculosis. Tuberculosis is a disease for which "eradication" has been promised so often in the past, but not delivered, that further use of the term seems inappropriately hyperbolic. The World Health Organization Tuberculosis Programme has set as its goals, with the use of existing technology, to detect 70% of all sputum smear-positive (infectious) cases and to increase cure rates to 85% by the year 2000 (World Health Organization, 1992). For the scientific community, perhaps the most realistic long-term goal for future tuberculosis research and control efforts, as formulated by the World Health Organization Leprosy Program in its approach to leprosy control, is to develop the means to "eliminate the disease as a public health problem."

REFERENCES

Ad Hoc Committee on Health Research Relating to Future Intervention Options (1996). "Investing in Health Research and Development." World Health Organization, Geneva: (Document TDR/Gen/96.1).

Addington, W. W. (1979). Patient compliance: The most serious remaining problem in the control of tuberculosis in the United States. *Chest* 76 (Suppl.), 741–743.

Aebischer, T. (1994). Recurrent cutaneous leishmaniasis: A role for persistent parasites? *Parasitol. Today* 10, 25–28.

Afonso, L. C. C., Scharton, T. M., Vieira, L. Q., Wysocka, M., Trinchieri, G., and Scott, P. (1994). The adjuvant effect of interleukin-12 in a vaccine against *Leishmania major*. *Science* 263, 235–237.

Aldovini, A., and Young, R. A. (1991). Humoral and cell-mediated immune responses to live recombinant BCG-HIV vaccines. *Nature (London)* 351, 479–482.

Alland, D., Kalkut, G. E., Moss, A. R., McAdam, R. A., Hahn, J. A., Bosworth, W., Drucker, E., and Bloom, B. R. (1994). Transmission of tuberculosis in New York City: An analysis by DNA fingerprinting and conventional epidemiologic methods. *N. Engl. J. Med.* 330, 1710–1716.

American Thoracic Society. (1980). Guidelines for short-course chemotherapy. *Am. Rev. Respir. Dis.* 121, 611–614.

Andersen, P. (1994). Effective vaccination of mice against *Mycobacterium tuberculosis* infection with a soluble mixture of secreted mycobacterial proteins. *Infect. Immun.* 62, 2536–2544.

Armstrong, J. A., and Hart, P. D. (1971). Response of cultured macrophages to *Mycobacterium tuberculosis*, with observations on fusion of lysosomes with phagosomes. *J. Exp. Med* 134, 713–740.

Armstrong, J. A., and Hart, P. D. (1975). Phagosome–lysosome interactions in cultured macrophages infected with virulent tubercle bacilli: Reversal of the usual fusion pattern and observations on bacterial survival. *J. Exp. Med.* 142, 1–16.

Arno, P. S., Murray, C. J., Bonuck, K. A., and Alcabes, P. (1993). The economic impact of tuberculosis in hospitals in New York City: A preliminary analysis. *J. Law, Med. Ethics* 21, 317–323.

Aronson, J. D., Aronson, C. F., and Taylon, H. C. (1958). A twenty-year appraisal of BCG vaccination in the control of tuberculosis. *Arch. Int. Med.* 101, 881–893.

Auerbach, O., Hobby, G. L., Small, M. J., Lenert, T. F., and Comer, J. V. (1955). The clinicopathologic significance of the demonstration of viable tubercle bacilli in resected lesions. *J. Thoracic Surg.* 29, 109–135.

Azad, A. K., Sirakova, T. D., Rogers, L. M., and Kolattukudy, P. E. (1996). Targeted replacement of the mycocerosic acid synthase gene in *Mycobacterium bovis* BCG produces a mutant that lacks mycosides. *Proc. Natl. Acad. Sci. U.S.A.* **93**, 4787–4792.

Baker, S. K., and Glassroth, J. (1994). Miliary tuberculosis. *In* "Tuberculosis" (W. N. Rom and S. Garay, eds.), pp. 493–511. Little, Brown and Co., Boston, MA.

Balasubramanian, V., Wiegeshaus, E. H., Taylor, B. T., and Smith, D. W. (1994). Pathogenesis of tuberculosis: Pathway to apical localization. *Tubercle Lung Dis.* **75**, 168–178.

Balasubramanian, V., Pavelka, M. S., Jr., Bardarov, S. S., Martin, J., Weisbrod, T. R., McAdam, R. A., Bloom, B. R., and Jacobs, W. R., Jr. (1996). Allelic exchange in *Mycobacterium tuberculosis* with long linear recombination substrates. *J. Bacteriol.* **178**, 273–279.

Banerjee, A., Dubnau, E., Quemard, A., Balasubramanian, V., Sun Um, K., Wilson, T., Collins, D., de Lisle, G., and Jacobs, W. R., Jr. (1994). *inhA*, a gene encoding a target for isoniazid and ethionamide in *Mycobacterium tuberculosis*. *Science* **263**, 227–230.

Bange, F.-C., Brown, A. M., and Jacobs, W. R., Jr. (1996). Leucine auxotrophy restricts growth of *Mycobacterium bovis* BCG in macrophages. *Infec. Immun.* **64**, 1794–1799.

Barnes, H. L., and Barnes, L. R. P. (1928). The duration of life in pulmonary tuberculosis with cavity. *Am. Rev. Tuberculosis* **18**, 412–424.

Barnes, P. F., Fong, S. J., Brennan, P. J., Twomey, P. E., Mazumder, A., and Modlin, R. L. (1990). Local production of tumor necrosis factor and IFN-γ in tuberculous pleuritis. *J. Immunol.* **145**, 149–154.

Barnes, P. F., Bloch, A. B., Davidson, P. T., and Snider, D. E. (1991). Tuberculosis in patients with human immunodeficiency virus infection. *N. Engl. J. Med.* **324**, 1644–1650.

Barnes, P. F., Lu, S., Abrams, J. S., Wang, E., Yamamura, M., and Modlin, R. L. (1993). Cytokine production at the site of disease in human tuberculosis. *Infect. Immun.* **61**, 3482–3489.

Barnes, P. F., Modlin, R. L., and Ellner, J. J. (1994). T-cell responses and cytokines. *In* "Tuberculosis: Pathogenesis, Protection, and Control" (B. R. Bloom, ed.), pp. 417–435. ASM Press, Washington, D.C.

Barnum, H. N. (1986). Cost savings from alternative treatments for tuberculosis. *Soc. Sci. Med.* **23**, 847–850.

Barry III, C. E., and Mdluli, K. (1996). Drug sensitivity and environmental adaptation of mycobacterial cell wall components. *Trends Microbiol.* **4**, 275–281.

Bates, J. H., and Stead, W. W. (1993). The history of tuberculosis as a global epidemic. *Med. Clin. North Am.* **77**, 1205–1217.

Baulard, A., Kremer, L., and Locht, C. (1996). Efficient homologous recombination in fast-growing and slow-growing mycobacteria. *J. Bacteriol.* **178**, 3091–3098.

Beck, F., and Yegian, D. (1952). A study of the tubercle bacillus in resected pulmonary lesions. *Am. Rev. Tuberculosis* **66**, 44–51.

Beckman, E. M., Porcelli, S. A., Morita, C. T., Behar, S. M., Furlong, S. T., and Brenner, M. B. (1994). Recognition of a lipid antigen by CD1-restricted αβ+ T cells. *Nature (London)* **372**, 691–694.

Besra, G. S., and Chatterjee, D. (1994). Lipids and carbohydrates of *Mycobacterium tuberculosis*. *In* "Tuberculosis: Pathogenesis, Protection, and Control" (B. R. Bloom, ed.), pp. 285–306. ASM Press, Washington, D.C.

Black, F. L. (1966). Measles endemicity in insular populations: Critical community size and its evolutionary implication. *J. Theor. Biol.* **11**, 207–211.

Black, F. L. (1975). Infectious diseases in primitive societies. *Science* **187**, 515–518.

Black, F. L., Hierholzer, W. J., Pinheiro, F. D., Evans, A. S., Woodall, J. P., Opton, E. M., Emmons, J. E., West, B. S., Edsall, G., Downs, W. G., and Wallace, G. D. (1974). Evidence for persistence of infectious agents in isolated human populations. *Am. J. Epidemiol.* **100**, 230–250.

Blackwell, J. M. (1996). Genetic susceptibility to leishmanial infections: Studies in mice and man. *Parasitology* **112** (Suppl.), S67–S74.

Bloch, A. B., Cauthen, G. M., Onorato, I. M., Dansbury, K. G., Kelly, G. D., Driver, C. R., and Snider, D. E., Jr. (1994). Nationwide survey of drug-resistant tuberculosis in the United States. *J. Am. Med. Assoc.* **271**, 665–671.

Bloom, B. R., and Bennett, B. (1966). Mechanism of a reaction in vitro associated with delayed-type hypersensitivity. *Science* **153**, 80–82.

Bloom, B. R., and Fine, P. E. M. (1994). The BCG experience: Implications for future vaccines against tuberculosis. *In* "Tuberculosis: Pathogenesis, Protection, and Control" (B. R. Bloom, ed.), pp. 531–557. ASM Press, Washington, D.C.

Bloom, B. R., and Murray, C. J. L. (1992). Tuberculosis: Commentary on a reemergent killer. *Science* **257**, 1055–1064.

Blower, S. M., McLean, A. R., Porco, T. C., Small, P. M., Hopewell, P. C., Sanchez, M. A., and Moss, A. R. (1995). The intrinsic transmission dynamics of tuberculosis epidemics. *Nature Medicine* **1**, 815–821.

Bobrowitz, I. D., and Robins, D. E. (1967). Ethambutol-isoniazid versus PAS-isoniazid in original treatment of pulmonary tuberculosis. *Am. Rev. Respir. Dis.* **96**, 428–438.

Boom, W. H., Wallis, R. S., and Chervenak, K. A. (1991). Human *Mycobacterium tuberculosis*-reactive CD4+ T-cell clones: Heterogeneity in antigen recognition, cytokine production, and cytotoxicity for mononuclear phagocytes. *Infect. Immun.* **59**, 2737–2743.

Braun, M. M., and Cauthen, G. (1992). Relationship of the human immunodeficiency virus epidemic to pediatric tuberculosis and Bacillus Calmette Guerin immunization. *Pediatr. Infect. Dis. J.* **11**, 220–227.

Brennan, P. J., and Draper, P. (1994). Ultrastructure of *Mycobacterium tuberculosis*. *In* "Tuberculosis: Pathogenesis, Protection, and Control" (B. R. Bloom, ed.), pp. 271–284. ASM Press, Washington, D.C.

Brewer, T. F., and Colditz, G. A. (1995). Relationship between bacille Calmette-Guerin (BCG) strains and the efficacy of BCG vaccine in the prevention of tuberculosis. *Clin. Infect. Dis.* **20**, 126–135.

Brudney, K., and Dobkin, J. (1991). Resurgent tuberculosis in New York City: Human immunodeficiency virus, homelessness, and the decline of tuberculosis control programs. *Am. Rev. Respir. Dis.* **144**, 745–749.

Burlein, J. E., Stover, K. C., Offutt, S., and Hanson, M. S. (1994). Expression of foreign genes in mycobacteria. *In* "Tuberculosis: Pathogenesis, Protection, and Control" (B. R. Bloom, Ed.), pp. 239–252. ASM Press, Washington, D.C.

Buschman, E., and Skamene, E. (1988). Genetic background of the host and expression of natural resistance and acquired immunity to *M. tuberculosis*. *In* "Mycobacterium tuberculosis: Interactions with the Immune System" (M. Bendinelli and H. Friedman, eds.), pp. 59–79. Plenum, New York.

Canetti, G. (1965). Present aspects of bacterial resistance in tuberculosis. *Am. Rev. Respir. Dis.* **92**, 687–703.

Cantwell, M. F., Snider, D. E., Cauthen, G. M., and Onorato, I. M. (1994). Epidemiology of tuberculosis in the United States, 1985 through 1992. *J. Am. Med. Assoc.* **272**, 535–539.

Cavalli-Sforza, L. L., and Cavalli-Sforza, F. (1995). "The Great Human Diasporas." Addison-Wesley, New York.

Chamberlayne, E. C., ed. (1972). "Status of Immunization of Tuberculosis in 1971." U.S. Department of Health, Education, and Welfare, Washington, D.C.

Chan, J., and Kaufmann, S. H. E. (1994). Immune mechanisms of protection. *In* "Tuberculosis: Pathogenesis, Protection, and Control" (B. R. Bloom, ed.), pp. 389–415. ASM Press, Washington, D.C.

Chan, J., Xing, Y., Magliozzo, R. S., and Bloom, B. R. (1992). Killing of virulent *Mycobacterium tuberculosis* by reactive nitrogen intermediates produced by activated murine macrophages. *J. Exp. Med.* **175**, 1111–1122.

Chan, J., Tanaka, K., Carroll, D., Flynn, J., and Bloom, B. R. (1995). Effects of nitric oxide synthase inhibitors on murine infection with *Mycobacterium tuberculosis*. *Infect. Immun.* **63**, 736–740.

Chase, M. W. (1945). The cellular transfer of cutaneous hypersensitivity to tuberculin. *Proc. Soc. Exp. Biol. Med.* **59**, 134–135.

Chaulk, C. P., Moore-Rice, K., Rizzo, R., and Chaisson, R. E. (1995). Eleven years of community-based directly observed therapy for tuberculosis. *J. Am. Med. Assoc.* **274**, 945–951.

Clemens, D. L. (1996). Characterization of the *Mycobacterium tuberculosis* phagosome. *Trends Microbiol.* **4**, 113–118.

Clemens, D. L., and Horwitz, M. A. (1995). Characterization of the *Mycobacterium tuberculosis* phagosome and evidence that phagosomal maturation is inhibited. *J. Exp. Med.* **181**, 257–270.

Cobbett, L. (1932). The rabbit as a means of distinguishing the human from the bovine type of tuberculosis. *J. Pathol. Bacteriol.* **35**, 681–699.

Coetzee, A. M., and Berjak, J. (1968). BCG in the prevention of tuberculosis in an adult population. *Proc. Mine Med. Officers Assoc.* **48**, 41–53.

Cohen, J. (1995). "How Many People Can the Earth Support?" Norton, New York.

Cohn, M. L., Oda, U., Kovitz, C., and Middlebrook, G. (1954a). Studies on isoniazid and tubercle bacilli. I. The isolation of isoniazid-resistant mutants *in vitro. Am. Rev. Tuberculosis* **70**, 465–475.

Cohn, M. L., Kovitz, C., Oda, U., and Middlebrook, G. (1954b). Studies on isoniazid and tubercle bacilli. II. The growth requirements, catalase activities, and pathogenic properties of isoniazid-resistant mutants. *Am. Rev. Tuberculosis* **70**, 641–664.

Collins, D. M. (1996). In search of tuberculosis virulence genes. *Trends Microbiol.* **4**, 426–430.

Collins, D. M., Kawakami, R. P., de Lisle, G. W., Pascopella, L., Bloom, B. R., and Jacobs, W. R., Jr. (1995). Mutation of the principal σ factor causes loss of virulence in a strain of the *Mycobacterium tuberculosis* complex. *Proc. Natl. Acad. Sci. U.S.A.* **92**, 8036–8040.

Colston, M. J., and Davis, E. O. (1994). Homologous recombination, DNA repair, and mycobacterial *recA. In* "Tuberculosis: Pathogenesis, Protection, and Control" (B. R. Bloom, ed.), pp. 217–226. ASM Press, Washington, D.C.

Comroe, J. H. (1978). Pay dirt: the story of streptomycin. *Am. J. Respir. Dis.* **117**, 773–781 and 957–968.

Comstock, G. W. (1978). Tuberculosis in twins: A re-analysis of the Prophit survey. *Am. Rev. Respir. Dis.* **117**, 621–624.

Comstock, G. W. (1982). Epidemiology of tuberculosis. *Am. Rev. Respir. Dis.* **125**, 8–16.

Comstock, G. W., and Webster, R. G. (1969). Tuberculosis studies in Muscogee County, Georgia. VII. A twenty-year evaluation of BCG vaccination in a school population. *Am. Rev. Respir. Dis.* **100**, 839–845.

Comstock, G. W., Livesay, V. T., and Woolpert, S. F. (1974a). The prognosis of a positive tuberculin reaction in childhood and adolescence. *Am. J. Epidemiol.* **99**, 131–138.

Comstock, G. W., Livesay, V. T., and Woolpert, S. F. (1974b). Evaluation of BCG vaccination among Puerto Rican children. *Am. J. Public Health* **64**, 283–291.

Comstock, G. W., Woolpert, S. F., and Livesay, V. T. (1976). Tuberculosis studies in Muscogee County, Georgia. Twenty-year evaluation of a community trial of BCG vaccination. *Public Health Rep.* **91**, 276–280.

Connell, N. D., Medina-Acosta, E., McMaster, W. R., Bloom, B. R., and Russell, D. G. (1993). Effective immunization against cutaneous leishmaniasis with recombinant bacille Calmette–Guerin expressing the *Leishmania* proteinase gp63. *Proc. Natl. Acad. Sci. U.S.A.* **90**, 11473–11477.

Cooper, A. M., and Flynn, J. L. (1995). The protective immune response to *Mycobacterium tuberculosis. Curr. Opin. Immunol.* **7**, 512–516.

Cooper, A. M., Dalton, D. K., Stewart, T. A., Griffin, J. P., Russell, D. G., and Orme, I. M. (1993). Disseminated tuberculosis in interferon γ gene-disrupted mice. *J. Exp. Med.* **178**, 2243–2247.

Cooper, A. M., Roberts, A. D., Rhoades, E. R., Callahan, J. E., Getzy, D. M., and Orme, I. M. (1995). The role of interleukin-12 in acquired immunity to *Mycobacterium tuberculosis* infection. *Immunology* **84**, 423–432.

Corper, H. J., and Cohn, M. L. (1933). The viability and virulence of old cultures of tubercle bacilli: Studies on twelve-year broth cultures maintained at incubator temperature. *Am. Rev. Tuberculosis Pulmonary Dis.* **28**, 856–874.

Corper, H. J., Lurie, M. B., and Uyei, N. (1927). The variability of localization of tuberculosis in the organs of different animals: III. The importance of the growth of tubercle bacilli as determined by gaseous tension. *Am. Rev. Tuberculosis* **15**, 65–87.

Crofton, J., and Mitchison, D. A. (1948). Streptomycin resistance in pulmonary tuberculosis. *Br. Med. J.* **2**, 1009–1015.

Crowle, A. J., and Elkins, N. (1990). Relative permissiveness of macrophages from black and white people for virulent tubercle bacilli. *Infect. Immun.* **58**, 632–638.

Crowle, A. J., Ross, E. J., and May, M. H. (1987). Inhibition by 1,25(OH)$_2$-vitamin D$_3$ of the multiplication of virulent tubercle bacilli in cultured human macrophages. *Infect. Immun.* **55**, 2945–2950.

Dahl, K. E., Shiratsuchi, H., Hamilton, B. D., Ellner, J. J., and Toossi, Z. (1996). Selective induction of transforming growth factor β in human monocytes by lipoarabinomannan of *Mycobacterium tuberculosis. Infect. Immun.* **64**, 399–405.

Dahlstrom, G. (1952). Tuberculosis in BCG vaccinated and non-vaccinated young adults: A comparative prognostic study. *Acta Tuberculosea Scand.* **32** (Suppl.), 1–138.

Dahlstrom, G., and Difs, H. (1951). The efficacy of BCG vaccination: A study of vaccinated and tuberculin-negative conscripts. *Acta Tuberculosea Scand.* **27** (Suppl.), 1.

Daniel, T. M., Bates, J. H., and Downes, K. A. (1994). History of tuberculosis. *In* "Tuberculosis: Pathogenesis, Protection, and Control" (B. R. Bloom, ed.), pp. 13–24. ASM Press, Washington, D.C.

Dannenberg, A. M., Jr., and Rook, G. A. W. (1994). Pathogenesis of pulmonary tuberculosis: An interplay of tissue-damaging and macrophage-activating immune responses—Dual mechanisms that control bacillary multiplication. *In* "Tuberculosis: Pathogenesis, Protection, and Control" (B. R. Bloom, ed.), pp. 459–483. ASM Press, Washington, D.C.

David, J. R. (1966). Delayed hypersensitivity in vitro: Its mediation by cell-free substances formed by lymphoid cell–antigen interaction. *Proc. Natl. Acad. Sci. U.S.A.* **56**, 72–77.

De Cock, K. M., Soro, B., Coulibaly, I. M., and Lucas, S. B. (1992). Tuberculosis and HIV infection in sub-Saharan Africa. *J. Am. Med. Assoc.* **268**, 1581–1587.

De Cock, K. M., Binkin, N. J., Zuber, P. L. F., Tappero, J. W., and Castro, K. G. (1996). Research issues involving HIV-associated tuberculosis in resource-poor countries. *J. Am. Med. Assoc.* **276**, 1502–1507.

De Maio, J., Zhang, Y., Ko, C., Young, D. B., and Bishai, W. R. (1996). A stationary-phase stress-response sigma factor from *Mycobacterium tuberculosis*. *Proc. Natl. Acad. Sci. U.S.A.* **93**, 2790–2794.

Denis, M. (1991a). Interferon-gamma-treated murine macrophages inhibit growth of tubercle bacilli via the generation of reactive nitrogen intermediates. *Cell. Immunol.* **132**, 150–157.

Denis, M. (1991b). Killing of *Mycobacterium tuberculosis* within human monocytes: Activation by cytokines and calcitriol. *Clin. Exp. Immunol.* **84**, 200–206.

Deretic, V., Philipp. W., Dhandayuthapani, S., Mudd, M. H., Curcic, R., Garbe, T., Heym, B., Via, L. E., and Cole, S. T. (1995). *Mycobacterium tuberculosis* is a natural mutant with an inactivated oxidative-stress regulatory gene: Implications for sensitivity to isoniazid. *Mol. Microbiol.* **17**, 889–900.

Dessen, A., Quemard, A., Blanchard, J. S., Jacobs, W. R., and Sacchettini, J. C. (1995). Crystal structure of the isoniazid target of *Mycobacterium tuberculosis*. *Science* **267**, 1638–1641.

DiMasi, J. A. (1995). Success rates for new drugs entering clinical testing in the United States. *Clin. Pharmacol. Ther.* **58**, 1–14.

Ding, A., Nathan, C., Grayca, J., Derynck, R., Stuehr, D. J., and Srimal, S. (1990). Macrophage deactivating factor and TGF-β1, -β2, and -β3 inhibit induction of macrophage nitrogen oxide synthesis by IFN$_\gamma$. *J. Immunol.* **145**, 940–944.

Draper, P. (1982). The anatomy of mycobacteria. *In* "The Biology of the Mycobacteria, Volume 1: Physiology, Identification and Classification (C. Ratledge and J. Stanford, eds.), pp. 9–52. Academic Press, London, 1982.

Drolet, G. J. (1934). The incidence of tuberculous infection among children in New York City: A survey of 14,699 children tuberculin-tested in hospitals or attending clinics, three-year period, 1930–1932. *Am. Rev. Tuberculosis* **30**, 1–32.

Drolet, G. J. (1938). Present trend of case fatality rates in tuberculosis. *Am. Rev. Tuberculosis* **37**, 125–151.

Drolet, G. J. (1946). Epidemiology of tuberculosis. *In* "Clinical Tuberculosis" (B. Goldberg, ed.), pp. A3–A61. Davis, Philadelphia.

Dubos, R. J. (1949). The tubercle bacillus and tuberculosis. *Am. Sci.* **37**, 353–370.

Dubos, R., and Dubos, J. (1952). "The White Plague: Tuberculosis, Man, and Society." Little, Brown, and Company, Boston.

Dubos, R. J., and Schaefer, W. B. (1956). Antituberculous immunity induced in mice by virulent primary infection: Its inhibition by chemoprophylaxis. *Am. Rev. Tuberculosis Pulmonary Dis.* **74**, 541–551.

Edwards, W. M., Cox, R. S., Jr., Cooney, J. P., and Crone, R. I. (1970). Active pulmonary tuberculosis with cavitation of forty-one years' duration. *Am. Rev. Respir. Dis.* **102**, 448–455.

Ehrlich, P. (1913). Address in pathology on chemotherapeutics: Scientific principles, methods, and results. *Lancet* **2**, 445–451.

Enarson, D. A., and Murray, J. F. (1996). Global epidemiology of tuberculosis. *In* "Tuberculosis" (W. N. Rom and S. Garay, eds.), pp. 57–75. Little, Brown, and Co., Boston, MA.

Engbaek, H. C. (1952). Growth of *Mycobacterium tuberculosis* determined by direct agar micros-copy. The initial growth of young and old cultures. *Acta Pathol. Microbiol. Scand.* **31**, 370–376.

Espersen, E. (1949). Morphological studies of the normal growth of a human tubercle strain, and the effects of some antibacterial substances on same. *Acta Pathol. Microbiol. Scand.* **26**, 178–204.

Everett, F. R. (1933). The pathological anatomy of pulmonary tuberculosis in the American negro and in the white race. *Am. Rev. Tuberculosis* **27**, 411–464.

Ewart, W. (1882). Pulmonary cavities: Their origin, growth, and repair. *Br. Med. J.* **1**, 333–337, 369–372, 415–418, 453–456, 493–496, 530–533, and 569–572.

Falk, A., Tucker, W. B., and Kaufman, J. E. (1954) A clinical and bacteriologic correlation of re-sected pulmonary tuberculosis lesions. *Am. Rev. Tuberculosis* **70**, 689–700.

Falkinham III, J. O., and Crawford, J. T. (1994). Plasmids. *In* "Tuberculosis: Pathogenesis, Protec-tion, and Control" (B. R. Bloom, ed.), pp. 185–198. ASM Press, Washington, D.C.

Fauci, A. S. (1996). Host factors and the pathogenesis of HIV-induced disease. *Nature (London)* **384**, 529–534.

Feldman, W. H. (1954). Streptomycin: Some historical aspects of its development as a chemothera-peutic agent in tuberculosis. *Am. Rev. Tuberculosis* **69**, 859–868.

Feldman, W. H., and Baggenstoss, A. H. (1938). The residual infectivity of the primary complex of tuberculosis. *Am. J. Pathol.* **14**, 473–490.

Feldman, W. H., and Baggenstoss, A. H. (1939). The occurrence of virulent tubercle bacilli in pre-sumably non-tuberculous lung tissue. *Am. J. Pathol.* **5**, 501–515.

Ferguson, R. G., and Simes, A. B. (1949). BCG vaccination of indian infants in Saskatchewan. *Tubercle* **30**, 5–11.

Filley, E. A., and Rook, G. A. W. (1991). Effect of mycobacteria on sensitivity to the cytotoxic effects of tumor necrosis factor. *Infect. Immun.* **59**, 2567–2572.

Filley, E. A., Bull, H. A., Dowd, P. M., and Rook, G. A. W. (1992). The effect of *Mycobacterium tuberculosis* on the susceptibility of human cells to the stimulatory and toxic effects of tumour necrosis factor. *Immunology* **77**, 505–509.

Fine, P. E. M. (1981). Immunogenetics of susceptibility to leprosy, tuberculosis, and leishmaniasis. An epidemiological perspective. *Int. J. Leprosy* **49**, 437–454.

Fine, P. E. M. (1989). The BCG story: Lessons from the past and implications for the future. *Rev. Infect. Dis.* **11**, S353–S359.

Fine, P. E. M., Sterne, J. A. C., Ponnighaus, J. M., and Rees, R. J. W. (1994). Delayed-type hyper-sensitivity, mycobacterial vaccines, and protective immunity. *Lancet* **344**, 1245–1249.

Finken, M., Kirschner, P., Meier, A., Wrede, A., and Bottger, E. C. (1993). Molecular basis of strep-tomycin resistance in *Mycobacterium tuberculosis*: Alterations of the ribosomal protein S12 gene and point mutations within a functional 16S ribosomal RNA pseudoknot. *Mol. Microbiol.* **9**, 1239–1246.

Fleming, A. (1929). On the antibacterial action of cultures of a *Penicillium*, with special reference to their use in the isolation of *B. influenzae*. *Br. J. Exp. Pathol.* **10**, 226–236.

Flesch, I. E., and Kaufmann, S. H. E. (1987). Mycobacterial growth inhibition by interferon-γ-activated bone marrow macrophages and differential susceptibility among strains of *Mycobac-terium tuberculosis*. *J. Immunol.* **138**, 4408–4413.

Flesch, I. E., and Kaufmann, S. H. E. (1991). Mechanisms involved in mycobacterial growth inhi-bition by gamma-interferon-activated bone marrow macrophages: Role of reactive nitrogen in-termediates. *Infect. Immun.* **59**, 3213–3218.

Flesch, I. E. A., Hess, J. H., Huang, S., Aguet, M., Rothe, J., Bluethmann, H., and Kaufmann, S. H. E. (1995). Early interleukin 12 production by macrophages in response to mycobacterial infection depends on interferon γ and tumor necrosis factor α. *J. Exp. Med.* **181**, 1615–1621.

Flynn, J. L., Goldstein, M. M., Triebold, K. J., Koller, B., and Bloom, B. R. (1992). Major histocom-patibility complex class I-restricted T cells are required for resistance to *Mycobacterium tuber-culosis* infection. *Proc. Natl. Acad. Sci. U.S.A.* **89**, 12013–12017.

Flynn, J. L., Chan, J., Triebold, K. J., Dalton, D. K., Stewart, T. A., and Bloom, B. R. (1993). An essential role for interferon γ in resistance to *Mycobacterium tuberculosis* infection. *J. Exp. Med.* **178**, 2249–2254.

Flynn, J. L., Goldstein, M. M., Triebold, K. J., Pfeffer, K., Lowenstein, C. J., Schreiber, R., Mak,

T. W., and Bloom, B. R.. (1995a). Tumor necrosis factor-α is required in the protective immune response against *Mycobacterium tuberculosis* in mice. *Immunity* 2, 561–572.

Flynn, J. L., Goldstein, M. M., Triebold, K. J., Sypek, J., Wolf, S., and Bloom, B. R. (1995b). IL-12 increases resistance of BALB/c mice to *Mycobacterium tuberculosis* infection. *J. Immunol.* 155, 2515–2524.

Fox, W., and Mitchison, D. A. (1975). Short-course chemotherapy for pulmonary tuberculosis. *Am. Rev. Respir. Dis.* 111, 325–353.

Frieden, T. R., Sterling, T., Pablos-Mendez, A., Kilburn, J. O., Cauthen, G. M., and Dooley, S. W. (1993). The emergence of drug-resistant tuberculosis in New York City. *N. Engl. J Med.* 328, 521–526.

Frieden, T. R., Fujiwara, P. I., Washko, R. M., and Hamburg, M. A. (1995). Tuberculosis in New York City—Turning the tide. *N. Engl. J. Med.* 333, 229–233.

Frimoldt-Moller, J., Acharyulu, G. S., and Pillai, K. K. (1973). Observations on the protective effect of BCG vaccination in a South Indian rural population: Fourth report. *Bull. Int. Union Against Tuberculosis* 48, 40–52.

Frost, W. H. (1939). The age selection of mortality from tuberculosis in successive decades. *Am. J. Hygiene* 30 (Section A), 91–96.

Fulton, S. A., Johnsen, J. M., Wolf, S. F., Sieburth, D. S., and Boom, W. H. (1996). Interleukin-12 production by human monocytes infected with *Mycobacterium tuberculosis:* Role of phagocytosis. *Infect. Immun.* 64, 2523–2531.

Geison, G. L. (1995). "The Private Science of Louis Pasteur." Princeton Univ. Press, Princeton, New Jersey.

Gercken, J., Pryjma, J., Ernst, M., and Flad, H.-D. (1994). Defective antigen presentation by *Mycobacterium tuberculosis*-infected monocytes. *Infect. Immun.* 62, 3472–3478.

Ghon, A. (1923). The primary complex in human tuberculosis and its significance. *Am. Rev. Tuberculosis* 7, 314–317.

Girling, D. J. (1989). The chemotherapy of tuberculosis. *In* "The Biology of the Mycobacteria, Volume 3: Clinical Aspects of Mycobacterial Disease" (C. Ratledge, J. Stanford, and J. M. Grange, eds.), pp. 285–323. Academic Press, London.

Goble, M., Iseman, M. D., Madsen, L. A., Waite, D., Ackerson, L., and Horsburgh, C. R., Jr. (1993). Treatment of 171 patients with pulmonary tuberculosis resistant to isoniazid and rifampin. *N. Engl. J. Med.* 328, 527–532.

Goletti, D., Weissman, D., Jackson, R. W., Graham, N. M. H., Vlahov, D., Klein, R. S., Munsiff, S. S., Ortona, L., Cauda, R., and Fauci, A. S. (1996). Effect of *Mycobacterium tuberculosis* on HIV replication: Role of immune activation. *J. Immunol.* 157, 1271–1278.

Goodwin, R. A., and Des Prez, R. M. (1983). Apical localization of pulmonary tuberculosis, chronic pulmonary histoplasmosis, and progressive massive fibrosis of the lung. *Chest* 83, 801–805.

Goren, M. B., D'Arcy Hart, P., Young, M. R., and Armstrong, J. A. (1976). Prevention of phagosome–lysosome fusion in cultured macrophages by sulfatides of *Mycobacterium tuberculosis*. *Proc. Natl. Acad. Sci. U.S.A.* 73, 2510–2514.

Grigg, E. R. N. (1958). The arcana of tuberculosis, with a brief epidemiologic history of the disease in the U.S.A. *Am. Rev. Tuberculosis Pulmonary Dis.* 78, 151–172, 426–453, and 583–603.

Grosset, J. (1978). The sterilizing value of rifampicin and pyrazinamide in experimental short-course chemotherapy. *Tubercle* 59, 287–297.

Guleria, I., Teitelbaum, R., McAdam, R. A., Kalpana, G., Jacobs, W. R., Jr., and Bloom, B. R. (1996). Auxotrophic vaccines for tuberculosis. *Nat. Med.* 2, 334–337.

Haas, F., and Haas, S. S. (1996). The origins of *Mycobacterium tuberculosis* and the notion of its contagiousness. *In* "Tuberculosis" (W. N. Rom and S. Garay, eds.), pp. 3–19. Little, Brown, Co., Boston, MA.

Hamburg, M. A. (1995). "Health of the City: Focus on Tuberculosis." New York City Department of Health, New York.

Hart, P. D. (1967). Efficacy and applicability of mass BCG vaccination in tuberculosis control. *Br. Med. J.* 1, 587–592.

Hart, P. D., Sutherland, I., and Thomas, J. (1967). The immunity conferred by effective BCG and vole bacillus vaccines, in relation to individual variations in induced tuberculin sensitivity and to technical variations in the vaccines, *Tubercle* 48, 201–210.

Hart, P. D., and Sutherland, I. (1977). BCG and vole bacillus vaccines in the prevention of tuberculosis in adolescence and early adult life. *Br. Med. J.* **2**, 293–295.

Hatfull, G. F., and Jacobs, W. R., Jr. (1994). Mycobacteriophages: Cornerstones of mycobacterial research. *In* "Tuberculosis: Pathogenesis, Protection, and Control" (B. R. Blood, ed.), pp. 165–183. ASM Press, Washington, D.C.

Heinzel, F. P., Schoenhaut, D. S., Rerko, R. M., Rosser, L. E., and Gately, M. K. (1993). Recombinant interleukin 12 cures mice infected with *Leishmania major. J. Exp. Med.* **177**, 1505–1509.

Helweg-Larsen, P., Hoffmeyer, H., Kieler, J., Hess Thaysen, E. H., Hess Thaysen, J., Thygesen, P., and Wulff, M. H. (1952). Famine disease in German concentration camps: Complications and sequels. Chapter XXI: Tuberculosis. *Acta Med. Scand.* Supplement 274, 330–361.

Heym, B., Honore, N., Truffot-Pernot, C., Banerjee, A., Schurra, C., Jacobs, W. R., Jr., van Embden, J. D. A., Grosset, J. H., and Cole, S. T. (1994). Implications of multidrug resistance for the future of short-course chemotherapy of tuberculosis: A molecular study. *Lancet* **344**, 293–298.

Hinshaw, H. C. (1954). Historical notes on earliest use of streptomycin in clinical tuberculosis. *Am. Rev. Tuberculosis* **70**, 9–14.

Hinshaw, H. C., Pfuetze, K., and Feldman, W. H. (1943). Treatment of tuberculosis with Promin: A progress report. *Am. Rev. Tuberculosis* **47**, 26–34.

Hinshaw, H. C., Pfuetze, K. H., and Feldman, W. H. (1944). Chemotherapy of clinical tuberculosis with Promin: p,p'-Diaminodiphenylsulfone-N,N'-didextrose sulfonate; a second report of progress. *Am. Rev. Tuberculosis* **50**, 52–57.

Hirsch, C. S., Yoneda, T., Averill, L., Ellner, J. J., and Toossi, Z. (1994). Enhancement of intracellular growth of *Mycobacterium tuberculosis* in human monocytes by transforming growth factor-β_1. *J. Infect. Dis.* **170**, 1229–1237.

Hirsch, C. S., Hussain, R., Toossi, Z., Dawood, G., Shahid, F., and Ellner, J. J. (1996). Cross-modulation by transforming growth factor β in human tuberculosis: Suppression of antigen-driven blastogenesis and interferon γ production. *Proc. Natl. Acad. Sci. U.S.A.* **93**, 3193–3198.

Hobby, G. L., Auerbach, O., Lenert, T. F., Small, M. J., Comer, J. V. (1954). The late emergence of M. tuberculosis in liquid cultures of pulmonary lesions resected from humans. *Am. Rev. Tuberculosis* **70**, 191–218.

Hobby, G. L., and Lenert, T. F. (1957). The *in vitro* of antituberculous agents against multiplying and non-multiplying microbial cells. *Am. Rev. Tuberculosis* **76**, 1031–1048.

Hoge, C. W., Fisher, L., Donnell, H. D., Jr., Dodson, D. R., Tomlinson, G. V., Jr., Breiman, R. F., Bloch, A. B., and Good, R. C. (1994). Risk factors for transmission of *Mycobacterium tuberculosis* in a primary school outbreak: Lack of racial difference in susceptibility to infection. *Am. J. Epidemiol.* **139**, 520–530.

Honda, M., Matsuo, K., Nakasone, T., Okamoto, Y., Yoshizaki, H., Kitamura, K., Sugiura, W., Watanabe, K., Fukushima, Y., Haga, S., Katsura, Y., Tasaka, H., Komuro, K., Yamada, T., Asano, T., Yamazaki, A., and Yamazaki, S. (1995). Protective immune responses induced by secretion of a chimeric soluble protein from a recombinant *Mycobacterium bovis* bacillus Calmette–Guerin vector candidate vaccine for human immunodeficiency virus type 1 in small animals. *Proc. Natl. Acad. Sci. U.S.A.* **92**, 10693–10697.

Honore, N., and Cole, S. T. (1994). Streptomycin resistance in mycobacteria. *Antimicrobial Agents Chemother.* **38**, 238–242.

Horwitz, M. A., Lee, B.-W. E., Dillon, B. J., and Harth, G. (1995). Protective immunity against tuberculosis induced by vaccination with major extracellular proteins of *Mycobacterium tuberculosis. Proc. Natl. Acad. Sci. U.S.A.* **92**, 1530–1534.

Hsu, H. S. (1965a). *In vitro* studies on the interactions between macrophages of rabbits and tubercle bacilli. I. Cellular basis of native resistance. *Am. Rev. Respir. Dis.* **91**, 488–498.

Hsu, H. S. (1965b). *In vitro* studies on the interactions between macrophages of rabbits and tubercle bacilli. II. Cellular and humoral aspects of acquired resistance. *Am. Rev. Respir. Dis.* **91**, 499–509.

Huygen, K., Content, J., Denis, O., Montgomery, D. L., Yawman, A. M., Deck, R. R., DeWitt, C. M., Orme, I. M., Baldwin, S., D'Souza, C., Drowart, A., Lozes, E., Vandenbussche, P., van Vooren, J.-P., Liu, M. A., and Ulmer, J. B. (1996). Immunogenicity and protective efficacy of a tuberculosis DNA vaccine. *Nat. Med.* **2**, 893–898.

Institute of Medicine (1997). "America's Vital Interest in Global Health: Protecting our People,

Promoting Economic Interests, and Projecting U. S. Influence Abroad." National Academy Press, Washington, D.C.

Iseman, M. D. (1993). Treatment of multidrug-resistant tuberculosis. *N. Engl. J. Med.* **329**, 784–791.

Iseman, M. D., Cohn, D. L., and Sharbaro, J. A. (1993). Directly observed treatment of tuberculosis: We can't afford not to try it. *N. Engl. J. Med.* **328**, 576–578.

Ivanyi, J., and Thole, J. (1994). Specificity and function of T- and B-cell recognition in tuberculosis. *In* "Tuberculosis: Pathogenesis, Protection, and Control" (B. R. Bloom, ed.), pp. 437–458. ASM Press, Washington, D.C.

Johansen, K. A., Gill, R. E., and Vasil, M. L. (1996). Biochemical and molecular analysis of phospholipase C and phospholipase D activity in mycobacteria. *Infect. Immun.* **64**, 3259–3266.

Johnson, B. J., and McMurray, D. N. (1994). Cytokine gene expression by cultures of human lymphocytes with autologous *Mycobacterium tuberculosis*-infected monocytes. *Infect. Immun.* **62**, 1444–1450.

Jones G. S., Amirault, H. J., and Anderson, B. R. (1990). Killing of *Mycobacterium tuberculosis* by neutrophils: a nonoxidative process. *J. Infect. Dis.* **162**, 700–704.

Jouanguy, E., Altare, F., Lamhamedi, S., Revy, P., Emile, J.-F., Newport, M., Levin, M., Blanche, S., Fischer, A., and Casanova, J.-L. (1996). Interferon-γ-receptor deficiency in an infant with fatal bacille Calmette–Guerin infection. *N. Engl. J. Med.* **335**, 1956–1961.

Kahn, M. C. (1929). A developmental cycle of the tubercle bacillus as revealed by single-cell studies. *Am. Rev. Tuberculosis* **20**, 150–200.

Kallmann, F. J., and Reisner, D. (1943). Twin studies on the significance of genetic factors in tuberculosis. *Am. Rev. Tuberculosis* **47**, 549–574.

Kanai, K., and Yanagisawa, K. (1955). Antibacterial action of streptomycin against tubercle bacilli of various growth phase. *Japan J. Med. Sci. Biol.* **8**, 63–76.

Kanai, K. (1966). Experimental studies on host-parasite equilibrium in tuberculous infection, in relation to vaccination and chemotherapy. *Japan J. Med. Sci. Biol.* **19**, 181–199.

Kanai, K., and Kondo, E. (1971). Limits in tuberculosis chemotherapy as revealed by experimental studies in mice. *Japan J. Med. Sci. Biol.* **24**, 313–321.

Kapur, V., Whittam, T. S., and Musser, J. M. (1994a). Is *Mycobacterium tuberculosis* 15,000 years old? *J. Infect. Dis.* **170**, 1348–1349.

Kapur, V., Li, L. L., Iordanescu, S., Hamrick, M. R., Wanger, A., Kreiswirth, B. N., and Musser, J. M. (1994b). Characterization by automated DNA sequencing of mutations in the gene (*rpoB*) encoding the RNA polymerase β subunit in rifampicin-resistant *Mycobacterium tuberculosis* strains from New York City and Texas. *J. Clin. Microbiol.* **32**, 1095–1098.

Karlson, A. G., Pfuetze, K. H., Carr, D. T., Feldman, W. H., and Hinshaw, C. H. (1949). The effect of combined therapy with streptomycin, para-aminosalicylic acid, and promin on the emergence of streptomycin-resistant strains of tubercle bacilli: a preliminary report. *Proc. Staff Meet. Mayo Clinic* **24**, 85–88.

Kasahara, K., Tobe, T., Tomita, M., Mukaida, N., Shao-Bo, S., Matsushima, K., Yoshida, T., Sugihara, S., and Kobayashi, K. (1994). Selective expression of monocyte chemotactic and activating factor/monocyte chemoattractant protein 1 in human blood monocytes by *Mycobacterium tuberculosis*. *J. Infect. Dis.* **170**, 1238–1247.

Keeling, M. J., and Grenfell, B. T. (1997). Disease extinction and community size: Modeling the persistence of measles. *Science* **275**, 65–67.

Kempner, W. (1939). Oxygen tension and the tubercle bacillus. *Am. Rev. Tuberculosis* **40**, 157–168.

Kereszturi, C., Park, W. H., and Logie, A. J. (1935). A study of 2,900 chest roentgenograms of BCG vaccinated and control infants in tuberculous families. *Trans. Natl. Tuberculosis Assoc.* **97**, 97–108.

King, C., Sahish, M., Crawford, J. T., and Shinnick, T. M. (1993). Expression of contact-dependent cytolytic activity of *Mycobacterium tuberculosis* and isolation of the locus encoding the activity. *Infect. Immun.* **61**, 2708–2712.

Koch, R. (1882). Die Aetiologie der Tuberkulose. *Ber. Klin. Wochenschrift.* **15**; Koch, R. (1932). The aetiology of tuberculosis. *Am. Rev. Tuberculosis* **25**, 298–323.

Kopanoff, D. E., Snider, D. E., and Johnson, M. (1988). Recurrent tuberculosis: Why do patients develop disease again? *Am. J. Public Health* **78**, 30–33.

Krause, A. K. (1932). Introduction to the aetiology of tuberculosis (translated from the German). *Am. Rev. Tuberculosis* **25**, 285–298.

Kudoh, S. (1956). The growth mode of mycobacterium under microculture. *Acta Tuberculosea Scand.* **32**, 74–87.

Kumararatne, D. S., Pithie, A. S., Drysdale, P., Gaston, J. S. H., Kiessling, R., Iles, P. B., Ellis, C. J., Innes, J., and Wise, R. (1990). Specific lysis of mycobacterial antigen-bearing macrophages by class II MHC-restricted polyclonal T cell lines in healthy donors or patients with tuberculosis. *Clin. Exp. Immunol.* **80**, 314–323.

Lancet (1890). Professor Koch's remedy for tuberculosis. *Lancet* **2**, 1118–1122, 1239–1143, 1282, 1290–1296, and 1347–1352.

Lancet (1949). Control of tuberculosis. *Lancet* **2**, 1225.

Langermann, S., Palaszynski, S., Sadziene, A., Stover, C. K., and Koenig, S. (1994). Systemic and mucosal immunity induced by BCG vector expressing outer-surface protein A of *Borrelia burgdorferi*. *Nature (London)* **372**, 552–555.

Laochumroonvorapong, P., Wang, J., Chau-Ching, L., Ye, W., Moreira, A. L., Elkon, K. B., Freedman, V. H., Kaplan, G. (1997). Perforin, a cytotoxic molecule which mediates cell necrosis, is not required for the early control of mycobacterial infection in mice. *Infect. Immun.* **65**, 127–132.

Leao, S. C., Rocha, C. L., Murillo, L. A., Parra, C. A., & Patarroyo, M. E. (1995). A species-specific nucleotide sequence of *Mycobacterium tuberculosis* encodes a protein that exhibits hemolytic activity when expressed in *Escherichia coli*. *Infect. Immun.* **63**, 4301–4306.

Lin, Y., Zhang, M., Hofman, F. M., Gong, J., and Barnes, P. F. (1996). Absence of a prominent Th2 cytokine response in human tuberculosis. *Infect. Immun.* **64**, 1351–1356.

Lindgren, I. (1961). Anatomical and roentgenologic studies of tuberculous infections in BCG-vaccinated and non-vaccinated subjects. *Acta Radiologica* Supplement 209, 7–101.

Lindgren, I. (1965). The pathology of tuberculous infection in BCG-vaccinated humans. *Ad. Tuberculosis Res.* **14**, 202–234.

Lister, J. (1890). Lecture on Koch's treatment of tuberculosis. *Lancet* **2**, 1257–1259.

Loebel, R. O., Shorr, E., and Richardson, H. B. (1933). The influence of adverse conditions upon the respiratory metabolism and growth of human tubercle bacilli. *J. Bacteriol.* **26**, 167–200.

Lorgat, F., Keraan, M. M., Lukey, P. T., and Ress, S. R. (1992). Evidence for in vivo generation of cytotoxic T cells: PPD-stimulated lymphocytes from tuberculous pleural effucions demonstrate enhanced cytotoxicity with accelerated kinetics of induction. *Am. Rev. Respir. Dis.* **145**, 418–423.

Lucas, S., and Nelson, A. M. (1994). Pathogenesis of tuberculosis in human immunodeficiency virus-infected people. *In* "Tuberculosis: Pathogenesis, Protection, and Control" (B. R. Blood, ed.), pp. 503–513. ASM Press, Washington, D.C.

Lurie, M. (1964). "Resistance to Tuberculosis: Experimental Studies in Native and Acquired Defensive Mechanisms." Harvard Univ. Press, Cambridge, Massachusetts.

McAdam, R. A., Guilhot, C., and Gicquel, B. (1994). Transposition in mycobacteria. *In* "Tuberculosis: Pathogenesis, Protection, and Control" (B. B. Bloom, ed.), pp. 199–216. ASM Press, Washington, D.C.

McAdam, R. A., Weisbrod, T. R., Martin, J., Scuderi, J. D., Brown, A. M., Cirillo, J. D., Bloom, B. R., and Jacobs, W. R., Jr. (1995). *In vivo* growth characteristics of leucine and methionine auxotrophic mutants of *Mycobacterium bovis* BCG generated by transposon mutagenesis. *Infect. Immun.* **63**, 1004–1012.

McCune, R. M., and Tompsett, R. (1956). Fate of *Mycobacterium tuberculosis* in mouse tissues as determined by the microbial enumeration technique. I. The persistence of drug-susceptible tubercle bacilli in the tissues despite prolonged antimicrobial therapy. *J. Exp. Med.* **104**, 737–762.

McCune, R. M., Tompsett, R., and McDermott, W. (1956). Fate of *Mycobacterium tuberculosis* in mouse tissues as determined by the microbial enumeration technique. II. The conversion of tuberculous infection to the latent state by the administration of pyrazinamide and a companion drug. *J. Exp. Med.* **104**, 763–801.

McCune, R. M., Feldmann, F. M., Lambert, H. P., and McDermott, W. (1966a). Microbial persistence. I. The capacity of tubercle bacilli to survive sterilization in mouse tissues. *J. Exp. Med.* **123**, 445–468.

McCune, R. M., Feldmann, F. M., and McDermott, W. (1966b). Microbial persistence. II. Characteristics of the sterile state of tubercle bacilli. *J. Exp. Med.* **123**, 469–486.

McDermott, W. (1969). The story of INH. *J. Infect. Dis.* **119**, 678–683.

McDermott, W., McCune, R. M., Jr., and Tompsett, R. (1956). Dynamics of antituberculous chemotherapy. *Am. Rev. Tuberculosis Pulmonary Dis.* **74** (Suppl.), S100–S108.

McDonough, K. A., and Kress, Y. (1995). Cyotoxicity for lung epithelial cells is a virulence-associated phenotype of *Mycobacterium tuberculosis*. *Infect. Immun.* **63**, 4802–4811.

McDonough, K. A., Kress, Y., and Bloom, B. R. (1993). Pathogenesis of tuberculosis: Interaction of *Mycobacterium tuberculosis* with macrophages. *Infect. Immun.* **61**, 2763–2773.

McFadden, J. (1996). Recombination in mycobacteria. *Mol. Microbiol.* **21**, 205–211.

McGrath, J. W. (1988). Social networks of disease spread in the Lower Illinois Valley: A simulation approach. *Am. J. Phys. Anthropol.* **77**, 483–496.

Mackaness, G. B., and Blanden, R. V. (1967). Cellular immunity. *Prog. Allergy* **11**, 89–140.

McNeil, M. R., and Brennan, P. J. (1991). Structure, function and biogenesis of the cell envelope of mycobacteria in relation to bacterial physiology, pathogenesis and drug resistance; some thoughts and possibilities arising from recent structural information. *Res. Microbiol.* **142**, 451–464.

McNeill, W. H. (1976). "Plagues and Peoples." Doubleday, New York.

McNeill, W. H. (1992). "The Global Condition." Princeton Univ. Press, Princeton, New Jersey.

Mazzaccaro, R. J., Gedde, M., Jensen, E. R., van Santen, H. M., Ploegh, H. L., Rock, K. L., and Bloom, B. R. (1996). Major histocompatibility class I presentation of soluble antigen facilitated by *Mycobacterium tuberculosis* infection. *Proc. Natl. Acad. Sci. U.S.A.* **93**, 11786–11791.

Medina, E., and North, R. J. (1996). Evidence inconsistent with a role for the *Bcg* gene (Nramp 1) in resistance of mice to infection with virulent *Mycobacterium tuberculosis*. *J. Exp. Med.* **183**, 1045–1051.

Medlar, E. M. (1940). Pulmonary tuberculosis in cattle: The location and type of lesions in naturally acquired tuberculosis. *Am. Rev. Tuberculosis* **41**, 283–306.

Medlar, E. M. (1948). The pathogenesis of minimal pulmonary tuberculosis: A study of 1,225 necropsies in cases of unexpected and sudden death. *Am. Rev. Tuberculosis* **58**, 583–611.

Medlar, E. M., and Sasano, K. T. (1936). A study of the pathology of experimental pulmonary tuberculosis in the rabbit. *Am. Rev. Tuberculosis* **34**, 456–476.

Meylan, P. R. A., Richman, D. D., and Kornbluth, R. S. (1992). Reduced intracellular growth of mycobacteria in human macrophages cultivated at physiologic oxygen pressure. *Am. Rev. Respir. Dis.* **145**, 947–953.

Middlebrook, G. (1954). Isoniazid-resistance and catalase activity of tubercle bacilli. *Am. Rev. Tuberculosis* **69**, 471–472.

Middlebrook, G., Cohn, M. L., and Schaefer, W. B. (1954). Studies on isoniazid and tubercle bacilli. III. The isolation, drug-susceptibility, and catalase-testing of tubercle bacilli from isoniazid-treated patients. *Am. Rev. Tuberculosis* **70**, 852–872.

Middlebrook, G., and Yegian, D. (1946). Certain effects of streptomycin on mycobacteria *in vitro*. *Am. Rev. Tuberculosis* **54**, 553–558.

Mitchison, D. A., and Selkon, J. B. (1956). The bactericidal activities of antituberculous drugs. *Am. Rev. Tuberculosis Pulmonary Dis.* **74** (Suppl.), S109–S123.

Molloy, A., Laochumroovorapong, P., and Kaplan, G. (1994). Apoptosis, but not necrosis, of infected monocytes is coupled with killing of intracellular bacillus Calmette–Guerin. *J. Exp. Med.* **180**, 1499–1509.

Moreno, C., Taverne, J., Mehlert, A., Bate, C. A. W., Brealey, R. J., Meager, A., Rook, G. A. W., and Playfair, J.H.L. (1989). Lipoarabinomannan from *Mycobacterium tuberculosis* induces the production of tumour necrosis factor from human and murine macrophages. *Clin. and Exp. Immunol.* **76**, 240–245.

MRC (1949). Treatment of pulmonary tuberculosis with *para*-aminosalicylic acid and streptomycin. *Br. Med. J.* **2**, 1521.

MRC (1956). BCG and vole bacillus vaccines in the prevention of tuberculosis in adolescents: First report to the Medical Research Council by their tuberculosis vaccines clinical trials committee. *Br. Med. J.* **1**, 413–427.

MRC (1959). BCG and vole bacillus vaccines in the prevention of tuberculosis in adolescents: Second report to the Medical Research Council by their tuberculosis vaccines clinical trials committee. *Br. Med. J.* **2**, 379–396.

MRC (1963). BCG and vole bacillus vaccines in the prevention of tuberculosis in adolescence and early adult life: Third report to the Medical Research Council by their tuberculosis vaccines clinical trials committee. *Br. Med. J.* **1**, 973–978.

MRC (1972). BCG and vole bacillus vaccines in the prevention of tuberculosis in adolescence and early adult life. *Bull. W.H.O.* **46**, 371–385.

Murray, C. J. L. (1994a). Issues in operational, social, and economic research on tuberculosis. *In* "Tuberculosis: Pathogenesis, Protection, and Control" (B. R. Bloom, ed.), pp. 583–622. ASM Press, Washington, D.C.

Murray, C. J. L. (1994b). Quantifying the burden of disease: the technical basis for disability-adjusted life years. *Bull. W.H.O.* **72**, 429–445.

Murray, C. J. L., DeJonghe, E., Chum, H. G., Nyangulu, D. S., Salomao, A., and Styblo, K. (1994). Cost-effectiveness of chemotherapy for pulmonary tuberculosis in three sub-Saharan African countries. *Lancet* **338**, 1305–1308.

Murray, C. J. L., Styblo, K., and Rouillon, A. (1990). Tuberculosis in developing countries: Burden, intervention and cost. *Bull. Int. Union Against Tuberculosis Lung Dis.* **65**, 6–23.

Murray, P. J., Aldovini, A., and Young, R. A. (1996). Manipulation and potentiation of antimycobacterial immunity using recombinant bacille Calmette–Guerin strains that secrete cytokines. *Proc. Natl. Acad. U.S.A.* **93**, 934–939.

Musser, J. M., Kapur, V., Williams, D. L., Kreiswirth, B. N., van Soolingen, D., and van Embden, J. D. (1996). Characterization of the catalase-peroxidase gene (*katG*) and *inhA* locus in isoniazid-resistant and -susceptible strains of *Mycobacterium tuberculosis* by automated DNA sequencing:Restricted array of mutations associated with drug resistance. *J. Infect. Dis.* **173**, 196–202.

Myrvik, Q. N., Leake, E. S., and Wright, M. J. (1984). Disruption of phagosomal membranes of normal alveolar macrophages by the H37Rv strain of *Mycobacterium tuberculosis*: A correlate of virulence. *Am. Rev. Respir. Dis.* **129**, 322–328.

Newport, M. J., Huxley, C. M., Huston, S., Hawrylowicz, C. M., Oostra, B. A., Williamson, R., and Levin, M. (1996). A mutation in the interferon-γ-receptor gene and susceptibility to mycobacterial infection. *N. Engl. J. Med.* **335**, 1941–1949.

Nicholson, S., da Gloria Bonecini-Almeida, M., Lapa e Silva, J. R., Nathan, C., Xie, Q.-W., Mumford, R., Weidner, J. R., Calaycay, J., Geng, J., Boechat, N., Linhares, C., Rom, W., and Ho, J. L. (1996). Inducible nitric oxide synthase in pulmonary alveolar macrophages from patients with tuberculosis. *J. Exp. Med.* **183**, 2293–2302.

Office of Technology Assessment. (1993). "Pharmaceutical R & D: Costs, Risks, and Rewards," (OTA-H-523). U.S. Government Printing Office, Washington, D.C.

Opie, E. L. (1922). Phthisiogenesis and latent tuberculous infection. *Am. Rev. Tuberculosis* **6**, 525–546.

Opie, E. L. (1930). Anatomical characteristics of tuberculosis in Jamaica. *Am. Rev. Tuberculosis* **22**, 613–625.

Opie, E. L., and Aronson, J. D. (1927). Tubercle bacilli in latent tuberculous lesions and in lung tissue without tuberculous lesions. *Arch. Pathol.* **4**, 1–21.

Pagel, W. (1935). On the endogenous origin of early pulmonary tuberculosis. The anatomic view of its clinical diagnosis. *Am. J. Med. Sci.* **189**, 253–270.

Pal, P. G., and Horwitz, M. A. (1992). Immunization with extracellular proteins of *Mycobacterium tuberculosis* induces cell-mediated immune responses and substantial protective immunity in a guinea pig model of pulmonary tuberculosis. *Infect. Immun.* **60**, 4781–4792.

Pancholi, P., Mirza, A., Bhardwaj, N., and Steinman, R. M. (1993). Sequestration from immune CD4+ T cells of mycobacteria growing in human macrophages. *Science* **260**, 984–986.

Pape, J. W., Jean, S. S., Ho, J. L., Hafner, A., and Johnson, W. D., Jr. (1993). Effect of isoniazid prophylaxis on incidence of active tuberculosis and progression of HIV infection. *Lancet* **342**, 268–272.

Peizer, L. R., Widelock, D., and Klein, S. (1954). Effect of isoniazid on the viability of isoniazid-susceptible and isoniazid-resistant cultures of *Mycobacterium tuberculosis*. *Am. Rev. Tuberculosis* **69**, 1022–1028.

Peizer, L. R., Chaves, A. D., and Widelock, D. (1957). The effects of early isoniazid treatment in experimental guinea pig tuberculosis. *Am. Rev. Tuberculosis Pulmonary Dis.* **76**, 732–751.

Perriens, J. H., St. Louis, M. E., Mukadi, Y. B., Brown, C., Prignot, J., Pouthier, F., Portaels, F., Williame, J.-C., Mandala, J. K., Kaboto, M., Ryder, R. W., Roscigno, G., and Piot, P. (1995). Pulmonary tuberculosis in HIV-infected patients in Zaire: A controlled trial of treatment for either 6 or 12 months. *N. Engl. J. Med.* **332**, 779–784.

Pfuetze, K. H., Pyle, M. M., Hinshaw, H. C., and Feldman, W. H. (1955). The first clinical trial of streptomycin in human tuberculosis. *Am. Rev. Tuberculosis Pulmonary Dis.* **71**, 752–754.

Pierce, C., Dubos, R. J., and Middlebrook, G. (1947). Infection of mice with mammalian tubercle bacilli grown in Tween–albumin liquid medium. *J. Exp. Med.* **86**, 159–174.

Pinner, M., and Kasper, J. A. (1932). Pathological peculiarities of tuberculosis in the American negro. *Am. Rev. Tuberculosis* **26**, 463–491.

Poccia, F., Boullier, S., Lecoeur, H., Cochet, M., Poquet, Y., Colizzi, V., Fournie, J. J., and Gougeon, M. (1996). Peripheral Vγ9/Vδ2 T cell deletion and anergy to nonpeptidic mycobacterial antigens in asymptomatic HIV-1-infected persons. *J. Immunol.* **157**, 449–461.

Porcelli, S., Morita, C. T., and Brenner, M. B. (1992). CD1b restricts the response of human CD4/8⁻ T lymphocytes to a microbial antigen. *Nature (London)* **360**, 593–597.

Quetel, C. (1992). "History of Syphilis." Johns Hopkins Univ. Press, Baltimore.

Rasmussen, K. N. (1957). The apical localization of pulmonary tuberculosis. *Acta Tuberculosea Scand.* **34**, 245–259.

Ratledge, C. (1982). Nutrition, growth and metabolism. *In* "The Biology of the Mycobacteria, Volume I: Physiology, Identification and Classification" (C. Ratledge and J. Stanford, eds.), pp. 185–271. Academic Press, London.

Raviglione, M. C., Narain, J. P., and Kochi, A. (1992). HIV-associated tuberculosis in developing countries: Clinical features, diagnosis, and treatment. *Bull. W.H.O.* **70**, 515–526.

Raviglione, M. C., Sudre, P., Rieder, H. L., Spinaci, S., and Kochi, A. (1993). Secular trends of tuberculosis in Western Europe. *Bull. W.H.O.* **71**, 297–306.

Raviglione, M. C., Snider, D. E., and Kochi, A. (1995). Global epidemiology of tuberculosis: Morbidity and mortality of a worldwide epidemic. *J. Am. Med. Assoc.* **273**, 220–226.

Rees, R. J. W., and D'Arcy Hart, P. (1961). Analysis of the host–parasite equilibrium in chronic murine tuberculosis by total and viable bacillary counts. *Br. J. Exp. Pathol.* **42**, 83–88.

Reichle, H. S., and Gallavan, M. (1937). Reactivation of a primary tuberculous complex as a source of tuberculous reinfection. *Arch. Pathol.* **24**, 201–214.

Reyrat, J.-M, Berthet, F.-X, and Gicquel, B. (1995). The urease locus of *Mycobacterium tuberculosis* and its utilization for the demonstration of allelic exchange in *Mycobacterium bovis* bacillus Calmette–Guerin. *Proc. Natl. Acad. Sci. U.S.A.* **92**, 8768–8772.

Rhines, C. (1935). The persistence of avian tubercle bacilli in soil and in association with soil microorganisms. *J. Bacteriol.* **29**, 299–311.

Rhodes, R. (1986). "The Making of the Atomic Bomb." Simon & Schuster, Inc., New York.

Rich, A. R. (1944). "The Pathogenesis of Tuberculosis." Charles C. Thomas, Springfield, Ilinois.

Rich, A. R., and Follis, R. H., Jr. (1938). The inhibitory effect of sulfanilamide on the development of experimental tuberculosis in the guinea pig. *Bull. Johns Hopkins Hospital* **62**, 77–84.

Rich, A. R., and Follis, R. H., Jr. (1942). The effect of low oxygen tension upon the development of experimental tuberculosis. *Bull. Johns Hopkins Hospital* **71**, 345–363.

Riley, R. L. (1957). Aerial dissemination of pulmonary tuberculosis. *Am. Rev. Tuberculosis Pulmonary Dis.* **76**, 931–941.

Riley, R. L. (1960). Apical localization of pulmonary tuberculosis. *Bull. Johns Hopkins Hospital* **106**, 232–239.

Riley, R. L., Mills, C. C., Nyka, W., Weinstock, N., Storey, P. B., Sultan, L. U., Riley, M. C., and Wells, W. F. (1959). Aerial dissemination of pulmonary tuberculosis: A two-year study of contagion in a tuberculosis ward. *Am. J. Hygiene* **70**, 185–196.

Riley, R. L., Mills, C. C., O'Grady, F., Sultan, L. U., Wittstadt, F., and Shivpuri, D. N. (1962). Infectiousness of air from a tuberculosis ward. Ultraviolet irradiation of infected air: Comparative infectiousness of different patients. *Am. Rev. Respir. Dis.* **85**, 511–525.

Ristow, M., Mohlig, M., Rifai, M., Schatz, H., Feldman, K., and Pfeiffer, A. (1996). New isoniazid/ethionamide resistance gene mutation and screening for multidrug-resistant *Mycobacterium tuberculosis* strains. *Lancet* **346**, 502–503.

Roberts, A. D., Sonnenberg, M. G., Ordway, D. J., Furney, S. K., Brennan, P. J., Belisle, J. T., and

Orme, I. M. (1995). Characteristics of protective immunity engendered by vaccination of mice with purified culture filtrate protein antigens of *Mycobacterium tuberculosis. Immunology* **85**, 502–508.

Robertson, H. E. (1933). The persistence of tuberculous infections. *Am. J. Pathol.* **9** (Suppl.), 711–718.

Robinson, D. S., Ying, S., Taylor, I. K., Wangoo, A., Mitchell, D. M., Kay, A. B., Hamid, Q., and Shaw, R. J. (1994). Evidence for a Th1-like bronchoalveolar T-cell subset and predominance of interferon-gamma gene activation in pulmonary tuberculosis. *Am. J. Respir. Crit. Care Med.* **149**, 989–993.

Rook, G. A. W., and Bloom, B. R. (1994). Mechanisms of pathogenesis in tuberculosis. *In* "Tuberculosis: Pathogenesis, Protection, and Control" (B. R. Bloom, ed.), pp. 485–501. ASM Press, Washington, D.C.

Rook, G. A. W., Steele, J., Ainsworth, M., and Champion, B. R. (1986a). Activation of macrophages to inhibit proliferation of *Mycobacterium tuberculosis:* Comparison of the effects of recombinant gamma-interferon on human monocytes and murine peritoneal macrophages. *Immunology* **59**, 333–338.

Rook, G. A. W., Steele, J., Fraher, L., Barker, S., Karmali, R., and O'Riordan, J. (1986b). Vitamin D_3, gamma interferon, and control of proliferation of *Mycobacterium tuberculosis* by human monocytes. *Immunology* **57**, 159–163.

Rosenthal, S. R., Loewinsohn, E., Graham, M. L., Liveright, D., Thorne, M. G., and Johnson, V. (1961). BCG vaccination against tuberculosis in Chicago: A twenty-year study statistically analyzed. *Pediatrics* **28**, 622–641.

Rothman, S. M. (1994). "Living in the Shadow of Death: Tuberculosis and the Social Experience of Illness in American History." Harper Collins, New York.

Rothschild, H., Friedenwald, J. S., and Bernstein, C. (1934). The relation of allergy to immunity in tuberculosis. *Bull. Johns Hopkins Hospital* **54**, 232–276.

Ryan, F. (1993). "The Forgotten Plague: How the Battle Against Tuberculosis Was Won—and Lost." Little, Brown & Company, Boston.

Saha, B., Das, G., Vohra, H., Ganguly, N. K., and Mishra, G. C. (1994). Macrophage–T cell interaction in experimental mycobacterial infection. Selective regulation of co-stimulatory molecules on *Mycobacterium*-infected macrophages and its implication in the suppression of cell-mediated immune response. *Eur. J. Immunol.* **24**, 2618–2624.

Salo, W. L., Aufderheide, A. C., Buikstra, J., and Holcomb, T. A. (1994). Identification of *Mycobacterium tuberculosis* DNA in pre-Columbian Peruvian mummy. *Proc. Natl. Acad. Sci. U.S.A.* **91**, 2091–2094.

Sanchez, F. O., Rodriguez, J. I., Agudelo, G., and Garcia, L. F. (1994). Immune responsiveness and lymphokine production in patients with tuberculosis and healthy controls. *Infect. Immun.* **62**, 5673–5678.

Schatz, A., Bugie, E., and Waksman, S. A. (1944). Streptomycin, a substance exhibiting antibiotic activity against gram-positive and gram-negative bacteria. *Proc. Soc. Exp. Biol. Med.* **55**, 66.

Schlesinger, L. S. (1993). Macrophage phagocytosis of virulent but not attenuated strains of *Mycobacterium tuberculosis* is mediated by mannose receptors in addition to complement receptors. *J. Immunol.* **150**, 2920–2930.

Schlesinger, L. S., Bellinger-Kawahara, C. G., Payne, N. R., and Horwitz, M. A. (1990). Phagocytosis of *Mycobacterium tuberculosis* is mediated by human monocyte complement receptors and complement component C3. *J. Immunol.* **144**, 2771–2780.

Schmidt, L. H. (1956). Some observations on the utility of simian pulmonary tuberculosis in defining the therapeutic potentialities of isoniazid. *Am. Rev. Tuberculosis Pulmonary Dis.* **74** (Suppl.), 138–159.

Scorpio, A., and Zhang, Y. (1996). Mutations in *pncA*, a gene encoding pyrazinamidase/nicotinamidase, cause resistance to the antituberculous drug pyrazinamide in tubercle bacillus. *Nat. Med.* **2**, 662–667.

Segal, W. (1984). Growth dynamics of *in vivo* and *in vitro* grown mycobacterial pathogens. *In* "The Mycobacteria: A Sourcebook" (G. P. Kubica and L. G. Wayne, eds.), pp. 547–573. Dekker, New York.

Segal, W., and Bloch, H. (1956). Biochemical differentiation of *Mycobacterium tuberculosis* grown *in vivo* and *in vitro. J. Bacteriol.* **72**, 132–141.

4

MOBILE GENETIC ELEMENTS AND THE EVOLUTION OF NEW EPIDEMIC STRAINS OF *VIBRIO CHOLERAE*

ERIC J. RUBIN
Department of Microbiology
and Molecular Genetics
Harvard Medical School
Boston, Massachusetts

MATTHEW K. WALDOR
Division of Geographic Medicine
and Infectious Diseases
Tupper Research Institute
New England Medical Center
Tufts University School of Medicine
Boston, Massachusetts

JOHN J. MEKALANOS
Department of Microbiology and Molecular Genetics
and Shipley Institute of Medicine
Harvard Medical School
Boston, Massachusetts

I. INTRODUCTION

The histories of modern man and cholera are intertwined. Although there are many bacterial causes of diarrhea, only cholera has caused repeated pandemics. Cholera has, at one time or another since the beginning of the nineteenth century, caused epidemic disease associated with enormous mortality on every continent (Barua, 1992). Despite the fact that the recognition of its infectious nature by John Snow led to the birth of epidemiology more than a century ago we are still unable to control cholera throughout the developing world. Even though there are safe antibiotics effective against the causative agent, *Vibrio cholerae,* cholera still causes millions of cases of diarrhea and thousands of deaths worldwide (Anonymous, 1995a,b).

Although cholera has most likely been present since antiquity, the pandemic spread of cholera outside of Asia is believed to have begun less than two centuries ago (Barua, 1992). The first cholera pandemic started in India in 1817 and swept through the Middle East and Asia. Since that time there have been six more pandemics, the last starting in 1961 and continuing to the present. Some would argue that the seventh pandemic has now given way to the eighth pandemic with the appearance of the new cholera serogroup O139 (see below). More recently the number of cholera cases worldwide has been increasing, owing in large part to the arrival of the seventh pandemic in Africa in 1970 (Swerdlow and Isaäcson, 1994), to the appearance of the pandemic strain in Peru in 1991 (Tauxe *et al.*, 1994), and to the emergence of O139 cholera in 1992 (Cholera Working Group, 1993). This has led to large numbers of deaths in some populations such as during the outbreak among Rwandan refugees in camps in the former Zaire (now the Democratic Republic of Congo) in 1994 (Anonymous, 1995b). Between these large outbreaks of disease cholera has remained present but largely confined to endemic foci.

What causes new pandemics? Clearly, a new outbreak of disease requires both a susceptible population and a virulent organism. Since infection with epidemic strains of cholera results in long-lasting (perhaps lifetime) immunity (Levine and Pierce, 1992), a susceptible population must contain a large enough number of individuals who have either never had cholera or at least never been infected with an antigenically related strain. In endemic areas the incidence of cholera is heavily skewed toward children (Glass and Black, 1992), who have not had the opportunity to develop protective immunity. In the current epidemic in Latin America, where cholera has not been present since well before the beginning of the current pandemic, both adults and children are being affected (Tauxe *et al.*, 1994).

Changes in the organisms that cause cholera have been linked with the onset of new outbreaks. The clearest example of this has been the recent outbreak of cholera which started on the coast of the Bay of Bengal (Cholera Working Group, 1993). A rise in the number of cases of cholera was also associated with an unusual number of cases among adults. Analysis of the epidemic strain revealed that it did not express the O1 serogroup antigen which had been characteristic of all previous epidemic strains. Instead, the strain contained a new serogroup antigen, now known as O139, which had not previously been described. This new serogroup spread rapidly throughout India and into the rest of Asia, in many areas completely supplanting the O1 strain which had been present since the 1960s (Nair *et al.*, 1994).

Vibrio cholerae, unlike many other infectious causes of diarrhea, does not require an association with a mammalian host for its survival. Rather, this bacterium inhabits estuarine environments often in association with organisms such as zooplankton (Colwell, 1996). It has been hypothesized that changes in global climate patterns which influence phytoplankton blooms have played important roles in recent cholera outbreaks (Colwell, 1996).

How do new epidemic strains of cholera arise? We now have several examples of genetic changes in *V. cholerae* that have been associated with cholera-causing strains. As discussed elsewhere, because bacteria lack the variation produced by sexual reproduction, rapid evolution of organisms requires the acquisition of new DNA by horizontal gene transfer (Cohan, 1996). We now have evidence that genes which are important in virulence can be transferred between *V. cholerae* strains. Here we review mechanisms of genetic exchange of known or potential virulence factors in *V. cholerae*. In addition, we discuss genetic elements that are likely to have arisen by genetic exchange, although the mechanism of transfer has not been elucidated. Although we briefly discuss virulence mechanisms, readers are referred to a comprehensive review of cholera genetics for a thorough treatment (Kaper *et al.*, 1995).

II. MECHANISMS OF GENETIC EXCHANGE IN *VIBRIO CHOLERAE*

A. Filamentous Bacteriophage

All epidemic strains of *V. cholerae* produce cholera toxin. The toxin, which consists of five copies of the B protein bound to a single copy of the A protein, is encoded by the genes *ctxA* and *ctxB*. The B subunits bind to the surface of target cells and facilitate the entry of the A subunit into the cytoplasm (Finkelstein, 1992). There the A subunit catalyzes the ADP-ribosylation of the eukaryotic protein G_s, resulting in marked increases in cellular cyclic AMP levels. The *ctxAB* genes lie within a region on the *V. cholerae* chromosome which is referred to as the CTX element (Mekalanos *et al.*, 1983; Pearson *et al.*, 1993). Adjacent to the structural genes for cholera toxin are two genes which had been postulated to encode proteins with entertoxic activity, *ace* (Trucksis *et al.*, 1993) and *zot* (Fasano *et al.*, 1991). Additionally, the element encodes a peptide, the product of the *cep* gene (Pearson *et al.*, 1993), which resembles genes which encode pilus subunits, and *orfU* (Trucksis *et al.*, 1993), a gene of unknown function. This group of genes is contained within a part of the CTX element known as the core (Pearson *et al.*, 1993). Adjacent to the core region is another set of genes known as the RS element (Pearson *et al.*, 1993). The RS element was known to encode a site-specific recombination system that catalyzes the integration of plasmids carrying this element into the *V. cholerae* genome. Thus, as a whole, the *ctx* element was thought to be a site-specific transposon.

To our surprise, we found that the *ctx* element was transmissible (Waldor and Mekalanos, 1996). Strains which contain a kanamycin resistance gene in place of the *ctx* structural genes were capable of transferring antibiotic resistance to other strains. This transfer did not require cell-to-cell contact. In fact, filtered culture supernatants from the donor strains were capable of mediating transfer of the phenotype. The addition of nucleases to supernatants did not inhibit this transfer. These experiments demonstrated that transduction was the mechanism of transfer of the kanamycin resistance gene in *ctx*. Electron microscopic studies showed that concentrates of culture supernatants from donor strains contained structures similar in morphology to filamentous bacteriophages. When partially purified these phages were found to contain a 6.9-kb single-stranded DNA genome encoding the *ctx* element. Therefore, the *ctx* element actually represents the genome of a filamentous bacteriophage, CTXφ (Fig. 1).

In fact, some of the genes in the *ctx* element resemble those of known filamentous phages. Koonin (1992) had noted that the predicted sequence of the protein encoded by the *zot* gene resembled the sequence of the gene I product of the *Escherichia coli* phage M13. The gene I product of M13 is an inner membrane protein required for phage assembly (Russel, 1995). The *cep* gene encodes a small protein which is similar in size and in the distribution of charged and nonpolar amino acids to the major coat proteins of other filamentous phages (such as the gene VIII product of M13) (Model and Russel, 1988). *ace* encodes a predicted product similar to that of the gene VI homolog of the *Pseudomonas aeruginosa* filamentous phage Pf1 (Hill *et al.*, 1991). *orfU* bears little resemblance to known proteins, but the size and relative position of this gene in

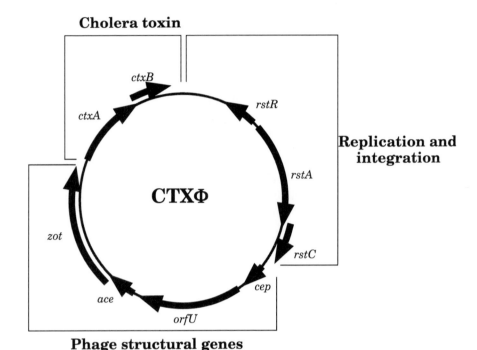

FIGURE I   Organization of the CTXφ genome. Open reading frames are designated with arrows showing the direction of transcription.

CTXφ is similar to gene III of M13 (Model and Russel, 1988). pIII is a minor coat protein which determines the binding specificity of the phage to the F pilus. All of the above-mentioned genes of M13 are required for proper assembly of infectious phage particles. Similarly, deletions in either *zot* or *orfU* abrogate transduction by CTXφ. Thus, CTXφ structurally and functionally resembles other filamentous phages.

However, the ability of the CTXφ genome to integrate into the host chromosome is very unusual. The only previously described examples of integrating filamentous phages are those of *Xanthomonas campestris* (Shieh *et al.*, 1995). Like temperate phages, CTXφ appears to choose between lysogeny and productive (though nonlytic) infection. This choice is determined by whether a phage attachment site (termed *attRS*) exists in the genome. In strains which lack *attRS* CTXφ DNA replicates as a double-stranded plasmid and high titers of phage are produced. However, in *attRS*-containing strains, the phage genome integrates (sometimes as a tandem duplication), no replicating phage DNA is detectable, and little or no phage is produced. Subclones of the CTXφ genome which contain only the RS region behave identically (with respect to replication and integration) to intact phage.

The RS element contains three open reading frames (Waldor *et al.*, 1997). *rstR* is transcribed in the opposite orientation from other phage genes. The sequence of its predicted product closely resembles a family of transcriptional repressors typified by the *Bacillus subtilis* protein Xre (McDonnell and McConnell, 1994). Indeed, the expression of *lacZ* fusions to the second gene in the

RS region, *rstA*, is repressed in the presence of *rstR*. *rstA* and *rstB*, the third RS-encoded gene, have little similarity to previously described proteins. Plasmids that contain insertion mutations in *rstA* neither replicate nor integrate in *V. cholerae*, whereas an insertion mutant in *rstB* is able to replicate but cannot integrate. The simplest explanation for these results is that RstA is a replication protein and RstB is an integrase. The reason why *rstA* insertion mutants do not integrate is unclear. It is possible that RstA is required for integration or that replication is required prior to integration. Alternatively, insertions in *rstA* may be polar on the downstream *rstB*.

Therefore, the CTXφ genome can be thought of as having three functional modules (Fig. 1). The first module, the RS element, is responsible for integration and replication and their regulation. The 5' end of the core region, which contains phage structural and morphogenesis genes, constitutes the second module. The third module, the *ctx* genes located at the 3' end of the core region, does not as yet have a role in the production of phage. This modular arrangement is probably not simply an artifact of laboratory investigation. The chromosomes of naturally occurring *V. cholerae* strains contain a variety of arrangements of CTXφ modules (Fig. 2). Many chromosomal elements contain two different RS elements which are tandemly arranged and are not identical. Bacteriophage genomes that have thus far been studied all contain the RS element, which is adjacent to the core genes in the bacterial chromosome (termed RS2). RS1, which does not appear to be incorporated into the phage particle, contains *rstR*, *rstA*, and *rstB* genes which are quite similar to those in RS2 and an additional open reading frame, *rstC*. The protein encoded by *rstC* does not appear similar to known proteins, and we do not yet know its function. RS1-containing subclones are able to replicate and integrate in much the same way as RS2-containing plasmids. These varying RS elements may constitute a toolbox for the construction of phages.

The receptor for CTXφ is the TCP pilus which is encoded by the large *tcp* operon (Waldor and Mekalanos, 1996). The TCP pilus is extremely important in pathogenesis by *V. cholerae*. In the infant mouse model of cholera (Taylor *et al.*, 1987), which essentially measures the ability of the organisms to colonize the gut, and in human volunteers TCP has been shown to be required for *V. cholerae* colonization of the small intestine (Herrington *et al.*, 1988). Mutations in *tcpA* (the major pilin gene) prevent CTXφ infection as does antiserum to the pilus. Phage DNA introduced into *tcp*-mutant strains by electroporation, however, is able to replicate and produce infectious phage. Finally, CTXφ can only infect cells grown under conditions that result in TCP expression.

The requirement for host strains to express TCP pilus in order to be infected by CTXφ has important implications. While classical biotype strains (which caused pandemics prior to the current pandemic) express TCP under a variety of specific laboratory conditions, El Tor strains (the causative agents of the current pandemic do not express TCP well under any *in vitro* conditions (Rhine and Taylor, 1994). Clearly, however, TCP must be expressed during infection. In fact, this can be demonstrated by coinfecting mice with a phage-producing donor strain and a recipient El Tor strain (Waldor and Mekalanos, 1996). El Tor recipients harvested from mouse intestine after such a coinfection are infected with CTXφ at a rate more than a millionfold higher than under *in vitro*

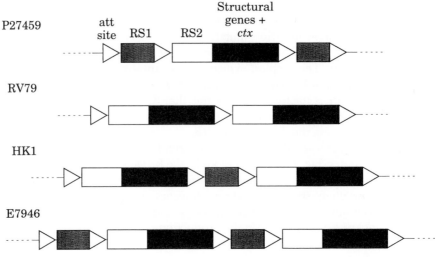

FIGURE 2 CTX element in various *V. cholerae* O1 El Tor biotype strains. Arrowheads indicate positions of the repeated sequences at the phage integration site.

conditions. The infected host is the only natural site currently known where TCP is efficiently expressed. Thus, the transmission of CTXφ may only occur in the human gut. The generation of new cholera-toxin producing strains may occur in human hosts coinfected with toxinogenic (CTXφ+) and nontoxinogenic (CTXφ−) *V. cholerae* strains. This is reminiscent of the reassortment of the influenza genome resulting in antigenic shift which occurs in multiply infected animals (Scholtissek, 1994).

Three more filamentous phages of *V. cholerae* have been reported. The most extensively characterized is VSK (Scholtissek, 1994). This phage, isolated from an O139 serogroup strain, shares several characteristics with CTXφ, although it appears to have a different restriction map. VSK has a 7-kb single-stranded DNA genome and replicates using a double-stranded DNA intermediate. Like CTXφ, VSK is not produced in measurable amounts from uninduced bacteria; however, treatment with mitomycin C does result in the production of phage. VSK is also capable of integrating into the genome. The receptor for VSK and any role it may have in virulence have not yet been reported.

Two more filamentous phages of *V. cholerae* have been isolated (Shimodori *et al.*, 1996). fs1, which has a 6.4-kb genome, and fs2 whose genome is 8.5 kb, have each been found in a variety of both serogroup O1 and O139 strains. Both phages have similar morphologies and contain single-stranded DNA, although they are genetically distinct. Unlike CTXφ, fs1 and fs2 both form plaques. It is not known whether their genomes integrate into the host chromosome. TCP pili do not act as a receptor for either of these phages. Nonlysogenic strains are difficult to infect with fs1 and fs2 after *in vitro* growth. However, some strains become susceptible after culture in rabbit ileal loops. As is the case for CTXφ, a signal present in the host may help create the necessary conditions for gene transfer. This may represent a common theme among pathogenic bacteria (Mel and Mekalanos, 1996).

There appear to be several filamentous phages present in various *V. cholerae* strains. When the DNA sequence of CTXϕ is compared with that of M13, a well-characterized phage of *E. coli*, several interesting differences emerge. CTXϕ contains the two additional modules described above, namely, the RS element and the cholera toxin structural genes. However, CTXϕ lacks genes required for M13 function. Some of these gene functions may be complemented by chromosomal loci. It is also possible that the filamentous phages of *V. cholerae* complement functions in each other and, therefore, constitute an interdependent system of otherwise defective phages.

B. Temperate Phage

Several virulence factors in both gram-positive and gram-negative bacteria are encoded in the genomes of lysogenizing bacteriophage (Betley *et al.*, 1986; Bishai and Murphy, 1988). Several temperate phages have been described in *V. cholerae* (Rowe and Frost, 1992). It has been difficult, however, to show a link between the presence of these phages and virulence. Most phage-encoded virulence genes have been identified only after recognizing their protein product and then finding that the coding sequence is linked to a phage.

A temperate phage-encoded gene has been shown to be associated with a change in virulence in *V. cholerae* (Reidl and Mekalanos, 1995). This phage, K139, is related to the kappa family of bacteriophages previously described in serogroup O1 *V. cholerae*. K139 was originally isolated from an O139 strain but can be found in a variety of virulent strains. Several virulent strains which do not produce phage contain kappa-related DNA in their genomes, possibly representing defective prophage. The phage particles contain a 35-kb double-stranded linear DNA genome which circularizes and integrates site-specifically into the *V. cholerae* chromosome.

A minitransposon was used to identify genes that encode secreted products in K139 phage. The transposon, *Tnbla*, contains a copy of the β-lactamase gene, *bla*, without its signal sequence. For the encoded enzyme to function (and, therefore, mediate resistance to ampicillin) it must be fused to a protein that provides a signal sequence which directs secretion into the bacterial periplasm. Therefore, mutants that contain an insertion in genes encoding secreted proteins can be selected with ampicillin. Similar strategies have been used to identify many genes encoding secreted products (Manoil and Beckwith, 1985), although *Tnbla* differs in that it is selectable. As several genes required to cause disease encode secreted proteins, methods which identify such genes have been useful for finding new virulence factors (Taylor *et al.*, 1989).

Four independent *Tnbla* insertions in K139 were all found to be in the same open reading frame. This open reading frame, known as *glo*, encodes a 137 amino acid protein with a predicted signal sequence (Fig. 3). Comparison with known proteins showed significant sequence similarity to a eukaryotic GTP-binding protein $G_s\alpha$, one of the regulatory subunits of adenylate cyclase and the substrate for ADP-ribosylation by cholera toxin. Significantly, the C terminus of the Glo protein contains a motif which is recognized in eukaryotic cells by enzymes that catalyze the isoprenylation of cysteine residues near the C termini (Willumsen *et al.*, 1984). No such system has been identified in bacteria. The

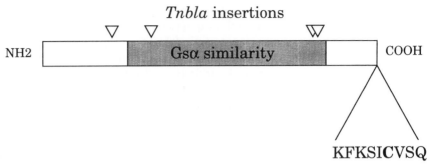

FIGURE 3   Map of the *glo* gene. The expanded sequence at the carboxyl terminus represents the region of homology with -CAAX box-containing GTP-binding proteins. The conserved cysteine residue is indicated in bold type (Reidl and Mekalanos, 1995).

similarity of Glo to $G_s\alpha$ is especially interesting given that cholera toxin itself targets $G_s\alpha$ and requires a second GTP-binding protein (ARF) for its activity (Finkelstein, 1992).

Is Glo important in virulence? When a virulent *V. cholerae* strain is lysogenized with *glo+* or *glo−* K139 phage there is no difference in colonization in infant mice. However, mice were found to have more diarrhea when infected with bacteria containing the wild-type phage. In addition, the number of bacteria required to kill infant mice was at least 10-fold higher for *glo− V. cholerae*. These results suggest that *glo* does play a role in the virulence of *V. cholerae*. As noted above, the infant mouse model of cholera has generally been used to study only colonization. It will be important to study the effect of *glo* mutations in other experimental models where the toxic effects of *V. cholerae* can be quantitated.

Other temperate phage-encoded virulence factors may certainly exist. The technique which identified *glo* in K139 phage has not been applied to other *V. cholerae* phages. In fact, it is possible that K139 phage itself may encode other virulence factors. Because *Tnbla* mutagenesis only identifies genes which encode secreted proteins, this method would not identify possible cytoplasmic virulence factors. In addition, a mutation in a virulence factor which was also necessary for production of infectious bacteriophage would not be detected after a transposon insertion disrupted the gene. Other techniques are likely to be required to identify all phage-encoded genes likely to be important in the ability of *V. cholerae* to cause disease.

C. Chromosomal Elements

When the O139 serogroup started causing disease in southeastern India in late 1992, isolates were found to be resistant to a number of antibiotics (Albert *et al.,* 1993). This pattern of resistance, to streptomycin, trimethoprim, sulfamethoxazole, and furazolidine, was not characteristic of the endemic O1 strains from the same area. Genetic analysis revealed that the genes encoding resistances to three of the antibiotics, namely, streptomycin, trimethoprim, and sulfamethoxazole, are physically linked (Waldor *et al.,* 1996). Furthermore, mating experi-

genes. By understanding the mechanisms of transfer we may also be able to prevent their dissemination.

ACKNOWLEDGMENTS

This work was supported by National Institutes of Health Grants AI02137 to E. J. R., AI01321 to M. K. W., and AI18045 to J. J. M. M. K. W. is also supported by a New England Medical Center Young Investigator Award and a Tupper Scholar Award.

REFERENCES

Albert, M. J., Siddique, A. K., Islam, M. S., Faruque, A. S. G., Ansaruzzaman, M., Faruque, S. M., and Sack, R. B. (1993). Large outbreak of clinical cholera due to *Vibrio cholerae* non-O1 in Bangladesh. *Lancet* **341,** 704.

Anonymous. (1995a). Cholera in 1994. Part I. *Weekly Epidemiological Record* **70,** 201–208.

Anonymous. (1995b). Cholera in 1994. Part II. *Weekly Epidemiological Record* **70,** 209–211.

Barua, B. (1992). History of cholera. *In* "Cholera" (D. Barua and W. B. Greenough III, eds.), pp. 1–36. Plenum, New York.

Betley, M. J., Miller, V. L., and Mekalanos, J. J. (1986). Genetics of bacterial enterotoxins. *Annu. Rev. Microbiol.* **40,** 577–605.

Bik, E. M., Bunschoten, A. E., Gouw, R. D., and Mooi, F. (1995). Genesis of the novel epidemic *Vibrio cholerae* O139 strain: Evidence for horizontal transfer of genes involved in polysaccharide synthesis. *EMBO J.* **14,** 209–216.

Bik, E. M., Bunschoten, A. E., Willems, R. J. L., Chang, A. C. Y., and Mooi, F. R. (1996). Genetic organization and functional analysis of the *otn* DNA essential for cell-wall polysaccharide synthesis in *Vibrio cholerae* O139. *Mol. Microbiol.* **20,** 799–811.

Bishai, W. R., and Murphy, J. R. (1988). Bacteriophage gene products that cause human disease. *In* "The Bacteriophages" (R. Calender, ed.), pp. 683–724. Plenum, New York.

Brown, R. C., and Taylor, R. K. (1995). Organization of *tcp, acf,* and *toxT* genes within a ToxT-dependent operon. *Mol. Microbiol.* **16,** 425–439.

Cholera Working Group. (1993). Large epidemic of cholera-like disease in Bangladesh caused by *Vibrio cholerae* O139 synonym Bengal. *Lancet* **342,** 387–390.

Cohan, F. M. (1996). The role of genetic exchange in bacterial evolution. *ASM News* **62,** 631–636.

Colwell, R. R. (1996). Global climate and infectious disease—The cholera paradigm. *Science* **274,** 2025–2031.

Comstock, L. E., Maneval, D., Panigrahi, P., Joseph, A., Levine, M. M., Kaper, J. B., Morris, J. G., and Johnson, J. (1995). The capsule and O antigen in *Vibrio cholerae* O139 Bengal are associated with a genetic region not present in *Vibrio cholerae* O1. *Infect. Immun.* **63,** 317–323.

Comstock, L. E., Johnson, J. A., Machalski, J. M., Morris, J. G., and Kaper, J. B. (1996). Cloning and sequence of a region encoding a surface polysaccharide of *Vibrio cholerae* O139 and characterization of the insertion site in the chromosome of *Vibrio cholerae* O1. *Mol. Microbiol.* **19,** 815–826.

DiRita, V. J. (1992). Co-ordinate expression of virulence genes by ToxR in *Vibrio cholerae. Mol. Microbiol.* **6,** 451–458.

Everiss, K. D., Hughes, K. J., and Peterson, K. M. (1996). The accessory colonization factor and the toxin-coregulated pilus gene clusters are physically linked on the *Vibrio cholerae* O395 genome. *DNA Sequence* **5,** 51–55.

Fasano, A., Baudry, B., Pumplin, D. W., Wasserman, S. S., Tall, B. D., Ketley, J. N., and Kaper, J. B. (1991). *Vibrio cholerae* produces a second enterotoxin which affects intestinal tight junctions. *Proc. Natl. Acad. Sci. U.S.A.* **88,** 5242–5246.

Finkelstein, R. A. (1992). Cholera enterotoxin (choleragen): A historical perspective. *In* "Cholera" (D. Barua and W. B. Greenough III, eds.), pp. 155–187. Plenum, New York.

Glass, R. I., and Black, R. E. (1992). The epidemiology of cholera. *In* "Current Topics in Infectious Disease: Cholera" (D. Barua and W. B. Greenough III, eds.), pp. 129–154. Plenum, New York.

Harkey, C. W., Everiss, K. D., and Peterson, K. M. (1995). Isolation and characterization of a *Vibrio cholerae* gene (*tagA*) that encodes a ToxR-regulated lipoprotein. *Gene* **153**, 81–84.

Herrington, D. A., Hall, R. H., Losonsky, G., Mekalanos, J. J., Taylor, R. K., and Levine, M. M. (1988). Toxin, toxin-coregulated pili, and the *toxR* regulon are essential for *Vibrio cholerae* pathogenesis in humans. *J. Exp. Med.* **168**, 1487–1492.

Hill, D. F., Short, N. J., Perham, R. N., and Petersen, G. B. (1991). DNA sequence of the filamentous bacteriophage Pf1. *J. Mol. Biol.* **218**, 349–363.

Hisatsune, K., Kondo, S., Isshiki, Y., Iguchi, T., Kawamata, Y., and Shimada, T. (1993). O-antigenic lipopolysaccharide of *Vibrio cholerae* O139 Bengal, a new epidemic strain for recent cholera in the Indian subcontinent. *Biochem. Biophys. Res. Commun.* **196**, 1309–1315.

Hughes, K. J., Everiss, K. D., Harkey, C. W., and Peterson, K. M. (1994). Identification of a *Vibrio cholerae* ToxR-activated gene (*tagD*) that is physically linked to the toxin-coregulated pilus (*tcp*) gene cluster. *Gene* **148**, 97–100.

Kaper, J. B., Morris, J. G., and Levine, M. M. (1995). Cholera. *Clin. Microbiol. Rev.* **8**, 48–86.

Karaolis, D. K., Lan, R., and Reeves, P. R. (1995). The sixth and seventh cholera pandemics are due to independent clones separately derived from environmental, nontoxigenic, non-O1 *Vibrio cholerae*. *J. Bacteriol.* **177**, 3191–3198.

Koonin, E. V. (1992). The second cholera toxin, Zot, and its plasmid-encoded and phage encoded homologues constitute a group of putative ATP-ases with an altered purine NTP-binding motif. *FEBS Lett.* **312**, 3–6.

Kovach, M. E., Shaffer, M. D., and Peterson, K. M. (1996). A putative integrase gene defines the distal end of large cluster of ToxR-regulated colonization genes in *Vibrio cholerae*. *Microbiology* **142**, 2165–2174.

Lee, C. A. (1996). Pathogenicity islands and the evolution of bacterial pathogens. *Infect. Agents Dis.* **5**, 1–7.

Levine, M. M., and Pierce, N. F. (1992). Immunity and vaccine development. *In* "Cholera" (D. Barua and W. B. Greenough III, eds.), pp. 285–328. Plenum, New York.

McDonnell, G. E., and McConnell, D. J. (1994). Overproduction, isolation, and DNA-binding characteristics of Xre, the repressor protein from the *Bacillus subtilis* defective prophage PBSX. *J. Bacteriol.* **176**, 5831–5834.

Manoil, C., and Beckwith, J. (1985). Tn*phoA:* A transposon probe for protein export signals. *Proc. Natl. Acad. Sci. U.S.A.* **82**, 8129–8133.

Mekalanos, J. J., Swartz, D. J., Pearson, G. D., Harford, N., Groyne, F., and deWilde, M. (1983). Cholera toxin genes: Nucleotide sequence, deletion analysis and vaccine development. *Nature (London)* **306**, 551–557.

Mel, S. F., and Mekalanos, J. J. (1996). Modulation of horizontal gene transfer in pathogenic bacteria by *in vivo* signals. *Cell (Cambridge, Mass.)* **87**, 795–798.

Model, P., and Russel, M. (1988). Filamentous bacteriophage. *In* "The Bacteriophages" (R. Calendar, ed.), pp. 375–456. Plenum, New York.

Nair, G. B., Ramamurthy, T., Bhattacharya, S. K., Mukhopadhyay, A. K., Garg, S., Bhattacharya, M. K., Takeda, T., Shimada, T., Takeda, Y., and Deb, B. C. (1994). Spread of *Vibrio cholerae* O139 Bengal in India. *J. Infect. Dis.* **169**, 1029–1034.

Pearson, G. D. N., Woods, A., Chiang, S. L., and Mekalanos, J. J. (1993). CTX genetic element encodes a site-specific recombination system and an intestinal colonization factor. *Proc. Natl. Acad. Sci. U.S.A.* **90**, 3750–3754.

Peterson, K. M., and Mekalanos, J. J. (1988). Characterization of the *Vibrio cholerae* ToxR regulon: Identification of novel genes involved in intestinal colonization. *Infect. Immun.* **56**, 2822–2829.

Reidl, J., and Mekalanos, J. J. (1995). Characterization of *Vibrio cholerae* bacteriophage K139 and use of a novel mini-transposon to identify a phage-encoded virulence factor. *Mol. Microbiol.* **18**, 685–701.

Rhine, J. A., and Taylor, R. K. (1994). TcpA pilin sequences and colonization requirements for O1 and O139 *Vibrio cholerae*. *Mol. Microbiol.* **13**, 1013–1020.

Rowe, B., and Frost, J. A. (1992). Vibrio phages and phage-typing. *In* "Cholera" (D. Barua and W. B. Greenough III, eds.), pp. 95–106. Plenum, New York.

Russel, M. (1995). Moving through the membrane with filamentous phages. *Trends Microbiol.* **3**, 223–228.

Salyers, A. A., Shoemaker, N. B., and Li, L.-Y. (1995). In the driver's seat: The *Bacteroides* conjugative transposons and the elements they mobilize. *J. Bacteriol.* **177**, 5727–5731.

Scholtissek, C. (1994). Source for influenza pandemics. *Eur. J. Epidemiol.* **10**, 455–458.

Shieh, G. J., Lin, C. H., Kuo, J. L., and Kuo, T. T. (1995). Characterization of an open reading frame involved in site-specific integration of filamentous phage Cf1t from *Xanthomonas campestris* pv. *citri. Gene* **158**, 73–76.

Shimodori, S., Kojima, F., Amako, K., Ehara, M., Ichinose, Y., Hirayama, T., Honma, Y., Iwanaga, M., and Albert, M. J. (1996). Filamentous phages of *Vibrio cholerae* O139 and O1. *In* "Thirty-second Joint Conference on Cholera and Related Diarrheal Diseases," pp. 34–35. The U.S.–Japan Cooperative Medical Science Program, Nagasaki.

Stroeher, U. H., Jedani, K. E., Dredge, B. K., Morona, R., Brown, M. H., Karageorgos, L. E., Albert, M. J., and Manning, P. A. (1995). Genetic rearrangements in the *rfb* regions of *Vibrio cholerae* O1 and O139. *Proc. Natl. Acad. Sci. U.S.A.* **84**, 2833–2837.

Swerdlow, D. L., and Isaäcson, M. (1994). The epidemiology of cholera in Africa. *In* "*Vibrio cholerae* and Cholera" (I. K. Wachsmuth, P. A. Blake, and Ø. Olsvik, eds.), pp. 297–308. ASM Press, Washington, D.C.

Tauxe, R., L., S., Tapia, R., and Libel, M. (1994). The Latin American epidemic. *In* "*Vibrio cholerae* and Cholera" (I. K. Wachsmuth, P. A. Blake, and Ø. Olsvik, eds.), pp. 321–344. ASM Press, Washington, D.C.

Taylor, R. K., Miller, V. L., Furlong, D. B., and Mekalanos, J. J. (1987). Use of *phoA* gene fusions to identify a pilus colonization factor coordinately regulated with cholera toxin. *Proc. Natl. Acad. Sci. U.S.A.* **84**, 2833–2837.

Taylor, R. K., Manoil, C., and Mekalanos, J. J. (1989). Broad-host-range vectors for delivery of TnphoA: Use in genetic analysis of secreted virulence determinants of *Vibrio cholerae. J. Bacteriol.* **171**, 1870–1878.

Trucksis, M., Galen, J., Michalski, J., Fasano, A., and Kaper, J. B. (1993). Accessory cholera enterotoxin (Ace), the third toxin of a *Vibrio cholerae* virulence cassette. *Proc. Natl. Acad. Sci. U.S.A.* **90**, 5267–5271.

Waldor, M. K., and Mekalanos, J. J. (1996). Lysogenic conversion by a filamentous phage encoding cholera toxin. *Science* **272**, 1910–1914.

Waldor, M. K., Rubin, E. J., Pearson, G. D. N., Kimsey, H., and Mekalanos, J. J. Regulation, replication, and integration functions of the *Vibrio cholerae* CTXϕ are encoded by region RS2. *Mol. Microbiol.* **24**:917–926.

Waldor, M. K., Tschape, H., and Mekalanos, J. J. (1996). A new type of conjugative transposon encodes resistance to sulfamethoxazole, trimethoprim, and streptomycin in *Vibrio cholerae* O139. *J. Bacteriol.* **178**, 4157–4165.

Willumsen, B. M., Norris, K., Papageorge, A. G., Hubbert, N. L., and Lowy, D. R. (1984). Harvey murine sarcoma virus p21[ras] protein: Biological and biochemical significance of the cysteine nearest the carboxy terminus. *EMBO J.* **3**, 2581–2585.

Yamasaki, S., Hoshino, K., Shimizu, T., Garg, S., Shimada, T., Ho, S., R. K., Bhadra, Nair, G. B., and Takeda, Y. (1996). Comparative analysis of the gene responsible for lipopolysaccharide synthesis of *Vibrio cholerae* O1 and O139 and those of non-O1 non-O139 *Vibrio cholerae. In* "Thirty-second Joint Conference on Cholera and Related Diarrheal Diseases," pp. 24–27. The U.S.–Japan Cooperative Medical Science Program, Nagasaki.

5

PATHOGENIC *ESCHERICHIA COLI* O157:H7: A MODEL FOR EMERGING INFECTIOUS DISEASES

THOMAS S. WHITTAM, ELIZABETH A. MCGRAW, AND SEAN D. REID

Institute of Molecular Evolutionary Genetics
Department of Biology
Pennsylvania State University
University Park, Pennsylvania

I. INTRODUCTION

The evolution and emergence of new bacterial pathogens is fundamentally a two-stage process. The first stage is the creation of genetic variation in virulence among strains in a population of bacteria by the processes of mutation, lateral gene transfer, or recombination. At this stage chance can play a major role—most new mutations in natural populations of bacteria are lost by random genetic drift, and genetic variation is purged by periodic selection (Levin, 1981; Milkman, 1997). The second stage is where natural selection dominates: the new virulent strain must spread and increase in frequency relative to other less virulent strains. This spreading of a new pathogen may be sufficient to increase disease incidence to noticeable levels. In addition, certain ecological opportunities, such as infiltration of a water or food distribution system, can expose large numbers of susceptible hosts to infection by the emerging pathogen and trigger conspicuous outbreaks of disease.

Escherichia coli O157:H7 exemplifies this two-stage process of evolution and emergence. *Escherichia coli* O157:H7 is a newly emerged pathogen, first incriminated in infectious disease in 1982 (Riley *et al.*, 1983), and is now recognized as a major cause of large-scale epidemics and thousands of sporadic cases of gastrointestinal illness in North America, Europe, and Japan (Feng, 1995; Griffin and Tauxe, 1991; Izumiya *et al.*, 1997). In the United States, *E. coli* O157:H7 infections are estimated to account for 20,000 illnesses and 250 deaths a year (Armstrong, *et al.*, 1996). *Escherichia coli* O157:H7 was originally implicated in human disease because it produced a severe bloody diarrhea (hemorrhagic colitis) that is clinically distinct from other types of diarrheal diseases caused by other enteric pathogens (Levine, 1987)

The crucial steps in the evolution and emergence of *E. coli* O157:H7 involved the acquisition of a suite of virulence determinants, including a pathogenicity island encoding factors for intestinal adherence, a large plasmid specifying hemolysin and other putative factors, and bacteriophage carrying genes encoding powerful cytotoxins. A critical element in the emergence of *E. coli* O157:H7 as a foodborne pathogen was the evolution of acid resistance, an attribute that promotes survival in acidic foods and results in efficient transmission with a low infective dose. The tolerance of highly acidic environments is one aspect of overall durability, that is, the ability of a pathogen to remain viable in the external environment outside of the host.

Our purpose in this chapter is to examine the emergence of new diseases and novel pathogens from an evolutionary perspective. This review begins with some background on how virulence can evolve and why high levels of pathogenicity might be expected for waterborne and foodborne transmitted diseases. We then focus on recent advances in elucidating the molecular mechanisms of pathogenesis of *E. coli* O157:H7 and in reconstructing the major steps in the evolution of this pathogen. We suggest that changing conditions in the secondary environment, for instance, acidification caused by acid rain, selects for enhanced durability and fosters the spread of durable mutants of pathogenic strains. These naturally occurring durable mutants have the potential to emerge as new waterborne or foodborne pathogens.

A. Evolutionary Perspective on Virulence

Evolutionary biologists have had a long-standing interest in host–parasite interactions, but the emergence of human immunodeficiency virus (HIV) and other infectious agents has drawn attention to fundamental questions about how parasite and pathogen virulence evolves (Bull, 1994; Lenski and May, 1994; Lipsitch and Nowak, 1995). To what extent is virulence an adaptation? Do pathogens inevitably evolve toward attenuation? If they do not, what determines the rate and direction of changes in virulence? How does the mode of transmission influence the evolution of virulence?

These questions and others are now being addressed by a synthetic view of the evolution of virulence that has developed since the late 1980s (Levin, 1996). This shift in paradigm has been stimulated mostly by mathematical population biology theory (Cohen and Newman, 1989; Levin *et al.*, 1982; May and Anderson, 1983; May and Nowak, 1995; Nowak and May, 1994). The conventional view (May and Anderson, 1983) predicted that parasites always evolve toward reduced virulence in order to ensure the survival of the host population, and hence their own. This view of inevitable attenuation through evolution garnered support from many natural situations. For example, Bluetongue, Rinderpest, Rift Valley Fever, and African Swine Fever viruses show evidence of having coevolved in a long-term association with their indigenous ungulate hosts. Domesticated animals, however, when introduced into areas of Africa where the viral diseases are endemic, were devastated by the diseases (Garnick, 1992). Other pathogens, such as Ebola virus and Hantavirus, demonstrate very high

PATHOGENIC E. COLI O157:H7

LIVERPOOL
JOHN MOORES UNIVERSITY
AVRIL ROBARTS LRC
TEL. 0151 231 4022

165

virulence in secondary hosts as well (Bull, 1994). Perhaps the most convincing example is the well-known case of the myxoma virus, which when introduced into Australia to control populations of rabbits experienced a rapid decrease in virulence (Lenski and May, 1994).

One of the troubling problems with the conventional view is that it relies on group selection, that is, the idea that natural selection operates on differences in the rates of origination and extinction of local populations rather than on the survival and reproduction of individuals. Moreover, there is no well-developed theoretical base to support an inevitable drive toward attenuation (Levin et al., 1982; Levin and Eden, 1990). Recent theory, however, indicates that there is no consistent outcome in the relationship between host and pathogen, and the evolution of virulence depends on the relationship between the parameters of the infection and transmission processes (Anderson and May, 1982, 1991; Bonhoeffer and Nowak, 1994; Lenski and May, 1994).

The essential features of the complexity of the problem are seen in the familiar equation for net growth rate of a pathogen (microparasite) that undergoes direct transmission in susceptible host populations (May and Anderson, 1983). The basic reproductive rate of a pathogen (R_o) is defined as

$$R_o = \beta H/(\alpha + b + \nu),$$

where α is equal to the infection-induced host mortality rate, b is equal to the disease-free mortality rate, ν is the host recovery rate, β is equal to the transmission coefficient which is the probability of infection per contact between an infected and susceptible host, and H is the number of susceptible hosts. R_o measures the fitness of the pathogen and can be interpreted as the average number of secondary infections due to a single infectious case when a pathogen is introduced into a fully susceptible host population. From the evolutionary perspective of the pathogen, natural selection should act to maximize R_o. An increase in reproductive rate will ensure a greater contribution to the next generation of pathogens. When these parameters are independent of one another, natural selection will favor reduced virulence (decreasing α) and greater transmissibility (increasing β) (Levin, 1996).

In many natural situations, however, the parameters of infection and disease are not independent so that, for example, virulence is coupled with transmission. This coupling occurs because disease processes, such as the induction of diarrhea by an enteric pathogen, can increase the probability of transmission to new susceptible hosts (Ewald, 1991). In such cases, the evolution of virulence traits are linked to fitness through transmissibility, and natural selection can maintain high levels of virulence (Ewald, 1996; Levin, 1996).

B. Waterborne and Foodborne Pathogens

From an evolutionary perspective, waterborne and foodborne transmission fosters high virulence because the pathogens do not rely on host mobility to be transmitted—there is no cost in terms of pathogen transmission for an organism causing severe disease that immobilizes the host (Ewald, 1983, 1991). In this case, the systems for distribution of water and food provide a "cultural vector" for delivery of contaminated material and successful spread of a pathogen. Ewald (1991) draws support for this hypothesis by showing that the virulence of gastrointestinal pathogens, including *Vibrio cholerae, Salmonella typhi,* and *Shigella dysenteriae,* is positively correlated with their tendency for waterborne transmission. Interestingly, this application of the evolutionary theory predicts that, if all things remain equal, diarrheal pathogens should evolve toward lower virulence as the purity of water supplies improves (Ewald, 1991).

Another reason why virulence could remain high in diarrheal pathogens is because natural selection in the external environment could favor the evolution of durable strains that have improved ability to tolerate nonspecific systems of host defense. Many natural populations of enteric bacteria have a general life cycle that involves transitions between two principal habitats: the lower intestine of warm-blooded animals and the water, sediment, and soil of the external environment. These habitats differ markedly in their biotic and abiotic features, including, for example, the availability of carbohydrates and amino acids, uniformity of temperature, and interactions with other microorganisms (Hartl and Dykhuizen, 1984; Mason and Richardson, 1981; Savageau, 1983). Although it is estimated that the half-life of *E. coli* in the secondary environment is on the order of a few days (Hartl and Dykhuizen, 1984), some strains can survive for months in freshwater (Flint, 1987; Rice *et al.*, 1992). For enteric organisms, it is possible that the sojourn in the external environment selects for durable strains that evolve resistance mechanisms to survive extreme aspects of the environments. Enhanced durability could improve the ability of a pathogen to tolerate gastric acidity, thus reducing the infective dose and increasing the probability that the pathogen successfully colonizes a susceptible host. Durable strains could also better survive the extremes and stress of food processing and storage.

II. EVOLUTION AND EMERGENCE OF *ESCHERICHIA COLI* O157:H7

Levine (1987) classifies O157:H7 strains along with *E. coli* of other serotypes associated with hemorrhagic colitis into a distinct group of pathogenic strains called the enterohemorrhagic *E. coli* or EHEC. *Escherichia coli* O157:H7 infections produce a wide range of disease severity from asymptomatic carriage and mild diarrhea to bloody diarrhea and associated extraintestinal complications such as hemolytic uremic syndrome (Tarr, 1995). The EHEC grouping highlights the fact that *E. coli* O157:H7 has a novel mechanism of pathogenesis involving the high expression of Shiga toxins (O'Brien and Holmes, 1987), the intimate adherence of bacteria to intestinal epithelial cells (Tesh and O'Brien, 1992), and the contribution of ancillary factors whose roles in high virulence of this organism are less clear.

A. What Makes *Escherichia coli* O157:H7 So Virulent?

Some of the primary virulence genes that have been identified in *E. coli* O157: H7 and implicated in pathogenesis are shown in Fig. 1. Most of these genes are absent from nonpathogenic *E. coli* although several occur in other pathogenic

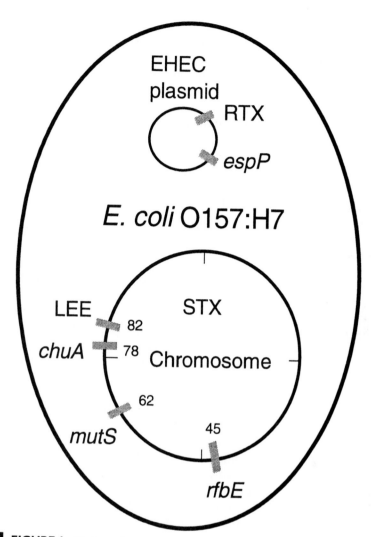

FIGURE 1 Virulence determinants of *E. coli* O157:H7. Chromosomal loci include LEE (locus of enterocyte effacement) inserted at *selC*; *chuA* (*E. coli* heme utilization gene); *mutS*; *rfb* gene cluster, specifies O-antigen; STX, Shiga toxin genes; RTX, repeat in toxin enterohemolysin; *espP*, excreted serine protease genes on 90-kbp EHEC plasmid. The genomic location of the Stx prophages has not been reported.

groups of *E. coli* where, in combination with additional factors, they contribute to different types of disease. For example, enteroaggregative *E. coli* are a pathogenic group of bacteria that cause persistent diarrhea in young children primarily in developing countries. These *E. coli* elaborate several enterotoxins including a 38-amino acid heat-stable toxin (EAST 1) that stimulates intestinal secretion in the rabbit ileum (Savarino *et al.*, 1991). The toxin gene has been detected in EHEC strains, so EAST 1 toxin expression potentially has a role in the diarrhea elicited by *E. coli* O157:H7 infection (Savarino *et al.*, 1996).

1. Shiga Toxins

The first characteristic of *E. coli* O157:H7 implicated in pathogenesis was a potent cytotoxic effect produced by the bacteria on Vero cells in tissue culture, an effect that could be neutralized by polyclonal antisera to Shiga toxin (O'Brien *et al.*, 1983). It is now known that *E. coli* O157:H7 elaborate one or two cytotoxins which act to inhibit protein synthesis in eukaryotic cells. These cytotoxins belong to the Shiga toxin (Stx) family, formerly referred to as Shiga-like toxins or Verotoxins (Calderwood *et al.*, 1996.)

Members of the Shiga toxin family have a common operon structure comprising two genes that encode the A and B polypeptide subunits (O'Brien and Holmes, 1987; O'Brien *et al.*, 1992). The mature holotoxin is composed of a single A subunit and five B subunits. The enzymatic A subunit cleaves the N-glycoside bond in a specific adenine residue of the 28 S rRNA in the 60 S ribosomal subunit. The removal of the adenine residue inhibits the elongation factor-dependent binding of aminoacyl-tRNAs to the ribosome, and thus truncates peptide elongation, suppressing overall protein synthesis and eventually killing the cell. The receptor-binding B subunit binds to glycolipids on the eukaryotic cell surface. Although the precise role of Shiga toxins in the pathogenicity of O157:H7 strains is not completely understood, it is clear that Stx damages vascular endothelial cells in certain organs and contributes to the extraintestinal complications of O157:H7 infection.

Escherichia coli O157:H7 often produces two antigenically distinct types of Shiga toxins, Stx1 and Stx2, which have about 60% sequence similarity. The Stx genes of O157:H7 are encoded by bacteriophages and have been transferred to nontoxigenic strains in laboratory conditions (Newland *et al.*, 1985; O'Brien *et al.*, 1984; Strockbine *et al.*, 1986). The Stx-converting phages are members of a diverse family of lambda-like phages that are widespread in nature (O'Brien and Holmes, 1987). In addition to O157:H7, strains of many different *E. coli* serotypes produce Shiga toxins (Karmali, 1989), suggesting that bacteriophages have disseminated Stx genes in the *E. coli* population in nature. Related cytotoxins have also been discovered in *Citrobacter freundii*, indicting that Stx genes can spread among different bacterial species (Schmidt *et al.*, 1993).

2. Locus of Enterocyte Effacement Pathogenicity Island

Escherichia coli O157:H7, like the enteropathogenic *E. coli* (EPEC) that cause infantile diarrhea, have the ability to disrupt the intestinal epithelium by intimately adhering to enterocytes (Levine, 1987). The close attachment of bacteria destroys the microvilli and creates a characteristic histopathology referred to as attaching and effacing (AE) lesions (Donnenberg and Kaper, 1992; Law, 1994; Moon *et al.*, 1983; Tesh and O'Brien, 1992). The ability to cause AE lesions is encoded by a 35-kb segment of DNA designated as the locus of enterocyte effacement or LEE (McDaniel *et al.*, 1995). In both O157:H7 and EPEC E2348/69, LEE is inserted at the selenocysteine tRNA gene (*selC*) at 82.6 minutes on the K-12 chromosome (Fig. 1) (McDaniel *et al.*, 1995). Interestingly, in unrelated *E. coli* strains, the exact *selC* site of insertion contains different DNA inserts including, in one case, a retron phage (Jin *et al.*, 1992) and, in another case, a block of virulence genes in a uropathogenic clone (Blum *et al.*, 1994).

The LEE encodes a variety of proteins whose functions in pathogenesis are now under intense scrutiny. The first gene *(eaeA)* discovered in LEE encodes intimin, a 94-kDa outer membrane protein that is required for production of AE lesions (Donnenberg and Kaper, 1992). Intimin is produced by both EPEC and EHEC strains, and it is homologous to proteins of *Citrobacter rodentium* and *Hafnia alvei* (Schauer and Falkow, 1993) and to the invasins of *Yersinia pseudotuberculosis* and *Yersinia enterocolitica* (Isberg *et al.*, 1987; Young *et al.*, 1990). In EPEC strain E2348/69, LEE contains genes that specify a type III secretion system *(sep* genes) which is necessary for the export of specific secreted proteins (Jarvis *et al.*, 1995; Kenny and Finlay, 1995). LEE also encodes an EPEC secreted protein, EspB, that binds to eukaryotic cells, activates signal transduction pathways in host cells, and induces tyrosine phosphorylation of host proteins (Donnenberg *et al.*, 1993; Kenny and Finlay, 1995). McDaniel and Kaper (1997) confirmed the role of LEE-encoded proteins in pathogenesis by cloning and transferring the entire pathogenicity island to an avirulent *E. coli* K-12 strain. The recipient strain gained the ability to secrete EPEC virulence proteins, induce host signal transduction pathways, and form AE lesions in cultured epithelial cells (McDaniel and Kaper, 1997).

Although LEE has been mainly studied in EPEC strain E2348/69, evidence suggests that it functions similarly in EHEC O157:H7. Jarvis and Kaper (1996) have shown that proteins secreted by O157:H7 react with rabbit antiserum raised against EPEC secreted proteins and with antiserum from a patient infected with O157:H7. A *sepB* mutant of O157:H7 strain 86-24 did not secrete smaller (<40 kDa) proteins and was complemented by an intact EPEC *sepB* gene (Jarvis and Kaper, 1996). In addition, a 37-kDa protein secreted by O157:H7 is recognized by antiserum to EPEC EspB and has the same four terminal amino acids. There is also a secreted 24-kDa polypeptide that may be homologous to EspA (Jarvis and Kaper, 1996). Together these results suggest that LEE encodes the molecular machinery necessary for AE lesion formation by *E. coli* O157:H7.

3. Enterohemolysin and the EHEC Plasmid

Escherichia coli O157:H7, as well as other serotypes (e.g., O26:H11 and O111:H8) associated with hemorrhagic colitis and hemolytic uremic syndrome, have a 90-kb plasmid (the EHEC plasmid) that carries genes that may encode other virulence factors (Levine *et al.*, 1987). Genes for an enterohemolysin (Bauer and Welch, 1996; Schmidt *et al.*, 1995, 1996), catalase peroxidase *(katP)* (Brunder *et al.*, 1996), and a secreted serine protease *(espP)* (Brunder *et al.*, 1997) have been characterized on the EHEC plasmid (Fig. 1), although at this point the precise role of these genes, if any, in EHEC pathogenesis is unknown.

The enterohemolysin belongs to the RTX (repeat in toxin) family of hemolysins but differs in its target cell specificity and lytic activity from the prototypic *E. coli* α-hemolysin (Hly) associated with extraintestinal infections (Bauer and Welch, 1996). The EHEC plasmid is not essential for the full expression of virulence by O157:H7 strains in gnotobiotic pigs (Tzipori *et al.*, 1987), but its presence is strongly correlated in strains of serotypes O157:H7, O26:H11, and O111:H8 that have been associated with hemorrhagic colitis (Levine *et al.*, 1987).

The EHEC plasmid also carries *espP*, a gene encoding a 104-kDa extracellular protein (EspP) that functions as a serine protease (Brunder *et al.*, 1997). EspP is homologous to the IgA1 proteases of the pathogenic *Neisseria* and *Haemophilus influenzae* and to the similar-sized EspC protein of EPEC. EspP cleaves both pepsin and human coagulation factor V *in vitro*. Immunoblot analysis of patient sera indicates that this protease is immunoreactive and presumably expressed and secreted during EHEC infections (Brunder *et al.*, 1997). Brunder *et al.* (1997) suggest that EspP degradation of human coagulation factor V contributes to the mucosal hemorrhaging associated with O157:H7 infection and aids in the dissemination of the organism.

4. Heme Receptor

Iron is a limiting resource for bacterial growth, and a variety of iron transport systems have evolved in different pathogenic bacteria. Mills and Payne (1995) identified and characterized a heme utilization locus *(shuA)* in *Shigella dysenteriae* and detected a homologous gene in *E. coli* O157:H7 by hybridization. The O157:H7 gene, called *chuA (E. coli* heme utilization gene), encodes a 69-kDa outer membrane protein that is synthesized in response to low iron conditions (Torres and Payne, 1997). Sequences homologous to *chuA* were not detected in other *Shigella* species or in nonpathogenic strains of *E. coli*. However, *chuA* sequences were found in O55:H7 strains and localized to 78 minutes (Fig. 1) on the chromosome (S. Payne, 1997, *personal communication)* suggesting that the most recent common ancestor of O157:H7 and O55:H7 produced the *chuA* heme receptor (see beyond).

Law and Kelly (1995) showed that heme and hemoglobin stimulated the *in vitro* growth of O157:H7 strains relative to non-O157 strains. The ability of O157:H7 to utilize heme or hemoglobin could promote rapid multiplication and allow the bacteria to invade and colonize new sites within a host (Torres and Payne, 1997).

5. Genetic Changes in the Vicinity of *mutS*

LeClerc *et al.* (1996) reported an unusual genomic region from an O157:H7 strain that contains a novel DNA sequence (~2750 bp) in the intergenic region between *mutS* and *rpoS* compared to the 6098 bp of DNA found in *E. coli* K-12 (Fig. 1). The novel intergenic sequence is found in O157:H7 strains with normally low mutation rates, as well as in closely related O55:H7 strains (LeClerc *et al.*, 1996). In contrast, O157:H7 mutator strains, those with elevated spontaneous mutation rates, have large deletions in the intergenic region which may influence the expression of *mutS* (LeClerc *et al.*, 1996).

6. O157 Somatic Antigen and the *rfb* Region

The genes involved in the synthesis of the lipopolysaccharide (LPS) O-antigen occur together in the *rfb* gene cluster at 45 minutes on the *E. coli* chromosome (Fig. 1). *Escherichia coli* is a highly antigenically variable species, and strains with particular O-antigens are prevalent in certain diseases of humans and domestic animals (Ørskov *et al.*, 1977). To begin to elucidate the role of the O157 O-antigen in disease, Bilge *et al.* (1996) used TnphoA mutagenesis to identify mutants deficient in the expression of O157 LPS. A mutant of

O157:H7 strain 86-24 that did not express O-antigen had the transposon inserted into an open reading frame (364 codons) in the *rfb* region that is 54% identical in amino acid sequence to perosamine synthetase (encoded by *rfbE*) of *Vibrio cholerae* O1. This homology may explain why humans immunized with *V. cholerae* O1 develop antibodies that cross-react with O157 LPS (Bilge *et al.*, 1996).

The O157 *rfbE* region was detected by hybridization in the chromosomal DNA of 30 O157 strains of a variety of H types including H antigens 3, 12, 16, 38, 43, and 45, as well as an Stx-producing nonmotile O157 strain (Bilge *et al.*, 1996). The homologous fragment was not found in O55:H7 strains, the closest non-O157 relative to O157:H7 strains (see below). Interestingly, the *rfbE* mutant O157 strain exhibited enhanced levels of adhesion to HeLa cells, suggesting that O157 LPS interferes with adherence to epithelial cells (Bilge *et al.*, 1996). The hypothesis that O-somatic antigens play critical functions in pathogenesis has long been suggested, and there is clearly a need for more investigation to elucidate the molecular basis of the contribution of O-somatic antigens in virulent *E. coli*.

The absence of the O157 *rfbE* region from O55:H7 supports the hypothesis that a lateral transfer and recombination event involving the part of the *rfb* complex and the adjacent *gnd* locus was a crucial step in the differentiation of O157:H7 from an O55:H7-like ancestor (Whittam, 1995).

B. Stepwise Evolution of *Escherichia coli* O157:H7

To understand the steps in the evolution of the virulence of *E. coli* O157:H7, we have used techniques from molecular evolutionary genetics (Nei, 1987) to study the genetic relationships among O157:H7 isolates and strains of other serotypes implicated in diarrheal disease (Whittam *et al.*, 1988, 1993; Whittam and Wilson, 1988). Our first studies, based on multilocus enzyme electrophoresis (Selander *et al.*, 1986), showed that O157:H7 isolates from recent epidemics of hemorrhagic colitis in North America belonged to a clone complex that was not closely allied to Stx-producing strains of other *E. coli* serotypes (Whittam *et al.*, 1988), many of which produce a clinically similar form of bloody diarrhea (Tzipori *et al.*, 1986, 1987). The O157:H7 clone was also found to be only distantly related, based on electrophoretic type (ET), to other O157 strains associated with enteric infections in animals (Whittam and Wilson, 1988). Further comparisons of O157:H7 strains to a diverse collection of isolates of serotypes associated with infectious diarrheal disease revealed that 72% of the isolates belong to 15 major ETs each of which marks a bacterial clone with a widespread geographic distribution (Whittam *et al.*, 1993).

Genetically, the O157:H7 clone is most closely related to an EPEC strain of serotype O55:H7, a non-Stx-producing clone associated with infantile diarrhea (Whittam *et al.*, 1993). Bacteria of the O55:H7 clone isolated from disease cases have the intimin *(eaeA)* gene but usually do not have the EPEC plasmid and do not display the localized adherence phenotype typical of EPEC (Bokete *et al.*, 1997; Rodrigues *et al.*, 1996; Whittam and McGraw, 1996). Interestingly, characterization of O55 strains isolated from children with diarrhea in Brazil showed that O55:H7 strains carried not only the intimin gene, but in some cases displayed novel combinations of other adherence factors (Rodrigues *et al.*, 1996). The results suggested that O55:H7 is a pathogenic clone with a propensity to acquire new virulence factors in nature (Rodrigues *et al.*, 1996).

The clonal nature of *E. coli* O157:H7 has facilitated its phenotypic identification (Feng, 1995). *Escherichia coli* O157:H7 strains do not ferment sorbitol rapidly (in contrast to about 95% of wild *E. coli* strains), which provides a useful phenotype for screening of O157:H7 strains based on sorbitol indicator medium (Farmer and Davis, 1985). Isolates of O157:H7 also do not exhibit β-glucuronidase (GUD) activity, although the *uidA* gene that encodes GUD is intact (Feng *et al.*, 1991), and they are thus consistently negative in the fluorogenic 4-methylumbelliferyl β–D-glucuronide assay (Doyle and Schoeni, 1984).

In 1990, Karch and colleagues discovered a nonmotile (H−) O157 clone that was implicated in an outbreak of hemolytic uremic syndrome in Upper Bavaria, Germany (Karch *et al.*, 1990). Isolates of the O157:H− clone possess the intimin gene *(eaeA)* and produce Stx2; however, phenotypically they are dissimilar from typical O157:H7 in that they are sorbitol-positive (Sor+). The Sor+ O157:H− strains had nearly identical *Xba*I restriction patterns in pulsed-field gel electrophoresis and were strikingly different from the patterns for sorbitol-negative O157:H7 (Karch *et al.*, 1993). From these observations, Karch and co-workers (1993) hypothesize that the sorbitol-fermenting O157: H− is a new clone with similar virulence properties to those of O157:H7.

Comparison of the multilocus enzyme genotypes of the Sor+ O157:H7 strains from Germany reveals that this distinct clone is part of the O157:H7 clone complex (Table I). The O157:H− clone can be distinguished from O157: H7 strains by mobility variants of two enzymes. The bacteria of this clone are similar to O55:H7 in that they ferment sorbitol and express β-glucuronidase (Feng *et al.*, 1997).

The clonal analysis suggests that the pathogenic O157:H7 emerged when an O55:H7-like progenitor, already possessing a mechanism for adherence to intestinal cells, acquired secondary virulence factors (Shiga toxins and the EHEC plasmid) via horizontal transfer and recombination. The working model is outlined in Fig. 2 as a branching phylogenetic tree with ancestral states to the left and contemporary states to the right. The evolutionary model begins at time T_1 when an ancestral *E. coli* acquired the LEE pathogenicity island into the *selC* site. This event gave rise in a single evolutionary step to a pathogen with the ability to adhere intimately and cause attaching–effacement. This attribute alone may be sufficient for bacteria to cause diarrheal disease in infants, as is the case of the contemporary EPEC O55:H7 clone.

In the time between T_1 and T_2, other transfer events occurred including the

TABLE I   Bacteria of the *Escherichia coli* O157:H7 Complex

Serotype	Shiga toxins	Common phenotype
O157:H7	*stx1, stx2*	Sor−, GUD−
O157:H−	*stx2*	Sor+, GUD+
O55:H7	None	Sor+, GUD+

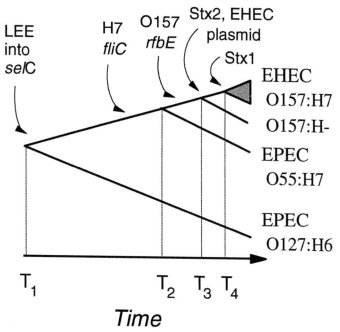

FIGURE 2 Evolutionary model of the stepwise acquisition of virulence factors of *E. coli* O157:H7.

evolution of the flagellin-encoding *fliC* gene and the *rfbE* gene associated with expression of O157 somatic antigen. At this point, the chuA heme receptor gene, which occurs in the O157:H7 complex, was most likely acquired by lateral transfer and recombination. In the transition from T_2 to T_3, the lineage acquired the *rfbE* gene associated with expression of O157 somatic antigen which occurs in all of the descendants. In this same period, this ancestor may have acquired an Stx2 gene and the EHEC plasmid which is characteristic of the derived O157:H7 lines. At time T_3, the branches of the Stx-2 producing O157 EHEC split. One branch gave rise to the nonmotile O157:H− clone discovered in Germany. This clone retained the ability to ferment sorbitol and express β-glucuronidase but lost the expression of H7 flagellar antigen (Feng *et al.*, 1997). The other branch lead to the sorbitol-negative O157:H7 clones that are common and widespread in North America and Europe. One O157:H7 clone acquired a second Stx gene (Stx1), and this double toxin producer was in particularly high frequency in the northwestern United States in the early 1980s (Tarr *et al.*, 1989).

The model presented for the evolution of the O157:H7 complex (Fig. 2) is testable in that it makes predictions about the order and timing of clonal divergence which can be tested by examining sequence divergence of individual genes. Comparative sequencing of the intimin *(eaeA)* and flagellin *(fliC)* genes in our laboratory suggests independent support for this model. The *eaeA* sequences of the O157:H7 complex are very closely related (Whittam and McGraw, 1996), with O55:H7 differing at only two codon positions from the other sequences (Table II). Mutations in the flagellin of the members of the

TABLE II   Polymorphic Codons among Intimin (*eaeA*) and Flagellin (*fliC*) Sequences in the *Escherichia coli* O157:H7 Clone Complex

Serotype (strain)	Codon in *eaeA*		Codon in *fliC*			
	911	928	44	45	235	576
O157:H7 (3a)	Arg	Gly	Asp	Ala	Asp	Arg
	CGT	GGG	GAC	GCC	GAT	CGA
O157:H7 (Cl8)	Arg	Gly	Asp	Ala	Asp	Arg
	CGT	GGG	GAC	GCC	GAT	CGA
O157:H− (3f)	Arg	Gly	His	Ser	Asn	Arg
	CGT	GGG	CAC	TCC	AAT	CGA
O55:H7 (5d)	Ser	Arg	Asp	Ala	Asn	Pro
	AGT	AGG	GAC	GCC	AAT	CCG

O157:H7 complex reflect the close relationship of O55:H7 to O157:H7, with only two codon differences. The nonmotile O157:H− *fliC* occupies an intermediate state sharing the ancestral Asn at codon 235 with O55:H7 and sharing the derived Arg at codon 576 with O157:H7. In addition, the *fliC* from O157:H− has two distinct amino acid changes that occurred at conserved positions in the protein. The presence of these amino acid replacements suggests that selection on the gene has been relaxed as a result of the long-term loss of flagellar expression.

C. Adaptation and Emergence

Most of the knowledge of the epidemiology of *E. coli* O157:H7 has accumulated from outbreak investigations and studies of the prevalence of the organism in the bovine reservoir (Armstrong *et al.*, 1996; Griffin, 1995). The various routes of transmission of the organism into humans is summarized in Fig. 3. *Escherichia coli* O157:H7 has been isolated from both healthy and ill cattle, although generally it is not pathogenic to adult cattle. Numerous studies have isolated the organism from healthy cattle, indicating that it is generally a commensal organism (Armstrong *et al.*, 1996; Griffin, 1995). The prevalence has been estimated to range from less than 1 to 10%. It is widespread in cattle geographically and is more common in calves than in older cattle. In addition, other ruminants, such as sheep, harbor *E. coli* O157:H7. In one study of sheep in Idaho, there was seasonal variation with no fecal shedding of the organism in November and a high prevalence of shedding (31%) in June (Kudva *et al.*, 1996).

Outbreak investigations have demonstrated that *E. coli* O157:H7 can be transmitted by food, raw milk, water, and by direct person-to-person spread (Armstrong *et al.*, 1996; Griffin, 1995). Outbreaks of foodborne illness have been most often traced to undercooked contaminated ground meat although cross-contamination of other foods by contaminated meat has occurred (Fig. 4). From the first outbreaks in the early 1980s it was suspected that O157:H7 has a low infectious dose (Neill, 1994), and the observation of secondary infections caused by person-to-person spread and transmission by water supports this

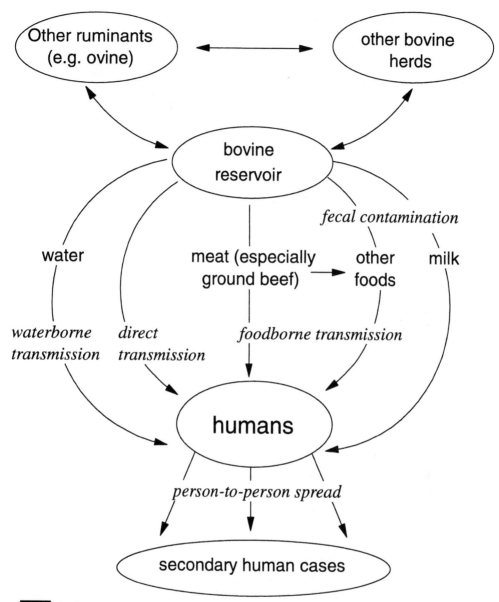

FIGURE 3   Reservoirs and modes of transmission of E. coli O157:H7. Modified from Armstrong et al. (1996).

idea (Griffin, 1995). In addition, O157:H7 has been implicated in several out-breaks in which acidic foods served as the vehicles for spread (Armstrong *et al.*, 1996), which has drawn attention to the unusual acid tolerance of this organism (see below).

One point to emphasize in the diagram in Fig. 3 is that E. *coli* O157:H7 must adapt to two distinct habitats. The primary habitat is the enteric environment of cattle, other ruminants, and humans (enclosed in circles in Fig. 3). The

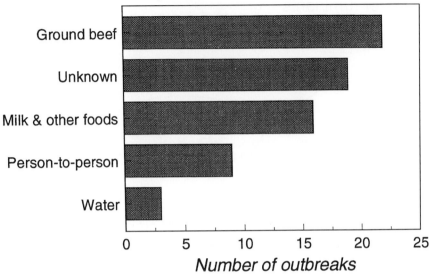

FIGURE 4   Outbreaks of diarrheal disease caused by *E. coli* O157:H7 in the United States from 1982 to 1994 (Armstrong *et al.,* 1996).

secondary habitat is the water, sediment, soil, and food of the environment external to the mammalian hosts. These habitats differ markedly in their biotic and abiotic features. Thus to successfully spread, *E. coli* must be able to survive a sojourn in the secondary habitat in order to colonize a new host in the reservoir or to infect a human.

1. *Escherichia Coli* O157:H7 Are Unusually Acid Tolerant

Acid tolerance refers to the ability of microorganisms to survive a defined period of time at a pH that is lethal. Waterman and Small (1996) found that 76% of 38 O157:H7 strains had greater than 10% survival after a 2-hr exposure to pH 2.5. Under laboratory conditions, the percentage log survival in synthetic gastric fluid (pH 1.5) for *E. coli* O157:H7 exceeds 95% for 1-hr acid exposures of stationary phase cultures (Arnold and Kaspar, 1995). Although there is some variation among O157:H7 strains, the average survival is greater than that observed in other pathogenic groups of *E. coli* and greatly exceeds nonpathogenic controls (Fig. 5). In another experiment, Leyer *et al.* (1995) found that *E. coli* O157:H7 cells that were exposed to acidic conditions (pH 5.0) for short periods were better able to survive in acidic food than unexposed cells.

Escherichia coli O157:H7 has enhanced survival in extreme acidic conditions at low temperatures (Clavero and Beuchat, 1996; Ragubeer *et al.,* 1995; Zhao *et al.,* 1993). For example, the pathogen has greater survival in processed salami stored at 5°C than salami stored at 20°C, regardless of the physiological conditions of the cells prior to inoculation (Clavero and Beuchat, 1996). It also

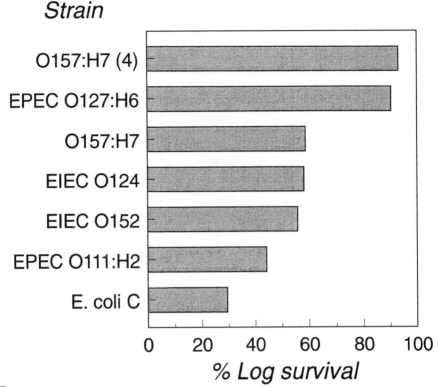

Strain

FIGURE 5 Level of acid resistance among different pathogenic strains of *E. coli*. Data from Arnold and Kaspar (1995).

can survive longer in salad dressing (Ragubeer *et al.*, 1995) and apple juice (pH <4.0) (Zhao *et al.*, 1993) when stored at lower temperatures.

How did extreme acid tolerance evolve in *E. coli* O157:H7? *Escherichia coli* in general has multiple genetic systems that respond to physical and chemical challenges, and a confer resistance to low and lethal pH levels (Slonczwewski and Foster, 1996). Some of the genetic systems also provide cross-protection against other environmental stresses. Homologous systems are found in *Salmonella enterica* (Slonczwewski and Foster, 1996). There are two pathways which impart resistance to normally lethal acidic conditions and are induced by prior exposure to moderate acidity (Slonczwewski and Foster, 1996; Small *et al.*, 1994). One system is induced at log phase and the other is induced in stationary phase. A third set of genes involves σ^{38}, the sigma factor encoded by *rpoS*, which is induced at stationary phase independent of pH. The *rpoS* system regulates a multitude (>50) of protective proteins that enhance resistance to a variety of extreme environmental conditions. Both inducible acid tolerance systems and *rpoS* regulated proteins have been shown to be important for the survival of O157:H7 at low pH and in acidic foods (Benjamin and Datta, 1995; Cheville *et al.*, 1996; Leyer *et al.*, 1995; Small *et al.*, 1994).

It is not surprising that enteric pathogens have complex mechanisms for tolerating extremes in pH within a host. Ingested bacteria encounter low pH in the stomach (pH <3.0), and within phagosomes and phagolysosomes of intestinal epithelial cells or macrophages. They are also exposed to alkaline secretions of the pancreas. What is problematic is why such unusually extreme acid tolerance has evolved in the recently emergent *E. coli* O157:H7.

2. The Rumen Hypothesis

One hypothesis to account for the unusual acid tolerance of *E. coli* O157: H7 is that it is an adaptation for this organism to survive in the rumen of cows (Armstrong *et al.*, 1996). Ruminants, including cows, sheep, goats, deer, and other herbivores, possess a four-chambered stomach. Most of the fermentation occurs in the largest chamber, the rumen, where special bacteria and protozoans break down complex polysaccharides into carbon dioxide, methane, and short-chain (volatile) fatty acids. The pH of the rumen varies from a low of around pH 5 for a well-fed animal, when the concentration of fatty acids is high, to a more neutral environment in a starving animal when fatty acid concentration is lower. It may be that *E. coli* O157:H7 has spread through ruminant populations because it has adapted and specialized to exploit the rumen environment.

3. The Acid Rain Hypothesis

A second hypothesis is that the acidification of natural waters, soil, and agricultural areas by acid rain has imposed a selective pressure on *E. coli* in the secondary environment that favors the spread of acid-tolerant mutants. Acid rain is precipitation that has been polluted by emissions of sulfur dioxide (SO_2) and oxides of nitrogen (NO_x) in the atmosphere. These oxides react with water vapor and form sulfuric acid and nitric acid. These components fall to earth in the form of rain or snow, dry deposition of SO_2, and sulfuric or nitrogen salts. Although SO_2 and NO_x are naturally found in the atmosphere, the combustion of fossil fuels in industry and automobiles has dramatically increased the presence of these substances.

III. LESSONS FROM *ESCHERICHIA COLI* O157:H7

In summary, we hypothesize that the emergence of *E. coli* O157:H7 as a new pathogen was dependent on two components. The first component involved the evolution of a highly virulent mechanism of pathogenesis. The mechanism built up in steps, first with the acquisition of LEE and the attaching–effacing phenotype to create an EPEC-like ancestor, and then with the acquisition of the EHEC plasmid and phage-encoded Shiga toxins to give rise to a novel pathogen that causes a new disease. The second, and equally important component, is the set of factors that favored the increase in frequency and spread of the new pathogen. The ability to cause diarrhea at various stages would enhance transmission and maintain the pathogen against competition from nonpathogenic strains in the *E. coli* population. Acid tolerance is likely to have been one of the determining factors that benefitted survival of bacteria in the external environment and bovine reservoir, especially with the increased acidic pollution caused by acid rain. Moreover, it is these very features of the organism that allow it both to survive the extreme conditions of food processing and to circumvent the host gastric acid defense.

What is the potential of bacteria with the same virulence factors as *E. coli* O157:H7 to evolve acid resistance and thus gain the ability to cause massive outbreaks of foodborne disease? How quickly can acid resistance or other aspects of durability evolve under direct selective pressure in extreme environments? To what extent does acid resistance evolve as a correlated or indirect response to selection for survival in harsh environmental conditions? Are there trade-offs between the adaptations for durability in the external environment versus growth rate in the enteric environment? To understand the conditions that promote emergence, future investigations need to focus not only on molecular mechanisms of virulence but also on the ecology and evolution of potential pathogens in their natural environments. For example, Ewald (1996) has proposed that measuring aspects of durability may serve as valuable indicators of the degree of virulence for directly transmitted pathogens. Finally, it should be noted that changes in the external environment imposed by human activities, such as acid pollution, are creating new selective pressures that foster the emergence of durable foodborne and waterborne pathogens.

REFERENCES

Anderson, R. M., and May, R. M. (1982). "Population Biology of Infectious Diseases." Springer-Verlag, New York.

Anderson, R. M., and May, R. M. (1991). "Infectious Diseases of Humans: Dynamics and Control." Oxford Univ. Press, New York.

Armstrong, G. L., Hollingsworth, J., and Morris, J. G., Jr. (1996). Emerging foodborne pathogens: *Escherichia coli* O157:H7 as a model of entry of a new pathogen into the food supply of the developed world. *Epidemiol. Rev.* **18**, 29–51.

Arnold, K. W., and Kaspar, C. W. (1995). Starvation- and stationary-phase-induced acid tolerance in *Escherichia coli* O157:H7. *Appl. Environ. Microbiol.* **61**, 2037–2039.

Bauer, M. E., and Welch, R. A. (1996). Characterization of an RTX toxin from enterohemorrhagic *Escherichia coli* O157:H7. *Infect. Immun.* **64**, 167–175.

Benjamin, M. M., and Datta, A. R. (1995). Acid tolerance of enterohemorrhagic *Escherichia coli*. *Appl. Environ. Microbiol.* **61**, 1669–1672.

Bilge, S. S., Vary, J. C., Dowell, S. F., and Tarr, P. I. (1996). Role of the *Escherichia coli* O157:H7 O side chain in adherence and analysis of an *rfb* locus. *Infect. Immun.* **64**, 4795–4801.

Blum, G., Ott, M., Lischewski, A., Ritter, A., Imrich, H., Tschäpe, H., and Hacker, J. (1994). Excision of large DNA regions termed pathogenicity islands from tRNA-specific loci in the chromosome of an *Escherichia coli* wild-type pathogen. *Infect. Immun.* **62**, 606–614.

Bokete, T. N., Whittam, T. S., Wilson, R. A., Clausen, C. R., O'Callahan, C. M., Moseley, S. L., Fritsche, T. R., and Tarr, P. I. (1997). Genetic and phenotypic analysis of *Escherichia coli* with enteropathogenic characteristics isolated from Seattle children. *J. Infect. Dis.* **175**, 1382–1389.

Bonhoeffer, S., and Nowak, M. A. (1994). Intra-host versus inter-host selection: Viral strategies of immune function impairment. *Proc. Natl. Acad. Sci. U.S.A.* **91**, 8062–8066.

Brunder, W., Schmidt, H., and Karch, H. (1996) KatP, a novel catalase–peroxidase encoded by the large plasmid of enterohaemorrhagic *Escherichia coli* O157:H7. *Microbiology* **142**, 3305–3315.

Brunder, W., Schmidt, H., and Karch, H. (1997). EspP, a novel extracellular serine protease of enterohemorrhagic *Escherichia coli* O157:H7 cleaves human coagulation factor V. *Mol. Microbiol.* **24**, 767–778.

Bull, J. J. (1994). Virulence. *Evolution* **48**, 1423–1437.

Calderwood, S. B., Acheson, D. W. K., Keusch, G. T., Barrett, T. J., Griffin, P. M., Strockbine, N. A., Swaminathan, B., Kaper, J. B., Levine, M. M., Kaplan, B. S., Karch, H., O'Brien, A. D., Obrig, T. G., Takeda, Y., Tarr, P. I., and Wachsmuth, I. K. (1996). Proposed new nomenclature for SLT (VT) family. *ASM News* **62**, 118–119.

Cheville, A. M., Arnold, K. W., Burchrieser, C., Cheng, C.-M., and Kaspar, C. W. (1996). *rpoS* regulation of acid, heat, and salt tolerance in *Escherichia coli* O157:H7. *Appl. Environ. Microbiol.* **62**, 1822–1824.

Clavero, M. R. S., and Beuchat, L. R. (1996). Survival of *Escherichia coli* O157:H7 in broth and processed salami as influenced by pH, water activity, and temperature and suitability of media for its recovery. *Appl. Environ. Microbiol.* **62**, 2735–2740.

Cohen, J. E., and Newman, C. M. (1989). Host–parasite relations and random zero-sum games: The stabilizing effect of strategy diversification. *Am. Nat.* **133**, 533–552.

Donnenberg, M. S., and Kaper, J. B. (1992). Enteropathogenic *Escherichia coli*. *Infect. Immun.* **60**, 3953–3961.

Donnenberg, M. S., Yu, J., and Kaper, J. B. (1993). A second chromosomal gene necessary for intimate attachment of enteropathogenic *Escherichia coli* to epithelial cells. *J. Bacteriol.* **175**, 4670–4680.

Doyle, M. P., and Schoeni, J. L. (1984). Survival and growth characteristics of *Escherichia coli* associated with hemorrhagic colitis. *Appl. Environ. Microbiol.* **48**, 855–856.

Ewald, P. W. (1983). Host–parasite relations, vectors, and the evolution of disease severity. *Annu. Rev. Ecol. System.* **14**, 465–485.

Ewald, P. W. (1991). Waterborne transmission and the evolution of virulence among gastrointestinal bacteria. *Epidemiol. Infect.* **106**, 83–119.

Ewald, P. W. (1996). Guarding against the most dangerous emerging pathogens: Insights from evolutionary biology. *Emerging Infect. Dis.* **4**, 245–269.

Farmer, J. J., and Davis, B. R. (1985). H7 antiserum–sorbitol fermentation medium: A single tube screening medium for detecting *Escherichia coli* O157:H7 associated with hemorrhagic colitis. *J. Clin. Microbiol.* **22**, 620–625.

Feng, P. (1995). *Escherichia coli* serotype O157:H7: Novel vehicles of infection and emergence of phenotypic variants. *Emerging Infect. Dis.* **1**, 47–52.

Feng, P., Lum, R., and Chang, G. W. (1991). Identification of *uidA* gene sequences in β-D-glucuronidase-negative *Escherichia coli*. *Appl. Environ. Microbiol.* **57**, 320–323.

Feng, P., Lampel, K. A., Karch, H., and Whittam, T. S. (1997). Sequential genetic and phenotypic changes in the emergence of *Escherichia coli* O157:H7. *J. Infect. Dis.* submitted.

Flint, K. P. (1987). The long-term survival of *Escherichia coli* in river water. *J. Appl. Bacteriol.* **63**, 261–270.

Garnick, E. (1992). Parasite virulence and parasite–host coevolution: A reappraisal. *J. Parasitol.* **78**, 381–386.

Griffin, P. M. (1995). *Escherichia coli* O157:H7 and other enterohemorrhagic *Escherichia coli*. *In* "Infections of the Gastrointestinal Tract" (M. J. Blaser, P. D. Smith, J. I. Ravdin, H. B. Greenberg, and R. L. Guerrant, eds.), pp. 739–761. Raven, New York.

Griffin, P. M., and Tauxe, R. V. (1991). The epidemiology of infections caused by *Escherichia coli* O157:H7, other enterohemorrhagic *E. coli*, and the associated hemolytic uremic syndrome. *Epidemiol. Rev.* **13**, 60–98.

Hartl, D. L., and Dykhuizen, D. E. (1984). The population genetics of *Escherichia coli*. *Annu. Rev. Genet.* **18**, 31–68.

Isberg, R. R., Vorrhis, D. L., and Falkow, S. (1987). Identification of invasin: A protein that allows enteric bacteria to penetrate cultured mammalian cells. *Cell (Cambridge, Mass.)* **50**, 769–778.

Izumiya, H., Terajima, J., Wada, A., Inagaki, Y., Itoh, K. I., Tamura, K., and Watanabe, H. (1997). Molecular typing of enterohemorrhagic *Escherichia coli* O157:H7 isolates in Japan using pulse-field gel electrophoresis. *J. Clin. Microbiol.* **35**, 1675–1680.

Jarvis, K. G., and Kaper, J. B. (1996). Secretion of extracellular proteins by enterohemorrhagic *Escherichia coli* via a putative type III secretion system. *Infect. Immun.* **64**, 4826–4829.

Jarvis, K. G., Giron, J. A., Jerse, A. E., McDaniel, T. K., Donnenberg, M. S., and Kaper, J. B. (1995). Enteropathogenic *Escherichia coli* contains a putative type III secretion system necessary for the export of proteins involved in attaching and effacing lesion formation. *Proc. Natl. Acad. Sci. U.S.A.* **92**, 7996–8000.

Jin, T., Rudd, K. E., and Inouye, M. (1992). The *nlpA* lipoprotein gene is located near the *selC* tRNA gene on the *Escherichia coli* chromosome. *J. Bacteriol.* **174**, 3822–3823.

Karch, H., Wiss, R., Gloning, H., Emmich, P., Aleksic, S., and Bockemühl, J. (1990). Hämolytisch–urämisches syndrome bei Kleinkindern durch Verotoxin-produzierende *Escherichia coli. Dtsch. Med. Wochensch.* **115**, 489–495.

Karch, H., Böhm, H., Schmidt, H., Gunzer, F., Aleksic, S., and Heesemann, J. (1993). Clonal structure and pathogenicity of Shiga-like toxin-producing, sorbitol-fermenting *Escherichia coli* O157:H–. *J. Clin. Microbiol.* **31**, 1200–1205.

Karmali, M. (1989). Infection by Verocytotoxin-producing *Escherichia coli. Clin. Microbiol. Rev.* **2**, 15–38.

Kenny, B., and Finlay, B. B. (1995). Protein secretion by enteropathogenic *Escherichia coli* is essential for transducing signal to epithelial cells. *Proc. Natl. Acad. Sci. U.S.A.* **92**, 7991–7995.

Kudva, I. T., Hatfield, P. G., and Hovde, C. J. (1996). *Escherichia coli* O157:H7 in microbial flora of sheep. *J. Clin. Microbiol.* **34**, 431–433.

Law, D. (1994). Adhesion and its role in the virulence of enteropathogenic *Escherichia coli. Clin. Microbiol. Rev.* **7**, 152–173.

Law, D. and Kelly, J. (1995). Use of heme and hemoglobin by *Escherichia coli* 0157 and other shiga-like-toxin-producing *E. coli* serogroups. *Infect. Immun.* **63**, 700–702.

LeClerc, J. E., Li, B., Payne, W. L., and Cebula, T. A. (1996). High mutation frequencies among *Escherichia coli* and *Salmonella* pathogens. *Science* **274**, 1208–1211.

Lenski, R. E., and May, R. M. (1994). The evolution of virulence in parasites and pathogens: Reconciliation between two competing hypotheses. *J. Theor. Biol.* **169**, 253–265.

Levin, B. L. (1981). Periodic selection, infectious gene exchange, and the genetic structure of *E. coli* populations. *Genetics* **99**, 1–23.

Levin, B. R. (1996). The evolution and maintenance of virulence in microparasites. *Emerging Infect. Dis.* **2**, 93–102.

Levin, B. R., and Eden, C. S. (1990). Selection and evolution of virulence in bacteria: An ecumenical excursion and modest suggestion. *Parasitology* **100**, S103–S115.

Levin, B. R., Allison, A. C., Bremermann, H. J., Cavalli-Sforza, L. L., Clarke, B. C., Frentzel-Beyme, R., Hamilton, W. D., Levin, S. A., May, R. M., and Thieme, H. R. (1982). Evolution of parasites and hosts. *In* "Population Biology of Infectious Diseases" (R. M. Anderson and R. M. May, eds.), pp. 213–243, Springer-Verlag, New York.

Levine, M. M. (1987). *Escherichia coli* that cause diarrhea: Enterotoxigenic, enteropathogenic, enteroinvasive, enterohemorrhagic, and enteroadherent. *J. Infect. Dis.* **155**, 377–389.

Levine, M. M., Xu, J., Kaper, J. B., Lior, H., Prado, V., Tall, B., Nataro, J., Karch, H., and Wachsmuth, K. (1987). A DNA probe to identify enterohemorrhagic *Escherichia coli* of O157:H7 and other serotypes that cause hemorrhagic colitis and hemolytic uremic syndrome. *J. Infect. Dis.* **156**, 175–182.

Leyer, G. J., Wang, L.-L., and Johnson, E. A. (1995). Acid adaptation of *Escherichia coli* O157:H7 increases survival in acidic food. *Appl. Environ. Microbiol.* **61**, 3752–3755.

Lipsitch, M., and Nowak, M. A. (1995). The evolution of virulence in sexually transmitted HIV/AIDS. *J. Theor. Biol.* **174**, 427–440.

McDaniel, T. K., Jarvis, K. G., Donnenberg, M. S., and Kaper, J. B. (1995). A genetic locus of enterocyte effacement conserved among diverse enterobacterial pathogens. *Proc. Natl. Acad. Sci. U.S.A.* **92**, 1664–1668.

McDaniel, T. K., and Kaper, J. B. (1997). A cloned pathogenicity island from enteropathogenic *Escherichia coli* confers the attaching and effacing phenotype on *E. coli* K-12. *Mol. Microbiol.* **23**, 399–407.

Mason, T. G., and Richardson, G. (1981). A review of *Escherichia coli* and the human gut: Some ecological considerations. *J. Appl. Bacteriol.* **1**, 1–16.

May, R. M., and Anderson, R. M. (1983). Parasite–host coevolution. *In* "Coevolution" (D. J. Futuyma and M. Slatkin, eds.), pp. 186–206, Sinauer, Sunderland, Massachusetts.

May, R. M., and Nowak, M. A. (1995). Coinfection and the evolution of parasite virulence. *Proc. R. Soc. London, Ser. B: Biol. Sci.* **261**(1361), 209–215.

Milkman, R. (1997). Recombination and DNA sequence variation in *E. coli. In* "Ecology of Pathogenic Bacteria: Molecular and Evolutionary Aspects" (B. A. M. van der Zeijst, W. P. M. Hoekstra, J. D. A. van Embdam, and A. J. W. van Alphen, eds.), pp. 177–189. North-Holland, Amsterdam.

Mills, M., and Payne, S. M. (1995). Genetics and regulation of heme iron transport in *Shigella dysenteriae* and detection of an analogous system in *Escherichia coli* O157:H7. *J. Bacteriol.* **177**, 3004–3009.

Moon, H. W., Whipp, S. C., Argenzio, R. A., Levine, M. M., and Giannella, R. A. (1983). Attaching and effacing activities of rabbit and human enteropathogenic *Escherichia coli* in pig and rabbit intestines. *Infect. Immun.* **41**, 1340–1351.

Nei, M. (1987). "Molecular Evolutionary Genetics" Columbia Univ. Press, New York.

Neill, M. A. (1994). *E. coli* O157:H7 time capsule: What do we know and when did we know it? *Dairy, Food, and Environmental Sanitation* **14**, 374–379.

Newland, J. W., Strockbine, N. A., Miller, S. F., O'Brien, A. D., and Holmes, R. K. (1985). Cloning of Shiga-like toxin structural genes from a toxin converting phage of *Escherichia coli*. *Science* **230**, 179–181.

Nowak, M. A., and May, R. M. (1994). Superinfection and the evolution of parasite virulence. *Proc. R. Soc. London, Ser. B: Biol. Sci.* **255**(1342), 81–89.

O'Brien, A. D., and Holmes, R. K. (1987). Shiga and Shiga-like toxins. *Microbiol. Rev.* **51**, 206–220.

O'Brien, A. D., Lively, T. A., Chen, M. E., Rothman, S. W., and Formal, S. B. (1983). *Escherichia coli* O157:H7 strains associated with haemorrhagic colitis in the United States produce a *Shigella dysenteriae* 1 (Shiga) like cytotoxin. *Lancet* **1**, 702.

O'Brien, A. D., Newland, J. W., Miller, S. F., Holmes, R. K., Smith, H. W., and Formal, S. B. (1984). Shiga-like toxin-converting phages from *Escherichia coli* strains that cause hemorrhagic colitis or infantile diarrhea. *Science* **226**, 694–696.

O'Brien, A. D., Tesh, V. L., Donohue-Rolfe, A., Jackson, M. P., Olsnes, S., Sandvig, K., Lindberg, A. A., and Keusch, G. T. (1992). Shiga toxin: Biochemistry, genetics, mode of action, and role in pathogenesis. *Curr. Top. Microbiol. Immunol.* **180**, 65–94.

Ørskov, I., Ørskov, F., Jann, B., and Jann, K. (1977). Serology, chemistry, and genetics of O and K antigens of *Escherichia coli*. *Bacteriol. Rev.* **41**, 667–710.

Ragubeer, R. V., Ke, J. S., Campbell, M. L., and Meyer, R. S. (1995). Fate of *Escherichia coli* O157:H7 and other coliforms in commercial mayonnaise and refrigerated salad dressing. *J. Food Protect.* **58**, 13–18.

Rice, E. W., Johnson, C. H., Wild, D. K., and Reasoner, D. J. (1992). Survival of *Escherichia coli* O157:H7 in drinking water associated with waterborne disease outbreak of hemorrhagic colitis. *Lett. Appl. Microbiol.* **15**, 38–40.

Riley, L. W., Remis, R. S., Helgerson, S. D., McGee, H. B., Wells, J. G., Davis, B. R., Hebert, R. J., Olcott, E. S., Johnson, L. M., Hargrett, N. T., Blake, P. A., and Cohen, M. L. (1983). Hemorrhagic colitis associated with a rare *Escherichia coli* serotype. *N. Engl. J. Med.* **308**, 681–685.

Rodrigues, J., Scaletsky, I. C. A., Campos, L. C., Gomes, T. A. T., Whittam, T. S., and Trabulsi, L. R. (1996). Clonal structure and virulence factors in strains of *Escherichia coli* of the classic serogroup O55. *Infect. Immun.* **64**, 2680–2686.

Savageau, M. A. (1983). *Escherichia coli* habitats, cell types, and molecular mechanism of gene control. *Am. Nat.* **122**, 732–744.

Savarino, S. J., Fasano, A., Robertson, D. C., and Levine, M. M. (1991). Enteroaggregative *Escherichia coli* elaborate a heat-stable enterotoxin demonstrable in an in vitro rabbit intestinal model. *J. Clin. Invest.* **87**, 1450–1455.

Savarino, S. J., McVeigh, A., Watson, J., Cravioto, A., Molina, J., Echeverria, P., Bhan, M. K., Levine, M. M., and Fasano, A. (1996). Enteroaggregative *Escherichia coli* heat-stable enterotoxin is not restricted to enteroaggregative *E. coli*. *J. Infect. Dis.* **173**, 1019–1022.

Schauer, D. B., and Falkow, S. (1993). Attaching and effacing locus of a *Citrobacter freundii* biotype that causes transmissible murine colonic hyperplasia. *Infect. Immun.* **61**, 2486–2492.

Schmidt, H., Montag, M., Bockemuhl, J., Heesemann, J., and Karch, H. (1993). Shiga-like toxin II related cytotoxins in *Citrobacter freundii* strains from human and beef samples. *Infect. Immun.* **61**, 534–545.

Schmidt, H., Beutin, L., and Karch, H. (1995). Molecular analysis of the plasmid-encoded hemolysin of *Escherichia coli* O157:H7 strain EDL 933. *Infect. Immun.* **63**, 1055–1061.

Schmidt, H., Kernbach, C., and Karch, H. (1996). Analysis of the EHEC *hly* operon and its location on the physical map of the large plasmid of enterohaemorrhagic *Escherichia coli* O157:H7. *Microbiology* **142**, 907–914.

Selander, R. K., Caugant, D. A., Ochman, H., Musser, J. M., Gilmour, M. H., and Whittam, T. S.

(1986). Methods of multilocus enzyme electrophoresis for bacterial population genetics and systematics. *Appl. Environ. Microbiol.* **51**, 873–884.

Slonczwewski, J. L., and Foster, J. W. (1996). pH-regulated genes and survival at extreme pH. *In* "*Escherichia coli* and *Salmonella*: Cellular and Molecular Biology," (F. C. Neidhardt, R. Curtis III, J. L. Ingraham, E. C. C. Lin, K. B. Low, B. Magasanik, W. S. Reznikoff, M. Riley, M. Schaechter, and H. E. Umbarger, eds.), 2nd Ed., pp. 1539–1549. American Society for Microbiology, Washington, D.C.

Small, P., Blackenhorn, D., Welty, D., Zinser, E., and Slonczewski, J. L. (1994). Acid and base resistance in *Escherichia coli* and *Shigella flexneri*: Role of *rpoS* and growth pH. *J. Bacteriol.* **176**, 1729–1737.

Strockbine, N. A., Marques, L. R. M., Newland, J. W., Smith, H. W., Holmes, R. K., and O'Brien, A. D. (1986). Two toxin-converting phages from *Escherichia coli* O157:H7 strain 933 encode antigenically distinct toxins with similar biologic activities. *Infect. Immun.* **53**, 135–140.

Tarr, P. I. (1995). *Escherichia coli* O157:H7: Clinical, diagnostic, and epidemiological aspects of human infection. *Clin. Infect. Dis.* **20**, 1–10.

Tarr, P. I., Neill, M. A., Clausen, C. R., Newland, J. W., Neill, R. J., and Moseley, S. L. (1989). Genotypic variation in pathogenic *Escherichia coli* O157:H7 isolated from patients in Washington, 1984–1987. *J. Infect. Dis.* **159**, 344–347.

Tesh, V. L., and O'Brien, A. D. (1992). Adherence and colonization mechanisms of enteropathogenic and enterohemorrhagic *Escherichia coli*. *Microb. Pathog.* **12**, 245–254.

Torres, A. G., and Payne, S. M. (1997). Haem iron-transport system in enterohaemorrhagic *Escherichia coli* O157:H7. *Mol. Microbiol.* **23**, 825–833.

Tzipori, S., Wachsmuth, I. K., Chapman, C., Birner, R., Brittingham, J., Jackson, C., and Hogg, J. (1986). The pathogenesis of hemorrhagic colitis caused by *Escherichia coli* O157:H7 in gnotobiotic piglets. *J. Infect. Dis.* **154**, 712–716.

Tzipori, S., Karch, H., Wachsmuth, K. I., Robins-Browne, R. M., O'Brien, A. D., Lior, H., Cohen, M. L., Smithers, J., and Levine, M. M. (1987). Role of a 60-megadalton plasmid and Shiga-like toxins in the pathogenesis of infection caused by enterohemorrhagic *Escherichia coli* O157:H7 in gnotobiotic piglets. *Infect. Immun.* **55**, 3117–3125.

Waterman, S. R., and Small, P. L. C. (1996). Characterization of the acid resistance phenotype and *rpoS* alleles of Shiga-like toxin-producing *Escherichia coli*. *Infect. Immun.* **64**, 2808–2811.

Whittam, T. S. (1995). Genetic population structure and pathogenicity in enteric bacteria. *In* "Population Genetics of Bacteria" (S. Baumberg, J. P. W. Young, E. M. H. Wellington, and J. R. Saunders, eds.), pp. 217–245. Cambridge Univ. Press, Cambridge.

Whittam, T. S., and McGraw, E. A. (1996). Clonal analysis of EPEC serogroups. *Rev. Microbiol. Sao Paulo* **27**, 7–16.

Whittam, T. S., and Wilson, R. A. (1988). Genetic relationships among pathogenic *Escherichia coli* of serogroup O157. *Infect. Immun.* **56**, 2467–2473.

Whittam, T. S., Wachsmuth, I. K., and Wilson, R. A. (1988). Genetic evidence of clonal descent of *Escherichia coli* O157:H7 associated with hemorrhagic colitis and hemolytic uremic syndrome. *J. Infect. Dis.* **157**, 1124–1133.

Whittam, T. S., Wolfe, M. L., Wachsmuth, I. K., Ørskov, F., Ørskov, I., and Wilson, R. A. (1993). Clonal relationships among *Escherichia coli* strains that cause hemorrhagic colitis and infantile diarrhea. *Infect. Immun.* **61**, 1619–1629.

Young, V. B., Miller, V. L., Falkow, S., and Schoolnik, G. K. (1990). Sequence, localization and function of the invasin protein of *Yersinia enterocolitica*. *Mol. Microbiol.* **4**, 119–128.

Zhao, T., Doyle, M. P., and Besser, R. (1993). Fate of enterohemorrhagic *Escherichia coli* O157:H7 in apple cider with and without preservatives. *Appl. Environ. Microbiol.* **59**, 2526–2530.

6

THE REVIVAL OF GROUP A STREPTOCOCCAL DISEASES, WITH A COMMENTARY ON STAPHYLOCOCCAL TOXIC SHOCK SYNDROME

JAMES M. MUSSER
Section of Molecular Pathobiology
Department of Pathology
Baylor College of Medicine
Houston, Texas

RICHARD M. KRAUSE
Fogarty International Center
National Institutes of Health
Bethesda, Maryland

I. COMMENTARY ON TOXIC SHOCK SYNDROME CAUSED BY *STAPHYLOCOCCUS AUREUS*

A. Background and Brief Overview

Staphylococcus aureus is a gram-positive bacterium that is the cause of numerous diseases in humans and animals, including food poisoning, endocarditis, sepsis, scalded-skin syndrome, furunculosis, and economically important mastitis in cows and sheep. The pathogen also is responsible for human toxic shock syndrome (TSS), which will be discussed here. Although sporadic cases of TSS have been recorded in the literature for many years, the sudden occurrence of hundreds of cases in the early 1980s was totally unexpected (reviewed in Todd, 1988). The subsequent research defining its epidemiology, bacteriology, and pathophysiology sets the stage for a more extensive discussion of streptococcal toxic shock syndrome.

Human TSS was initially described in the recent literature in 1978 as a severe acute illness (characterized by high fever, erythematous rash, hypotension or shock, multiorgan involvement, and desquamation of the skin) of young children associated with infection with *S. aureus* (Todd *et al.*, 1978). Two years later, TSS occurred as a geographically widespread disease affecting mainly young, healthy, menstruating women, especially those using certain high absorbency tampons. Most vaginal isolates of *S. aureus* recovered from patients with TSS produced a chromosomally encoded toxin, eventually designated toxic shock syndrome toxin-1 (TSST-1) (Blomster-Hautamaa et al., 1986). Evidence has accumulated implicating TSST-1 as a major virulence factor in the pathogenesis of TSS in most patients. For example, almost all strains recovered from patients with menstrual TSS, which account for about 90% of the cases, synthesize TSST-1, whereas only 50–60% of isolates from nonmenstrual cases of TSS and 5–25% of strains causing other diseases produce this protein. When injected into animals, purified TSST-1 produces many signs and symptoms characteristic of human TSS. Moreover, isogenic strains of S. aureus with an inactivated TSST-1 coding gene fail to cause TSS-like disease in animals (de Azavedo et al., 1985; Sloane et al., 1991). These data and other research have unambiguously demonstrated that this toxin is causally involved in the pathogenesis of many cases of TSS.

B. Generally Applicable Lessons

Research on TSS also taught us several lessons that are generally applicable to many other emerging infectious diseases. For example, study of the medical bacteriology and evolutionary genetics of TSST-1-producing S. aureus strains revealed that the organisms responsible for most cases of TSS in females are members of a single distinctive clone designated as electrophoretic type (ET) 41 (Musser et al., 1990). This finding provides an explanation for the observation that most isolates recovered from patients with TSS share many phenotypic and genotypic traits (Todd, 1988), and is important new evidence in support of the hypothesis that a clone of a particular microbe is the cause of an outbreak or an epidemic. This idea was first developed to explain the repeated association of Escherichia coli strains expressing only a few of the many O and H antigen serotypes with outbreaks of infantile diarrhea in diverse geographic localities that occurred years apart (Orskov et al., 1976).

The study of TSS has also provided new information on the critical role of bacterial superantigens in mediating human disease. TSST-1 is a member of the class of structurally related molecules designated "superantigens" (reviewed in Thibodeau and Sekaly, 1995). TSST-1 and other superantigens bind to the major histocompatibility complex (MHC) class II molecules of antigen-presenting cells. Unlike conventional antigens, these molecules do not require proteolytic digestion. They stimulate proliferation of a large proportion of T cells in a Vβ-specific manner, and thereby profoundly alter immune system homeostasis. This T-cell proliferation may result in enhanced production of cytokines including several interleukins, tumor necrosis factors (TNF-α and -β), and interferon-γ, leading to a cascade of events that results in acute systemic illness, shock, and sometimes death. In addition to promoting T-cell proliferation, superantigens can also downregulate the immune response. Superantigen–T-cell interaction can induce anergy and/or deletion of these cells. Investigation of humans with TSS has shown selective expansion of T cells bearing particular Vβ regions of the T-cell receptor (TCR) (Choi et al., 1990).

Genetic studies of TSST-1-producing S. aureus underscored the important role of allelic variation in microbial virulence genes in mediating disease manifestations, and host susceptibility. For example, TSST-producing strains are a major cause of sheep mastitis. However, these isolates are genotypically distinct from the major human TSS clone (Musser and Selander, 1990), that is, they represent a separate and distinct phylogenetic lineage of the species. This observation suggested that sheep isolates have a TSST-1 gene (*tst-1*) allelic variant differing in nucleotide and amino acid sequence from that made by urogenital TSS-associated organisms. This view was subsequently borne out by the demonstration that the sheep isolates produce a toxin variant termed TSST-O (for ovine). TSST-O differs from TSST-1 by 9 amino acid residues, including 7 (T19A, A55T, T57S, T69I, Y80W, E132K, and I140T) over the 194-amino acid length of the mature protein (Lee *et al.*, 1992). Compared to TSST-1, TSST-O is only weakly mitogenic for rabbit or mouse splenocytes. Sophisticated genetic, biochemical, and immunologic studies have documented that amino acids E132 and I140 are critical residues for mediating the mitogenic activity of TSST-1 for human T cells (Deresiewicz *et al.*, 1994; Murray *et al.*, 1994). These residues are located on the α-2 helix in the C-terminal domain of TSST-1 implicated as important in TCR binding (Prasad *et al.*, 1993). Hence, some of the TSST-O amino acid changes have functional consequences which imply host adaptation, a major step leading to disease specialization.

C. The Adapted Clone Hypothesis

The detailed molecular insights noted above do not provide a satisfactory explanation for the occurrence of the particular clone of *S. aureus* that caused the marked increase in frequency of TSS episodes in geographically widespread areas in the late 1970s and early 1980s. One possibility is an episode of periodic selection involving a fitness mutation occurring in a cell giving rise to the major TSS clone (ET 41). However, this hypothesis is not favored because TSS increased in frequency in many different geographic regions at the same time. Moreover, studies on strains in collections isolated years ago have shown that TSST-1-positive *S. aureus* strains of ET 41 existed well before the emergence of TSS in the late 1970s. Alternatively, the increase in number of TSS cases may have been due to changes in lifestyle of the susceptible population so that the fitness of the ET 41 TSST-1-positive genotype that existed in an endemic state was enhanced relative to that of other genetic lines of staphylococci. According to this hypothesis, any alteration of the ecological niche occupied by toxin-positive cells already widely dispersed in human populations that results in a growth advantage or increased toxin production, or both, could produce a bloom of TSS cases. All of the evidence suggests that this is what occurred.

The evidence has implicated introduction of superabsorbent materials in catemenial products, perhaps associated with decreased or decreasing levels of anti-TSST-1 antibody in human populations, as a major factor in the rapid increase in disease frequency. Serological studies demonstrated that although most adults have preexisting antibody to TSST-1, most women who developed TSS associated with the menses have little or no detectable antibodies (Bonventre *et al.*, 1984; Todd, 1988). These data are consistent with the epidemiological observation that children and young adults are at highest risk for developing menstrual TSS. Moreover, patients have been described with recurrent TSS, and most of these individuals fail to develop significant levels of anti-TSST-1 antibody. Considerable research effort has been expended to investigate the influence of tampon absorbency, composition, and other parameters on level of

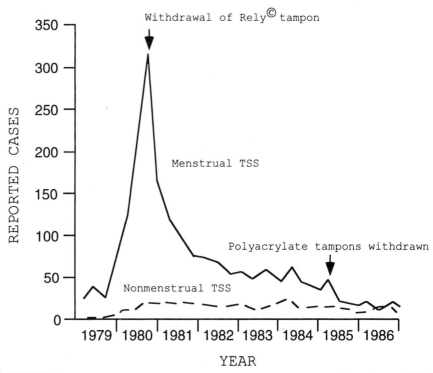

FIGURE I   Number of reported cases of toxic shock syndrome in the United States, by year (1979 1986), based on passive surveillance. Based on Broome (1989).

TSST-1 production by *S. aureus*. The toxic strains multiply within the mois fabric of the tampon and secrete the toxin which the host then absorbs. Th consequence is the generalized toxemia of TSS. Withdrawal of one brand o high-absorbency tampon from the market, which led to a subsequent rapid de crease in TSS cases in the United States (Fig. 1), is consistent with the thesis tha certain tampons were an important participant in TSS pathogenesis.

Several noteworthy bacteriologic and clinical observations about TSS de serve further comment. There remains no molecular explanation for the fac that the vast majority of menstrual TSS is caused by isolates of a clonal lineag of *S. aureus* that is distinct from strains causing nosocomial infections, or othe diseases caused by this pathogen. Production of TSST-1 alone by TSS strain is not a sufficient explanation for the occurrence of TSS because many other *S aureus* clones can express this toxin (Musser *et al.*, 1990; Musser and Selander 1990). One possibility is that the cervicovaginal milieu favors the expression o TSST-1 (or other microbial products critical for virulence) by ET 41 organism over that of isolates of other clones. For example, inasmuch as the expression o virulence factors and other extracellular proteins by *S. aureus* is regulated b positive transcriptional factors (Kornblum *et al.*, 1990; Cheung *et al.*, 1992), i is reasonable to speculate that the regulation of TSST-1 production by ET 4 organisms is distinct from that of other TSST-1-positive strains. Recent evidenc has been presented that supports this idea (Ji *et al.*, 1997).

An alternative but not mutually exclusive hypothesis is that isolates o

ET 41 are more prone to colonize the human vagina and, hence, are widely dispersed in an ecological niche of great consequence in TSS. This idea has been presented by one of us as the "adapted clone" hypothesis (Musser *et al.*, 1990). According to this hypothesis, isolates of ET 41 are responsible for most vaginal cases of TSS because this clone has special affinity for the cervicovagina, perhaps as a consequence of variation in regulation of toxin or other virulence gene expression. Therefore, the chance of colonization of this anatomic site by an isolate of ET 41 is much greater than that for other clones.

The adapted clone hypothesis is consistent with the observation that TSST-1-positive ET 41 strains existed well before 1978, which means that the *tst* gene did not recently evolve, nor recently associate with the chromosomal genotype marked by ET 41. Second, 28% of isolates of *S. aureus* recovered from the introitus, vagina, or cervix of random healthy carriers or women with non-TSS urogenital symptoms were ET 41 or closely allied clones, and yet many of these organisms produce TSST-1 (Musser *et al.*, 1990). Third, the adapted clone hypothesis does not require a precipitous drop in herd immunity to TSST-1 to explain the rapid rise in frequency of occurrence of TSS cases. An additional point worth noting is that all TSST-1-positive isolates of ET-41 produce staphylococcal enterotoxin A, another superantigen, but its role in pathogenesis has not been examined.

Taken together, these epidemiological and bacteriologic studies on *S. aureus* TSS enhanced our understanding of host–pathogen interactions at the molecular level, and they provided an important genetic framework for the analysis of other emerging microbial pathogens. In particular, the lessons learned from study of *S. aureus* TSS were invaluable in formulating strategies to investigate toxic shock type diseases caused by Group A Streptococcus that reemerged later in the 1980s.

Although not directly relevant to a discussion on the emergence of staphylococcal TSS, it should be noted that concurrent studies on staphylococcal immunity by Karakawa and colleagues (reviewed by Fattom and Naso, 1997) demonstrated that *S. aureus* can be classified on the basis of eight antigenically distinct carbohydrate capsules. Antibodies to these antigens promote type-specific phagocytosis and protective immunity in animals. A vaccine composed of capsular types 5 and 8 is now being tested in Phase I and II clinical trials, for the prevention of *S. aureus* infections in high risk populations, such as those that are particularly susceptible to nosocomial infection (Fattom and Naso, 1997).

II. REVIVAL OF GROUP A STREPTOCOCCAL DISEASES

The return of Group A Streptococcus (GAS) as a cause of serious infections in the 1980s was a sobering reminder that medical science knows embarrassingly little about the evolutionary and epidemiological forces that drive temporal variation in bacterial disease frequency and severity. The GAS rapidly emerged from an "endangered species" to an organism that once again was the cause of rheumatic fever outbreaks and toxic shock syndrome that required the close attention of pediatricians, internists, and surgeons. If emergency treatment was not vigorous and prompt, death was likely to occur.

The death of the popular creator of the Muppets Jim Henson in May, 1990, due to a toxic shock syndrome, highlighted the fact that no one is isolated from the apparent randomness of severe GAS infections. Later in the 1990s Lucien Bouchard, leader of the secession movement in Quebec, Canada, also had a near-fatal encounter with invasive GAS disease. Henson's death and Bouchard's illness heightened public awareness in North America about unusually virulent bacterial infections, and the concept of emerging microbial diseases. These and other events such as increased publicity about antibiotic-resistant bacteria led to numerous lay press accounts of virulent "germs," or "superbugs." This anxiety was fed by the dissemination of multidrug-resistant strains of *Mycobacterium tuberculosis* in New York City and elsewhere (Bifani *et al.*, 1996; Agerton *et al.*, 1997). Although calls of gloom and doom have largely been drowned by voices of calm and reason, it is clear from the weight of box office receipts, and the number and size of book contracts, that tales of infectious agents have captivated the imagination of the lay public.

A. Introduction to Group A Streptococcus

1. Diseases

Group A Streptococcus is a gram-positive bacterium that is the etiological agent of numerous diseases, including pharyngitis and/or tonsillitis, skin infections (impetigo, erysipelis, and other forms of pyoderma), scarlet fever, severe sepsis, meningitis, necrotizing fasciitis, and a toxic shock syndrome; and the non-infectious complications acute rheumatic fever, and acute glomerulonephritis. For reasons that remain obscure, despite 50 years of research, acute rheumatic fever and acute glomerulonephritis can occur as late noninfectious sequelae. Globally, these diseases cause immense human morbidity and mortality. For example, direct costs associated with pharyngitis in the United States have been estimated to be in the range of 1 billion dollars. The upside is that antibiotic treatment has essentially eliminated rheumatic fever and subsequent rheumatic heart disease in the United States and Europe. These crippling forms of heart disease, however, are still very common in the developing world and in certain indigenous populations in developed countries (Carapetis *et al.*, 1996; Martin and Sriprakash, 1996).

2. Putative and Proven Virulence Factors: An Embarrassment of Riches

Although there is a large body of literature on the pathogenesis of diseases due to GAS, this review must be selective rather than comprehensive. Unlike several pathogenic microbes with a relatively narrow disease spectrum and limited strain diversity, such as *Bordetella pertussis* and *Shigella sonnei* (Go *et al.*, 1996), GAS shows immensely complicated pathogenesis. Wild-type strains isolated worldwide possess substantial chromosomal and allelic diversity and cause an unusually diverse range of infection types (Kehoe *et al.*, 1996). In part due to this diversity, we lack a comprehensive understanding of the molecular pathogenesis of GAS virulence.

A variety of adherence molecules are involved in the initial host cell–GAS interaction that results in colonization. The relative contribution of each adherence molecule appears to depend, in part, on the target host cell used in the *in vitro* assay (reviewed in Hasty *et al.*, 1992). The following outer surface GAS products have been implicated in adherence: M protein, several M protein-like molecules that bind immunoglobulins, lipoteichoic acid, several fibronectin (FN)- and vitronectin (VN)-binding proteins, and C-carbohydrate. Host molecules implicated as receptors for GAS adherence include FN, fibrinogen, VN, and galactose- and fucose-containing glycoproteins. Based on genetic strategies, there is strong evidence that a FN-binding protein (protein F encoded by the *prt* gene; also known as the Sfb protein) mediates the adherence of some GAS to respiratory epithelial cells. Expression of protein F by some strains is stimulated by increased O_2 tension, suppressed by anaerobiosis, and controlled by a gene (*rofA*) encoding a trans-acting positive regulator of *prtF*. However, it is important to realize that many GAS strains—including a large percentage of organisms causing contemporary episodes of severe invasive disease—lack the *prt* gene and, hence, apparently employ other mechanisms for initial interaction with host tissues.

Following host tissue attachment, GAS may either remain at the site of colonization or initiate a more invasive extracellular infection involving tissue destruction and/or intravascular entry and dissemination. The tissue destruction in many patients is extensive, rapidly progressive, and characterized by a virtual total destruction of the extracellular matrix. This histopathology occurs at many distant sites where bacteria are not present (Jevon *et al.*, 1994). Numerous extracellular substances are produced by GAS. Many of these are highly toxic and are undoubtedly the cause of tissue necrosis associated with either localized pharyngitis or STSS. The molecular mechanism(s) responsible for tissue destruction is unknown. Microbial invasion can generally be thought of as either an "active," aggressive process mediated by (for example) degradative enzymes or a "passive" process that can involve subversion of normal host cell processes. Although both strategies probably operate in GAS infections, the exact molecular pathway(s) involved is poorly understood.

Evidence suggests that the following streptococcal molecules contribute to one or more phases of GAS pathogenesis: M protein, M-like protein, hyaluronic acid capsule, C5a peptidase, streptokinase, streptolysin O, and hyaluronidase. Evidence has also been presented indicating that a highly conserved potent extracellular cysteine protease also is an important Group A *Streptococcus* virulence factor (Lukomski *et al.*, 1997, 1998; Musser, 1997) (Fig. 2). A gene (*mga*) has also been identified that encodes a trans-acting positive regulator of expression of M protein and several other virulence factors (Podbielski *et al.*, 1995). Moreover, a functional *mga* gene is required for virulence in mice and resistance to phagocytosis by human phagocytes (Table I). The role of M protein in mouse virulence and resistance to phagocytosis was first shown by Lancefield in the 1920s (1941). M type-specific antibody protects mice and promotes phagocytosis of GAS. Another recent finding that may bear on the issue of GAS invasion—although the mechanism is unknown—is that stationary phase cultures of certain strains are efficiently internalized by cultured A549 human lung carcinoma epithelial cell monolayers (LaPenta *et al.*, 1994).

Several "superantigens" have been described; most are produced to a greater or lesser extent by all GAS isolates (reviewed in Musser, 1997). Superantigens are powerful stimulators of specific subsets of T cells that express cer-

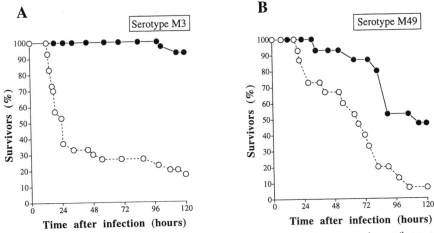

FIGURE 2 Streptococcal pyrogenic exotoxin B (extracellular cysteine protease) contributes to Group A *Streptococcus*-mediated death. Kaplan–Meier survival plots show significantly decreased mouse lethality of the cysteine protease-inactivated isogenic Group A *Streptococcus* strains (○, wild-type parental strain; ●, *speB* isogenic mutant derivative strain). Lethality studies were conducted with adult (18–20 g) male outbred CD-1 Swiss mice, as described in Lukomski et al. (1997). Inocula (10^6 colony-forming units) were injected intraperitoneally into groups of 15 or 30 mice (time = 0 hr). (A) M3 isogenic strains: $n = 30$ mice per group; $\chi^2 = 57.9$, $p < 0.00001$. (B) M49 isogenic strains, experiment 1; $n = 15$ mice per group; $\chi^2 = 24.3$, $p < 0.00001$. Modified from Lukomski et al. (1997).

tain Vβ regions of the T-cell receptor (TCR) in a novel fashion, a process that can have dramatic detrimental effects on the host (Thibodeau and Sekaly, 1995). Streptococcal pyrogenic exotoxin A (SpeA) has been given most attention because of a strong significant association between SpeA production and severe disease (Musser *et al.*, 1991).

Very little is known about the molecular pathogenesis of acute rheumatic fever. It has long been presumed that the disease has an autoimmune component, and the organism is known to express several products that elicit antibodies that cross-react with human tissue (Zabriskie, 1985; Cunningham, 1992; Cunningham *et al.*, 1997).

B. Recent Epidemiological Observations

1. Acute Rheumatic Fever

In early 1985, there was recorded an unanticipated and substantial increase in cases of acute rheumatic fever (ARF) in Salt Lake City and surrounding communities, resulting in 136 reported cases in Utah in 1985 and 1986 (Veasy *et al.*, 1987). The great majority of cases met the Jones' criteria for guidance in the diagnosis of rheumatic fever. Additional outbreaks of ARF that comprised more than several hundred patients in five separate states were subsequently reported in other regions of the United States (Table II) (Bisno, 1991). This startling reappearance led to detailed epidemiological investigations showing that most patients were white, lived in suburban or rural settings, and had access to medical care. The demographic features were particularly surprising because rheumatic fever had become a sporadic disease of inner city populations. Although many

▬ **TABLE I** *Streptococcus pyogenes* **Factors Involved in Host–Pathogen Interactions Defined by Use of Isogenic Strains**

Molecule	Assay	Phenotype of isogenic mutant	Representative references
M protein	Epithelial cell attachment	Decreased epithelial cell adherence	Courtney *et al.*, 1994; Dale *et al.*, 1996;
	Phagocytosis	Increased phagocytosis	Moses *et al.*, 1997
	Mouse virulence	Decreased mouse virulence	
C repeat domain	Human keratinocyte adherence	Decreased adherence	Perez-Casal *et al.*, 1993b
	Complement factor H binding	Decreased binding	
	Survival in human blood	No alteration	
M-related protein (MRP)	Phagocytosis by human granulocytes	Increased phagocytosis	Podbielski *et al.*, 1996
Protein F	Human respiratory epithelial cell adherence	Decreased adherence	Hanski and Caparon, 1992
Hyaluronic acid capsule	Phagocytosis	Decreased mouse soft tissue lesion size	Wessels *et al.*, 1991; Wessels and Bronze, 1994;
	Mouse virulence	Enhanced efficiency of keratinocyte entry	Wessels *et al.*, 1994; Schrager *et al.*, 1996;
	Internalization by human keratinocytes		Dale *et al.*, 1996; Moses *et al.*, 1997
Mga	Survival in human blood	Decreased survival	Perez-Casal *et al.*, 1993a
C5a peptidase	Mouse virulence	Altered organ trafficking	Ji *et al.*, 1996
Pyrogenic exotoxin B	Mouse virulence	Decreased mouse lethality	Lukomski *et al.*, 1997

European countries recorded increased invasive disease episodes, no increase in the numbers of ARF cases has been documented.

Historically, GAS has caused many large outbreaks of streptococcal pharyngitis and rheumatic fever (Quinn, 1989; Katz and Morens, 1992). However, from the 1940s through the mid-1980s, the incidence of these diseases declined dramatically in the United States and most developed nations, presumably as a consequence of antibiotic treatment, better public health measures, and other reasons that may be related to changes in the properties of GAS (Fig. 3).

For example, Stollerman (1997) has speculated that there has been a decline during the last four decades in the virulence of GAS isolated from endemic cases of streptococcal disease, in contrast to those highly virulent strains of GAS that were isolated during the epidemics in the military and civilian populations during World War II and the Korean War. Strains with diminished virulence have a small capsule and sparse M-protein, whereas the strains isolated during epidemics that resulted in rheumatic fever were rich in M-protein and had a bulky capsule. The latter strains are highly contagious and rapidly transmitted from

TABLE II Countries Reporting Problems with Increased Frequency of Group A Streptococcus Infections

Locality	Disease	Implicated serotypes	Study years	Comment	Representative reference
North America United States	Invasive	M1T1, M3T3, M28T28	1986–1988	20 patients; mortality 30%	Stevens *et al.*, 1989
	Invasive	M1, M3, M18	1972–1988	>5000 strains referred to CDC	Schwartz *et al.*, 1990
	Invasive	Most MNT	1983–1990	North Carolina; increase noted beginning 1987	Givner *et al.*, 1991
	Invasive	M1, M12	1980–1990	Denver, Colorado; 34 cases	Wheeler *et al.*, 1991
	Invasive	M1	1985–1990	Pima County, Arizona; increased disease severity noted in late 1980s	Hoge *et al.*, 1993
	Invasive	Unknown	1977–1993	Disease increase in 1990s associated with varicella	Doctor *et al.*, 1995
	Invasive	Unknown	1993–1995	14 children in Pacific Northwest with necrotizing fasciitis complicating varicella	Brogan *et al.*, 1995
	Invasive	M1, M3	1994	24 children in southern California; disease associated with varicella	Vugia *et al.*, 1996
	Invasive	M3	1994–1995	Two clustered outbreaks in Akron, Ohio	DiPersio *et al.*, 1996
	Invasive	M3	1995	Clustered outbreak of 7 cases in southeastern Minnesota	Cockerill *et al.*, 1997
	Invasive	M1, M3	1987–1995	North Carolina,	Kiska *et al.*, 1997
United States	ARF	M1, M3, M18, M78	1985–1986	Utah, Idaho, Nevada, Wyoming	Veasy *et al.*, 1987
	ARF	M18	1984–1986	Columbus, Ohio	Hosier *et al.*, 1987
	ARF	M1, M5, M6, M18	1986–1987	Akron, Ohio	Congeni *et al.*, 1987

Country	Disease	M types	Years	Description	Reference
	ARF	Unknown	1985–1986	Western Pennsylvania, Ohio	Wald et al., 1987
	ARF	Unknown	1985–1988	New York City	Griffiths and Gersony, 1990
	ARF	M18	1987–1988	Nashville, Tennessee	Westlake et al., 1990
	ARF	M11/12	1987–1988	Memphis, Tennessee	Leggiadro et al., 1990
	ARF	M3, M18	1987–1988	Ft. Leonard Wood, Missouri	Centers for Disease Control, 1988
Canada	Invasive	M1, M12, M4, M28, M3	1992–1993	323 patients	Davies et al., 1996
	Necrotizing fasciitis	M1, M3 common	1991–1995	77 patients studied	Kaul et al., 1997
Europe United Kingdom	Invasive	M1, M3	1980–1987	Dramatic rise in M1 strains 1984–1987	Gaworzewska and Colman, 1988
	Necrotizing fasciitis	M1, M3, M5, MNT	1994	Six temporally linked cases	Cartwright et al., 1995
Sweden	Invasive	M1T1	1988–1989 (winter)	Countrywide outbreak	Strömberg et al., 1991
Norway	Invasive	M1T1	1987–1988 (winter)	Countrywide outbreak	Martin and Høiby, 1990; Bucher et al., 1992
	Necrotizing fasciitis	M1, M3, M6, M28	1992–1994	13 patients	Chelsom et al., 1994
Finland	Invasive	M1T1	1988–1995	Disease peaks in 1989–1990	Muotiala et al., 1997
New Zealand	Invasive	M1T1	1980–1991	Disease peaks in 1984–1985, 1989	Martin and Single, 1993
Australia	Invasive	M1T1	1982–1993	Disease peaks in 1989, 1993	Carapetis et al., 1995
Japan	Invasive	M1, M3	1992–1994	Countrywide survey	Nakashima et al., 1997
	Invasive	M1T1, M3T3	1990–1995	Rapid increase in M3T3 isolates	Inagaki et al., 1997

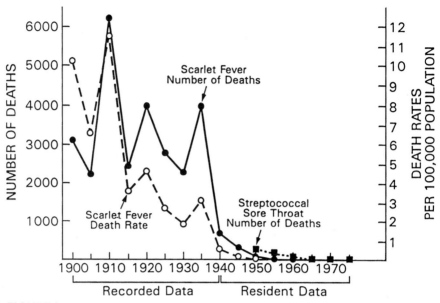

FIGURE 3 Decline in number of deaths from scarlet fever and streptococcal sore throat in the United States. The graph illustrates that the decline in streptococcal disease started well before the introduction and widespread use of penicillin. Reprinted from Quinn (1982).

person to person, and it is these strains that have made a reappearance in recent years as the cause of STSS and ARF.

Much has been learned recently about the molecular epidemiology of GAS and rheumatic fever, particularly in regard to the molecular structure and genetics of M-protein. Bessen *et al.* (1989) have evidence for two distinct classes of M-protein, class 1 and class 2. Patients with rheumatic fever had a higher than normal serum concentration of IgG antibodies with specificity for the class 1 epitope (Bessen *et al.*, 1995). The sera of patients who did not develop rheumatic fever had a lower than normal antibody concentration with the class 2 specific epitope. Finally, Bessen *et al.* (1996) showed that the genes of M-protein (*emm*) have four sub-families of nucleotide sequences arranged in five distinctive chromosomal patterns. These patterns clearly differentiate rheumatogenic pharyngitis strains from impetigo strains, and though less distinctly, from non-rheumatogenic pharyngitis strains. Inasmuch as these studies had been done with strains of GAS recovered in the United States and Western Europe, it is unknown if these general conclusions about the molecular biology of M protein apply to strains that will be isolated in worldwide locales.

A potential and practical fallout of this work may be the development of a successful vaccine for GAS. A major stumbling block thus far has been the large number of M-protein serotypes that can cause a serious infection and the M type-specific nature of protective immunity. Bessen and Fischetti (1990) have determined that the class 1 epitope of M-protein can stimulate immunity to heterologous M serotypes in mouse protection tests. For example, mice immunized with the M type 6 C repeat region epitope of M protein are immune to challenge with the heterologous type 14 M protein that shares a C repeat region with the type 6 M protein. In recent studies, Fischetti (1996) has employed non-infectious commensal bacteria to develop a recombinant strain that produces this repeat region epitope of M protein. A vaccine of this recombinant stimulates mucosal immunity following pharyngeal administration in the mouse. Both IgG and IgA antibodies are produced. The mice are protected from colonization when challenged with both homologous and heterologous streptococci that share the same C region repeat in their M proteins. Recent M-protein genetics research suggests that there are a limited number of C region repeat peptides among the wild M type strains isolated from rheumatic fever patients. A recombinant vaccine consisting of these several peptides may be effective in stimulating the immunity to the strains that are most likely to produce acute rheumatic fever. An alternative strategy has been pursued by Dale and Chiang (1995) using a recombinant Group A streptococcal M protein fragment fused to the B subunit of *Escherichia coli* labile toxin. It has also been demonstrated that immunization of mice with a highly conserved extracellular cysteine protease will protect mice from lethal intraperitoneal challenge with virulent GAS (Kapur *et al.*, 1994).

2. Pharyngitis Treatment Failures

Acute pharyngitis caused by GAS is a major health problem in children worldwide. Although no definite data are available, there has also been an increased number of reports of GAS pharyngitis that fail to be eradicated by treatment with standard drug regimens (Gray *et al.*, 1991; Pichichero, 1991). The failure of eradication may be accompanied by persistence or recrudescence of clinical illness, that is, clinical failure. The causes of these failures are obscure but may include patient noncompliance, antimicrobial tolerance, and coinfection with other pathogens. None of the strains isolated from these patients are resistant to penicillin *in vitro*. Some investigators have reported sharp increases in the rate of resistance to erythromycin, an alternative antibiotic used in treatment (Seppälä *et al.*, 1992). Because of space limitations, issues associated with GAS pharyngitis treatment failures are not discussed further in this chapter.

3. Invasive Episodes and Rise of Streptococcal Toxic Shock Syndrome

a. Introduction

For unknown reasons, the number of deaths due to severe or septic scarlet fever declined in the United States between 1910 and 1940, well before introduction and widespread use of antibiotics (Fig. 3). Scarlet fever is streptococcal pharyngitis with a rash followed by the characteristic desquamation in a week to 10 days. Madsen (1973) also reported similar trends in Norway occurring from 1870 to the late 1940s. These declines continued well into the 1980s, not only in these two countries, but in most developed nations as well. Since the late 1950s there has also been a progressive decline in the frequency and the severity of streptococcal pharyngitis. For this reason, the medical community was surprised and largely unprepared when several outbreaks of streptococcal disease characterized by unusual severity and high mortality rates were reported in European countries and the United States late in the 1980s (Table II).

Stevens *et al.* (1989) described 20 patients from the Rocky Mountain region (Utah, Idaho, Montana, and Nevada) with an unusually severe form of GAS infection that was characterized by extensive local tissue destruction and signs and symptoms of systemic toxicity. Nineteen patients had shock, 80% had renal impairment, about one-half had acute respiratory distress syndrome, and six

■■■ **TABLE III   Case Definition of Streptococcal Toxic-Shock Syndrome and Necrotizing Fasciitis**[a]

I. Streptococcal TSS
  A. Isolation of group A *Streptococcus* from
    1. A sterile site
    2. A nonsterile body site
  B. Clinical signs of severity
    1. Hypotension
    2. Clinical and laboratory abnormalities (two or more of the following required):
      a. Renal impairment
      b. Coagulopathy
      c. Liver abnormalities
      d. Acute respiratory distress syndrome
      e. Extensive tissue necrosis, i.e., necrotizing fasciitis
      f. Erythematous rash

Definite case = A1 + B(1 + 2)
Probable case = A2 + B(1 + 2)

II. Necrotizing fasciitis
  A. Definite case
    1. Necrosis of soft tissue with involvement of the fascia
  PLUS
    2. Serious systemic disease, including one or more of the following:
      a. Death
      b. Shock (systolic blood pressure <90 mm Hg)
      c. Disseminated intravascular coagulopathy
      d. Failure of organ systems
        (i)   Respiratory failure
        (ii) Liver failure
        (iii) Renal failure
    3. Isolation of group A *Streptococcus* from a normally sterile body site
  B. Suspected case
    1. 1 + 2 and serological confirmation of group A streptococcal infection by a 4-fold rise against
      a. streptolysin O
      b. DNase B
    2. 1 + 2 and histological confirmation: gram-positive cocci in a necrotic soft tissue infection

[a] Streptococcal toxic shock syndrome (streptococcal TSS) is defined as any Group A streptococcal infection associated with the early onset of shock and organ failure. Definitions describing criteria for shock, organ failure, definite cases, and probable cases are included. Source: The Working Group on Severe Streptococcal Infections, 1993; Stevens, 1995.

died. GAS was recovered from skin and wound cultures taken from virtually all patients, and the majority of patients had positive blood cultures. The case definition of streptococcal toxic shock syndrome (Stevens *et al.,* 1989; The Working Group on Severe Streptococcal Infections, 1993) (Table III) includes signs and symptoms that are due to multiorgan failure and are similar to those of staphylococcal toxic shock syndrome. In some patients necrotizing fasciitis was a prominent feature. Virtually all GAS isolates recovered from such patients produced one or more pyrogenic exotoxins that have significant amino acid sequence homology and functional similarity with TSST-1 produced by *S. aureus.* Significantly, many of the isolates produced streptococcal pyrogenic exotoxin A (SpeA), also known as erythrogenic toxin A, which has been rarely recorded in recent years as an extracellular product of wild type isolates of GAS (Stevens *et al.,* 1989). Outbreaks or cases of streptococcal TSS and other severe invasive diseases have now been reported throughout the United States (Table II).

The Centers for Disease Control reported that the rate and severity of GAS infections in the United States increased dramatically since the mid-1980s. This rise was associated predominantly with recovery of isolates expressing M protein serotypes M1, M3, and M18 from the patients (Schwartz *et al.,* 1990). The increase in invasive infections caused by type M1 organisms was most marked, with these strains accounting for 50% of bacteremias in 1980–1988 and 64% of septic episodes in 1989–1990. Observations on streptococcal TSS have now been made in several countries that have broadened our knowledge of its epidemiology and pathogenesis. This work is summarized next.

b. STSS in Other Countries

(i) **Canada** An increase in the frequency and severity of GAS infections has also been recorded in parts of Canada. Beginning in 1990, an apparent increase in invasive disease cases in Ontario led to initiation of a well-designed and executed prospective, population-based surveillance study of invasive GAS cases in the province (Demers *et al.,* 1993; McGeer *et al.,* 1995; Davies *et al.,* 1996; Kaul *et al.,* 1997). During 1992 and 1993, 323 patients with invasive GAS infections were identified, and an annual incidence of 1.5 cases per 100,000 population was recorded (Davies *et al.,* 1996). The attack rates were highest in children less than 10 years old and adults greater than 60 years old. This suggests that an absence of immunity at these age groups is a factor in the occurrence of STSS. Underlying chronic conditions were identified in 56% of patients, and risk factors for disease included cancer, diabetes, alcohol abuse, varicella, and, interestingly, human immunodeficiency virus (HIV) infection. Roughly one-half of the patients had soft tissue infections. Necrotizing fasciitis occurred in 6% of patients, and toxic shock in 13%. Importantly, there was substantial transmission of invasive streptococcal infection in households and health care institutions. The majority (80%) of all infections were caused by strains expressing M1, M3, M12, M4, and M28 proteins. As reported in other studies, many of the M1 strains produced exotoxin A. More recent data from Ontario has shown that a major new peak of disease occurred in the winter of 1994–1995 and was associated with a significant rise in cases of necrotizing fasciitis and toxic shock (McGeer *et al.,* 1995; Kaul *et al.,* 1997). Moreover, the percentage of infections caused by M3 organisms harboring the *speA* gene increased dramatically during this period. The Ontario study has thoroughly documented the rapid shifts in disease spectrum and strain M serotype associations that can occur in a relatively local area.

(ii) **European Countries** The occurrence of changes in the character and severity of infections caused by GAS strains also was observed in the 1980s in certain areas of Europe, specifically in the United Kingdom, Norway, Sweden, Denmark, Finland, Germany, former Czechoslovakia, and former Yugoslavia (Table II). In the United Kingdom, for example, a survey of 12,469 Group A isolates cultured from patients in 1980 through June, 1987, revealed the following epidemiological features (Gaworzewska and Colman, 1988): (1) the percentage of disease caused by isolates of serotypes M1, M3, and M28 increased, (2) a significant increase in the number of serious disease episodes and deaths occurred, and, (3) 80% of strains received between 1985 and 1987 from fatal infections were serotype M1 protein.

(iii) **New Zealand**   The Streptococcus Reference Laboratory, Communicable Disease Center, Wellington, has conducted extensive epidemiological surveillance of streptococcal infections in New Zealand since 1980 (Martin and Sriprakash, 1996). The data show the striking rapid shifts in prevalence of certain M types that can occur. On the basis of analysis of approximately 12,000 isolates, Group A types M1T1 (M protein type 1, T protein type 1) was the predominant serotype during three periods, 1983–1986, 1987–1989, and 1992–1996 (Fig. 4). The number of sterile site M1T1 isolates reflected the total number of M1T1 isolates recovered during these periods. Molecular typing by restriction fragment length polymorphism (RFLP) analysis showed that a total of six distinct RFLP types were present among the M1 strains. A majority of these M1 strains isolated before 1983 were arbitrarily designated RFLP type 1b. Since 1983, RFLP type 1b GAS has been replaced by GAS RFLP type 1a. The rise of RFLP type 1a was subsequently responsible for the majority of M1 infections in New Zealand. These results were important because they provided additional evidence that genotypic heterogeneity exists among strains classified on the basis of M serotyping.

(iv) **Australia**   To determine if the pattern of invasive GAS infections in Australia had changed, Carapetis *et al.* (1995) conducted a 12-year review of all cases of Group A streptococcal bacteremia seen at the Royal Children's Hospital of Melbourne between 1982 and 1993. A trend toward increased incidence of infection was found, as was a clear increase in disease severity during the study period. As described in other studies, an M1 clone that harbored the gene encoding pyrogenic exotoxin A was the most frequent cause of invasive disease episodes. Hence, in general, there was a striking parallel between the findings reported from this Australian study and those published from several Northern Hemisphere sites.

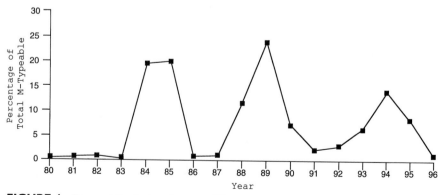

FIGURE 4   Temporal distribution of serotype M1 strains in New Zealand, 1980 to 1996. The data are presented as percentages of all M typeable organisms. Based in part on Martin and Single (1993).

C. Pathogenic Properties of GAS

1. Value of Phenotype Analysis: Importance of Strain Serotype

Until 1989, most epidemiological information for GAS was generated by assigning strains to one of approximately 80 M-protein serotypes, first described by Lancefield in the 1920s (reviewed in Fischetti, 1989). A second surface antigen known as T protein has also been employed in sero-epidemiology. In addition, strains have been classified into two categories (opacity factor positive and negative) on the basis of the ability of culture supernatants to make serum opaque. For many decades, determination of M serotype of clinical isolates has been used to define many aspects of the epidemiology of GAS. Most important was the discovery by Lancefield that M protein was a virulence factor that impaired phagocytosis, and antibodies to M protein promoted phagocytosis. Such serological analysis in and of itself has not, however, revealed the special characteristics of GAS that produce streptococcal TSS.

2. Population Genetic Framework Provided by Multilocus Enzyme Electrophoresis

The great appeal of serological classification is the identification of a small number of phenotypic surface properties that have important relationships to epidemiology and virulence. Although M typing has been particularly useful to examine phagocytosis-mediated immunity, the present convention of classifying isolates on the basis of one or several surface antigens is less satisfactory for the study of genomic relationships because too few phenotypic traits are detected in this way. Such serologic methods do not reflect the complexity of the genetic structure of GAS. Thus, for example, genetic studies have shown that not all isolates of a particular M-protein serotype are closely related to one another; moreover, different isolates of closely similar chromosomal genotypes may differ in serotypic profiles (Kehoe *et al.*, 1996). In short, serotyping techniques have provided little basis for understanding the genetic variations in population structure occurring in connection with temporal changes in the frequency or character of invasive streptococcal diseases.

To determine the genetic diversity and clonal relationships among GAS isolated from patients with streptococcal TSS or other invasive diseases in the United States, 108 organisms were studied by multilocus enzyme electrophoresis and analyzed for pyrogenic exotoxin A, B, and C synthesis (Musser *et al.*, 1991). Multilocus enzyme electrophoresis is a convenient technique for estimating levels of chromosomal diversity and relationships among strains, and it has been the primary research tool used to examine the population genetics of emerging bacterial pathogens (Musser, 1996). By this method the 108 strains were classified into 33 electrophoretic types (ETs). Nearly one-half of the 108 strains, including more than two-thirds from streptococcal TSS, were identified as related clones ET-1 and ET-2 (Musser *et al.*, 1991) (Fig. 5).

3. Expression of Virulence Factors

Several studies of GAS isolates recovered from invasive disease episodes have reported results on virulence factor expression. Most of the studies have concentrated on analysis of production of pyrogenic exotoxins A, B, and C, because of their presumed importance in scarlet fever and possible role in strep-

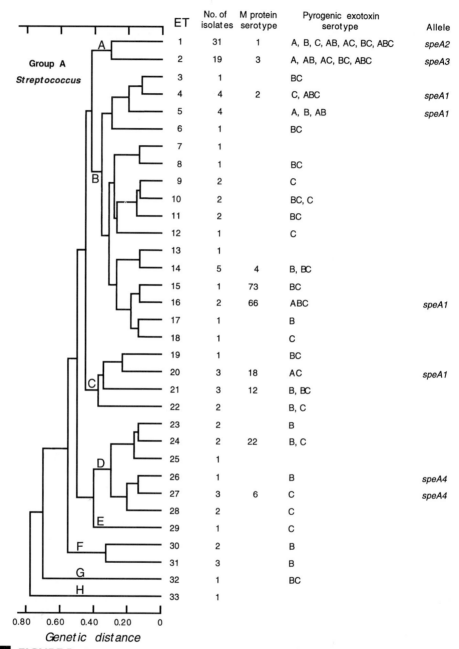

FIGURE 5   Dendrogram showing estimates of genetic relationships among 33 ETs of Group A *Strep-tococcus*, based on allele profiles at 12 enzyme loci. Eight major lineages are identified by letters A–H. Pyrogenic exotoxin serotypes, M protein serotypes, and *speA* alleles of isolates of each ET are indicated. Modified from Musser *et al.* (1991).

tococcal TSS. The results of the first study on 108 isolates recovered in the United States showed that 98 (91%) expressed one or more exotoxins. Exotoxins B and C, alone or in combination, were produced by strains in nearly all of

LIVERPOOL
JOHN MOORES UNIVERSITY
AVRIL ROBARTS LRC
TEL. 0151 231 4022

REVIVAL OF GROUP A STREPTOCOCCAL DISEASES

203

the 33 ETs in the sample. In contrast, expression of exotoxin A was confined to a limited number of distinct electrophoretic types. Importantly, there was a relatively high frequency of exotoxin A expression among isolates recovered from streptococcal TSS patients; 28 of 31 isolates (90%) expressed exotoxin A, whereas only 12 of 45 isolates causing less severe diseases produced this exotoxin ($p < 0.001$) (Musser et al., 1991). Isolates from STSS patients produced pyrogenic exotoxin A, either alone or in combination with other pyrogenic exotoxins. This significant association was present with isolates of the same clone, as well as those of distantly related phylogenetic lineages, an observation interpreted as strong circumstantial evidence that exotoxin A itself, or possibly the product of a gene linked to it, is a factor in streptococcal TSS pathogenesis.

4. Genotype Analysis of Group A Streptococcus Strains Causing Streptococcal Toxic Shock Syndrome

a. Pulsed-Field Gel Electrophoresis and Other Chromosomal Analyses

Analysis of RFLPs generated by pulsed-field gel electrophoresis was employed for epidemiological studies in an attempt to enhance the genetic resolution of GAS strains from several geographic areas. Strains expressing serotype M1 and M3 have been examined most intensively because of their frequent recovery from invasive infections occuring worldwide (Table IV). Analysis of 90 serotype M1 strains from patients in New Zealand identified six RFLP subtypes, with most isolates belonging to a single type, designated 1a. Subsequently, a sample of 126 M1 strains from 13 countries on five continents was studied by a variety of genetic strategies, including pulsed-field gel electrophoresis analysis. Substantial levels of chromosomal diversity were identified among the isolates. The primary finding was that most invasive disease episodes were caused by two subclones as defined by a combination of RFLP type and multilocus enzyme electrophoretic type. One of these two subclones (ET 1/ RFLP type 1a) had the speA gene and was recovered worldwide.

Evidence consistent with the theme that a distinct M1 subclone is responsible for many of the contemporary episodes of invasive disease has been presented by Muotiala et al. (1997). These investigators studied 115 isolates of serotype M1T1 cultured in Finland and Norway during 1988 to 1995 from patients with pharyngitis and invasive diseases. The organisms were analyzed by restriction endonuclease analysis of chromosomal DNA, rRNA gene polymorphism (ribotyping), and random amplified polymorphic DNA analysis (RAPD). Representative genotypes revealed by these methods were then examined for presence of the speA gene by polymerase chain reaction (PCR). The speA gene was detected in all of these strains. Moreover, two-thirds of the organisms were of the same M1T1 genotype as determined by these methods. This is the same M1T1 genotype identified in the earlier study by Musser et al. (1995). The common M1T1 genotype cultured from invasive cases was identical to that recovered from patients with pharyngitis. This observation was important because it provided additional strong evidence that host factors play an important role in determining the outcome of host–pathogen interactions. Several other studies have supported the notion that a distinct speA-positive M1T1 subclone has produced streptococcal TSS worldwide.

In the case of M3T3 strains, analysis of isolates from invasive disease also

TABLE IV Studies Documenting Association of a Distinct *speA*-Positive M1 or M3 Subclone with Invasive Episodes

Locality	M type	Analytical method	Reference
United States	M1	Multilocus enzyme electrophoresis Exotoxin gene probing Toxin expression	Musser *et al.*, 1991
United States	M1	Pulsed-field gel electrophoresis Exotoxin gene probing	Cleary *et al.*, 1992
North America, Europe	M1, M3	Multilocus enzyme electrophoresis Sequencing of *speA, ska, speB*	Musser *et al.*, 1993

Scandinavia	M1	Pulsed-field gel electrophoresis
		Chromosomal restriction endonuclease
		  analysis
		RNA gene polymorphisms (ribotyping)
		Random amplified polymorphic DNA analysis
		  (RAPD)
		Exotoxin gene PCR
			Norgren *et al.*, 1992;
			Seppälä *et al.*, 1994a,b;
			Mylvaganam *et al.*, 1994

Australia	M1	Pulsed-field gel electrophoresis Exotoxin gene PCR	Carapetis *et al.*, 1995
Global	M1	Multilocus enzyme electrophoresis Pulsed-field gel electrophoresis Automated DNA sequencing of *emm, speA,* *speB, ska,* and *scp*	Musser *et al.*, 1995
United States	M3	Multilocus enzyme electrophoresis Pulsed-field gel electrophoresis Exotoxin gene sequencing *emm* sequencing	Cockerill *et al.*, 1997

Scotland	M3	Pulsed-field gel electrophoresis
		Chromosomal restriction endonuclease
		  analysis
		Ribotyping
		Exotoxin gene polymerase chain
		  reaction (PCR)
		emm sequencing
			Upton *et al.*, 1995, 1996

revealed a distinct multilocus enzyme genotype and pulsed-field electrophoretic type, and possession of the *speA* gene (Upton *et al.*, 1995, 1996; Cockerill *et al.*, 1997). In addition, analogous to the finding with the M1T1 strains, it has been shown in two studies that the common genotype of M3T3 strains or other organisms cultured from patients with invasive episodes were genetically indistinguishable from those isolated from patients with only pharyngitis or from asymptomatic contacts (Musser *et al.*, 1994; Cockerill *et al.*, 1997).

b. DNA Sequence of Genes Encoding Virulence Factors

Allelic variation of several genes encoding putative virulence factors may be related to the increase in disease frequency and severity. Initially the *speA* gene was chosen for study based on the observation of a strong statistical association of SpeA production with streptococcal TSS (Musser *et al.*, 1991). The *speA* gene was examined first in a sample of 20 strains selected on the basis of the population genetic framework provided by multilocus enzyme electrophoresis. Surprisingly, the analysis identified four alleles of *speA* in natural populations, designated *speA1* through *speA4* (Nelson *et al.*, 1991). The *speA1* allele was found in association with strains of several distinct chromosomal backgrounds, implying from an evolutionary point of view that *speA1* is the ancestral allele. Two other alleles (*speA2* and *speA3*), characterized solely by a single nucleotide change resulting in a single amino acid change, were each identified in single clones (M1/ET1 and M3/ET2, respectively). GAS possessing these *speA* alleles have caused the majority of the current streptococcal TSS cases (Fig. 5).

Two subsequent studies investigated this interesting observation in detail. Sequence analysis of *speA* in 90 ET1/M1 and ET2/M3 strains revealed that all contemporary *speA*-positive strains have the *speA2* and *speA3* allelic variants, respectively. In striking contrast, ET1/M1 and ET2/M3 strains from the 1920s and 1930s contained the *speA1* allele. The data suggested that there is temporal variation in the occurrence of clone–*speA* allele combinations, an observation that may in part explain fluctuations in disease frequency, severity, and character. Kline and Collins (1996) discovered that the SpeA3 toxin variant is significantly more active as a superantigen compared to SpeA1 and SpeA2, a result first suggested by the population genetic data presented previously (Nelson *et al.*, 1991).

Because it is likely that several streptococcal products are important in host–parasite interaction, in a subsequent study serotype M1 strains were examined for sequence variation in genes encoding additional putative virulence factors, including M protein, streptokinase, streptococcal pyrogenic exotoxin B, and C5a peptidase. A major finding from this study was the identification of seven variants of the M1 protein (Fig. 6). There was strong association of ET, RFLP type, and *emm* allele. For example, virtually all strains of ET 1/RFLP type 1a/*speA*-positive bacteria had allele *emm1.0*, whereas ET1/RFLP type 1k/ *speA*-negative isolates usually had the *emm1.3* allele. The only reasonable interpretation of these DNA sequence data on several distinct loci is that the recent increase in invasive disease episodes in the United States, Europe, and elsewhere

```
              10         20         30         40         50         60         70         80         90
M1.0   TVLGAGFANQ TEVKANGDGN PREVIEDLAA NNPAIQNIRL RHENKDLKA- ------RLEN AMEVAGRDFK RAEELEKAKQ ALEDQRKDLE
M1.1   .......T.. ......N..R S.D.T.EI.. ..TTV..... .N...N...- ------.... ..N....... .......... ..........
M1.1*  .......T.. ......N..R S.D.T.EI.. ..TTV..... .N...N..K NEDLEA.... ..N....... .......... ..........
M1.3   .......... .........S .......... .......... .......... ------.... .......... .......... ..........
M1.4   ..V....... ......ENVG ..D.VKE.VE KD.VL..K.. .S..QK..E- ------S.... .RD....... .......... ...A......
M1.5   .......... .......... ......G... .......... .......... ------.... .......... .......... ..........
M1.6   .......... .......... .......... .......... .Y.......- ------.... .......... .......... ..........


              10         20         30         40         50         60         70         80         90
M3.0   DARS----VN GEFPRHVKLK NE--IENLLD -----QVTQL YNKHNSNYQQ Y----SAQAG RLDLRQKAEY LKGLNDWAER LLQELNGEDV
M3.2   ....DARS.. .......... ...--..... -----..... .T........ .NAQDN.... .......... .......... ..........
M3.3   ....DARS.. .......... ...--..... -----..... .T........ .----N.... .......... .......... ..........
M3.6   ....----.. .......... ...--..... -----..... .A........ .----NT..R .......... .......... ..........
M3.14  ...N----.. .......... .......... .......... .......... .----..... .......... .......... ......S....
M3.25  ....----.. .......... ...--..... ENSLD..... .T........ .----N.... .......... .......... ..........
M3.26  ....----.. .......... ...IE...... -----..... .T........ .----N.... .......... .......... ..........
```

FIGURE 6 Alignment of deduced N-terminal amino acid sequences of seven alleles of *emm* from Group A *Streptococcus* serotype M1 isolates and seven alleles of *emm* from serotype M3 isolates. The region shown represents amino acids beginning after cleavage of a 26-amino acid leader peptide sequence. Amino acid residues that are identical to those encoded by *emm1.0* and *emm3.0* are represented by dots, and gaps (−) are inserted to maximize amino acid identity among the M1 and M3 variants. The M3 sequences are representative variants identified among a sample of several hundred serotype M3 strains recovered from global sources and characterized in the laboratory of J. M. Musser. Based in part on M1 data of Musser *et al.* (1995).

has been caused, in part, by a distinct subclone expressing serotype M1 protein. This implies large interclonal variance in traits such as the degree of virulence and disease association. Similar studies conducted with samples of strains expressing serotype M3 have also revealed considerable levels of chromosomal heterogeneity and *emm* sequence diversity (Fig. 6). Although these studies on the microdissection of allelic variation in the M1 gene have been useful in identifying the major clone responsible for streptococcal TSS, they do not implicate the M protein directly as the cause of the recent disease surge.

D. Host Factors in the Occurrence of Disease

One of the particularly elusive goals in studies on streptococcal diseases is the elucidation of risk factors that identify those individuals who are unusually susceptible to the occurrence of severe infection such as STSS or sequelae such as rheumatic fever and acute nephritis. Despite decades of streptococcal research, our understanding of host risk factors at the molecular level is limited.

1. Acute Rheumatic Fever

Epidemiological investigations on the startling reappearance of ARF in the United States in the 1980s led to the findings that most patients were white, lived in suburban or rural settings, and had access to medical care. These results were unexpected because recent dogma held that ARF patients (in the USA, at least) were members of ethnic minorities, lived in cramped urban environments, and had limited health care access. It is safe to say that we have no significant insight into the molecular mechanisms responsible for these fluctuations in the occurrence of ARF.

With the development of molecular strategies designed to permit large-scale automated interrogation of sequence diversity in the human genome, analyses of the genetics of susceptibility to infectious agents is now beginning in earnest (Dean *et al.*, 1996; Jouanguy *et al.*, 1996). It has long been speculated that there is a human genetic component to streptococcal disease susceptibility, most notably rheumatic fever (Unny and Middlebrooks, 1983; Zabriskie, 1985). This idea developed in part out of the results of studies conducted at Warren Air Force Base. The attack rate of rheumatic fever following streptococcal pharyngitis in recruits who did not receive antibiotic treatment was roughly 3% (Denny *et al.*, 1950). The observation that the ARF attack rate was constant in the population, despite the fact that several different M serotypes were prevalent at the base during the study period, led to the ideas that (i) rheumatic fever was not specific to any single M protein serotype, (ii) all pharyngeal infections were potentially rheumatogenic, (iii) curtailing of an antigenic stimulus, thus blunting the immune response, would prevent the disease, and (iv) host factors play an important role in development of this disease.

Most work in the area of host susceptibility has been directed to analysis of major histocompatibility complex markers in the context of rheumatic fever (reviewed in Unny and Middlebrooks, 1983). However, the data have been inconsistent, sometimes conflicting, and inconclusive. For example, studies of identical twins showed a relatively low concordance compared with several other infectious diseases that have a genetic predisposition (Taranta *et al.*, 1959).

More recently, nine patients with acute rheumatic fever were studied to determine whether the Vβ repertoire was altered in peripheral blood T cells (Abbott *et al.*, 1996). The analysis was undertaken because there is evidence that activated T cells are involved in the pathogenesis of ARF, and patients with severe invasive GAS disease have abnormalities in the Vβ repertoire of peripheral blood T cells. These changes are thought to be caused by superantigens expressed by many GAS. However, the mean Vβ repertoire of peripheral blood T cells in the nine patients was similar to that of 34 controls and, moreover, did not change during 6 months of follow-up. The results suggested that a superantigen is not involved in ARF pathogenesis in a direct way. Although clearly an important issue, study of the host genetics of susceptibility to streptococcal infection has not been examined thoroughly with contemporary biomedical methods.

2. Invasive Infections

Several reports on a large number of patients have identified predisposing factors that contribute to the occurrence of GAS invasive diseases. In the initial description of 20 patients from the Rocky Mountain region by Stevens *et al.* (1989), about one-half had no apparent predisposing factors. In Sweden, a country that experienced a doubling of the invasive infection case rate in the winter of 1988 to 1989, study of 79 hospitalized patients found that two-thirds had predisposing conditions, including cardiovascular disease, arthritis, trauma or skin ulcer, alcoholism or drug abuse, malignancy, use of corticosteroids, diabetes mellitus, or recent varicella (Strömberg *et al.*, 1991). A substantial increase in invasive disease during the same time period was reported in Denmark. As in Sweden, many patients had underlying medical conditions that apparently predisposed to disease. A nearly 3-fold increase in occurrence of GAS bacteremia compared with previous years was reported for Norway (Martin and Høiby, 1990). The increase was most marked in older children and adults 30–50 years old. A study of seven epidemiologically linked patients in rural Minnesota reported that five had underlying medical conditions or predisposing factors that included most of those described in the Scandinavian cases.

Numerous reports have demonstrated that children with varicella, or a history of recent varicella, are clearly at enhanced risk (Cowan *et al.*, 1993; Doctor *et al.*, 1995; Vugia *et al.*, 1996). The skin lesions are a likely site for infection with GAS. This association has been known since early in the twentieth century (Bullowa and Wishik, 1935). The death of five children with streptococcal TSS associated with varicella in Los Angeles, and two postpartum women in Buffalo, has heightened the concern of public health officials. As this chapter was being finished in early 1997, an ongoing outbreak of GAS disease in Rochester, New York, had claimed five lives. The molecular underpinnings of these important observations regarding host risk factors are unknown.

Because so many aspects of the molecular pathogenesis of streptococcal disease remain obscure, we are largely forced to speculate about the causes of the reemergence of ARF and streptococcal TSS. It is reasonable to presume that population changes in host immune status (herd immunity) are involved. It is important to differentiate among the mechanisms that lead to changes in herd immunity. For example, it is possible that herd immunity can be altered simply by the passage of time, as, for example, might occur with an aging cohort. Pro-

TABLE V   Median Levels of Antibody Titers in
Acute-Phase and Convalescent-Phase Sera from
Patients with Group A Streptococcal Infections[a]

| | Disease course | | |
| | Death | Bacteremia | Uncomplicated |
Serum phase	($n = 6$)	($n = 18$)	($n = 11$)
Acute			
ET-A	36	43	49
ET-B	7[b]	61	37
ET-C	23	45	11
M1 antigen	8[c]	11	48
Convalescent			
ET-A		110	78
ET-B		151	91
ET-C		134	115
M1 antigen		93	35

[a] Reprinted from Holm *et al.* (1992). Antibody titers were measured by enzyme-linked immunosorbent assay (ELISA). ET, erythrogenic toxin (types A–C).

[b] Significantly different ($p < 0.05$) from ET-B antibody titers in bacteremia and uncomplicated infections.

[c] Significantly different ($p < 0.05$) from patients with uncomplicated infections.

tective levels of antibody may simply decrease among members in the cohort to a point below which many individuals are now susceptible to colonization and to invasion into deeper tissues. One mechanism postulated to contribute to changes in herd immunity is widespread use of antibiotics to treat children with group A streptococcal pharyngitis. This idea grew out of the unambiguous demonstration that penicillin treatment of patients with pharyngitis blunted the host immune response (Table VI).

Other superb immunologic studies conducted by Holm's group (Holm *et al.*, 1992; Norrby-Teglund *et al.*, 1994, 1996) have contributed very important information on the relationship between serum antibody levels and the risk

TABLE VI   Anti-Streptolysin O Response to
Type 12 Group A Streptococcal Infections:
Effect of Treatment on Response[a]

Treatment	Number of patients	Mean ASO rise[b]
None	108	148
Penicillin	44	72
γ-Globulin	32	246

[a] Modified from Stetson *et al.* (1955).

[b] Calculated as mean convalescent anti-streptolysin O (ASO) titer minus mean initial titer.

of STSS. Serologic studies were performed on a group of patients with invasive streptococcal disease in Sweden during 1988–1989, when a countrywide outbreak caused mainly by M1T1 strains occurred. All isolates produced high amounts of streptococcal pyrogenic exotoxins B and C. Importantly, patients with uncomplicated infections had high serum antibody levels to these toxins, and to the M1 antigen (Table V), whereas patients with bacteremia, or fatal infections, had low antibody levels to these substances. The most striking observation was that individuals with fatal infection also had very low antibody levels to streptococcal pyrogenic exotoxin B, a potent cysteine protease (Musser, 1997). The level of antibody directed against exotoxins A and C was apparently not an important variable in these patients. The authors proposed that a combination of production of large amounts of exotoxin B and low antibody titers to it, and to M antigen of the infecting strain, is the determining factor for the outcome of the infection. Inasmuch as antibody directed against M protein is well known to elicit type-specific immunity, a protective role for antibody directed against this molecule is not surprising. However, the implication that serum antibody directed against SpeB may also participate in host protection was unexpected, and it has led, in part, to renewed interest in this molecule (Musser, 1997). Immunization of mice with purified cysteine protease elicits significant protective immunity against intraperitoneal challenge with live, virulent streptococci that produce this enzyme (Kapur *et al.*, 1994). Holm's data also imply that decreased levels of antibody directed against M antigen or SpeB are important factors in the increases in streptococcal disease frequency and severity.

Changes in herd immunity were thought to play a role in the occurrence of a 1966 epidemic of poststreptococcal nephritis due to type M49 GAS among Chippewa Indians living on the Red Lake Indian Reservation in northern Minnesota (Anthony *et al.*, 1967). It was noted that an M49-associated nephritis epidemic also had occurred 13 years previously in 1953, and that in the intervening years, despite intensive surveillance, M49 strains were rare or absent. Because all patients in the 1966 epidemic were too young to have acquired type-specific immunity as a result of the 1953 epidemic, it was suggested that the affected children were members of an immunologically virgin cohort.

It is also possible that a precipitous change in the genetic makeup of the bacterium (as might occur with acquisition or creation of one or more new virulence factors, or structural variants of a virulence factor) results in the *de novo* genesis of a streptococcal variant with increased fitness. Such a process would thrust a potentially devastating pathogen into a naive population, in a process that resembles the well-known antigenic shifts described in influenza virus epidemiology. From the standpoint of host–pathogen interactions, the introduction of a preexisting strain to a new geographic area would be functionally equivalent to *de novo* origin of a streptococcal variant. The recent introduction model has the advantage that one need not invoke pathogen genetic gymnastics. Recent analysis of a gene (*sic*) encoding a hypervariable protein that inhibits human complement function suggests that allelic variation may contribute to temporal variation in disease frequency (Stockbauer *et al.*, 1997). A study of strains recovered from two disease peaks of M1 disease in the former East Germany found that a single amino acid substitution in the Sic protein was associated with a striking increase in GAS infections in that country.

III. CONCLUSIONS

The molecular epidemiology reviewed here has revealed that the resurgence of streptococcal disease is associated with two distinct GAS clones that have been responsible for a substantial proportion of the invasive disease episodes. These clones possess the gene encoding exotoxin A, and represent distinct genotypes expressing serotype M1 and M3 proteins. These data add to the important concept that an outbreak or an epidemic of an infection is due to a microbial clone or cell line. Because there is significant chromosomal variation among strains assigned to the same M serologic type, future studies on the epidemiology of GAS must employ genetic markers in addition to antigenic identification. This is an especially important issue because evidence has been presented that human M type specific opsonic antibody may have greater reactivity against the homologous strain, whereas there is less reactivity for other strains of the same M type (de Malmanche and Martin, 1994). This strain variability appears to be related to the structural variation in M proteins that can occur among isolates of the same M type (Harbaugh *et al.*, 1993; Musser *et al.*, 1995).

Serotype M1 and M3 strains recovered from recent patients with invasive disease have single mutations in the *speA* gene that result in single amino acid replacements in an area of the molecule that has functional importance. One of these toxin variants (SpeA3) has significantly increased mitogenic activity and affinity for MHC class II compared with that of SpeA1 (Kline and Collins, 1996). Clearly, more must be learned about the diversity of nucleotide and amino acid sequences of microbial products that may be involved in streptococcal virulence and pathogenesis, as has been accomplished with other pathogens (Kawaoka and Webster, 1988).

One of the unanticipated findings from the study of the molecular evolutionary genetics of GAS is that the population structure derives from recombination events rather than clone dominance (Kehoe *et al.*, 1996). Most pathogenic bacteria evolve largely as distinct chromosomal lineages with little or no effective lateral transfer of chromosomal genes. This is not the case in GAS. Population genetic analyses have identified a relative lack of linkage disequilibrium in this species. Linkage disequilibrium refers to the occurrence of nonrandom association of alleles over loci. A lack of linkage disequilibrium is very strong evidence that effective horizontal gene transfer is a relatively frequent occurrence. Stated another way, alleles are reassorting more or less at random over loci, which implies that recombination is generating new associations of virulence genes and other genes. The occurrence of significant levels of recombination means that this pathogen can accelerate the overall rate of evolution in the species, in excess of that due to simple nucleotide substitutions.

Horizontal gene transfer processes are playing a fundamental role in generating diversity in GAS (Table VII). The discovery of significant levels of chromosomal gene recombination among isolates of GAS was unexpected because it was the first time such a finding has been made for a microbe in which DNA transformation is not known to occur. This finding has major implications for understanding how new, unusually virulent cell lines of the species are generated. Extrapolating to disease pathogenesis, such genetic and molecular processes may be driving variation over time in disease frequency and severity.

■ **TABLE VII** Genes Encoding Putative or Proven Group A Streptococcus Virulence Factors Implicated in Horizontal Transfer in Natural Populations

Gene	Product	Genomic location	Representative reference
speA	Pyrogenic exotoxin A	Bacteriophage	Nelson *et al.*, 1991
speC	Pyrogenic exotoxin C	Bacteriophage	Kapur *et al.*, 1992
emm	M protein	Chromosomal	Whatmore *et al.*, 1994
skn	Streptokinase	Chromosomal	Kapur *et al.*, 1995
enn	M-like protein	Chromosomal	Whatmore *et al.*, 1995
ssa	Streptococcal superantigen A	Bacteriophage?	Reda *et al.*, 1996
hyl	Hyaluronidase	Bacteriophage	Marciel *et al.*, 1997

Although there is no direct evidence that pyrogenic exotoxins are directly involved in the resurgence of invasive streptococcal disease including STSS, several lines of evidence support the idea that they contribute, in part, to pathogenesis. The evidence is most compelling for involvement of SpeA and SpeB. SpeA presumably acts as a superantigen to elicit enhanced cytokine production and shock, whereas the action of SpeB may be initiated through a direct destruction of host macromolecules (Musser, 1997). The involvement of exotoxins in the pathogenesis of rheumatic fever has not been established.

Evidence exists that serum antibody directed against M protein and SpeB is involved in protection against severe invasive group A streptococcal infection. Although it is probable that a decline in herd immunity has been an important factor in disease resurgence, no direct evidence exists to prove this notion. It is possible that aggressive early treatment of streptococcal pharyngitis has contributed to altered host immune status, thereby creating a large cohort of individuals at risk for developing disease when confronted by a virulent organism. The host genetic factors contributing to development of pharyngitis, invasive disease, and rheumatic fever are largely unknown.

At the close of the twentieth century, ARF has been essentially banished from the developed world, except for the unexpected, but limited, outbreaks reported here. The demise of ARF is due largely to primary prevention with penicillin treatment of GAS infections, and the prevention of secondary attacks with prolonged penicillin prophylaxis in patients who had a previous attack. Such a state of affairs would have never been thought possible prior to 1950, when there were a number of hospitals in the United States that were devoted entirely to the care of patients with rheumatic fever and rheumatic heart disease.

And yet, for much of the world, ARF and RHD remain a common affliction, particularly among the majority of the population who lack access to adequate medical care. These millions remain at risk of developing ARF following GAS pharyngitis. A vaccine to prevent infection is the only reasonable long term solution to the GAS problem. Until recently, a vaccine seemed an insurmountable challenge. However, recent work on several GAS target molecules provides hope for a future successful vaccine.

ACKNOWLEDGMENTS

We thank D. R. Martin and D. E. Low for sharing unpublished data, and D. Meyer for excellent assistance in preparation of the manuscript. Research in the laboratory of J. M. M. is supported by U.S. Public Health Service Grants AI-33119, AI-37004, AI-41168, and DA-09238, and by Grant 004949-016 from the Texas Advanced Research/Advanced Technology Program. J. M. M. is an Established Investigator of the American Heart Association.

REFERENCES

Abbott, W. G. H., Skinner, M. A., Voss, L., Lennon, D., Tan, P. L. J., Fraser, J. D., Simpson, I. J., Ameratunga, R., and Geursen, A. (1996). Repertoire of transcribed peripheral blood T-cell-receptor beta chain variable-region genes in acute rheumatic fever. *Infect. Immun.* **64**, 2842–2845.

Agerton, T., Valway, S., Gore, B., Pozsik, C., Plikaytis, B., Woodley, C., and Onorato, I. (1997). Transmission of a highly drug-resistant strain (strain W1) of *Mycobacterium tuberculosis*. Community outbreak and nosocomial transmission via a contaminated bronchoscope. *JAMA* **278**, 1073–1077.

Anthony, B. F., Kaplan, E. L., Chapman, S. S., Quie, P. G., and Wannamaker, L. W. (1967). Epidemic acute nephritis with reappearance of type-49 streptococcus. *Lancet* **2**, 787–790.

Bessen, D., and Fischetti, V. A. (1990). Synthetic peptide vaccine against mucosal colonization by group A streptococci. I. Protection against a heterologous M serotype with shared C repeat region epitopes. *J. Immunol.* **145**, 1251–1256.

Bessen, D., Jones, K. F., and Fischetti, V. A. (1989). Evidence for two distinct classes of streptococcal M protein and their relationship to rheumatic fever. *J. Exp. Med.* **169**, 269–283.

Bessen, D. E., Sotir, C. M., Readdy, T. L., and Hollingshead, S. K. (1996). Genetic correlates of throat and skin isolates of group A streptococci. *J. Infect. Dis.* **173**, 896–900.

Bessen, D. E., Veasy, L. G., Hill, H. R., Augustine, N. H., and Fischetti, V. A. (1995). Serologic evidence for a class I group A streptococcal infection among rheumatic fever patients. *J. Infect. Dis.* **172**, 1608–1611.

Bifani, P., Plikaytis, B. B., Kapur, V., Stockbauer, K., Pan, X., Lutfey, M. L., Moghazeh, S. L., Eisner, W., Daniel, T. M., Kaplan, M. H., Crawford, J. T., Musser, J. M., and Kreiswirth, B. N. (1996). Origin and interstate spread of a New York City multidrug-resistant *Mycobacterium tuberculosis* clone family. *JAMA* **275**, 452–457.

Bisno, A. L. (1991). Group A streptococcal infections and acute rheumatic fever. *N. Engl. J. Med.* **325**, 783–793.

Blomster-Hautamaa, D. A., Kreiswirth, B. N., Kornblum, J. S., Novick, R. P., and Schlievert, P. M. (1986). The nucleotide and partial amino acid sequence of toxic shock syndrome toxin-1. *J. Biol. Chem.* **261**, 15783–15786.

Bonventre, P. F., Linnemann, C., Weckbach, L. S., Staneck, J. L., Buncher, C. R., Vigdorth, E., Ritz, H., Archer, D., and Smith, B. (1984). Antibody responses to toxic-shock-syndrome (TSS) toxin by patients with TSS and by healthy staphylococcal carriers. *J. Infect. Dis.* **150**, 662–666.

Brogan, T. V., Nizet, V., Waldhausen, J. H. T., Rubens, C. E., and Clarke, W. R. (1995). Group A streptococcal necrotizing fasciitis complicating primary varicella: A series of fourteen patients. *Pediatr. Infect. Dis. J.* **14**, 588–594.

Broome, C. V. (1989). Epidemiology of toxic shock syndrome in the United States: Overview. *Rev. Infect. Dis.* **11**, S14–S21.

Bucher, A., Martin, P. R., Høiby, E. A., Halstensen, A., Odegaard, A., Hellum, K. B., Westlie, L., and Hallan, S. (1992). Spectrum of disease in bacteraemic patients during a *Streptococcus pyogenes* serotype M-1 epidemic in Norway in 1988. *Eur. J. Clin. Microbiol. Infect. Dis.* **11**, 416–426.

Bullowa, J. G. M., and Wishik, S. M. (1935). Complications of varicella: I. Their occurrence among 2534 patients. *Am. J. Dis. Child.* **49**, 923–926.

Carapetis, J., Robins-Browne, R., Martin, D., Shelby-James, T., and Hogg, G. (1995). Increasing severity of invasive group A streptococcal disease in Australia: Clinical and molecular epidemiological features and identification of a new virulent M-nontypeable clone. *Clin. Infect. Dis.* **21**, 1220–1227.

Carapetis, J. R., Wolff, D. R., and Currie, B. J. (1996). Acute rheumatic fever and rheumatic heart disease in the Top End of Australia's Northern Territory. *Med. J. Aust.* **164**, 146–149.

Cartwright, K., Logan, M., McNulty, C., Harrison, S., George, R., Efstratiou, A., McEnvoy, M., and Begg, N. (1995). A cluster of cases of streptococcal necrotizing fasciitis in Gloucestershire. *Epidemiol. Infect.* **115**, 387–397.

Centers for Disease Control. (1988). Acute rheumatic fever among Army trainees—Fort Leonard Wood, Missouri, 1987–1988. *Morbidity and Mortality Weekly Report* **37**, 519–522.

Chelsom, J., Halstensen, A., Haga, T., and Høiby, E. A. (1994). Necrotising fasciitis due to group A streptococci in western Norway: Incidence and clinical features. *Lancet* **344**, 1111–1115.

Cheung, A. L., Koomey, J. M., Butler, C. A., Projan, S. J., and Fischetti, V. A. (1992). Regulation of exoprotein expression in *Staphylococcus aureus* by a locus (*sar*) distinct from *agr. Proc. Natl. Acad. Sci. U.S.A.* **89**, 6462–6466.

Choi, Y., Lafferty, J. A., Clements, J. R., Todd, J. K., Gelfand, E. W., Kappler, J., Marrack, P., and Kotzin, B. L. (1990). Selective expansion of T cells expressing Vβ2 in toxic shock syndrome. *J. Exp. Med.* **172**, 981–984.

Cleary, P. P., Kaplan, E. L., Handley, J. P., Wlazlo, A., Kim, M. H., Hauser, A. R., and Schlievert, P. M. (1992). Clonal basis for resurgence of serious *Streptococcus pyogenes* disease in the 1980s. *Lancet* **339**, 518–521.

Cockerill III, F. R., MacDonald, K. L., Thompson, R. L., Roberson, F., Kohner, P. C., Besser-Wiek, J., Manahan, J. M., Musser, J. M., Schlievert, P. M., Talbot, J., Frankfort, B., Steckelberg, J. M., Wilson, W. R., Osterholm, M. T., and the Investigation Team. (1997). An outbreak of severe invasive group A streptococcal disease associated with high carriage rates of the invasive clone among school-aged children. *JAMA* **277**, 38–43.

Congeni, B., Rizzo, C., Congeni, J., and Sreenivasan, V. V. (1987). Outbreak of acute rheumatic fever in northeast Ohio. *J. Pediatr.* **111**, 176–179.

Courtney, H. S., Bronze, M. S., Dale, J. B., and Hasty, D. L. (1994). Analysis of the role of M24 protein in group A streptoccal adherence and colonization by use of Ω-interposon mutagenesis. *Infect. Immun.* **62**, 4868–4873.

Cowan, M. R., Primm, P. A., Scott, S. M., Abramo, T. J., and Wiebe, R. A. (1993). Serious group A β-hemolytic streptococcal infections complicating varicella. *Ann. Emerg. Med.* **23**, 818–823.

Cunningham, M. W. (1992). Group A streptococci: Molecular mimicry, autoimmunity and infection. In "Microbial Adhesion and Invasion" (M. Hook and L. Switalski, eds.), pp. 149–169. Springer-Verlag, New York.

Cunningham, M. W., Antone, S. M., Smart, M., Liu, R., and Kosanke, S. (1997). Molecular analysis of human cardiac myosin-cross-reactive B- and T-cell epitopes of the group A streptococcal M5 protein. *Infect. Immun.* **65**, 3913–3923.

Dale, J. B., and Chiang, E. C. (1995). Intranasal immunization with recombinant group A streptococcal M protein fragment fused to the B subunit of *Escherichia coli* labile toxin protects mice against systemic challenge infections. *J. Infect. Dis.* **171**, 1038–1041.

Dale, J. B., Washburn, R. G., Marques, M. B., and Wessels, M. R. (1996). Hyaluronate capsule and surface M protein in resistance to opsonization of group A streptococci. *Infect. Immun.* **64**, 1495–1501.

Davies, H. D., McGeer, A., Schwartz, B., Green, K., Cann, D., Simor, A. E., Low, D. E., and The Ontario Group A Streptococcal Study Group. (1996). Invasive group A streptococcal infections in Ontario, Canada. *N. Engl. J. Med.* **335**, 547–554.

Dean, M., Carrington, M., Winkler, C., Huttley, G. A., Smith, M. W., Allikmets, R., Goedert, J. J., Buchbinder, S. P., Vittinghoff, E., Gomperts, E., Donfield, S., Vlahov, D., Kaslow, R., Saah, A., Rinaldo, C., Detels, R., Hemophilia Growth and Development Study, Multicenter AIDS Cohort Study, Multicenter Hemophilia Cohort Study, San Francisco City Cohort, ALIVE Study, and O'Brien, S. J. (1996). Genetic restriction of HIV-1 infection and progression to AIDS by a deletion allele of the *CKR5* structural gene. *Science* **273**, 1856–1862.

de Azavedo, J. C., Foster, T. J., Hartigan, P. J., Arbuthnott, J. P., O'Reilly, M., Kreiswirth, B. N., and Novick, R. P. (1985). Expression of the cloned toxic shock syndrome toxin 1 gene (*tst*) in vivo with a rabbit uterine model. *Infect. Immun.* **50**, 304–309.

de Malmanche, S. A., and Martin, D. R. (1994). Protective immunity to the group A *Streptococcus* may be only strain specific. *Med. Microbiol. Immunol.* **183**, 299–306.

Demers, B., Simor, A. E., Vellend, H., Schlievert, P. M., Bryne, S., Jamieson, F., Walmsley, S., and Low, D. E. (1993). Severe invasive group A streptococcal infections in Ontario, Canada: 1987–1991. *Clin. Infect. Dis.* **16**, 792–800.

Denny, F. W., Wannamaker, L. W., Brink, W. R., Rammelkamp, C. H., Jr., and Custer, E. A. (1950). Prevention of rheumatic fever. Treatment of the preceding streptococcic infection. *JAMA* **143**, 151–153.

Deresiewicz, R. L., Woo, J. H., Chan, M., Finberg, R. W., and Kasper, D. L. (1994). Mutations affecting the activity of toxic shock syndrome toxin-1. *Biochemistry* **33**, 12844–12851.

DiPersio, J. R., File, T. M., Jr., Stevens, D. L., Gardner, W. G., Petropoulos, G., and Dinsa, K. (1996). Spread of serious disease-producing M3 clones of group A streptococcus among family members and health care workers. *Clin. Infect. Dis.* **22**, 490–495.

Doctor, A., Harper, M. B., and Fleisher, G. R. (1995). Group A β-hemolytic streptococcal bacteremia: Historical overview, changing incidence, and recent association with varicella. *Pediatrics* **96**, 428–433.

Fattom, A. F., and Naso, R. B. (1997). Vaccines for *Staphylococcus aureus* infections. *In* "New Generation Vaccines" (M. M. Levine, G. C. Woodrow, J. B. Kaper, G. S. Cobon, Eds.) pp. 979–988. Marcel Dekker, Inc., New York.

Fischetti, V. A. (1989). Streptococcal M protein: Molecular design and biological behavior. *Clin. Microbiol. Rev.* **2**, 285–314.

Fischetti, V. A., Medaglini, D., and Pozzi, G. (1996). Gram-positive commensal bacteria for mucosal vaccine delivery. *Curr. Opin. Biotech.* **7**, 659–666.

Gaworzewska, E., and Colman, G. (1988). Changes in the pattern of infection caused by *Streptococcus pyogenes*. *Epidemiol. Infect.* **100**, 257–269.

Givner, L. B., Abramson, J. S., and Wasilauskas, B. (1991). Apparent increase in the incidence of invasive group A beta-hemolytic streptococcal disease in children. *J. Pediatr.* **118**, 341–346.

Go, M. F., Kapur, V., Graham, D. Y., and Musser, J. M. (1996). Population genetic analysis of *Helicobacter pylori* by multilocus enzyme electrophoresis: Extensive allelic diversity and recombinational population structure. *J. Bacteriol.* **178**, 3934–3938.

Gray, G. C., Escamilla, J., Hyams, K. C., Struewing, J. P., Kaplan, E. L., and Tupponce, A. K. (1991). Hyperendemic *Streptococcus pyogenes* infection despite prophylaxis with penicillin G benzathine. *N. Engl. J. Med.* **325**, 92–97.

Griffiths, S. P., and Gersony, W. M. (1990). Acute rheumatic fever in New York City (1969 to 1988): A comparative study of two decades. *J. Pediatr.* **116**, 882–887.

Hanski, E., and Caparon, M. (1992). Protein F, a fibronectin-binding protein, is an adhesin of the group A streptococcus *Streptococcus pyogenes*. *Proc. Natl. Acad. Sci. U.S.A.* **89**, 6172–6176.

Harbaugh, M. P., Podbielski, A., Hugl, S., and Cleary, P. P. (1993). Nucleotide substitutions and small-scale insertion produce size and antigenic variation in group A streptococcal M1 protein. *Mol. Microbiol.* **8**, 981–991.

Hasty, D. L., Ofek, I., Courtney, H. S., and Doyle, R. J. (1992). Multiple adhesins of streptococci. *Infect. Immun.* **60**, 2147–2152.

Hoge, C. W., Schwartz, B., Talkington, D. F., Breiman, R. F., MacNeill, E. M., and Englender, S. J. (1993). The changing epidemiology of invasive group A streptococcal infections and the emergence of streptococcal toxic shock-like syndrome. A retrospective population-based study. *JAMA* **269**, 384–389.

Holm, S. E., Norrby, A., Bergholm, A.-M., and Norgren, M. (1992). Aspects of pathogenesis of serious group A streptococcal infections in Sweden, 1988–1989. *J. Infect. Dis.* **166**, 31–37.

Hosier, D. M., Craenen, J. M., Teske, D. W., and Wheller, J. J. (1987). Resurgence of acute rheumatic fever. *Amer. J. Dis. Child.* **141**, 730–733.

Inagaki, Y., Konda, T., Murayama, S., Yamai, S., Matsushima, A., Gyobu, Y., Tanaka, D., Tamaru, A., Katsukawa, C., Katayama, A., Tomita, M., Fuchi, Y., Hoashi, K., Watanabe, H., and The Working Group for Group A Streptococci in Japan. (1997). Serotyping of *Streptococcus pyogenes* isolated from common and severe invasive infections in Japan, 1990–5: implication of the T3 serotype strain-expansion in TSLS. *Epidemiol. Infect.* **119**, 41–48.

Jevon, G. P., Dunne, W. M., Jr., Hawkins, H. K., Armstrong, D. L., and Musser, J. M. (1994). Fatal group A streptococcal meningitis and toxic shock-like syndrome: Case report. *Clin. Infect. Dis.* **18**, 91–93.

Ji, G., Beavis, R., and Novick, R. P. (1997). Bacterial interference caused by autoinducing peptide variants. *Science* **276**, 2027–2030.

Ji, Y., McLandsborough, L., Kondagunta, A., and Cleary, P. P. (1996). C5a peptidase alters clearance and trafficking of group A streptococci by infected mice. *Infect. Immun.* **64**, 503–510.

Jouanguy, E., Altare, F., Lamhamedi, S., P. Revy, J.-F. Emile, M. Newport, M. Levin, S. Blanche, E. Seboun, A. Fischer, and J.-L. Casanova. (1996). Interferon-γ-receptor deficiency in an infant with fatal Bacille Calmette–Guerin infection. *N. Engl. J. Med.* **335**, 1956–1961.

Kapur, V., Nelson, K., Schlievert, P. M., Selander, R. K., and Musser, J. M. (1992). Molecular population genetic evidence of horizontal spread of two alleles of the pyrogenic exotoxin C gene (*speC*) among pathogenic clones of *Streptococcus pyogenes*. *Infect. Immun.* **60**, 3513–3517.

Kapur, V., Maffei, J. T., Greer, R. S., Li, L.-L., Adams, G. J., and Musser, J. M. (1994). Vaccination with streptococcal extracellular cysteine protease (interleukin-1β convertase) protects mice against challenge with heterologous group A streptococci. *Microb. Pathog.* **16**, 443–450.

Kapur, V., Kanjilal, S., Hamrick, M. R., Li, L.-L., Whittam, T. S., Sawyer, S. A., and Musser, J. M. (1995). Molecular population genetic analysis of the streptokinase gene of *Streptococcus pyogenes*: Mosaic alleles generated by recombination. *Mol. Microbiol.* **16**, 509–519.

Katz, A. R., and Morens, D. M. (1992). Severe streptococcal infections in historical perspective. *Clin. Infect. Dis.* **14**, 298–307.

Kaul, R., McGeer, A., Low, D. E., Green, K., Schwartz, B., Ontario Group A Streptococcal Study, and Simor, A. E. (1997). Population-based surveillance for group A streptococcal necrotizing fasciitis: Clinical features, prognostic indicators, and microbiologic analysis of seventy-seven cases. *Am. J. Med.* **103**, 18–24.

Kawaoka, Y., and Webster, R. G. (1988). Molecular mechanism of acquisition of virulence in influenza virus in nature. *Microb. Pathog.* **5**, 311–318.

Kehoe, M. A., Kapur, V., Whatmore, A., and Musser, J. M. (1996). Horizontal gene transfer among group A streptococci: Implications for pathogenesis and epidemiology. *Trends Microbiol.* **4**, 436–443.

Kiska, D. L., Thiede, B., Caracciolo, J., Jordan, M., Johnson, D., Kaplan, E. L., Gruninger, R. P., Lohr, J. A., Gilligan, P. H., and Denny, Jr., F. W. (1997). Invasive group A streptococcal infections in North Carolina: Epidemiology, clinical features, and genetic and serotype analysis of causative organisms. *J. Infect. Dis.* **176**, 992–1000.

Kline, J. B., and Collins, C. M. (1996). Analysis of the superantigenic activity of mutant and allelic forms of streptococcal pyrogenic exotoxin A. *Infect. Immun.* **64**, 861–869.

Kornblum, J., Kreiswirth, B., Projan, S. J., Ross, H., and Novick, R. P. (1990). *Agr*: a polycistronic locus regulating exoprotein synthesis in *Staphylococcus aureus*. In "Molecular Biology of the Staphylococci" (R. N. Novick and R. A. Skurray, eds.), pp. 373–402. VCH, New York.

Lancefeld, R. C. (1941). Specific relationship of cell composition to biological activity of hemolytic streptococci. Harvey Lectures (1940–1941), **36**, 251–290.

LaPenta, D., Rubens, C., Chi, E., and Cleary, P. P. (1994). Group A streptococci efficiently invade human respiratory epithelial cells. *Proc. Natl. Acad. Sci. U.S.A.* **91**, 12115–12119.

Lee, P. K., Kreiswirth, B. N., Deringer, J. R., Projan, S. J., Eisner, W., Smith, B. L., Carlson, E., Novick, R. P., and Schlievert, P. M. (1992). Nucleotide sequences and biologic properties of toxic shock syndrome toxin 1 from ovine- and bovine-associated *Staphylococcus aureus*. *J. Infect. Dis.* **165**, 1056–1063.

Leggiadro, R. J., Birnbaum, S. E., Chase, N. A., and Myers, L. K. (1990). A resurgence of acute rheumatic fever in a mid-South children's hospital. *S. Med. J.* **83**, 1418–1420.

Lukomski, S., Sreevatsan, S., Amberg, C., Reichardt, W., Woischnik, M., Podbielski, A., and Musser, J. M. (1997). Inactivation of *Streptococcus pyogenes* extracellular cysteine protease significantly decreases mouse lethality of serotype M3 and M49 strains. *J. Clin. Invest.* **99**, 2574–2580.

Lukomski, S., Burns, E. H., Jr., Wyde, P. R., Podbielski, A., Rurangirwa, J., Moore-Poveda, D. K., and Musser, J. M. (1998). Genetic inactivation of the extracellular cysteine protease (SpeB) expressed by *streptococcus pyogenes* decreases resistance to phagocytosis and dissemination to organs. *Infect. Immun.* In press.

McGeer, A., Green, K., Cann, D., Schwartz, B., Kaul, R., Fletcher, A., Matsumura, S., Ontario Group A Streptococcal Study, and Low, D. E. (1995). Changing epidemiology of invasive group A streptococcal infection—Population based surveillance, Ontario Canada, 1992–5. 35th Intersciences Conference on Antimicrobial Agents and Chemotherapy Annual Meeting, San Francisco, California (Abstract K135).

Madsen, S. T. (1973). Scarlet fever and erysipelas in Norway during the last hundred years. *Infection* **1**, 76–81.

Marciel, A. M., Kapur, V., and Musser, J. M. (1997). Molecular population genetic analysis of a *Streptococcus pyogenes* bacteriophage-encoded hyaluronidase gene: Recombination contributes to allelic diversity. *Microb. Pathog.* **22**, 209–217.

Martin, D. R., and Single, L. A. (1993). Molecular epidemiology of group A streptococcus M type 1 infections. *J. Infect. Dis.* **167**, 1112–1117.

Martin, D. R., and Sriprakash, K. S. (1996). Epidemiology of group A streptococcal disease in Australia and New Zealand. *Rec. Adv. Microbiol.* **4**, 1–40.

Martin, P. R., and Høiby, E. A. (1990). Streptococcal serogroup A epidemic in Norway 1987–1988. *Scand. J. Infect. Dis.* **22**, 421–429.

Moses, A. E., Wessels, M. R., Zalcman, K., Albertí, S., Natanson-Yaron, S., Menes, T., and Hanski, E. (1997). Relative contributions of hyaluronic acid capsule and M protein to virulence in a mucoid strain of the group A *Streptococcus*. *Infect. Immun.* **65**, 64–71.

Muotiala, A., Seppälä, H., Huovinen, P., and Vuopio-Varkila, J. (1997). Molecular comparison of group A streptococci of T1M1 serotype from invasive and noninvasive infections in Finland. *J. Infect. Dis.* **175**, 392–399.

Murray, D. L., Prasad, G. S., Earhart, C. A., Leonard, B. A. B., Kreiswirth, B. N., Novick, R. P., Ohlendorf, D. H., and Schlievert, P. M. (1994). Immunological and biochemical properties of mutants of toxic shock syndrome toxin-1. *J. Immunol.* **152**, 87–95.

Musser, J. M. (1996). Molecular population genetic analysis of emerged bacterial pathogens: Selected insights. *Emerging Infect. Dis.* **2**, 1–17.

Musser, J. M. (1997), Streptococcal superantigen, mitogenic factor, and pyrogenic exotoxin B expressed by *Streptococcus pyogenes*: Structure and function. *In* "Superantigens: Structure, Biology, and Relevance to Human Disease" (D. Y. M. Leung, B. T. Huber, and P. M. Schlievert, eds.), pp. 281–310. Dekker, New York.

Musser, J. M., and Selander, R. K. (1990). Molecular population genetics of *Staphylococcus aureus*. *In* "Molecular Biology of the Staphylococci" (R. N. Novick and R. A. Skurray, eds.), pp. 59–67. VCH, New York.

Musser, J. M., Schlievert, P. M., Chow, A. W., Ewan, P., Kreiswirth, B. N., Rosdahl, V. T., Naidu, A. S., Witte, W., and Selander, R. K. (1990). A single clone of *Staphylococcus aureus* causes the majority of cases of toxic shock syndrome. *Proc. Natl. Acad. Sci. U.S.A.* **87**, 225–299.

Musser, J. M., Hauser, A. R., Kim, M. H., Schlievert, P. M., Nelson, K., and Selander, R. K. (1991). *Streptococcus pyogenes* causing toxic-shock-like syndrome and other invasive diseases: Clonal diversity and pyrogenic exotoxin expression. *Proc. Natl. Acad. Sci. U.S.A.* **88**, 2668–2672.

Musser, J. M., Kapur, V., Kanjilal, S., Shah, U., Musher, D. M., Barg, N. L., Johnston, K. H., Schlievert, P. M., Henrichsen, J., Gerlach, D., Rakita, R. M., Tanna, A., Cookson, B. D., and Huang, J. C. (1993). Geographic and temporal distribution and molecular characterization of two highly pathogenic clones of *Streptococcus pyogenes* expressing allelic variants of pyrogenic exotoxin A (scarlet fever toxin). *J. Infect. Dis.* **167**, 337–346.

Musser, J. M., Kapur, V., Peters, J. E., Hendrix, C. W., Drehner, D., Gackstetter, G. D., Skalka, D. R., Fort, P. L., Maffei, J. T., Li, L.-L., and Melcher, G. P. (1994). Real-time molecular epidemiologic analysis of an outbreak of *Streptococcus pyogenes* invasive disease in US Air Force trainees. *Arch. Pathol. Lab. Med.* **118**, 128–133.

Musser, J. M., Kapur, V., Szeto, J., Pan, X., Swanson, D. S., and Martin, D. R. (1995). Genetic diversity and relationships among *Streptococcus pyogenes* strains expressing serotype M1 protein: Recent intercontinental spread of a subclone causing episodes of invasive disease. *Infect. Immun.* **63**, 994–1003.

Mylvaganam, H., Bjorvatn, B., Hofstad, T., Hjetland, R., Høiby, E. A., and Holm, S. E. (1994). Small-fragment restriction endonuclease analysis in epidemiological mapping of group A streptococci. *J. Med. Microbiol.* **40**, 256–260.

Nakashima, K., Ichiyama, S., Iinuma, Y., Hasegawa, Y., Ohta, M., Ooe, K., Shimizu, Y., Igarashi, H., Murai, T., and Shimokata, K. (1997). A clinical and bacteriologic investigation of invasive streptococcal infections in Japan on the basis of serotypes, toxin production, and genomic DNA fingerprints. *Clin. Infect. Dis.* **25**, 260–266.

Nelson, K., Schlievert, P. M., Selander, R. K., and Musser, J. M. (1991). Characterization and clonal distribution of four alleles of the *speA* gene encoding pyrogenic exotoxin A (scarlet fever toxin) in *Streptococcus pyogenes*. *J. Exp. Med.* **174**, 1271–1274.

Norgren, M., Norrby, A., and Holm, S. E. (1992). Genetic diversity in T1M1 group A streptococci in relation to clinical outcome of infection. *J. Infect. Dis.* **166**, 1014–1020.

Norrby-Teglund, A., Pauksens, K., Holm, S. E., and Norgren, M. (1994). Relation between low capacity of human sera to inhibit streptococcal mitogens and serious manifestation of disease. *J. Infect. Dis.* **170**, 585–591.

Norrby-Teglund, A., Kaul, R., Low, D. E., McGeer, A., Newton, D. W., Andersson, J., Andersson, U., and Kotb, M. (1996). Plasma from patients with severe invasive group A streptococcal infections treated with normal polyspecific IgG inhibits streptococcal superantigen-induced T cell proliferation and cytokine production. *J. Immunol.* **156**, 3057–3064.

Orskov, F., Orskov, I., Evans, D. J., Sack, R. B., and Wadstrom, T. (1976). Special *Escherichia coli* serotypes among enterotoxigenic strains from diarrhoea in adults and children. *Med. Microbiol. Immunol.* **162**, 73–80.

Perez-Casal, J. F., Dillon, H. F., Husmann, L. K., Graham, B., and Scott, J. R. (1993a). Virulence of two *Streptococcus pyogenes* strains (types M1 and M3) associated with toxic-shock-like syndrome depends on an intact *mry*-like gene. *Infect. Immun.* **61**, 5426–5430.

Perez-Casal, J., Price, J. A., Maguin, E., and Scott, J. R. (1993b). An M protein with a single C repeat prevents phagocytosis of *Streptococcus pyogenes:* Use of a temperature-sensitive shuttle vector to deliver homologous sequences to the chromosome of *S. pyogenes. Mol. Microbiol.* **8**, 809–819.

Pichichero, M. E. (1991). The rising incidence of penicillin treatment failures in group A streptococcal tonsillopharyngitis: An emerging role for the cephalosporins? *Pediatr. Infect. Dis. J.* **10**, S50–S55.

Podbielski, A., Flosdorff, A., and Weber-Heynemann, J. (1995). The group A streptococcal *virR49* gene controls expression of four structural *vir* regulon genes. *Infect. Immun.* **63**, 9–20.

Podbielski, A., Schnitzler, N., Beyhs, P., and Boyle, M. D. P. (1996). M-related protein (Mrp) contributes to group A streptococcal resistance to phagocytosis by human granulocytes. *Mol. Microbiol.* **19**, 429–441.

Prasad, G. S., Earhart, C. A., Murray, D. L., Novick, R. P., Schlievert, P. M., and Ohlendorf, D. H. (1993). Structure of toxic shock syndrome toxin 1. *Biochemistry* **32**, 13761–13766.

Quinn, R. W. (1982). Streptococcal infections. *In* "Bacterial Infections of Humans: Epidemiology and Control" (A. S. Evans and H. A. Feldman, eds.), pp. 525–552. Plenum Medical, New York.

Quinn, R. W. (1989). Comprehensive review of morbidity and mortality trends for rheumatic fever, streptococcal disease, and scarlet fever: The decline of rheumatic fever. *Rev. Infect. Dis.* **11**, 928–953.

Reda, K. B., Kapur, V., Goela, D., Lamphear, J. G., Musser, J. M., and Rich, R. R. (1996). Phylogenetic distribution of streptococcal superantigen SSA allelic variants provides evidence for horizontal transfer of *ssa* within *Streptococcus pyogenes. Infect. Immun.* **64**, 1161–1165.

Schrager, H. M., Rheinwald, J. G., and Wessels, M. R. (1996). Hyaluronic acid capsule and the role of streptococcal entry into keratinocytes in invasive skin infection. *J. Clin. Invest.* **98**, 1954–1958.

Schwartz, B., Facklam, R. R., and Breiman, R. F. (1990). Changing epidemiology of group A streptococcal infection in the USA. *Lancet* **336**, 1167–1171.

Seppälä, H., Nissinen, A., Järvinen, H., Huovinen, S., Henriksson, T., Herva, E., Holm, S. E., Jahkola, M., Katila, M.-L., Klaukka, T., Kontiainen, S., Liimatainen, O., Oinonen, S., Passi-Metsomaa, L., and Huovinen, P. (1992). Resistance to erythromycin in group A streptococci. *N. Engl. J. Med.* **326**, 292–297.

Seppälä, H., Vuopio-Varkila, J., Österblad, M., Jahkola, M., Rummukainen, M., Holm, S. E., and Huovinen, P. (1994a). Evaluation of methods for epidemiologic typing of group A streptococci. *J. Infect. Dis.* **169**, 519–525.

Seppälä, H., He, Q., Österblad, M., and Huovinen, P. (1994b). Typing of group A streptococci by random amplified polymorphic DNA analysis. *J. Clin. Microbiol.* **32**, 1945–1948.

Sloane, R., de Azavedo, J. C., Arbuthnott, J. P., Hartigan, P. J., Kreiswirth, B., Novick, R., and Foster, T. J. (1991). A toxic shock syndrome toxin mutant of *Staphylococcus aureus* isolated by allelic replacement lacks virulence in a rabbit uterine model. *FEMS Microbiol. Lett.* **78**, 239–244.

Stegmayr, B., Björck, S., Holm, S., Nisell, J., Rydvall, A., and Settergren, B. (1992). Septic shock induced by group A streptococcal infection: Clinical and therapeutic aspects. *Scand. J. Infect. Dis.* **24**, 589–597.

Stetson, C. A., Rammelkamp, C. H., Jr., Krause, R. M., Kohen, R. J., and Perry, W. D. (1955). Epidemic acute nephritis: Studies on etiology, natural history and prevention. *Medicine* **34**, 431–450.

Stevens, D. L. (1995). Streptococcal toxic-shock syndrome: Spectrum of disease, pathogenesis, and new concepts in treatment. *Emerging Infect. Dis.* **1**, 69–78.

Stevens, D. L., Tanner, M. H., Winship, J., Swarts, R., Ries, K. M., Schlievert, P. M., and Kaplan, E. (1989). Severe group A streptococcal infections associated with a toxic shock-like syndrome and scarlet fever toxin A. *N. Engl. J. Med.* **321**, 1–7.

Stockbauer, K. E., Grigsby, D., Pan, X., Perea Mejia, L. M., Cravioto, A., and Musser, J. M. (1997). Hypervariability generated by natural selection in an extracellular complement inhibiting protein made by serotype M1 strains of group A *Streptococcus. Proc. Natl. Acad. Sci. USA.* In press.

Stollerman, G. H. (1997). Rheumatic fever. *Lancet* **349**, 935–942.

Strömberg, A., Romanus, V., and Burman, L. G. (1991). Outbreak of group A streptococcal bacteremia in Sweden: An epidemiologic and clinical study. *J. Infect. Dis.* **164**, 595–598.

Taranta, A., Torosdag, S., Metrakos, J. D., Jegier, W., and Uchida, I. (1959). Rheumatic fever in monozygotic and dizygotic twins. *Circulation* **20**, 778.

The Working Group on Severe Streptococcal Infections. (1993). Defining the group A streptococcal toxic shock syndrome. Rationale and consensus definition. *JAMA* **269**, 390–391.

Thibodeau, J., and Sekaly, R.-P. (1995). "Bacterial Superantigens: Structure, Function and Therapeutic Potential." R. G. Landes, Austin, Texas.

Todd, J. K. (1988). Toxic shock syndrome. *Clin. Microbiol. Rev.* **1**, 432–446.

Todd, J., Fishaut, M., Kapral, F., and Welch, T. (1978). Toxic shock syndrome associated with phage-group-I staphylococci. *Lancet* **2**, 1116–1118.

Unny, S. K., and Middlebrooks, B. L. (1983). Streptococcal rheumatic carditis. *Microbiol. Rev.* **47**, 97–120.

Upton, M., Carter, P. E., Morgan, M., Edwards, G. F. S., and Pennington, T. H. (1995). Clonal structure of invasive *Streptococcus pyogenes* in Northern Scotland. *Epidemiol. Infect.* **115**, 231–241.

Upton, M., Carter, P. E., Orange, G., and Pennington, T. H. (1996). Genetic heterogeneity of M type 3 group A streptococci causing severe infections in Tayside, Scotland. *J. Clin. Microbiol.* **34**, 196–198.

Veasy, L. G., Wiedmeier, S. E., Orsmond, G. S., Ruttenberg, H. D., Boucek, M. M., Roth, S. J., Tait, V. F., Thompson, J. A., Daly, J. A., Kaplan, E. L., and Hill, H. R. (1987). Resurgence of acute rheumatic fever in the intermountain area of the United States. *N. Engl. J. Med.* **316**, 421–427.

Vugia, D. J., Peterson, C. L., Meyers, H. B., Kim, K. S., Arrieta, A., Schlievert, P. M., Kaplan, E. L., Werner, S. B., and Mascola, L. (1996). Invasive group A streptococcal infections in children with varicella in southern California. *Pedriatr. Infect. Dis. J.* **15**, 146–150.

Wald, E. R., Dashefsky, B., Feidt, C., Chiponis, D., and Byers, C. (1987). Acute rheumatic fever in western Pennsylvania and the tristate area. *Pediatrics* **80**, 371–374.

Wessels, M. R., and Bronze, M. S. (1994). Critical role of the group A streptococcal capsule in pharyngeal colonization and infection in mice. *Proc. Natl. Acad. Sci. U.S.A.* **91**, 12238–12242.

Wessels, M. R., Moses, A. E., Goldberg, J. B., and DiCesare, T. J. (1991). Hyaluronic acid capsule is a virulence factor for mucoid group A streptococci. *Proc. Natl. Acad. Sci. U.S.A.* **88**, 8317–8321.

Wessels, M. R., Goldberg, J. B., Moses, A. E., and DiCesare, T. J. (1994). Effects on virulence of mutations in a locus essential for hyaluronic acid capsule expression in group A streptococci. *Infect. Immun.* **62**, 433–441.

Westlake, R. M., Graham, T. P., and Edwards, K. M. (1990). An outbreak of acute rheumatic fever in Tennessee. *Pediatr. Infect. Dis. J.* **9**, 97–100.

Whatmore, A. M., Kapur, V., Sullivan, D. J., Musser, J. M., and Kehoe, M. A. (1994). Noncongruent relationships between variation in *emm* gene sequences and the population genetic structure of group A streptococci. *Mol. Microbiol.* **14**, 619–631.

Whatmore, A. M., Kapur, V., Musser, J. M., and Kehoe, M. A. (1995). Molecular population genetic analysis of the *enn* subdivision of group A streptococcal *emm*-like genes: Horizontal gene transfer and restricted variation among *enn* genes. *Mol. Microbiol.* **15**, 1039–1048.

Wheeler, M. C., Roe, M. H., Kaplan, E. L., Schlievert, P. M., and Todd, J. K. (1991). Outbreak of group A streptococcus septicemia in children. *JAMA* **266**, 533–537.

Zabriskie, J. B. (1985). Rheumatic fever: The interplay between host, genetics, and microbe. *Circulation* **71**, 1077–1086.

7

▇ LYME DISEASE

ALLEN C. STEERE
Tufts University School of Medicine
New England Medical Center
Boston, Massachusetts

I. INTRODUCTION

Lyme disease or Lyme borreliosis, which is caused by the tick-borne spirochete *Borrelia burgdorferi*, occurs in temperate regions of North America, Europe, and Asia. It is now the most common vector-borne disease in the United States (Centers for Disease Control, 1996). Since the Centers for Disease Control and Prevention began surveillance for this infection in 1982, the number of cases has increased dramatically, and more than 10,000 new cases have been reported each summer during the 1990s (Fig. 1; Centers for Disease Control, 1996). Although sporadic cases have been noted in 48 states, most cases have clustered in the northeastern United States from Massachusetts to Maryland, and in the midwest in Wisconsin and Minnesota (Fig. 2; Centers for Disease Control, 1996). The infection also occurs in most European countries, particularly in northern parts of the continent, including Germany, Austria, and Sweden (Stanek *et al.*, 1988). In the former Soviet Union, a central area is affected from the Baltic Sea to the Pacific Ocean (Korenberg *et al.*, 1993). Cases have also been reported in China (Ai *et al.*, 1990) and Japan (Carlberg and Naito, 1991).

During the last several decades, Lyme disease has spread in the United States and has caused focal outbreaks, particularly in the northeastern United States. In addition to rural areas and coastal islands, suburban locations have been affected near large northeastern cities including Boston, New York, and Philadelphia (Steere, 1994). In these locations, the risk is greatest during the summer months in persons engaged in routine activities on suburban residential

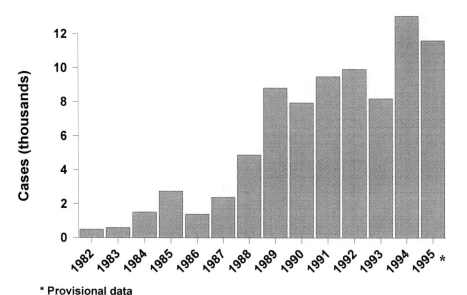

* **Provisional data**

FIGURE I   Number of cases of Lyme disease in the United States reported to the Centers for Disease Control and Prevention, by year from 1982 to 1995. Centers for Disease Control (1996), *Morbidity & Mortality Weekly Report* **45,** 481–484.

properties and in adjacent woodlands, but occupational and recreational exposures are also important. Currently, there are no practical methods to control enzootic Lyme disease or to prevent its spread.

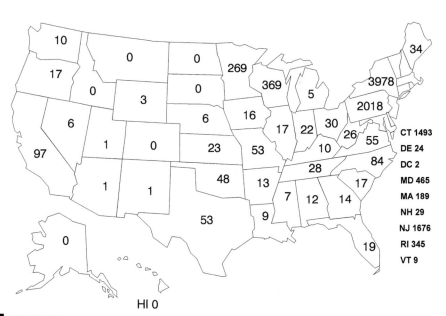

FIGURE 2   Locations of cases of Lyme disease in the United States reported to the Centers for Disease Control and Prevention in 1995. Centers for Disease Control (1996), *Morbidity & Mortality Weekly Report* **45,** 481–484.

II. ETIOLOGIC AGENT

Borrelia species, along with the *Leptospira* and *Treponema*, belong to the eubacterial phylum of spirochetes (Barbour and Hayes, 1986). Like all spirochetes, the *Borrelia* spp. have a protoplasmic cylinder that is surrounded first by a cell membrane, next by flagella, and finally by another, outer membrane, but the borrelia are longer and more loosely coiled than the other spirochetes (Barbour and Hayes, 1986). The unusual feature that distinguishes the borrelial genome from most other spirochetal or bacterial genomes is that its chromosome and some of its plasmids are linear DNA molecules (Saint Girons and Davidson, 1994).

Of the *Borrelia* spp., *Borrelia burgdorferi* is the longest (20–30 µm) and narrowest (0.2–0.3 µm); it has fewer flagella (7 to 11), and the G + C content of its genome is low (28 to 30.5 mol%) (Barbour and Hayes, 1986; Saint Girons and Davidson, 1994). It may be cultured in a complex, liquid medium called Barbour, Stoenner, Kelly (BSK) medium (Barbour, 1984). The organism contains at least 30 different proteins with as many as 10 outer surface proteins, including at least 5 outer surface lipoproteins (Brandt *et al.*, 1990). The genes encoding many of the outer surface proteins are located on circular or small linear plasmids (Barbour and Hayes, 1986; Saint Girons and Davidson, 1994). The 50-kb linear plasmid produces from a common promoter the two major outer surface proteins of the spirochete, OspA and OspB (Howe *et al.*, 1986). Similarly, the *OspE* and *OspF* genes are structurally arranged in tandem as one transcriptional unit under the control of a common promoter (Lam *et al.*, 1994). The *OspC* gene, which is located on a 26-kb circular plasmid (Marconi *et al.*, 1993), has sequence similarities with major variable protein of relapsing fever borrelia. Although considerable variation has been noted in *OspC* sequences from different isolates (Fung *et al.*, 1994), this outer membrane protein does not seem to undergo antigenic variation within the host during the infection.

Three genomic groups of *B. burgdorferi* sensu lato have now been identified (Baranton *et al.*, 1992), and more may exist. To date, all North American strains have belonged to the first group, *B. burgdorferi* sensu stricto (Baranton *et al.*, 1992). Although all three groups have been found in Europe, most isolates there have been group 2 (*Borrelia garinii*) or group 3 strains (*Borrelia afzelii*) (Baranton *et al.*, 1992; Canica *et al.*, 1993). These differences may well account for regional variations in the clinical picture of Lyme borreliosis. A fourth group, *Borrelia japonica* which was isolated in Japan, is not known to cause human disease (Kawabata *et al.*, 1993).

III. DISEASE VECTOR AND ANIMAL HOSTS

The vectors of Lyme borreliosis are several closely related ixodid ticks that are part of the *Ixodes ricinus* complex. In the northeastern and midwestern United States, *Ixodes dammini* (also called *Ixodes scapularis*) is the vector, and *Ixodes pacificus* is the vector in the West. In Europe, *I. ricinus* is the primary vector, and *Ixodes persulcatus* is in Asia. In the United States, the preferred host for both the larval and nymphal stages of *I. dammini* is the white-footed mouse, *Peromyscus leucopus* (Levine *et al.*, 1985). It is critical that both of the immature stages of the tick feed on the same host since the life cycle of the spirochete depends on horizontal transmission from infected nymphs to mice during early summer, who, in turn, infect larvae during late summer, which then molt to infected nymphs to begin the cycle again the next summer (Matschka and Spielman, 1986). White-tailed deer, which are not involved in the life cycle of the spirochete, are the preferred host for the adult stage of *I. dammini* and seem to be critical for the survival of the tick (Wilson *et al.*, 1985). In the northeastern United States, infection rates in ticks range from 10 to over 50% (Bosler *et al.*, 1984; Magnarelli *et al.*, 1984).

The vector ecology of *B. burgdorferi* is different on the west coast. This is because the nymphal stage of *I. pacificus* prefers to feed on lizards rather than mice, and lizards are not susceptible to infection with *B. burgdorferi* (Manweiler *et al.*, 1992). There, the spirochete is maintained in nature in a horizontal cycle between the dusky-footed woodrat *Neotoma fuscipes* and *Ixodes neotomae*, a tick that does not feed on humans (Brown and Lane, 1992). Only the relatively few larval and nymphal *I. pacificus* ticks that feed on infected woodrats rather than lizards are responsible for transmitting the spirochete to humans when they molt to the next stage. Accordingly, the infection rate of *I. pacificus* ticks is only 1 to 3%, and the number of cases on the west coast is much less than on the east coast.

IV. EMERGENCE OF LYME DISEASE IN THE LATE TWENTIETH CENTURY

The primary areas now affected by Lyme disease in the United States are near the terminal moraine of the glaciers formed 15,000 years ago (Davis, 1983; Spielman *et al.*, 1993). At that time, the northeastern and upper midwestern parts of what is now the United States were covered by tundra. As the glaciers retreated, forests grew in these areas, which eventually became populated with large numbers of deer. Early descriptions of colonial New England include comments about the abundance of deer and annoying ticks (Severinghaus and Brown, 1983). During the eighteenth and nineteenth centuries, forests were destroyed in New England to make farms, and deer were hunted practically to extinction. Deer are thought to have survived in only a few isolated locations, such as Naushon Island near Cape Cod and the eastern end of Long Island, New York (Spielman *et al.*, 1993). Analysis of museum tick collections by polymerase chain reaction (PCR) documented the presence of *B. burgdorferi*-infected *I. scapularis* ticks on Montauk Point, Long Island, in the 1940s (Persing *et al.*, 1990).

The emergence of Lyme disease in the United States in the late twentieth century is thought to have occurred primarily because of ecological conditions favorable for deer (Wilson *et al.*, 1985). As farmland in the northeast reverted to woodland, the habitat for deer improved: their natural predators were gone, the number of deer increased dramatically, they migrated to new areas, and federal programs protected them. With the advent of the automobile and superhighways, rural and suburban areas, where deer now lived, became populated with large numbers of susceptible suburbanites who had never been exposed to the spirochete. Lyme disease and the deer tick do not, however, occur in all areas where deer are found. Other factors, such as rodent populations, vegetation, temperature, and humidity, must also play an important role in the ecology of the disease.

Lyme disease was recognized as a separate entity in 1975–1976 because of geographic clustering of children in Lyme, Connecticut, who were thought to have juvenile rheumatoid arthritis (Steere *et al.,* 1977a). The rural setting of the case clusters and the identification of erythema migrans as a feature of the illness suggested that the disorder was transmitted by an arthropod. It soon became apparent that Lyme disease was a multisystem illness that affected primarily the skin, nervous system, heart, or joints (Steere *et al.,* 1977b). Erythema migrans also linked Lyme disease with certain previously described syndromes in Europe, including acrodermatitis chronica artophicans and Bannwarth's syndrome (Herxheimer and Hartmann, 1902; Bannwarth, 1944). These various syndromes were brought together conclusively in 1982 when Burgdorfer and Barbour isolated a previously unrecognized spirochete, now called *B. burgdorferi,* from *I. dammini* ticks (Burgdorfer *et al.,* 1982). This spirochete was then recovered from patients with Lyme disease in the United States (Benach *et al.,* 1983; Steere *et al.,* 1983a) and from those with erythema migrans, Bannwarth's syndrome, or acrodermatitis in Europe; in addition, the immune responses in affected patients were linked conclusively with this organism. Although there are regional variations, the basic outlines of the disorder are similar worldwide, and the commonest name for the overall disease is Lyme disease or Lyme borreliosis.

Clinically, this borrelial infection is most like syphilis in its multisystem involvement, occurrence in stages, and mimicry of other diseases. Whereas the various manifestations of syphilis were recognized to be a single entity in the nineteenth century, the protean manifestations of Lyme disease were not recognized to be part of single infection until late in the twentieth century.

V. CLINICAL MANIFESTATIONS, PATHOGENESIS, AND IMMUNITY

A. Early Infection (Stage 1: Localized Infection, Stage 2: Disseminated Infection)

After injection by the tick, *B. burgdorferi* first causes local infection of the skin (stage 1) in about 80% of patients (Steere *et al.,* 1983b; Steere, 1989; Nadelman *et al.,* 1996). Because of the small size of nymphal *I. dammini,* most people do not remember the tick bite. Several days later, the organisms begin to spread in the skin, resulting in a characteristic skin lesion, erythema migrans; within days to weeks (stage 2), they may disseminate to many sites. The spirochete has been cultured regularly from erythema migrans skin lesions (Berger *et al.,* 1992), and small numbers of organisms have been seen in patient specimens of myocardium, retina, muscle, bone, synovium, spleen, liver, meninges, and brain (Duray and Steere, 1988). Bacterial spread within the host is probably facilitated by the ability of the spirochetes to bind human plasminogen and urokinase-type plasminogen activator to its surface (Klempner *et al.,* 1995). Plasmin, the activated form of plasminogen, is a potent protease that could promote tissue invasion. *Borrelia burgdorferi* may then adhere to many different mammalian cell types. To date, two binding mechanisms have been identified. First, the spirochete attaches to several members of the integrin family of receptors, including a platelet-specific integrin receptor (Coburn *et al.*, 1993). A second pathway for cell attachment is mediated by host cell sugars, particularly glycosaminoglycans (Leong *et al.*, 1995).

Although spirochetes may disseminate widely, they seem to survive primarily in the skin, nervous system, heart, or joints; disseminated infection is often associated with characteristic symptoms in these systems (Steere *et al.*, 1983b; Nadelman *et al.*, 1996). Secondary annular skin lesions, which occur in about half of patients, look similar to the primary erythema migrans lesion, but they are generally smaller and migrate less. Excruciating headache and mild neck stiffness are common, but they typically occur in short attacks lasting only hours. The musculoskeletal pain of Lyme disease is generally migratory in joints, bursae, tendons, muscle, or bone, lasting only hours or days in a given location. At this stage, patients often appear quite ill and frequently have debilitating malaise and fatigue. Except for fatigue, the symptoms are typically intermittent and changing.

The specific immune response in Lyme disease develops gradually (Steere *et al.*, 1983a; Dressler *et al.*, 1993). Within several weeks, patients' peripheral blood mononuclear cells begin to have heightened responsiveness to *B. burgdorferi* antigens or mitogens (Sigal *et al.*, 1986). The specific immunoglobulin M (IgM) response peaks between the third and sixth week of infection (Steere *et al.*, 1983a) and often is associated with polyclonal activation of B cells, including elevated total serum IgM levels (Steere *et al.*, 1979), circulating immune complexes (Hardin *et al.*, 1979), and cryoglobulins (Steere *et al.*, 1979). Membrane lipoproteins, including OspA, are mitogenic for B cells (Schoenfeld *et al.*, 1992; Ma and Weis, 1993). The specific IgG response develops gradually over weeks to months to an increasing array of spirochetal polypeptides (Dressler *et al.*, 1993) and nonprotein antigens (Wheeler *et al.*, 1993). Histologically, all affected tissues show an infiltration of lymphocytes, plasma cells, or macrophages (Duray and Steere, 1988). Some degree of vascular damage, including mild vasculitis or hypercellular vascular occlusion, may be seen in multiple sites (Duray and Steere, 1988), suggesting that the spirochete may have been present in and around blood vessels.

After several weeks or months, as the infection begins to localize, about 15 to 20% of patients in the United States develop frank neurological involvement (Pachner and Steere, 1985). The characteristic manifestations are meningitis, cranial neuritis, and peripheral neuropathy, alone or in combination (Pachner and Steere, 1985). In patients with meningitis, cerebrospinal fluid (CSF) typically shows a lymphocytic pleocytosis of about 100 cells/mm³, often with elevated protein but normal glucose. Encephalitic signs, if present, are subtle; these include somnolence, poor memory, and mood change. Unilateral or bilateral facial palsy is the commonest cranial neuropathy, and it may be the only neurological abnormality (Clark *et al.*, 1985). The peripheral neuritis is usually an asymmetric motor, sensory, or mixed radiculoneuropathy of the limbs or trunk (Halperin *et al.*, 1987). Electrophysiological studies of affected extremities suggest primarily axonal nerve involvement with some demyelination of both proximal and distal nerve segments (Vallat *et al.*, 1987). Peripheral nerve lesions show predominantly axonal injury with perivascular infiltration of lymphocytes and plasma cells around epineural blood vessels (Vallat *et al.*, 1987). Spirochetes have not been seen in these lesions. Therefore, it has been postulated that certain neurological manifestations, such as cranial and radiculoneuritis, might result from immune-mediated phenomena. Antibody to an epitope of the 41-kDa flagellar antigen of *B. burgdorferi* cross-reacts with a human axonal antigen identified as chaperonin-Hsp60 (Dai *et al.*, 1993). However, it is not been shown that this example of molecular mimicry results in neuronal injury. Stage 2 neurological abnormalities usually last for weeks or months, but they may recur or become chronic.

Also within the period of several weeks after disease onset, about 4 to 8% of patients develop cardiac involvement (Steere *et al.*, 1980). The commonest abnormality is fluctuating degrees of atrioventricular block (first degree, Wenckebach, or complete heart block), but some patients have acute myopericarditis, mild left ventricular dysfunction, or, rarely, cardiomegaly or fatal pancarditis (Marcus *et al.*, 1985). The duration of cardiac abnormalities is usually brief (3 days to 6 weeks) (Steere *et al.*, 1980); complete heart block rarely persists for more than 1 week, and permanent pacemaker insertion is not necessary (McAlister *et al.*, 1989).

B. Late Infection (Stage 3: Persistent Infection)

A mean of 6 months after disease onset (range, 2 weeks to 2 years), within the context of strong cellular and humoral immune responses to *B. burgdorferi,* about 60% of patients in the United States begin to have brief attacks of asymmetric, oligoarticular arthritis, primarily in large joints, especially the knee (Steere *et al.*, 1987). Some attacks may affect periarticular structures, including peripheral entheses (Lawson and Steere, 1985; Steere *et al.*, 1987). Joint fluid white cell counts range from 500 to 110,000 cells/mm^3 and consist of predominantly polymorphonuclear leukocytes. Although the spirochete has been cultured from the joint fluid of only two patients with Lyme arthritis (Snydman *et al.*, 1986; Schmidli *et al.*, 1988), *B. burgdorferi* DNA may be detected by PCR in the synovial tissue or joint fluid of most patients (Bradley *et al.*, 1994; Nocton *et al.*, 1994). The synovial lesion, which is similar to that seen in other forms of chronic inflammatory arthritis, shows synovial hypertrophy, vascular proliferation, and an infiltration of mononuclear cells, primarily lymphocytes, plasma cells, and macrophages, with intense class II major histocompatibility molecule (HLA-DR) expression throughout the lesion (Johnston *et al.*, 1985; Steere *et al.*, 1988). T-cell reactivity to an array of spirochetal proteins is concentrated in the joint (Sigal *et al.*, 1986; Yoshinari *et al.*, 1991), and an antigen-specific T helper 1 response is dominant (Yssel *et al.*, 1991). The characteristics of the arthritis are similar in the United States and Europe, but arthritis seems to be a less frequent manifestation of the disease in Europe.

The number of patients who have recurrent attacks of arthritis decreases by about 10 to 20% each year (Steere *et al.*, 1987). In some patients, however, episodes of arthritis become longer during the second and third year of the illness, and chronic arthritis, defined as 1 year or more of continuous joint inflammation, characteristically begins in a small percentage of patients during this period. Typically, only one or a few large joints are affected, most commonly the knee. However, even patients with chronic arthritis rarely have continual joint inflammation for longer than several years.

Although most patients with either acute or chronic Lyme arthritis respond to antibiotic treatment, a small percentage have persistence of joint inflammation for months or even several years after therapy (Steere *et al.*, 1994). In these patients, *B. burgdorferi* DNA can usually not be detected in synovial tissue or joint fluid after antibiotic treatment (Nocton *et al.*, 1994). This outcome is associated with certain immunogenetic and immune markers, including HLA-DR4 alleles and humoral and cellular immunity to OspA (Lengl-Jannsen *et al.*, 1994; Steere *et al.*, 1994). Thus, in genetically susceptible individuals, it is possible that an immune response to OspA in the inflamed joint may cause tissue-specific autoimmunity that persists after the eradication of the spirochete from the joint with antibiotic therapy.

Although less common, chronic neurological involvement may also occur months to years after the onset of illness, sometimes following long periods of latent infection (Halperin *et al.*, 1987, 1989; Logigian *et al.*, 1990; Logigian and Steere, 1992). In the United States, the most common forms of chronic neuroborreliosis are a subtle encephalopathy affecting memory, mood, or sleep and an axonal polyneuropathy manifested as either distal paresthesias or spinal radicular pain. Patients with encephalopathy often have evidence of memory impairment, which can be demonstrated on neuropsychological tests, as well as abnormal CSF analyses. Single photon emission computed tomography (SPECT) scanning of the brain shows hypoperfusion of subcortical and cortical structures (Logigian *et al.*, 1997), but magnetic resonance imaging (MRI) scanning is usually normal (Logigian *et al.*, 1990). In those with polyneuropathy, electromyography generally shows extensive abnormalities of proximal and distal nerve segments (Logigian and Steere, 1992).

In Europe, Ackermann reported a series of 48 patients with a severe syndrome of chronic, progressive encephalomyelitis accompanied by intrathecal antibody synthesis to *B. burgdorferi* (Ackermann *et al.*, 1988). These patients developed spastic parapareses, bladder dysfunction, ataxia, seventh or eighth cranial nerve deficits, or cognitive impairment, including dementia. Only a few patients have been reported with borrelial encephalomyelitis in the United States (Halperin *et al.*, 1989; Logigian *et al.*, 1990). It is now known that neuroborreliosis in Europe is most often caused by infection with *B. garinii*, a species of the spirochete that has not been found in the United States. This may explain why encephalomyelitis seems to be less common in the United States than in Europe.

Acrodermatitis chronica atrophicans, the late skin manifestation of Lyme borreliosis, has been associated primarily with *B. afzelii* infection in Europe (Asbrink and Hovmark, 1988). This species of the spirochete has not been found in the United States, and only a few cases of acrodermatitis have been reported in the United States (Kaufman *et al.*, 1989). The skin lesions of acrodermatitis usually begin insidiously with bluish-red discoloration and swollen skin on one extremity. Erythema migrans may have been present years earlier.

The inflammatory phase of acrodermatitis may persist for many years, and it gradually leads to atrophy of the skin. *Borrelia afzelii* has been cultured from such lesions as long as 10 years after their onset (Asbrink and Hovmark, 1985). In some patients, scleroderma-like skin lesions can occur concomitantly with acrodermatitis (Asbrink and Hovmark, 1988).

VI. CONGENITAL INFECTION

During the 1980s, two provocative case reports described transplacental transmission of *B. burgdorferi* in infants whose mothers acquired Lyme borreliosis during the first trimester of pregnancy (Schlesinger *et al.*, 1985; Weber *et al.*, 1988). In both instances, spirochetes were seen in various fetal tissues, using the Dieterle silver stain, but cultures and serological testing were not done. In a subsequent, retrospective review of 19 cases of Lyme disease during pregnancy, 5 were associated with adverse fetal outcomes (Markowitz *et al.*, 1986). Because all of the outcomes differed, however, they could not be linked conclusively to maternal Lyme disease. A study of over 5000 infants from endemic or nonendemic areas found no association between congenital malformations and the presence of detectable antibody to *B. burgdorferi* in cord blood, and no infant had IgM antibody to the spirochete (Williams *et al.*, 1995). Although the Lyme disease spirochete can probably cause adverse fetal outcomes, it seems to be a rare event.

VII. ANIMAL MODELS OF LYME DISEASE

Animal models of Lyme disease have been developed in rodents, dogs, and nonhuman primates. Particularly important has been the murine model of Lyme disease in inbred C3H/HeJ mice (Barthold *et al.*, 1993); this model has been used to explore aspects of pathogenesis and to identify immune responses that are protective against reinfection. C3H mice develop acute arthritis and carditis 2 to 4 weeks after inoculation with *B. burgdorferi*; they do not develop erythema migrans or neurological abnormalities. Macrophages are the primary infiltrating cell in cardiac lesions (Ruderman *et al.*, 1995), whereas polymorphonuclear leukocytes and then lymphocytes are found in synovial tissue (Barthold *et al.*, 1993). Cellular immunity is critical for control of carditis, whereas antibody is crucial for resolution of arthritis (Barthold *et al.*, 1996). Despite the resolution of the acute lesions, spirochetes persist throughout the life of the mouse, primarily in the skin. Recurrent waves of spirochetemia from that site may lead to recurrent attacks of arthritis. However, this murine strain does not develop the equivalent of human, chronic Lyme arthritis.

Comparison of the infection in different inbred murine strains has shown the importance of genetic susceptibility, spirochetal burden, and the early immune response in the variability of subsequent arthritis. C3H/HeJ mice, which carry *H-2^k* alleles, develop severe arthritis when infected with *B. burgdorferi*, whereas BALB/c mice, which carry *H-2^d* alleles, develop only mild arthritis (Keane-Myers and Nickell, 1995). C3H/HeJ mice have a greater burden of spirochetes in their joints than BALB/c mice (Yang *et al.*, 1994). As an explanation for these findings, one study shows that C3H/HeJ mice produce interferon-γ (IFN-γ; Th1 response) when infected with the spirochete (Keane-Myers and Nickell, 1995). Although *B. burgdorferi*-infected BALB/c mice initially produce IFN-γ, they switch to interleukin-4 (IL-4) production (Th2 response). In this way, BALB/c mice seem to reduce the numbers of spirochetes initially and dampen the inflammatory response of the subsequent arthritis.

As with neurosyphilis, only human and nonhuman primates are known to be susceptible to neuroborreliosis. In one report, peripheral neuropathy was observed in all five rhesus monkeys that were infected with *B. burgdorferi* (Roberts *et al.*, 1995). On postmortem examination, nerve lesions were observed, including nerve sheath fibrosis and focal demyelinization.

VIII. DIAGNOSIS

A. Serology

Lyme disease is usually diagnosed by recognition of a characteristic clinical picture with serological confirmation. Although serological testing may be negative during the first several weeks of infection, most patients have a positive antibody response to *B. burgdorferi* after that time (Dressler *et al.*, 1993). The limitation of serological tests is that they do not clearly distinguish between active and inactive infection. Patients with previous Lyme disease, particularly if the disease progressed to late stages, often remain seropositive for years, even after adequate antibiotic treatment. In addition, some patients are seropositive because of asymptomatic infection. If these individuals subsequently develop another illness, the positive serological test for Lyme disease may cause diagnostic confusion. On the other hand, a small number of patients who receive inadequate antibiotic therapy during the first several weeks of infection still develop subtle joint or neurological symptoms, but are seronegative (Dattwyler *et al.*, 1988). The important point is that seronegative Lyme disease is usually a mild, attenuated disease that responds well to antibiotic therapy.

For serological testing in Lyme disease, the Centers for Disease Control and Prevention recommends a two-step approach in which samples are tested first by enzyme-linked immunosorbent assay (ELISA) and those with equivocal or positive results are tested by western blotting (Centers for Disease Control, 1994). During the first month of infection, both IgM and IgG responses to the spirochete should be determined, preferably in acute and convalescent serum samples. Approximately 20 to 30% of patients have a positive response in acute samples, whereas even after antibiotic treatment, about 70 to 80% have a positive response during convalescence 2 to 4 weeks later (Dressler *et al.*, 1993). After that time, the great majority of patients have a positive IgG antibody response, and a single test is usually sufficient. In persons with illness persisting longer than 1 month, a positive IgM test alone is likely to be a false-positive test result. According to current criteria adopted by the Centers for Disease Control and Prevention (1994), an IgM western blot is considered positive if two of the following three bands are present: 23, 39, and 41 kDa; however, the combination of the 23- and 41-kDa bands may still be a false-positive result. An IgG blot is considered positive if five of the following ten bands are present: 18, 23, 28, 30, 39, 41, 45, 58, 66, and 93 kDa.

B. Polymerase Chain Reaction

Because serological testing does not distinguish between active and inactive infection, a test is needed that identifies the spirochete itself. In most bacterial infections, culture serves this purpose. However, *B. burgdorferi* has only been cultured regularly from skin biopsy samples of erythema migrans early in the infection (Berger *et al.*, 1992). Detection of *B. burgdorferi* DNA from joint fluid by PCR has shown promise as a substitute for culture in patients with Lyme arthritis. In one study, 70 of 73 joint fluid samples (96%) had positive PCR results using primer–probe sets that detect OspA gene segments (Nocton *et al.*, 1994), and in another study, 6 of 7 patients had positive results using a primer–probe set that targets chromosomal DNA of the spirochete (Bradley *et al.*, 1994). However, PCR testing is not so sensitive in detecting spirochetal DNA in CSF samples in patients with neuroborreliosis. In one report, 6 of 16 patients (38%) with acute neuroborreliosis and 11 of 44 patients (25%) with chronic neuroborreliosis had positive PCR results using OspA primer–probe sets (Nocton *et al.*, 1996). The great problem with PCR is the risk of exogenous contamination causing false-positive results. For this reason, PCR results must be interpreted with caution.

IX. TREATMENT

The various manifestations of Lyme disease can usually be treated successfully with oral antibiotic therapy, except for objective neurological abnormalities which seem to require intravenous therapy (Table I). For early localized or disseminated Lyme disease, doxycycline is recommended in adult men and in nonpregnant women (Steere, 1989; Rahn and Malawista, 1991). An advantage of this regimen is that it is also effective treatment for the newly recognized agent of human granulocytic ehrlichiosis (Aguero-Rosenfeld *et al.*, 1996), which is transmitted by the same tick that transmits the Lyme disease agent. Amoxicillin, cefuroxime acetil, and erythromyin are second-, third-, and fourth-choice alternatives. In children, however, amoxicillin should be prescribed; in cases of penicillin allergy, cefuroxime axetil or erythromycin may be used. For patients with infection localized to the skin, 10 days of therapy is generally sufficient; in patients with disseminated infection, 20 to 30 days of therapy is recommended. Approximately 15% of patients experience a Jarisch–Herxheimer-like reaction during the first 24 hr of therapy. Although objective evidence of relapse is rare after these regimens, subjective symptoms, such as arthralgias, malaise, and fatigue, are not uncommon and may persist for months afterward.

These oral antibiotic regimens, when given for 30 to 60 days, are usually effective for the treatment of Lyme arthritis (Steere *et al.*, 1994). However, the response to therapy may be slow, and joint inflammation may take months to subside. A small percentage of patients with arthritis, particularly those with the HLA-DR4 allele and an immune response to OspA, may have persistent

TABLE I   Treatment Regimens for Lyme Disease[a]

Stage	Treatment
Early infection (local or disseminated)	
Adults	Doxycycline, 100 mg orally 2 times/day for 10 to 30 days[b]
	Amoxicillin, 500 mg orally 3 times/day for 10 to 30 days[b]
	Alternatives in case of doxycycline or amoxicillin allergy: Cefuroxime axetil, 500 mg orally twice daily for 10 to 30 days[b]
	Erythromycin, 150 mg orally 4 times a day for 10 to 30 days[b]
Children (age 8 or less)	Amoxicillin, 250 mg orally 3 times a day or 20 mg/kg/day in divided doses for 10 to 30 days[b]
	Alternatives in case of penicillin allergy: Cefuroxime axetil, 125 mg orally twice daily for 10 to 30 days[b]
	Erythromycin, 250 mg orally 3 times a day or 30 mg/kg/day in divided doses for 10 to 30 days[b]
Arthritis (intermittent or chronic)	Doxycycline 100 mg orally 2 times/day for 30 to 60 days
	Amoxicillin 500 mg orally 4 times/day for 30 to 60 days or
	Ceftriaxone 2 g i.v. once a day for 30 days
	Penicillin G, 20 million U i.v. in four divided doses daily for 30 days
Neurological abnormalities	Ceftriaxone 2 g i.v. once a day for 30 days
Early or late	Penicillin G, 20 million U i.v. in four divided doses daily for 30 days
	Alternative in cases of ceftriaxone or penicillin allergy: Doxycycline, 100 mg orally 3 times a day for 30 days[c]
Facial palsy alone	Oral regimens may be adequate
Cardiac abnormalities	
First-degree AV block (PR interval > 0.3 sec)	Oral regimens, as for early infection
High-degree AV block	Ceftriaxone, 2 g i.v. once a day for 30 days[d]
	Penicillin G, 20 million U i.v. in four divided doses daily for 30 days[d]

[a] Treatment failures have occurred with any of the regimens given, and a second course of therapy may be necessary. i.v., intravenous; AV, atrioventricular.

[b] For localized infection of the skin 10 days of therapy is generally adequate, but for disseminated infection 20 to 30 days of treatment is recommended.

[c] In the author's experience, this regimen is ineffective for the treatment of late neurological abnormalities of Lyme disease.

[d] Once the patient has stabilized, the course may be completed with oral therapy.

joint inflammation for months or even several years despite treatment with either oral or intravenous antimicrobial therapy (Steere *et al.*, 1994). If patients have persistent arthritis despite 2 months of oral therapy or 1 month of intravenous therapy, the author treats such patients with antiinflammatory agents. If this approach fails, synovectomy may be successful (Schoen *et al.*, 1991).

For objective neurological abnormalities, with the possible exception of facial palsy alone, parenteral antibiotic therapy is favored by most experts in the field (Logigian *et al.*, 1990; Rahn and Malawista, 1991). A 2- or 4-week course of intravenous ceftriaxone is most commonly used, but intravenous cefotaxime or intravenous penicillin G for the same duration may also be effective. The symptoms of early, acute neuroborreliosis often improve rapidly, whereas those of chronic neurological involvement subside more slowly over a period of months.

In patients with high-degree atrioventricular block or a PR interval of greater than 0.3 sec, intravenous therapy for at least part of the course and cardiac monitoring are recommended. In patients with complete heart block or congestive heart failure, glucocorticoids may be of benefit if the patient does not improve on antimicrobial therapy alone within 24 hr. These recommendations are based on expert opinion; a controlled trial has never been done.

It is unclear how and whether asymptomatic infection should be treated, but such patients are often given a course of oral antibiotics. The appropriate treatment for Lyme disease during pregnancy is also unclear. Because the risk of maternal–fetal transmission seems to be very low, standard therapy for the stage and manifestation of the illness may be sufficient. Relapse may occur with any of the antibiotic regimens for Lyme disease, and a second course of therapy may be necessary. However, there is currently no evidence that many courses of antibiotics are necessary for the treatment of Lyme disease. In addition, in patients who develop chronic fatigue syndrome or fibromyalgia after Lyme disease, further antibiotic therapy does not seem to be of benefit (Dinerman and Steere, 1992; Hsu *et al.*, 1993).

X. PREVENTION

A. Personal Protection Measures

The risk of tick bites in high risk areas can be reduced by wearing long trousers tucked into socks and by checking for ticks after exposure in wooded areas. In addition, although repellents containing DEET (*N,N*-diethylmetatoluamide) or permethrin effectively deter ticks (Schreck *et al.*, 1968), permethrin can only be applied on clothing and DEET may cause serious side effects when excessive amounts are applied directly to the skin (Oransky *et al.*, 1989). These measures have been largely ineffective in reducing the risk of Lyme disease presumably because they are difficult to carry out day after day in highly endemic residential areas.

B. Prophylactive Antibiotics for Tick Bites

Should patients with tick bites receive prophylactic antibiotic therapy? In a study in the highly endemic Lyme, Connecticut, region, the risk of Lyme disease in subjects with recognized tick bites was only 1.2 percent (Shapiro *et al.*, 1992) perhaps because 24 to 48 hr of tick attachment is often required for transmission of the spirochete (Piesman *et al.*, 1987). Although the risk of infection with

B. burgdorferi after a recognized tick bite is quite low, the physician may prefer to treat if the tick is engorged, if follow-up is difficult, or if the patient is quite anxious. Amoxicillin or doxycycline therapy for 10 days appears to prevent the occurrence of Lyme disease.

C. Public Health Measures

Environmental control of ticks over widespread areas is difficult. Eradication of deer (Wilson *et al.*, 1988), area application of insecticides (Curran *et al.*, 1993), habitat destruction by burning (Mather *et al.*, 1993), and distribution of permethrin-treated cotton balls (Stafford, 1991) have been tried. These measures have either been impractical or ineffective, or their acceptance has been limited by ecological consequences. Currently, there is no known way to control enzootic Lyme disease.

D. Vaccine Development

Because the risk of Lyme disease in endemic areas is great and because personal protection and environmental control measures are impractical, it has been hoped that a safe and effective vaccine could be developed to prevent this infection. In experimental animal models of Lyme disease, mice vaccinated with recombinant OspA have been shown to be protected from infection with *B. burgdorferi*, both by antibody-mediated killing of the spirochete within the host and by destruction of the organism within the tick prior to disease transmission (Fikrig *et al.*, 1990, 1992; Simon *et al.*, 1991). A recombinant OspA vaccine for human Lyme disease is currently being tested. In phase I and II trials, this vaccine has been shown to be immunogenic and safe (Keller *et al.*, 1994; Schoen *et al.*, 1995). Two manufacturers are currently completing phase III trials to assess the efficacy and safety of the vaccine in large numbers of individuals.

XI. FUTURE RESEARCH DIRECTIONS

By any measure, the elucidation of Lyme disease over the past 20 years has been a success, but many problems remain to be solved. First, more knowledge is needed regarding *B. burgdorferi* spirochetes and how they cause Lyme disease. The development of a gene transfer system for *B. burgdorferi* and studies in animal models will certainly help, but further studies of affected human patients are also vitally important. Basic research about the etiology and pathogenesis of one disease often has implications for other diseases. For example, not only may research in the pathogenesis of Lyme arthritis help in the understanding of that illness, but it may also give clues for research in other forms of chronic inflammatory arthritis, including rheumatoid arthritis, which are currently of unknown cause. Serological tests for Lyme disease have been standardized using crude spirochetal extracts, but these are first-generation tests that can be improved with further research, perhaps by the use of recombinant borrelial proteins. Regardless, tests are greatly needed that identify the spirochete directly.

PCR testing shows promise in this regard, but more research is needed. Although much is known about antibiotic treatment for Lyme disease, some patients continue to have symptoms after treatment. It is important to learn whether these patients have persistent infection, immune phenomena that continue after the eradication of the spirochete, or residual deficit from past infection. Finally, the tick that transmits Lyme disease continues to spread relentlessly. It is hoped that successful vaccine development will provide an important new public health approach for the control of the disease. Even so, the Lyme disease tick transmits other infectious agents, including *Babesia microti* and the newly recognized agent of human granulocytic ehrlichiosis (Goodman *et al.*, 1996). Thus, a foremost need are creative strategies to control the increasing problem of tick-borne diseases.

REFERENCES

Ackermann, R., Rehse-Kupper, B., Gollmer, E., and Schmidt, R. (1988). Chronic neurologic manifestations of erythema migrans borreliosis. *Ann. N.Y. Acad. Sci.* **539**, 16–23.

Aguero-Rosenfeld, M. E., Horowitz, H. W., Wormser, G. P., McKenna, D. F., Nowakowski, J., Munoz, J., and Dumler, J. S. (1996). Human granulocytic ehrlichiosis: A case series from a medical center in New York state. *Ann. Intern. Med.* **125**, 904–908.

Ai, C., Hu, R., Hyland, K. E., Wen, Y., Zhang, Y., Qui, Q., Liu, X., Shi, Z., Zhao, J., and Cheng, D. (1990). Epidemiological and aetiological evidence for transmission of Lyme disease by adult *Ixodes persulcatus* in an endemic area in China. *Int. J. Epidemiol.* **19**, 1061–1065.

Asbrink, E., and Hovmark, A. (1985). Successful cultivation of spirochetes from skin lesions of patients with erythema chronica migrans afzelius and acrodermatitis chronica atrophicans. *Acta Pathol. Microbiol. Immunol. Scand.* **93**, 161–163.

Asbrink, E., and Hovmark, A. (1988). Early and late cutaneous manifestations of *Ixodes*-borne borreliosis (erythema migrans borreliosis, Lyme borreliosis). *Ann. N.Y. Acad. Sci.* **539**, 4–15.

Bannwarth, A. (1944). Zur Klinik und Pathogenese der "chronischen lymphocytaren Meningitis." *Arch. Psychiatr. Nervenkrankh.* **117**, 161–185.

Baranton, G., Postic, D., Saint-Girons, I., Boerlin, P., Piffaretti, J. C., Assous, M., and Grimont, P. A. (1992). Delineation of *Borrelia burgdorferi* sensu stricto, *Borrelia garinii* sp. nov., and group VS461 associated with Lyme borreliosis. *Int. J. Syst. Bacteriol.* **42**, 378–383.

Barbour, A. G. (1984). Isolation and cultivation of Lyme disease spirochetes. *Yale J. Biol. Med.* **57**, 521–525.

Barbour, A. G., and Hayes, S. F. (1986). Biology of *Borrelia* species. *Microbiol. Rev.* **50**, 381–400.

Barthold, S. W., DeSouza, M. S., Janotka, J. L., Smith, A. L., and Persing, D. H. (1993). Chronic Lyme borreliosis in the laboratory mouse. *Am. J. Pathol.* **143**, 959–971.

Barthold, S. W., DeSouza, M., and Feng, S. (1996). Serum-mediated resolution of Lyme arthritis in mice. *Lab. Invest.* **74**, 57–67.

Benach, J. L., Bosler, E. M., Hanrahan, J. P., Coleman, J. L., Habicht, G. S., Bast, T. F., Cameron, D. J., Ziegler, J. L., Barbour, A. G., Burgdorfer, W., Edelman, R., and Kaslow, R. A. (1983). Spirochetes isolated from the blood of two patients with Lyme disease. *N. Engl. J. Med.* **308**, 740–742.

Berger, B. W., Johnson, R. C., Kodner, C., and Coleman, L. (1992). Cultivation of *Borrelia burgdorferi* from erythema migrans lesions and perilesional skin. *J. Clin. Microbiol.* **30**, 359–361.

Bosler, E. M., Ormiston, B. G., Coleman, J. L., Hanrahan, J. P., and Benach, J. L. (1984). Prevalence of the Lyme disease spirochete in populations of white-tailed deer and white-footed mice. *Yale J. Biol. Med.* **57**, 651–659.

Bradley, J. F., Johnson, R. C., and Goodman, J. L. (1994). The persistence of spirochetal nucleic acids in active Lyme arthritis. *Ann. Intern. Med.* **120**, 487–489.

Brandt, M. E., Riley, B. S., Radolf, J. D., and Norgard, M. V. (1990). Immunogenic integral membrane proteins of *Borrelia burgdorferi* are lipoproteins. *Infect. Immun.* 58, 983–991.

Brown, R. N., and Lane, R. S. (1992). Lyme disease in California: A novel enzootic transmission cycle of *Borrelia burgdorferi*. *Science* 256, 1439–1442.

Burgdorfer, W., Barbour, A. G., Hayes, S. F., Benach, J. L., Grunwaldt, E., and Davis, J. P. (1982). Lyme disease—A tick-borne spirochetosis? *Science* 216, 1317–1319.

Canica, M. M., Nato, F., Du Merle, L., Mazie, J. C., Baranton, G., and Postic, D. (1993). Monoclonal antibodies for identification of *Borrelia afzelii* sp. nov. associated with late cutaneous manifestations of Lyme borreliosis. *Scand. J. Infect. Dis.* 25, 441–448.

Carlberg, H., and Naito, S. (1991). Lyme borreliosis—A review and present situation in Japan. *J. Dermatol.* 18, 125–142.

Centers for Disease Control, ed. (1994). Proceedings of the Second National Conference on Serologic Diagnosis of Lyme Disease. 1st Ed., pp. 1–111. Association of State and Territorial Public Health Laboratory Directors, Washington, D.C.

Centers for Disease Control (1996). Lyme disease—United States, 1995. *Morbidity & Mortality Weekly Report* 45, 481–484.

Clark, J. R., Carlson, R. D., Sasaki, C. T., Pachner, A. R., and Steere, A. C. (1985). Facial paralysis in Lyme disease. *Laryngoscope* 95, 1341–1345.

Coburn, J., Leong, J. M., and Erban, J. K. (1993). Integrin $\alpha_{IIb}\beta_3$ mediates binding of *Borrelia burgdorferi* to human platelets. *Proc. Natl. Acad. Sci. U.S.A.* 90, 7059–7063.

Curran, K. L., Fish, D., and Piesman, J. (1993). Reduction of nymphal *Ixodes dammini* (Acari: Ixodidae) in a residential suburban landscape by area application of insecticides. *J. Med. Entomol.* 30, 107–113.

Dai, Z., Lackland, H., Stein, S., Li, Q., Radziewicz, R., Williams, S., and Sigal, L. H. (1993). Molecular mimicry in Lyme disease: Monoclonal antibody H9724 to *B. burgdorferi* flagellin specifically detects chaperonin-HSP60. *Biochim. Biophys. Acta* 1181, 97–100.

Dattwyler, R. J., Volkman, D. J., Luft, B. J., Halperin, J. J., Thomas, J., and Golightly, M. G. (1988). Seronegative Lyme disease: Dissociation of the specific T- and B-lymphocyte responses to *Borrelia burgdorferi*. *N. Engl. J. Med.* 319, 1441–1446.

Davis, M. B. (1983). Holocene vegetational history of the Eastern United States. *In* "Late Quaternary Environments of the U.S." (H. E. Wright, ed.), Vol. 2, pp. 166–179. University of Minnesota, Minneapolis, Minnesota.

Dinerman, H., and Steere, A. C. (1992). Lyme disease associated with fibromyalgia. *Ann. Intern. Med.* 117, 281–285.

Dressler, F., Whalen, J. A., Reinhardt, B. N., and Steere, A. C. (1993). Western blotting in the serodiagnosis of Lyme disease. *J. Infect. Dis.* 167, 392–400.

Duray, P. H., and Steere, A. C. (1988). Clinical pathologic correlations of Lyme disease by stage. *Ann. N.Y. Acad. Sci.* 539, 65–79.

Fikrig, E., Barthold, S. W., Kantor, F. S., and Flavell, R. A. (1990). Protection of mice against the Lyme disease agent by immunizing with recombinant OspA. *Science* 250, 553–556.

Fikrig, E., Telford III, S. R., Barthold, S. W., Kantor, F. S., Spielman, A., and Flavell, R. A. (1992). Elimination of *Borrelia burgdorferi* from vector ticks feeding on OspA-immunized mice. *Proc. Natl. Acad. Sci. U.S.A.* 89, 5418–5421.

Fung, B. P., McHugh, G. L., Leong, J. M., and Steere, A. C. (1994). Humoral immune response to outer surface protein C of *Borrelia burgdorferi* in Lyme disease: Role of the immunoglobulin M response in the serodiagnosis of early infection. *Infect. Immun.* 62, 3213–3221.

Goodman, J. L., Nelson, C., Vitale, B., Madigan, J. E., Dumler, J. S., Kurtti, T. J., and Munderloh, U. G. (1996). Direct cultivation of the causative agent of human granulocytic ehrlichiosis. *N. Engl. J. Med.* 334, 209–215.

Halperin, J. J., Little, B. W., Coyle, P. K., and Dattwyler, R. J. (1987). Lyme disease: Cause of a treatable peripheral neuropathy. *Neurology* 37, 1700–1706.

Halperin, J. J., Luft, B. J., Anand, A. K., Roque, C. T., Alvarez, O., Volkman, D. J., and Dattwyler, R. J. (1989). Lyme neuroborreliosis: Central nervous system manifestations. *Neurology* 39, 753–759.

Hardin, J. A., Steere, A. C., and Malawista, S. E. (1979). Immune complexes and the evolution of Lyme arthritis: Dissemination and localization of abnormal C1q binding activity. *N. Engl. J. Med.* 301, 1358–1363.

Herxheimer, K., and Hartmann, K. (1902). Ueber Acrodermatitis chronica atrophicans. *Arch. Dermatol. Syph.* **61**, 57–76, 255–300.

Howe, T. R., LaQuier, F. W., and Barbour, A. G. (1986). Organization of genes encoding two outer membrane proteins of the Lyme disease agent *Borrelia burgdorferi* within a single transcriptional unit. *Infect. Immun.* **54**, 207–212.

Hsu, V. M., Patella, S. J., and Sigal, L. H. (1993). "Chronic Lyme disease" as the incorrect diagnosis in patients with fibromyalgia. *Arthritis Rheum.* **36**, 1493–1500.

Johnston, Y. E., Duray, P. H., Steere, A. C., Kashgarian, M., Buza, J., Malawista, S. E., and Askenase, P. W. (1985). Lyme arthritis: Spirochetes found in synovial microangiopathic lesions. *Am. J. Pathol.* **118**, 26–34.

Kaufman, L. D., Gruber, B. L., Phillips, M. E., and Benach, J. L. (1989). Late cutaneous Lyme disease: Acrodermatitis chronica atrophicans. *Am. J. Med.* **86**, 828–830.

Kawabata, H., Masuzawa, T., and Yanagihara, Y. (1993). Genomic analysis of *Borrelia japonica* sp. *nov.* isolated from *Ixodes ovatus* in Japan. *Microbiol. Immunol.* **37**, 843–848.

Keane-Myers, A., and Nickell, S. P. (1995). T cell subset-dependent modulation of immunity to *Borrelia burgdorferi* in mice. *J. Immunol.* **154**, 1770–1776.

Keller, D., Koster, F. T., Marks, D. H., Hosbach, P., Erdile, L. F., and Mays, J. P. (1994). Safety and immunogenicity of a recombinant outer surface protein A Lyme vaccine. *J. Am. Med. Assoc.* **271**, 1764–1768.

Klempner, M. S., Noring, R., Epstein, M. P., McCloud, B., Hu, R., Limentani, S. A., and Rogers, R. A. (1995). Binding of human plasminogen and urokinase-type plasminogen activator to the Lyme disease spirochete, *Borrelia burgdorferi. J. Infect. Dis.* **171**, 1258–1265.

Korenberg, E. I., Kryuchechnikov, V. N., and Kovalevsky, Y. V. (1993). Advances in investigations of Lyme borreliosis in the territory of former USSR. *Eur. J. Epidemiol.* **9**, 86–91.

Lam, T. T., Nguyen, T. P., Montgomery, R. R., Kantor, F. S., Fikrig, E., and Flavell, R. A. (1994). Outer surface proteins E and F of *Borrelia burgdorferi*, the agent of Lyme disease. *Infect. Immun.* **62**, 290–298.

Lawson, J. P., and Steere, A. C. (1985). Lyme arthritis: Radiologic findings. *Radiology* **154**, 37–43.

Lengl-Janssen, B., Strauss, A. C., Steere, A. C., and Kamradt, T. (1994). The T helper cell response in Lyme arthritis: Differential recognition of *Borrelia burgdorferi* outer surface protein A (OspA) in patients with treatment-resistant or treatment-responsive Lyme arthritis. *J. Exp. Med.* **180**, 2069–2078.

Leong, J. L., Morrissey, P. E., Ortega-Barria, E., Pereira, M. E. A., and Coburn, J. (1995). Hemagglutination and proteoglycan binding by the Lyme disease spirochete, *Borrelia burgdorferi. Infect. Immun.* **63**, 874–883.

Levine, J. F., Wilson, M. L., and Spielman, A. (1985). Mice as reservoirs of the Lyme disease spirochete. *Am. J. Trop. Med. Hyg.* **34**, 355–360.

Logigian, E. L., and Steere, A. C. (1992). Clinical and electrophysiologic findings in chronic neuropathy of Lyme disease. *Neurology* **42**, 303–311.

Logigian, E. L., Kaplan, R. F., and Steere, A. C. (1990). Chronic neurologic manifestations of Lyme disease. *N. Engl. J. Med.* **323**, 1438–1444.

Logigian, E. L., Johnson, K. A., Kijewski, M. F., Kaplan, R. K., Becker, J. A., Jones, K. J., Garada, B. M., Holman, B. L., and Steere, A. C. (1997). Reversible cerebral hypoperfusion in Lyme encephalopathy. *Neurology* (in press).

Ma, Y., and Weis, J. J. (1993). *Borrelia burgdorferi* outer surface lipoproteins OspA and OspB possess B cell mitogenic and cytokine stimulatory properties. *Infect. Immun.* **61**, 3843–3853.

McAlister, H. F., Klementowicz, P. T., Andrews, C., Fisher, J. D., Feld, M., and Furman, S. (1989). Lyme carditis: An important cause of reversible heart block. *Ann. Intern. Med.* **110**, 339–345.

Magnarelli, L. A., Anderson, J. F., and Chappell, W. A. (1984). Antibodies to spirochetes in white-tailed deer and prevalence of infected ticks from foci of Lyme disease in Connecticut. *J. Wildl. Dis.* **20**, 21–26.

Manweiler, S. A., Lane, R. S., and Tempelis, C. H. (1992). The western fence lizard *Sceloporus occidentalis*: Evidence of field exposure to *Borrelia burgdorferi* in relation to infestation by *Ixodes pacificus* (Acari: Ixodidae). *Am. J. Trop. Med. Hyg.* **47**, 328–336.

Marconi, R. T., Samuels, D. S., and Garon, C. F. (1993). Transcriptional analyses and mapping of the ospC gene in Lyme disease spirochetes. *J. Bacteriol.* **175**, 926–932.

Marcus, L. C., Steere, A. C., Duray, P. H., Anderson, A. E., and Mahoney, E. B. (1985). Fatal pancarditis in a patient with coexistent Lyme disease and babesiosis: Demonstration of spirochetes in the myocardium. *Ann. Intern. Med.* **103**, 374–376.

Markowitz, L. E., Steere, A. C., Benach, J. L., Slade, J. D., and Broome, C. V. (1986). Lyme disease during pregnancy. *J. Am. Med. Assoc.* **255**, 3394–3396.

Mather, T. N., Duffy, D. C., and Campbell, S. R. (1993). An unexpected result from burning vegetation to reduce Lyme disease transmission risks. *J. Med. Entomol.* **30**, 642–645.

Matuschka, F. R., and Spielman, A. (1986). The emergence of Lyme disease in a changing environment in North America and central Europe. *Exp. Appl. Acarol.* **2**, 337–353.

Nadelman, R. B., Nowakowski, J., Forseter, G., Goldberg, N. S., Bittker, S., Cooper, D., Aguero-Rosenfeld, M., and Wormser, G. (1996). The clinical spectrum of early Lyme borreliosis in patients with culture-confirmed erythema migrans. *Am. J. Med.* **100**, 502–508.

Nocton, J. J., Dressler, F., Rutledge, B. J., Rys, P. N., Persing, D. H., and Steere, A. C. (1994). Detection of *Borrelia burgdorferi* DNA by polymerase chain reaction in synovial fluid in Lyme arthritis. *N. Engl. J. Med.* **330**, 229–234.

Nocton, J. J., Bloom, B. J., Rutledge, B. J., Persing, D. H., Logigian, E. L., Schmid, C. H., and Steere, A. C. (1996). Detection of *Borrelia burgdorferi* DNA by polymerase chain reaction in cerebrospinal fluid in patients with Lyme neuroborreliosis. *J. Infect. Dis.* **174**, 623–627.

Oransky, S., Roseman, B., Fish, D., *et al.* (1989). Seizures temporarily associated with the use of DEET insect repellent—New York and Connecticut. *Morbidity & Mortality Weekly Report* **38**, 678.

Pachner, A. R., and Steere, A. C. (1985). The triad of neurologic manifestations of Lyme disease: Meningitis, cranial neuritis, and radiculoneuritis. *Neurology* **35**, 47–53.

Persing, D. H., Telford, S. R., Rys, P. N., Dodge, D. E., White, T. J., Malawista, S. E., and Spielman, A. (1990). Detection of *Borrelia burgdorferi* DNA in museum specimens of *Ixodes dammini* ticks. *Science* **249**, 1420–1423.

Piesman, J., Mather, T. N., Sinsky, R. J., and Spielman, A. (1987). Duration of tick attachment and *Borrelia burgdorferi* transmission. *J. Clin. Microbiol.* **25**, 557–558.

Rahn, D. W., and Malawista, S. E. (1991). Lyme disease: Recommendations for diagnosis and treatment. *Ann. Intern. Med.* **114**, 472–481.

Roberts, E. D., Bohm, R. P., Jr., Cogswell, F. B., Lanners, H. N., Lowrie, R. C., Jr., Povinelli, L., Piesman, J., and Philipp, M. T. (1995). Chronic Lyme disease in the rhesus monkey. *Lab. Invest.* **72**, 146–160.

Ruderman, E. M., Kerr, J. S., Telford III, S. R., Spielman, A., Glimcher, L. H., and Gravallese, E. M. (1995). Early murine Lyme carditis has a macrophage predominance and is independent of major histocompatibility complex class II–CD4+ T cell interactions. *J. Infect. Dis.* **171**, 362–370.

Saint Girons, I. G. O., and Davidson, B. E. (1994). Molecular biology of the *Borrelia*, bacteria with linear replicons. *Microbiology* **140**, 1803–1816.

Schlesinger, P. A., Duray, P. H., Burke, B. A., Steere, A. C., and Stillman, M. T. (1985). Maternal-fetal transmission of the Lyme disease spirochete, *Borrelia burgdorferi*. *Ann. Intern. Med.* **103**, 67–69.

Schmidli, J., Hunziker, T., Moesli, P., and Schaad, U. B. (1988). Cultivation of *Borrelia burgdorferi* from joint fluid three months after treatment of facial palsy due to Lyme borreliosis. *J. Infect. Dis.* **158**, 905–906 [Correspondence].

Schoen, R. T., Aversa, J. M., Rahn, D. W., and Steere, A. C. (1991). Treatment of refractory chronic Lyme arthritis with arthroscopic synovectomy. *Arthritis Rheum.* **34**, 1056–1060.

Schoen, R. T., Meurice, F., Brunet, C. M., Cretella, S., Krause, D. S., Craft, J. E., and Fikrig, E. (1995). Safety and immunogenicity of an outer surface protein A vaccine in subjects with previous Lyme disease. *J. Infect. Dis.* **172**, 1324–1329.

Schoenfeld, R., Araneo, B., Ma, Y., Yang, L. M., and Weis, J. J. (1992). Demonstration of a B-lymphocyte mitogen produced by the Lyme disease pathogen, *Borrelia burgdorferi*. *Infect. Immun.* **60**, 455–464.

Schreck, C. E., Snoddy, E. L., and Spielman, A. (1968). Pressurized sprays of permethrin or DEET on military clothing for personal protection against *Ixodes dammini* (Acari: Ixodidae). *J. Med. Entomol.* **23**, 396.

Severinghaus, C. W., and Brown, C. Y. (1983). *In* "Changes in the Land: Indians, Colonists, and the Ecology of New England" (W. Cronon, ed.), p. 241. Hill & Wang, New York.

Shapiro, E. D., Gerber, M. A., Holabird, N. B., Berg, A. T., Feder, H. M., Jr., Bell, G. L., Rys, P. N., and Persing, D. H. (1992). A controlled trial of antimicrobial prophylaxis for Lyme disease after deer-tick bites. *N. Engl. J. Med.* **327,** 1769–1773.

Sigal, L. H., Steere, A. C., Freeman, D. H., and Dwyer, J. M. (1986). Proliferative responses of mononuclear cells in Lyme disease: Reactivity to *Borrelia burgdorferi* antigens is greater in joint fluid than in blood. *Arthritis Rheum.* **29,** 761–769.

Simon, M. M., Schaible, U. E., Kramer, M. D., Eckerskorn, C., Museteanu, C., Muller-Hermelink, H. K., and Wallich, R. (1991). Recombinant outer surface protein A from *Borrelia burgdorferi* induces antibodies protective against spirochetal infection in mice. *J. Infect. Dis.* **164,** 123–132.

Snydman, D. R., Schenkein, D. P., Berardi, V. P., Lastavica, C. C., and Pariser, K. M. (1986). *Borrelia burgdorferi* in joint fluid in chronic Lyme arthritis. *Ann. Intern. Med.* **104,** 798–800.

Spielman, A., Telford III, S. R., and Pollack, R. J. (1993). The origins and course of the present outbreak of Lyme disease. *In* "Ecology and Environmental Management of Lyme Disease" (H. S. Ginsberg, ed.), 1st Ed., Vol. 1, pp. 83–96. Rutgers University Press, New Brunswick, New Jersey.

Stafford III, K. C. (1991). Effectiveness of host-targeted permethrin in the control of *Ixodes dammini* (Acari: Ixodidae). *J. Med. Entomol.* **28,** 611–617.

Stanek, G., Pletschette, M., Flamm, H., Hirschl, A. M., Aberer, E., Kristoferitsch, W., and Schmutzhard, E. (1988). European Lyme borreliosis. *Ann. N.Y. Acad. Sci.* **539,** 274–282.

Steere, A. C. (1989). Lyme disease. *N. Engl. J. Med.* **321,** 586–596.

Steere, A. C. (1994). Lyme disease: A growing threat to urban populations. *Proc. Natl. Acad. Sci. U.S.A.* **91,** 2378–2383.

Steere, A. C., Malawista, S. E., Snydman, D. R., Shope, R. E., Andiman, W. A., Ross, M. R., and Steele, F. M. (1977a). Lyme arthritis: An epidemic of oligoarticular arthritis in children and adults in three Connecticut communities. *Arthritis Rheum.* **20,** 7–17.

Steere, A. C., Malawista, S. E., Hardin, J. A., Ruddy, S., Askenase, W., and Andiman, W. A. (1977b). Erythema chronicum migrans and Lyme arthritis: The enlarging clinical spectrum. *Ann. Intern. Med.* **86,** 685–698.

Steere, A. C., Hardin, J. A., Ruddy, S., Mummaw, J. G., and Malawista, S. E. (1979). Lyme arthritis: Correlation of serum and cryoglobulin IgM with activity, and serum IgG with remission. *Arthritis Rheum.* **22,** 471–483.

Steere, A. C., Batsford, W. P., Weinberg, M., Alexander, J., Berger, H. J., Wolfson, S., and Malawista, S. E. (1980). Lyme carditis: Cardiac abnormalities of Lyme disease. *Ann. Intern. Med.* **93,** 8–16.

Steere, A. C., Grodzicki, R. L., Kornblatt, A. N., Craft, J. E., Barbour, A. G., Burgdorfer, W., Schmid, G. P., Johnson, E., and Malawista, S. E. (1983a). The spirochetal etiology of Lyme disease. *N. Engl. J. Med.* **308,** 733–740.

Steere, A. C., Bartenhagen, N. H., Craft, J. E., Hutchinson, G. J., Newman, J. H., Rahn, D. W., Sigal, L. H., Spieler, P. H., Stenn, K. S., and Malawista, S. E. (1983b). The early clinical manifestations of Lyme disease. *Ann. Intern. Med.* **99,** 76–82.

Steere, A. C., Schoen, R. T., and Taylor, E. (1987). The clinical evolution of Lyme arthritis. *Ann. Intern. Med.* **107,** 725–731.

Steere, A. C., Duray, P. H., and Butcher, E. C. (1988). Spirochetal antigens and lymphoid cell surface markers in Lyme synovitis: Comparison with rheumatoid synovium and tonsillar lymphoid tissue. *Arthritis Rheum.* **31,** 487–495.

Steere, A. C., Levin, R. E., Molloy, P. J., Kalish, R. A., Abraham III, J. H., Liu, N. Y., and Schmid, C. H. (1994). Treatment of Lyme arthritis. *Arthritis Rheum.* **37,** 878–888.

Vallat, J. M., Hugon, J., Lubeau, M., Leboutet, M. J., Dumas, M., and Desproges-Gotteron, R. (1987). Tick-bite meningoradiculoneuritis: Clinical, electrophysiologic, and histologic findings in 10 cases. *Neurology* **37,** 749–753.

Weber, K., Bratzke, H. J., Neubert, U., Wilske, B., and Duray, P. H. (1988). *Borrelia burgdorferi* in a newborn despite oral penicillin for Lyme borreliosis during pregnancy. *Pediatr. Infect. Dis.* **7,** 286–289.

Wheeler, C. M., Garcia Monco, J. C., Benach, J. L., Golightly, M. G., Habicht, G. S., and Steere, A. C. (1993). Nonprotein antigens of *Borrelia burgdorferi. J. Infect. Dis.* **167,** 665–674.

Williams, C. L., Stobino, B., Weinstein, A., Spierling, P., and Medici, F. (1995). Maternal Lyme disease and congenital malformations: A cord blood serosurvey in endemic and control areas. *Paediatric & Perinatal Epidemiology* **9,** 320–330.

Wilson, M. L., Adler, G. H., and Spielman, A. (1985). Correlation between abundance of deer and that of the deer tick, *Ixodes dammini* (Acari: Ixodidae). *Ann. Entomol. Soc. Am.* **78,** 172–176.

Wilson, M. L., Telford III, S. R., Piesman, J., and Spielman, A. (1988). Reduced abundance of immature *Ixodes dammini* (Acari: Ixodidae) following elimination of deer. *J. Med. Entomol.* **25,** 224–228.

Yang, L., Weis, J. H., Eichwald, E., Kolbert, C. P., Persing, D. H., and Weis, J. J. (1994). Heritable susceptibility to severe *Borrelia burgdorferi*-induced arthritis is dominant and is associated with persistence of large numbers of spirochetes in tissues. *Infect. Immun.* **62,** 492–500.

Yoshinari, N. H., Reinhardt, B. N., and Steere, A. C. (1991). T cell responses to polypeptide fractions of *Borrelia burgdorferi* in patients with Lyme arthritis. *Arthritis Rheum.* **34,** 707–713.

Yssel, H., Shanafelt, M. C., Soderberg, C., Schneider, P. V., Anzola, J., and Peltz, G. (1991). *Borrelia burgdorferi* activates a T helper type 1-like T cell subset in Lyme arthritis. *J. Exp. Med.* **174,** 593–601.

8

ANTIBIOTIC RESISTANCE IN BACTERIA

JULIAN DAVIES and VERA WEBB
Department of Microbiology and Immunology
University of British Columbia
Vancouver, British Columbia, Canada

We must swim with the microbes and study their survival and adaptation to new habitats.
                                        Richard M. Krause (1994)

I. INTRODUCTION

The development of antibiotic resistance can be viewed as a global problem in microbial genetic ecology. It is a very complex problem to contemplate, let alone solve, due to the geographic scale, the variety of environmental factors, and the enormous number and diversity of microbial participants. In addition, the situation can only be viewed retrospectively, and what has been done was uncontrolled and largely unrecorded. Simply put, since the introduction of antibiotics for the treatment of infectious diseases in the late 1940s, human and animal microbial ecology has been drastically disturbed. The response of microbes to the threat of extinction has been to find genetic and biochemical evolutionary routes that led to the development of resistance to every antimicrobial agent used. The result is a large pool of resistance determinants in the environment. The origins, evolution, and dissemination of these resistance genes is the subject of this review.

II. MECHANISMS OF RESISTANCE: BIOCHEMISTRY AND GENETICS

The use of antibiotics should have created a catastrophic situation for microbial populations; however, their genetic flexibility allowed bacteria to survive (and even thrive) in hostile environments. The alternatives for survival for threatened microbial populations were either mutation of target sites or acquisition of

TABLE I   Transferable Antibiotic Resistance Mechanisms in Bacteria

Mechanism	Antibiotic
Reduced uptake into cell	Chloramphenicol
Active efflux from cell	Tetracyclines
Modification of target to eliminate or reduce binding of antibiotic	β-Lactams Erythromycin Lincomycin Mupirocin
Inactivation of antibiotic by enzymatic modification	
Hydrolysis	β-Lactams Erythromycin
Derivatization	Aminoglycosides Chloramphenicol Lincomycin
Sequestration of antibiotic by protein binding	β-Lactams Fusidic acid
Metabolic bypass of inhibited reaction	Trimethoprim Sulfonamides
Binding of specific immunity protein	Bleomycin
Overproduction of antibiotic target (titration)	Trimethoprim Sulfonamides

novel biochemical functions (also known as resistance determinants or R determinants).[1] Table I lists antibiotic resistance mechanisms that are transferable among bacteria. Mutation and the acquisition of R determinants are not mutually exclusive resistance strategies. Under the selective pressure of the antibiotic, mutation can lead to "protein engineering" of the acquired resistance determinant which may expand its substrate range to include semisynthetic molecules designed to be refractory to the wild-type enzyme (see Section II,A,2,a).

Theoretically, the ideal target for a chemotherapeutic agent is a constituent which is present in the target cell (e.g., bacterial pathogen, cancer cell) and not present in the host cell. The first antibiotics to be employed generally, the penicillins, targeted the synthesis of peptidoglycan, a component unique to the bacterial cell wall. Antibiotics have been found that inhibit the synthesis or interfere with the function of essentially all cellular macromolecules. Table II lists some of the common antimicrobial drugs used clinically and their targets within the bacterial cell. Our discussion of the mechanisms of resistance will be organized according to the targets of antibiotic activity.

---

[1] A note concerning terminology: "determinant" refers to the genetic element which encodes a "mechanism" or biochemical activity which confers resistance.

■ **TABLE II   Antimicrobial Drugs: Mechanisms of Action**

Target	Drug
Protein synthesis	
30 S subunit	Tetracyclines
	Aminoglycosides: streptomycin, amikacin, apramycin, gentamicin, kanamycin, tobramycin, netilmicin, isepamicin
50 S subunit	Chloramphenicol
	Fusidic acid
	Macrolides: erythromycin, streptogramin B
	Lincosamides
tRNA synthetase	Mupirocin
Cell wall synthesis	
Penicillin binding proteins	β-Lactams
	Penicillins: ampicillin, methicillin, oxacillin
	Cephalosporins: cefoxitin, cefotaxime
	Carbapenems: imipenen
	β-Lactamase inhibitors: clavulanic acid, sulbactam
D-Ala-D-Ala binding	Cyclic glycopeptides: vancomycin, teicoplanin, avoparcin
Muramic acid biosynthesis	Fosfomycin
Nucleic acid synthesis	
DNA	Quinolones and fluoroquinolones: ciprofloxicin, sparfloxacin
RNA	Rifampicin
Folic acid metabolism	Trimethoprim
	Sulfonamides

A. Targets and Specific Mechanisms

1. Protein Synthesis

Protein synthesis involves a number of components: the ribosome, transfer RNA (tRNA), messenger RNA (mRNA), numerous ancillary proteins, and other small molecules. When protein synthesis is inhibited by the action of an antibiotic, the ribosome is usually the target. The bacterial ribosome is composed of two riboprotein subunits. The small, 30 S subunit consists of approximately 21 ribosomal proteins (rprotein) and the 16 S ribosomal RNA (rRNA) molecule (about 1500 nucleotides), whereas the large, 50 S subunit contains approximately 34 rproteins and the 23 S and 5 S rRNA molecules (about 3000 and 120 nucleotides, respectively). The complexity of the ribosome structure and the redundancy of many of the genes encoding ribosomal components in most bacterial genera makes resistance due to point mutation an unlikely event.

Generally, resistance to antibiotics which inhibit protein synthesis is mediated by R determinants.

a. Aminoglycosides

Aminoglycosides, which are broad-spectrum antibiotics, are composed of three or more aminocyclitol units. They bind to the 30 S subunit and prevent the transition from the initiated complex to the elongation complex; they also interfere with the decoding process. As noted above, target site mutations of ribosomal components resulting in antibiotic resistance are rare; however, they do occur. For example, in the slow-growing *Mycobacterium tuberculosis*, mutants resistant to streptomycin appear more frequently than they do in *Escherichia coli*. Mutations leading to resistance result from an altered S12 rprotein or 16 S rRNA such that the ribosome has reduced affinity for the antibiotic. Most fast-growing bacteria have multiple copies of the rRNA genes, and because resistance is genetically recessive to antibiotic sensitivity, only rare mutations in the gene for protein S12 are isolated under normal situations. However, because the slow-growing mycobacteria possess only single copies of the rRNA genes, streptomycin resistance can arise by mutational alteration of either 16 S rRNA or ribosomal protein S12. Both types of mutations have been identified in *M. tuberculosis* (Finken *et al.*, 1993).

The introduction and therapeutic use of a series of naturally occurring and semisynthetic aminoglycosides over a 20-year period (1968–1988) led to the appearance of multiresistant strains resulting from selection and dissemination of a variety of aminoglycoside resistance determinants. For example, in 1994 the Aminoglycoside Resistance Study Group examined of the occurrence of aminoglycoside-resistance mechanisms in almost 2000 aminoglycoside-resistant *Pseudomonas* isolates from seven different geographic regions. In their study 37% of the isolates overall had at least two different mechanisms of resistance (Aminoglycoside Resistance Study Group, 1994).

A dozen different types of modifications are known to be responsible for resistance to the aminoglycosides. When one considers that each of these enzymes has a number of isozymic forms, there are at least 30 different genes implicated in bacterial resistance to this class of antibiotics. Different proteins in the same functional class may show as little as 44% amino acid similarity. The phylogenic relationships between different aminoglycoside-modifying enzymes were the subject of an excellent review by Shaw and co-workers, who have collated the nucleotide and protein sequences of the known aminoglycoside acetyltransferases, phosphotransferases, and adenylyltransferases responsible for resistance in both pathogenic bacteria and antibiotic-producing strains (Shaw *et al.*, 1993).

Rarely do a few point mutations in the aminoglycoside resistance gene arise sufficient to generate a modified enzyme with an altered substrate range that would lead to a significant change in the antibiotic resistance spectrum (Rather *et al.*, 1992; Kocabiyik and Perlin, 1992); in other words, these genes do not appear to undergo facile mutational changes that generate enzymes with altered substrate activity. *In vitro* mutagenesis studies have failed to generate extended-spectrum resistance to this class of antibiotics, and, so far, such changes have not been identified in clinical isolates, in contrast to the situation with the

col resistance, a second type causing the active efflux of chloramphenicol from *Pseudomonas* cells has been described (Bissonnette *et al.*, 1991). The *cml* determinant is encoded on integrons (see Section IV,A) and, like the tetracycline efflux pump, is a member of the MF family of efflux systems (see Section II,B,2, and Table V). The presence of the cloned *cml* gene in *E. coli* leads to, in addition to active pumping of the drug from the cell, a reduction in outer membrane permeability to chloramphenicol by repressing the synthesis of a major porin protein (Bissonnette *et al.*, 1991). This effect has also been reported for the homolog of the *cml* gene in *Haemophilus influenzae,* a major causative agent of meningitis.

d. Macrolides, Lincosamides, and Streptogramins

The macrolide–lincosamide–streptogramin (MLS) group of antibiotics have been used principally for the treatment of infections caused by gram-positive bacteria. The macrolides, especially erythromycin and its derivatives, have been used extensively and may be employed for the treatment of methicillin-resistant *Staphylococcus aureus* (MRSA), although the multiple drug resistance of the latter often includes the MLS class. Extensive studies on the mechanism of action of erythromycin identify the peptidyltransferase center as the target of the drug although there are clearly some subtleties in mechanism that remain to be resolved.

The principal mechanism of resistance to macrolide antibiotics involves methylation of the 23 S rRNA of the host giving the erm^R phenotype. In clinical isolates, enzymatic modification of rRNA, rendering the ribosome refractory to inhibition, is the most prevalent mechanism and, worldwide, compromises the use of this class of antibiotic in the treatment of gram-positive infections (Leclercq and Courvalin, 1991). Mutation of rRNA has also been shown to be important in some clinical situations: nucleotide sequence comparisons of clarithromycin-resistant clinical isolates of *Helicobacter pylori* revealed that all resistant isolates had a single base pair mutation in the 23 S rRNA (Debets-Ossenkopp *et al.*, 1996). In addition, a number of different mechanisms for the covalent modification of the MLS group have been described. For example, O-phosphorylation of erythromycin has been identified in a number of bacterial isolates (O'Hara *et al.*, 1989), and hydrolytic cleavage of the lactone ring of this class of antibiotics has also been described. The lincosamides (lincomycin and clindamycin) have been shown to be inactivated by enzymatic O-nucleotidylation in gram-positive bacteria. Macrolides are also inactivated by esterases and acetyltransferases. For macrolides and lincosamides and the related streptogramins (the MLS group) (Arthur *et al.*, 1987; Brisson-Noël *et al.*, 1988), the latter forms of antibiotic inactivation seem (for the moment) to be a relatively minor mechanism of resistance.

The macrolide antibiotics, especially the erythromycin group (14-membered lactones), contain a number of semisynthetic derivatives that have improved pharmacological characteristics. This includes activity against some resistant strains: however, no derivative has been found with effective potency against erm^R strains. The 23 S rRNA methyltransferases from pathogenic gram-positive cocci remain the most significant problem. To our knowledge, no useful inhibitor of these enzymes has been identified; with the availability of ample quantities of the purified enzymes it would not be surprising if rationally designed, specific inhibitors of some of the *erm* methyltransferases may eventually be developed.

The control of expression of the *erm* methyltransferases has been studied extensively by Weisblum and by Dubnau and their colleagues (Weisblum, 1995; Monod *et al.*, 1987). The majority of the resistant strains possess inducible resistance which is due to a novel posttranslational process in which the ribosome stalls on a 5'-leader sequence in the presence of low concentrations of antibiotic. The biochemistry of this process has been analyzed extensively. Point mutations or deletions that disrupt the secondary structure of the leader protein sequence generate constitutive expression, and such mutants have been identified clinically. These strains are resistant to the majority of MLS antibiotics. The regulation of antibiotic resistance gene expression takes many forms (see tetracycline and vancomycin for other examples); the leader-control process seen with the *erm* methylases is reminiscent of the control of amino acid biosynthesis by attenuation, which has been extensively studied. One cannot help but marvel at the simplicity and "cheapness" of this form of control of gene expression—all that is required is a few extra bases flanking the 5' end of the gene with no requirement of additional regulatory genes.

The streptogramins (especially virginiamycins) contain two components, a macrocyclic lactone and a depsipeptide which have synergistic activity (one such combination has been named Synercid®). These compounds have long been used as animal feed additives with the result that bacteria resistant to the MLS class of antibiotics are commonly found in the flora of farm animals. This nontherapeutic application is likely to compromise the newer MLS antibiotics being developed for human therapy. For example, derivatives of the pristinamycins (quinupristin and dalfopristin) will shortly be recommended for the treatment of vancomycin-resistant enterococci, against which they have potent activity. Regrettably, due to animal applications of MLS antibiotics a significant gene pool of resistance determinants is already widely disseminated.

e. Mupirocin

Mupirocin is an example of how quickly resistance can develop to a new antibiotic. Introduced in 1985, it has been used solely as a topical treatment for staphylococcal skin infections and as a nasal spray against commensal MRSA. Mupirocin, also known as pseudomonic acid A, is produced by the gram-negative bacterium *Pseudomonas fluorescens*. It acts by inhibiting isoleucyl-tRNA synthetase, which results in the depletion of charged $tRNA^{Ile}$, amino acid starvation, and ultimately the stringent response. The first reports of mupirocin resistance (Mu^R) appeared in 1987; while not yet a widespread problem, the Mu^R phenotype in coagulase-negative staphylococci and MRSA among others is emerging in many hospitals all over the world (Cookson, 1995; Zakrzewska-Bode *et al.*, 1995; Udo *et al.*, 1994). Resistance is due to the plasmid-encoded *mup*A gene, a mupirocin-resistant isoleucyl-rRNA synthetase (Noble *et al.*, 1988). Transfer has been demonstrated by filter mating and probably accounts for the rapid spread of resistance (Rahman *et al.*, 1993). An analysis of plasmids from clinical isolates of *S. aureus* has shown that mupirocin resistance can be found on multiple resistance, high copy number plasmids (Needham *et al.*, 1994).

2. Cell Wall Synthesis

The synthesis and integrity of the bacterial cell wall have been the focus of much attention as targets for antimicrobial agents. This is principally because this structure and its biosynthesis is unique to bacteria and also because inhibitors of cell well synthesis are usually bacteriocidal.

a. β-Lactams

In the years following the introduction of the β-lactam antibiotics (penicillins and cephalosporins) for the treatment of gram-negative and gram-positive infections, there has been a constant tug-of-war between the pharmaceutical industry and the bacterial population: the one to produce a novel β-lactam effective against the current epidemic of resistant bugs in hospitals, and the other to develop resistance to the newest "wonder" drug.

The role of mutation is especially important in the evolution or expansion of resistance in the case of β-lactams. In a 1992 review Neu showed the "phylogeny" of development of β-lactam antibiotics in response to the evolution of bacterial resistance to this class of antibiotics (Neu, 1992). Of the several known mechanisms of resistance to the β-lactam antibiotics (Table I), the most elusive target is hydrolytic inactivation by β-lactamases. A single base change in the gene for a β-lactamase can change the substrate specificity of the enzyme (Jacoby and Archer, 1991). Such changes occur frequently, especially in the Enterobacteriaceae, and it is frightening to realize that one single base change in a gene encoding a bacterial β-lactamase may render useless $100 million worth of pharmaceutical research effort.

The cycle of natural protein engineering in response to changing antibiotic-selection pressure has been demonstrated especially for the TEM β-lactamase (penicillinase and cephalosporinase) genes. The parental genes appear to originate from a variety of different (and unknown) sources (Couture *et al.*, 1992). The β-lactamase families differ by a substantial number of amino acids, as is the case for other antibiotic resistance genes. Sequential expansion of their substrate range to accommodate newly introduced β-lactam antibiotics is a special case and occurs by a series of point mutations at different sites within the gene that change the functional interactions between the enzyme and its β-lactam substrate. More than 30 of the so-called extended-spectrum β-lactamases have been identified (and more will come). The fact that the β-lactamase genes so readily undergo mutational alterations in substrate recognition could have several explanations, one being that the β-lactamases, like the related proteases, have a single active site that does not require interaction with any cofactors. Other antibiotic-modifying enzymes often have two active binding sites. The pharmacokinetic characteristics of the different classes of antibiotics (e.g., dose regimen, active concentration, and route of excretion) also may favor the pathway of mutational alteration in the development of resistance. One aspect of the mutational variations of the β-lactamase genes might involve the presence of mutator genes on the R plasmids in the bacterial hosts (LeClerc *et al.*, 1996). These genes increase mutation frequency severalfold and could explain the facile evolution of the TEM-based enzymes. This characteristic of bacterial pathogens has received comparatively little study.

As one approach to counteracting the destructive activity of β-lactamases, a series of effective inhibitors of these enzymes has been employed. The inhibitors are structural analogs of β-lactams that are, in most cases, dead-end irreversible inhibitors of the enzyme. Several have been used in combination with a β-lactam antibiotic for the treatment of infections from resistant microbes, for example, the successful combination of amoxicillin (antibiotic) and clavulanic acid (inhibitor). However, the wily microbes are gaining the upper hand once again by producing mutant β-lactamases that not only are capable of hydrolyzing the antibiotic but concomitantly become refractory to inhibition (Blazquez et al., 1993).

In addition to active site mutation, other changes in β-lactamase genes have evolved in response to continued β-lactam use. In some cases, increased resistance results from increased expression of the gene through an up-regulating promoter mutation (Chen and Clowes, 1987; Mabilat et al., 1990); alternatively, chromosomal β-lactamase genes can be overexpressed in highly resistant strains as a result of other changes in transcriptional regulation (Honoré et al., 1986).

Drug inactivation is not the only mechanism of resistance to the β-lactams. In fact, mutations that alter access to the target (penicillin binding proteins, pbp) of the drug through porin channels have been widely reported (Nikaido, 1994). Methicillin resistance in S. aureus (MRSA) is due to an unusual genetic complex which replaces the normal pbp2 with the penicillin refractory pbp2A. The origin of this resistance determinant is unknown, but the consequences of the clonal distribution of MRSA is well documented. It would appear that most MRSA are close relatives of a small number of parental derivatives that have been disseminated by international travel. Alterations of other pbp's in different bacterial pathogens have been responsible for widespread resistance to β-lactam antibiotics; in the case of Streptococcus pneumoniae and Neisseria gonorrhoeae this has occurred by interspecific recombination leading to the formation of mosaic genes that produce pbp's with markedly reduced affinity for the drugs.

b. Glycopeptides

The glycopeptide antibiotics vancomycin and teichoplanin were first discovered in the 1950s but have only come into prominence since the late 1980s, being the only available class of antibiotic effective for the treatment of MRSA and methicillin-resistant enterococci (MRE). However, numerous outbreaks of vancomycin-resistant enterococci (VRE) have been reported in hospitals around the world (VRE are essentially untreatable by any approved antibiotic), and there is great concern that vancomycin-resistance determinants will be transferred to pathogenic staphylococci; the resulting methicillin, vancomycin resistant S. aureus (MVRSA) will be the "Superbug," the "Andromeda" strain that infectious disease experts fear most (at least according to the newspapers).

The glycopeptides block cell wall synthesis by binding to the peptidoglycan precursor dipeptide D-alanyl-D-alanine (D-Ala-D-Ala) and preventing its incorporation into the macromolecular structure of the cell wall. The most common type of resistance is the vanA type found principally in *Enterococcus faecalis* (Walsh *et al.,* 1996; Arthur *et al.,* 1996). Resistance due to the VanA phenotype results from the substitution of the depsipeptide D-alanyl-D-lactate (D-Ala-D-Lac) for D-Ala-D-Ala residues, thereby reducing the binding of the antibiotic by eliminating a key hydrogen bond in the D-Ala-D-Ala complex. The introduction of D-Lac is encoded by the nine genes of the VanA cluster and is associated with a mechanism to prevent the formation of native D-Ala-D-Ala containing peptide in the same host; thus, the resistance is dominant. Vancomycin resistance is inducible by a two-component regulatory system; the inducers are not the antibiotics, but rather the accumulated peptidoglycan fragments produced by the initial inhibitory action of the glycopeptides. The VanA cluster is normally found on a conjugative transposon related to Tn*1546,* and it is probably responsible for the widespread dissemination of glycopeptide resistance among the enterococci. A plasmid carrying the VanA resistance determinants has been transferred from enterococci to staphylococci under laboratory conditions (Noble *et al.,* 1992), and there is apprehension that this will occur in clinical circumstances. Regrettably a glycopeptide antibiotic, avoparcin, has been employed extensively as a feed additive for chickens and pigs in certain European countries. This has led to the appearance of a high proportion glycopeptide-resistant enterococci in natural populations. This feeding practice has now been banned by the European Union (EU), but is it too late? Only time will tell.

c. Fosfomycin

A widely used mechanism for the detoxification of cell poisons in eukaryotes is the formation of glutathione adducts; for example, this mechanism is commonly used for herbicide detoxification in plants (Timmermann, 1988). However, in spite of the fact that many microbes generate large quantities of this important thiol, only one example of an antibiotic resistance mechanism of this type has so far been identified in bacteria, namely, that of fosfomycin. Fosfomycin, produced by a streptomycete, is an analog of phosphoenol pyruvate and interferes with bacterial cell wall synthesis by inhibiting the formation of N-acetylmuramic acid, a unique component of bacterial cell walls. It is employed in the treatment of sepsis, both alone and in combination with other antimicrobial agents. In gram-negative bacteria transmissible resistance is due to a plasmid-encoded glutathione S-transferase that catalyzes the formation of an inactive fosfomycin–glutathione adduct (Arca *et al.,* 1990). Two independent genes for fosfomycin resistance have been cloned and sequenced, one from *Serratia marcescens* (Suárez and Mendoza, 1991) and the other from *Staphylococcus aureus* (Zilhao and Courvalin, 1990); the two genes are unrelated at the sequence level. It is unlikely that the gram-positive gene encodes an enzyme involved in the production of a glutathione adduct, because *S. aureus* does not contain glutathione!

3. DNA and RNA Synthesis

Nucleic acid metabolism has attracted much attention as a potential target for antimicrobial drugs; the strategy of hitting at the "heart" of the microbe seemed the most likely to lead to effective bactericidal agents. Unfortunately, the ubiquity of DNA and RNA and the failure to identify discriminating target differences in the biosynthetic enzymes made this search unproductive until relatively recently when the fluoroquinolones (a class of synthetic drugs with no known natural analogs) were introduced. However, resistance mechanisms were not long in appearing, and the fluoroquinolones instead of being the "superdrugs" needed, are already limited by resistance.

a. Fluoroquinolones

Nalidixic acid, the prototype quinolone antibiotic discovered in 1962, had limited use, principally for urinary tract infections caused by gram-negative bacteria. Resistance developed by mutation, and the nal[R] phenotype proved to be the first useful marker for *gyrA* the gene encoding the A subunit of DNA gyrase (topoisomerase I). The development of resistance to nalidixic acid occurred solely by this type of mutation, and plasmid-determined resistance has never been reliably identified. This is not surprising, given that nalidixic acid is a purely synthetic chemical and no natural analog has been identified.

In the late 1970s the first fluoroquinolone antimicrobials were introduced; these proved vastly superior to nalidixic acid and are among the most potent antimicrobial agents known. A number of fluoroquinolone antibiotics have now been introduced as antiinfectives and most show good, broad-spectrum activity. In laboratory studies, mutations to high level resistance occurred at relatively low frequency, and genetic studies identified a number of different DNA replication-associated targets. As with nalidixic acid, topoisomerase I is the principal target, but other targets associated with bacterial DNA replication which give the FQ[R] phenotype have been identified in different bacterial species. In clinical use, resistance to the newer fluoroquinolones has been found to develop quite rapidly as a result of one or more mutations. In addition to target mutation, active efflux of the drug is also an important mechanism of resistance; for example, the *norA* mutation identifies a multiple drug resistance (mdr) system in *S. aureus* (see Section II,B,2 and Table V). Strains resistant to high levels of the drug have been identified frequently during the course of treatment especially with *P. aeruginosa* and *S. aureus* infections. Resistance is due to multiple mutations leading to increased efflux and alteration of components of the DNA synthetic apparatus. Clonal dissemination of FQ[R] strains appears to be quite common in nosocomial *P. aeruginosa* infections. No plasmid-mediated resistance to fluoroquinolones has been identified to date, although possible mechanisms leading to dominant resistance genes can be envisaged.

b. Rifampicin

Rifampicin is the only inhibitor of bacterial (and mitochondrial) RNA polymerases that has ever been used in the treatment of infectious disease. Mutations in the gene encoding the β subunit of RNA polymerase (*rpoB*) give high level resistance to the drug; the study of these mutations has provided important information on the structure and function of the RNA polymerase proteins in bacteria. Rifampicin is used in the treatment of gram-positive bacteria and is effective against mycobacterial and staphylococcal infections. Because it is lipophilic and thus diffuses rapidly across the hydrophobic cell envelop of mycobacteria, rifampicin is one of the frontline drugs for the treatment of tuberculosis and leprosy. However, the appearance of resistant strains as a result of *rpoB* mutations is quite common in multiple drug-resistant strains (Cole, 1994). In addition, inactivation of rifampicin in fast growing mycobacterial strains by phosphorylation, glucosylation, and ribosylation has been reported (Dabbs *et al.*, 1995).

4. Folic Acid Biosynthesis

As the above discussion illustrates, the biosynthesis and function of cellular macromolecules are the principal targets for the majority of antimicrobial agents. However, interference with the activity of enzymes involved in intermediary metabolism is also an effective strategy. Although many potential antimicrobial targets have been identified and tested, to date only folic acid biosynthesis has been successfully exploited in the development of useful drugs.

Folic acid is involved in the transfer of one-carbon groups utilized in the synthesis and metabolism of amino acids such as methionine and glycine and in the nucleotide precursors adenine, guanine, and thymine. Folic acid is converted in two reduction steps to tetrahydrofolate (FH_4), which serves as the intermediate carrier of hydroxymethyl, formyl, or methyl groups in a large number of enzymatic reactions in which "one-carbon" groups are transferred from one metabolite to another or are interconverted. The synthetic antibacterial agents sulfonamides and trimethoprim inhibit specific steps in the biosynthesis of FH_4. The current state of resistance to sulfonamides and trimethoprim in major bacterial pathogens and the mechanisms of sulfonamide and trimethoprim resistance have been reviewed (Huovinen *et al.*, 1995).

a. Sulfonamides

The sulfonamides were first discovered in 1932 and introduced into clinical practice in the late 1930s. They have a wide spectrum of activity and have been used in urinary tract infections due to the Enterobacteriaceae, in respiratory tract infections due to *Streptococcus pneumoniae* and *Haemophilus influenzae*, in skin infections due to *S. aureus*, and in gastrointestinal tract infections due to *E. coli* and *Shigella* spp. The wide range of clinical indications and low production costs maintain the popularity of the sulfonamides in Third World countries. The enzyme dihydropteroate synthase (DHPS) catalyzes the formation of dihydropteric acid, the immediate precursor of dihydrofolic acid. DHPS found in bacteria and some protozoan parasites, but not in human cells, is the target of sulfonamides. These drugs are structural analogs of *p*-aminobenzoic acid, the normal substrate of DHPS, and act as competitive inhibitors for the enzyme, thus blocking folic acid biosynthesis in bacterial cells. A large number of sulfonamides have been synthesized that show wide variations in therapeutic activity! One class, the dapsones, remains an effective anti-leprosy drug.

b. Trimethoprim

Trimethoprim was first introduced in 1962, and since 1968 it has been used (often in combination with sulfonamides due to a supposed synergistic effect) for numerous clinical indications. Like the sulfonamides, trimethoprim has a wide spectrum of activity and low cost. The target is the enzyme dihydrofolate reductase (DHFR), which is essential in all living cells. Trimethoprim is a struc-

tural analog of dihydrofolic acid and acts as a competitive inhibitor of the reductase. The human DHFR is naturally resistant to trimethoprim, which is the
basis for its use.

Resistance has been reported to both trimethoprim and sulfonamides since
their respective introductions into clinical practice. Although the principal form
of resistance is plasmid mediated, a clinical isolate of *E. coli* was described in
which the chromosomal DFHR was overproduced several hundredfold, leading
to very high trimethoprim resistance (minimum inhibitory concentration > 1 g/
liter). Sul[R] and Tmp[R] strains carry plasmid-encoded *dhps* and *dfhr* genes that
may be up to 100 times less susceptible to the inhibitors. Extensive genetic and
enzymatic studies, principally by Sköld and collaborators, have characterized
resistance in the Enterobacteriaceae. The *sul* gene is part of the 3' conserved
region of integron structures and may have been the first resistance determinant
acquired (see Section IV,A). Less information is available for resistant gram-
positive pathogens. Nonetheless, the origins of Sul[R] and Tmp[R] genes remains a
mystery.

B. Broad-Spectrum Resistance Systems

In the first part of this section we examined bacterial targets for antibiotic activity and mechanisms that specifically provide resistance to those antibiotics.
The mechanisms included modification of the target (e.g., vancomycin resistance), modification of the antibiotic (e.g., aminoglycoside methyltransferases),
overproduction of the target (e.g., trimethoprim resistance), and extrusion of
the drug from the cell (e.g., TetA-type tetracycline resistance). However, even as
the first antimicrobial agents were being tested, researchers noticed "intrinsic" differences in sensitivity among the target organisms to a wide range of
compounds.

I. Membranes and Cell Surfaces

Initially, intrinsic differences in resistance to antibiotics and other chemotherapeutic agents were attributed to structures such as the gram-negative
outer membrane and the mycobacterial cell surface. Intrinsic drug resistance
in mycobacteria has been reviewed by Nikaido and collaborators (Jarlier and
Nikaido, 1994). They suggest that lipophilic molecules are slowed by the low
fluidity of the lipid bilayer surrounding the cell wall and that hydrophobic molecules enter the mycobacterial cell slowly because the porins are inefficient and
few in number. They note that although these surface features strongly contribute the high level of natural resistance in mycobacteria, other factors are also
involved such as pbp's with low affinity for penicillin and the presence of
β-lactamases.

2. Multidrug Efflux Pumps

A cell has a number of different ways to export material across its
membrane. All of them involve the expenditure of energy. The most well-
characterized multiple drug resistance pumps in mammalian cells are the ABC
(ATP binding cassette) transporters. In bacteria, the ABC transporters are primarily seen in the translocation of virulence factors such as hemolysin in *E. coli*
and cyclolysin in *Bordetella pertussis*. A second type of active export system has

been characterized in which the energy to drive the pump comes from the proton motive force (PMF) of the transmembrane electrochemical proton gradient. These multidrug efflux systems have been the subject of a number of reviews (Lewis, 1994; Nikaido, 1996; Paulsen *et al.*, 1996). There are three families of PMF multidrug efflux pumps: the major facilitator superfamily (MFS), the staphylococcal (or small) multidrug resistance (SMR) family, and the resistance/nodulation/cell division (RND) family. Table V lists examples from each of these families and the types of compounds they pump out of the cell. In addition to these multisubstrate pumps, proton motive force efflux pumps for specific antibiotic resistance, such as the TetK and TetL (MFS-type) pumps, have also been described (see Section II,A,1,b).

The Acr multidrug efflux pump found in *E. coli* is one of the most well characterized of the RND-type pumps. Expression of the Acr efflux pump is controlled by the marA protein. The multiple antibiotic resistance (mar) phenotype was first described in 1983 by Levy and co-workers when they plated *E. coli* on medium containing either tetracycline or chloramphenicol and obtained mutants that were also resistant to the other antibiotic. Further analysis showed that these mutants had additional resistances to β-lactams, puromycin, rifampicin, and nalidixic acid. The most striking aspect of this multiple drug resistance phenotype was the wide range of structurally unrelated compounds with which it was observed. Subsequent studies have found that the mar phenotype is part of a complex stress response system which includes the superoxide response locus *soxRS* (Miller and Sulavik, 1996). Rather than encoding

TABLE V **Multidrug Efflux Pumps**[a]

Protein	Organism	Substrate
MFS		
QacA	*Staphylococcus aureus*	Mono- and divalent organic cations
NorA	*S. aureus*	Acriflavin, cetyltrimethylammonium bromide, fluoroquinolones, chloramphenicol, rhodamine 6G
EmrB	*Escherichia coli*	CCCP, nalidixic acid, organomercurials, tetrachlorosalicylanilide, thiolactomycin
Bcr	*E. coli*	Bicyclomycin, sulfathiazole
Blt	*Bacillus subtilis*	Similar range as NorA
Bmr	*B. subtilis*	Similar range as NorA
SMR		
Smr	*S. aureus*	Monovalent cations, e.g., cetyltrimethylammonium bromide, crystal violet, ethidium bromide
EmrE	*E. coli*	Monovalent cations, e.g., cetyltrimethylammonium bromide, crystal violet, ethidium bromide, methyl viologen, tetracycline
QacE	*Klebsiella pneumonia*	Similar range to Smr
RND		
AcrAB TolC	*E. coli*	Acriflavin, crystal violet, detergents, decanoate, ethidium bromide, erythromycin, tetracycline, chloramphenicol, β-lactams, nalidixic acid
AcrEF	*E. coli*	Acriflavin, actinomycin D, vancomycin
MexAB OprM	*Pseudomonas aeruginosa*	Chloramphenicol, β-lactams, fluoroquinolones, tetracycline

[a] Summarized from Paulsen *et al.* (1996).

the structural components responsible for the mar phenotype, the *mar* locus encodes a regulatory system. The marA protein is a positive regulator which controls expression of at least two loci, the *acr* locus, which encodes the genes for a multidrug efflux pump, and the *micF* locus, which encodes an antisense repressor of the outer membrane protein ompF. Expression from the *mar* locus is tightly controlled by the first gene in the operon, *marR*, which encodes a repressor protein (Miller and Sulavik, 1996).

As noted above, the *mar* locus is part of a complex stress response system. Levy and co-workers (Goldman *et al.*, 1996) have reported that mutations of the marR repressor protein protected *E. coli* from rapid killing by fluoroquinolones. They hypothesize that such protection may allow cells time to develop mutations that lead to higher levels of fluoroquinolone resistance, and may thus explain the increasing frequency of occurrence of resistant clinical isolates. Such mutations have been found among clinical strains of fluoroquinolone-resistant *E. coli*, suggesting that mutations at the *mar* locus may be the first step in clinically significant fluoroquinolone resistance. The latter may have been the case for all clinically significant antibiotic resistance: a mutational event leading to a low-level increase in drug efflux, followed by the acquisition of a heterologous resistance determinant, leading to high-level antibiotic resistance.

3. Other Types of Natural Resistance

Other types of natural resistance in bacteria will depend on the organism and the drug in question. For example, three species of enterococci, *Enterococcus gallinarum*, *Enterococcus casseliflavus*, and *Enterococcus flavescens*, produce peptidoglycan precursors which end in D-serine residues and are intrinsically resistant to low levels of vancomycin. Similarly, genera from the lactic acid bacteria *Lactobacillus*, *Leuconostoc*, and *Pediococcus* are resistant to high levels of glycopeptides because their cell wall precursors end with D-lactate. As noted above, mycobacteria species have pbp's which have a low affinity for penicillin. In addition, some mycobacteria have been reported to have ribosomes that have a lower affinity for macrolides than the ribosomes of *S. aureus* (Jarlier and Nikaido, 1994).

III. GENE TRANSFER

All available evidence suggests that the acquisition and dissemination of antibiotic resistance genes in bacterial pathogens has occurred since the late 1940s. The best support for this notion comes from studies of Hughes and Datta (Datta and Hughes, 1983; Hughes and Datta, 1983) who examined the "Murray Collection," a collection of (mostly) gram-negative pathogens obtained from clinical specimens in the preantibiotic era. None of these strains show evidence of resistance to antibiotics in current use. Given the critical role of gene exchange in bacterial evolution, it is self-evident that extensive interspecific and intergeneric gene transfer must have occurred during the golden age of antibiotics. This subject has been reviewed many times, and a variety of mechanisms of gene transfer have been invoked in the process of resistance determinant acquisition and dispersion (see Table VI) and characterized in laboratory studies. It can be assumed that these (and probably other processes) all occur in nature.

TABLE VI Gene Transfer Processes

Process	Components	Required genetic determinants		Observed in		DNA form[b]	DNA size	Host range	Comments
		Donor	Recipient[a]	Laboratory	Nature				
Conjugation	Cell/cell	Transfer genes	May require receptor	+	+	ss	≥4 Mb	Very broad interspecific, intergeneric	Can be of very high efficiency; can reduce problem of restriction in recipient
Fusion	Cell/cell	?	?	+ (rare)	−	ds	Unlimited	?	Likely to involve partners with damaged wall or membranes (protoplasts or spheroplasts)
Transduction	Bacteriophage/cell	Phage receptor	Phage receptor	+	+	ds	≤50 kb	Limited to closely related species	Very efficient
Transformation[c]	DNA/cell	−	a. Competence determinants b. Chemical or physical changes	+	+	ss/ds	≤50 kb	Very broad	Chemically and electrically induced competence required; efficiency variable

[a] All DNA exchange processes are subject to the negative effects of restriction (nuclease action) in the recipient.
[b] All processes, in principle, may take place with transfer of intact plasmids, chromosomes, or linear fragments. The requirement for recombination in the recipient depends on the nature of transferred DNA and its properties. ss, Single stranded; ds, double stranded.
[c] Many bacterial species are genetically non-competent, but may be converted to a competent state by laboratory processes. Electrotransformation is a good example.

A. Transduction

Transduction is the exchange of bacterial genes mediated by bacteriophage, or phage. When a phage infects a cell, the phage genes direct the takeover of the host DNA and protein synthesizing machinery so that new phage particles can be made. Transducing particles are formed when plasmid DNA or fragments of host chromosomal DNA are erroneously packaged into phage particles during the replication process. Transducing particles (those carrying nonphage DNA) are included when phage are liberated from the infected cell to encounter another host and begin the next round of infection.

Although there are many laboratory studies of transduction of antibiotic resistance, this mechanism has been considered less important in the dissemination of antibiotic resistance genes because phage generally have limited host ranges; they can infect only members of the same or closely related species, and the size of DNA transferred does not usually exceed 50 kb. However, phage of extraordinarily broad host specificity have been described. For example, phage PRR1 and PRD1 will infect any gram-negative bacterium containing the resistance plasmid RP1 (Olsen and Shipley, 1973; Olsen et al., 1974) and could, in principle, transfer genetic information between unrelated bacterial species.

Transduction has been documented in at least 60 species of bacteria found in a wide variety of environments (Kokjohn, 1989). Although the actual level of intraspecies and interspecies transduction are unknown, the potential for transductional gene exchange is likely to be universal among the eubacteria. A study of the presence of bacteriophage in aquatic environments has shown that there may be as many as 108 phage particles per milliliter (Bergh et al., 1989); the authors calculate that at such concentrations one-third of the total bacterial population is subject to phage attack every 24 hr. In transduction the DNA is protected from degradation within the phage particle, and Stotzky has suggested that transduction may be as important as conjugation or transformation as a mechanism of gene transfer in natural habitats (Stotzky, 1989).

B. Conjugation

Conjugation is the process in which DNA is transferred during cell-to-cell contact. It has long been considered the most important mechanism for the dissemination of antibiotic resistance genes. During an epidemic of dysentery in Japan in the late 1950s increasing numbers of *Shigella dysenteriae* strains were isolated that were resistant to up to four antibiotics simultaneously. It soon became clear that the emergence of multiply resistant strains could not be attributed to mutation. Furthermore, both sensitive and resistant *Shigella* could be isolated from a single patient, and the *Shigella* sp. and *E. coli* obtained from the same patients often exhibited the same multiple resistance patterns. These finding led to the discovery of resistance transfer factors and were also an early indication of the contribution of conjugative transfer to the natural evolution of new bacterial phenotypes.

In addition to plasmid-mediated conjugal transfer, another form of conjugation has been reported to take place in gram-positive organisms. Conjugative transposons were first reported in *Streptococcus pneumoniae* when the transfer of antibiotic resistance determinants occurred in the absence of plasmids (Shoemaker *et al.,* 1980). Salyers has suggested that conjugative transposons may be more important that conjugative plasmids in broad host-range gene transfer between some species of bacteria (Salyers, 1993). Conjugative transposons are not considered typical transposons (Scott, 1992). As of 1993, three different families of conjugative transposons had been found: (a) the Tn*916* family (originally found in streptococci but now known also to occur in gram-negative bacteria such as *Campylobacter*) (Salyers, 1993), (b) the *S. pneumoniae* family, and (c) the *Bacteroides* family. Conjugative transposons range in size from 15 to 150 kb, and, in addition to other resistance genes, most encode tetracycline-resistance determinants [e.g., Tn*916* encodes the TetM determinant, and TetQ (Salyers, 1993) is found on the conjugative transposon from the *Bacteroides* group].

Does conjugation take place in the environment? The ideal site for gene transfer is the warm, wet, nutrient-abundant environment of the mammalian intestinal tract with its associated high concentration of bacteria. The resident microflora is believed to serve as a reservoir for genes encoding antibiotic resistance which could be transferred not only to other members of this diverse bacterial population, but also to transient colonizers of the intestine, such as soil or water microbes or human pathogens. Using oligonucleotide probes having DNA sequence similarity to the hypervariable regions of the TetQ determinant, Salyers and co-workers provided evidence that gene transfer between species of *Bacteroides,* one of the predominant genera of the human intestine, and *Prevotella* sp., one of the predominant genera of livestock rumen, has taken place under physiological conditions (Nikolich *et al.,* 1994).

C. Transformation

Natural transformation is a physiological process characteristic of many bacterial species in which the cell takes up and expresses exogenous DNA. Although natural transformation has been reported to occur only in a limited number of genera, these include many pathogenic taxa such as *Haemophilus, Mycobacterium, Streptococcus, Neisseria, Pseudomonas,* and *Vibrio* (Stewart, 1989). Initial studies suggested that natural transformation in some of these genera was limited to DNA from that particular species. For example, an 11-base pair recognition sequence permits *Haemophilus influenzae* to take up its own DNA preferentially compared to heterologous DNAs (Kahn and Smith, 1986). Given such specificity one could ask if natural transformation is an important mechanism in the transfer of antibiotic resistance genes. Spratt and co-workers have reported the transfer of penicillin resistance between *Streptococcus pneumoniae* and *Neisseria gonorrhoeae* by transformation. Further, Roberts reported that when the TetB determinant (which is conferred by conjugation in other gram-negative groups) is present in *Haemophilus* species and highly tetracycline-resistant *Moraxella catarrhalis,* it is disseminated by transformation (Roberts *et al.,* 1991).

D. General Considerations

There is a world of difference between laboratory and environmental studies. While the isolation of pure cultures is an important component of bacterial strain identification, the use of purified bacterial species in gene transfer studies does little to identify the process and probably bears no relationship to the processes of genetic exchange that take place in the complex microbial populations of the gastrointestinal tract (for example). When antibiotic-resistant bacteria are isolated from diseased tissue and identified as the responsible pathogen, this is the identification of the final product of a complex and poorly understood environmental system. Although gene exchange may under normal circumstances be rare in stable microbial microcosms, the intense selective pressure of antibiotic usage is likely to have provoked cascades of antibiotic resistance gene transfer between unrelated microbes. These transfers must involve different biochemical mechanisms during which efficiency is not a critical factor since the survival and multiplication of a small number of resistant progeny suffices to create a clinically problematic situation.

It should be apparent that a great deal of additional study using modern molecular and amplification methods with complex microbial communities is necessary before the parameters of natural antibiotic resistance gene transfer can be defined properly.

IV. EVOLUTION OF RESISTANCE DETERMINANT PLASMIDS

A. The Integron Model

Studies by Hughes and Datta of plasmids they isolated from the Murray collection (Hughes and Datta, 1983; Datta and Hughes, 1983) suggest that the appearance of resistance genes is a recent event, that is, the multiresistance plasmids found in pathogens must have been created since the 1940s. What really takes place when a new antimicrobial agent is introduced and plasmid-determined resistance develops within a few years? The most significant component of the process of antibiotic resistance flux in the microbial population is gene pickup, which has now been emulated in the laboratory.

Largely due to the studies of Hall and co-workers (Stokes and Hall, 1989; Collis *et al.*, 1993; Recchia and Hall, 1995), we have a good idea of the way in which transposable elements carrying multiple antibiotic resistance genes might be formed. From their studies of the organization of transposable elements, these researchers have identified a key structural constituent of one class of transposon that they named an "integron."

The integron is a mobile DNA element with a specific structure consisting of two conserved segments flanking a central region in which "cassettes" that encode functions such antibiotic resistance can be inserted. The 5′ conserved region encodes a site-specific recombinase (integrase) and a strong promoter or promoters that ensure expression of the integrated cassettes. The integrase is responsible for the insertion of antibiotic resistance gene cassettes downstream of the promoter; ribosome binding sites are conveniently provided. More than one promoter sequence exists on the element (Lévesque *et al.*, 1994); transcription initiation is very efficient and functions in both gram-negative and gram-positive bacteria. The 3′ conserved region carries a gene for sulfonamide resistance (*sul*) and two open reading frames of unknown function. Probably the ancestral integron encoded no antibiotic resistance (Fig. 1). The ubiquitous

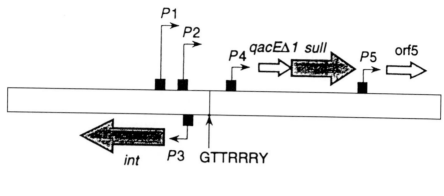

FIGURE I The general structure of an integron. Integrons consist of a 5′ conserved sequence that encodes an integrase and contains promoter sequences (P) responsible for the transcription of inserted gene cassettes. The 3′ conserved sequence encodes resistance to sulfonamide drugs (sul). The insertion site for gene cassettes (GTTRRRY) is indicated.

presence of *sul* in an element of this type might be surprising, although sulfon-amides (see above) have been employed since the mid-1930s and (apart from mercury salts) are the longest used agents for the treatment of infectious diseases. The resistance gene cassettes are integrated into a specific insertion site in the integron. Typically, in the case of Tn21-related transposons, each antibiotic resistance cassette is associated with one of a functional family of closely related, palindromic 59-base pair elements (or recombination hot spots) located to the 3′ side of the resistance gene (Fig. 1). Integrase-catalyzed insertion of resistance gene cassettes into resident integrons has been demonstrated (Collis *et al.*, 1993; Martinez and de la Cruz, 1990). In addition, site-specific deletion and rearrangement of the inserted resistance gene cassettes can result from integrase-catalyzed events (Collis and Hall, 1992).

Francia *et al.* (1993) have expanded our understanding of the role of inte-grons in gene mobilization by showing that the Tn21 integrase can act on sec-ondary target sites at significant frequencies and so permit the fusion of two R plasmids by interaction between the recombination hot spot of one plasmid and a secondary integrase target site on a second plasmid. The secondary sites are characterized by the degenerate pentanucleotide sequence Ga/tTNa/t. Though the details of the mechanism by which new integrons are then generated from the fusion structure are not established, the use of secondary integration sites could explain how new genes may be inserted into integrons without the neces-sity for a 59-bp element, as the authors point out.

B. Other Multiple Resistance Plasmids

Analyses of the integron-type transposons provide a good model for the way in which antibiotic resistance genes from various (unknown) sources may be in-

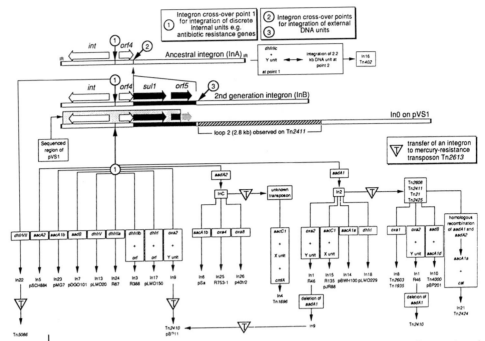

FIGURE 2 The diversity of antibiotic resistance integrons. The diagram illustrates the insertion of antibiotic resistance gene cassettes into the basic integron structure (Fig. 1). It should be noted that the antibiotic resistance gene cassettes are usually inserted in tandem array and more than five genes may be found in a single integron structure.

corporated into an integron by recombination events into mobile elements and hence into bacterial replicons, providing the R plasmids that we know today (Fig. 2) (Bissonnette and Roy, 1992). However, in bacterial pathogens a variety of transposable elements have been found that undergo different processes of recombinational excision and insertion. It is not known what evolutionary mechanisms are implicated or whether some form of integron-related structure is present in all cases. For the type of integron found in the Tn21 family, we have plausible models, supported by *in vivo* and *in vitro* studies, to provide a modus operandi by which antibiotic resistance genes were (and are) molecularly cloned in the evolution of R plasmids. A large number of transposable elements carrying virtually all possible combinations of antibiotic resistance genes have been identified (Berg, 1989), and nucleotide sequence analysis of multiresistant integrons shows that the inserted resistance gene cassettes differ markedly in codon usage, indicating that the antibiotic resistance determinants are of diverse origins. Microbes are masters at genetic engineering, and heterologous expression vectors of broad host range in the form of integrons were present in bacteria long before they became vogue for biotechnology companies in the 1980s.

V. ORIGINS

We have described how the majority of antibiotic resistance genes found in microbes have been acquired by their hosts. The important question is, From where did they acquire these genetic determinants? The integron model defines a mechanism by which antibiotic genes can be procured by members of the Enterobacteriaceae and pseudomonads. This mechanism requires the participation of extrinsic resistance genes (or cassettes); however, the origins of these open reading frames are a mystery. The same questions can be applied to any of the resistance genes found in pathogenic bacteria. Available evidence concerning some of the origins is discussed below.

A. Antibiotic Producers

Antibiotic-producing microbes are the prime suspects for the maintenance of a pool of resistance genes in nature. Any organism producing a toxic molecule must, by definition, possess a mechanism(s) to survive this potentially suicidal situation. Because the majority of antibiotics are produced by bacteria (principally the actinomycete group), one would expect these organisms to have mechanisms of protection against the antibiotics they make. These mechanisms take various forms, and it is significant that the mechanisms of resistance for the known antibiotics in producing organisms and clinical isolates are biochemically identical (see Table VII). The gene clusters for antibiotic biosynthesis in producing organisms almost invariably comprise one or more genes that encode resistance to the antibiotic produced. Many of these genes have been cloned and sequenced. Sequence comparisons of the genes from the producer and the clinical isolate often show very high degrees of similarity. Thus, the homologous biochemical mechanisms and the relatedness of the gene sequences support the hypothesis that producing organisms are likely to be the source of most resistance genes. However, other sources of resistance genes are not excluded, and some of these are discussed below.

TABLE VII Resistance Determinants with Homologs in Antibiotic Producing Organisms

Antibiotic	Mechanism
Penicillins	β-Lactamases
Cephalosporins	Penicillin binding proteins
Aminoglycosides	Phosphotransferases, acetyltransferases, nucleotidyltransferases
Chloramphenicol	Acetyltransferases
Tetracyclines	Ribosomal protection, efflux
Macrolides	rRNA methylation
Streptogramins	Esterases
Lincosamides	Phosphotransferases, acetyltransferases
Phosphonates	Phosphorylation, glutathionylation (?)
Bleomycin	Acetyltransferases, immunity protein

I. Tetracyclines

The first tangible evidence of resistance gene transfer involving antibiotic-producing streptomycetes in a clinical setting has come from Pang *et al.* (1994). These researchers analyzed human infections of nontuberculous mycobacteria and *Streptomyces* spp.; the infections did not respond to treatment with tetracyclines. Both microbial species contained resistance genes (*tetK* and *tetL*) known to be the basis of tetracycline resistance in gram-positive bacteria (see Section II,A,1,b on tetracycline resistance determinants). These resistance determinants promote efflux of the drug and are typically transposon-associated. Surprisingly, the mycobacteria and the streptomycetes both had the tetracycline resistance genes (*otrA* and *otrB*) previously identified in the tetracycline-producing strain *Streptomyces rimosis* (Davies, 1992; Doyle *et al.*, 1991). Reciprocally, the streptomycetes had acquired the *tetK* and *tetL* genes. Because the latter are clearly "foreign" genes (*tetK* and *tetL* have a G + C content different from those of streptomycete and mycobacterial chromosomal DNA), they must have been acquired as the result of a recent gene transfer. Although this evidence is consistent with resistance gene transfer between the streptomycetes and other bacteria, it is not known which is donor and which is recipient, nor whether the newly acquired tetracycline resistance genes are plasmid or chromosomally encoded.

2. Aminoglycosides

Covalent modification of their inhibitory biochemical products is very common in antibiotic-producing bacteria. It was the discovery of antibiotic modification as a means of self-protection in the streptomycetes that led to the proposal that antibiotic-producing microbes were the origins of the antibiotic resistance determinants found in other bacteria (Benveniste and Davies, 1973; Walker and Skorvaga, 1973). Support for this hypothesis has been provided by nucleic acid and protein sequence comparisons of aminoglycoside resistance determinants from producing organisms and clinical isolate sources (Shaw, 1984; Davies, 1992). As mentioned above, producing organisms are not the only potential source of antibiotic resistance mechanisms. The proposal that the enzymes that modify aminoglycosides evolved from such "housekeeping" genes as the sugar kinases and acyltransferases has been made by a number of groups (Udou *et al.*, 1989; Shaw *et al.*, 1992; Rather *et al.*, 1993).

3. Macrolides

The principal mechanism of resistance to the macrolide antibiotics involves methylation of the 23 S rRNA of the host. Methylation occurs on a specific adenine residue in the rRNA, and mono- and dimethylation has been described. The N-methyltransferase genes responsible for encoding this protective modification have been studied in detail from both the producing organisms and clinical isolates (Rather *et al.*, 1993). Although no direct transfer of resistance between producer and clinical isolates has been found, the identity of the biochemical mechanism and its regulation make for a compelling evolutionary relationship.

4. Other Antibiotics

Perhaps the most straightforward path toward the development of a drug resistance mechanism is via the major facilitator superfamily (MFS; see Section II,B,2) of transporters. The MFS is found in all organisms involved and consists of membrane transport systems involved in the symport, antiport, or uniport of various substrates. Other examples of MFS pumps include sugar uptake systems, phosphate ester/phosphate antiport, and oligosaccharide uptake. One might expect that, in addition to methods for the import of nutrients, a cell would have methods for the removal of harmful substances.

The β-lactamases are an interesting case because they are widely distributed among the bacterial kingdom. The ubiquity of β-lactamases suggests that these enzymes may play a part in normal metabolic or synthetic processes, but an essential role has not been shown. Examination of the sequences of β-lactamases by Bush *et al.* (1995) permitted the establishment of extensive phylogenetic relationships, which includes those from microbes employed in the commercial production of β-lactam antibiotics.

In the case of the glycopeptide antibiotics (vancomycin and teichoplanin), resistance in the enterococci is due to the acquisition of a cluster of genes that encode a novel cell wall precursor (see Section II,A,3). A comparison of D-Ala-D-Ala ligases from different sources has shown that the *vanA* gene is dissimilar to the other known genes, suggesting a divergent origin (Arthur *et al.*, 1996). The *vanA* homolog from the producing organism, *Streptomyces toyocaensis*, has been cloned and sequenced. The predicted amino acid sequence was compared with the D-Ala-D-Ala ligase from *Enterococcus* and is greater than 60% similar (G. D. Wright et al., 1997).

B. Unknown Origin

There are gaps in our understanding of the origin of resistance determinants. For example, the aminoglycoside nucleotidyltransferases have been found only in clinical isolates and have no known relatives; the potential origins of the chloramphenicol acetyltransferase genes are still unclear (see Section II,A,1,c); and the sources for sulfonamide and trimethoprim resistance are still not known (see Section II,A,4). There may be examples of "housekeeping" functions of resistance genes. For example, *Providencia stuartii* has a chromosomally encoded aminoglycoside acetyltransferase that may play a role in cell wall peptidoglycan formation (Payie *et al.*, 1995).

VI. MAINTENANCE OF ANTIBIOTIC RESISTANCE

A variety of surveys have indicated that normal healthy humans (who are not pursuing a course of antibiotic therapy) carry antibiotic-resistant enteric species in their intestinal tract; a substantial proportion are found to contain transmissible antibiotic resistance plasmids. Studies have demonstrated that a lack of antibiotic selective pressure, for example, removing antimicrobials from cattle feed, can lead to a gradual decrease in the percentage of resistance genes and resistance bacteria found in a population (Langlois *et al.*, 1986; Hinton *et al.*, 1985).

Although the results of these studies are encouraging, other studies indicate that once resistance cassettes have been developed it is unlikely that they will disappear completely from an environment where antibiotics are routinely used. In a sense "the cat is out of the bag." Chemostat studies (Chao *et al.*, 1983; Hartl *et al.*, 1983) suggest that insertion elements may themselves provide their hosts with a selective advantage independent of the resistance determinants they carry. Roberts has postulated that commensal bacteria can be reservoirs for tetracycline resistance determinants. When bacteria from the urogenital tracts of females who had not taken antibiotics for 2 weeks previously were examined, 82% of the viridans-type streptococci hybridized with at least one Tet determinant (Roberts *et al.*, 1991).

A. Multiple Antibiotic Resistance and Mercury Resistance

Although the cooccurrence of antibiotic resistance and resistance to heavy metals such as mercury has long been known, its implications for public health are only now becoming clear. In 1964 the cotransduction of genes encoding resistance to penicillin and mercury by a staphylococcal phage was reported (Richmond and John, 1964). Ten years later it was found that 25% of the antibiotic resistance plasmids isolated from enteric bacteria in Hammersmith Hospital also carried mercury resistance (Schottel *et al.*, 1974). DNA sequence analyses has shown that the Tn*21*-type transposons carry both a copy of the *mer* locus and an integron (see above) (Stokes and Hall, 1989; Grinsted *et al.*, 1990). Summers and co-workers (1993) have observed that resistance to mercury occurs frequently in human fecal flora and is correlated with the occurrence of multiple antibiotic resistance. How does this phenomenon become a public health concern? In the same report Summers' group found that the mercury released from the amalgam in "silver" dental fillings in monkeys led to the rapid enrichment of many different mercury-resistant bacteria in the oral and fecal flora. They suggest that in humans this chronic and biologically significant exposure to mercury may foster the persistence of multidrug-resistant microbes through selection of linked markers.

B. Other Examples

There may well be other examples of this phenomenon, where subclinical concentrations of an antibiotic could serve as selection for the maintenance (and propagation) of genetic elements and their resident resistance genes. There is also maintenance by constant selection pressure for other phenotypes, a type of "linked" selection (we have mentioned the role of mercury in this respect). How else does one explain the fact that streptomycin and chloramphenicol resistance can be still found on plasmids in hospitals, even though these antibiotics are no longer used? Are there other positive selective functions carried by plasmids? Resistance to ultraviolet light or other physical or chemical toxins (e.g., detergents, disinfectants) would be a possibility. Is it also conceivable that plasmids improve the fitness of their hosts under "normal" conditions? Evidence for this comes from the work of Lenski (personal communication, 1996).

VII. CONCLUDING REMARKS—FOR NOW AND THE FUTURE

It should be apparent from the foregoing discussion of antibiotic modification that there must be a substantial pool of antibiotic resistance genes (or close relatives of these genes) in nature. Gene flux between bacterial replicons and their hosts is likely to be the rule rather than the exception, and it appears to respond quickly to environmental changes (Levy and Novick, 1986; Levy and Miller, 1989; Hughes and Datta, 1983). This gene pool is readily accessible to bacteria when they are exposed to the strong selective pressure of antibiotic usage—in hospitals, for veterinary and agricultural purposes, and as growth promoters in animal and poultry husbandry. It is a life-or-death situation for microbes, and they have survived. A better knowledge of the components of this gene pool, particularly with respect to what might happen on the introduction of a new chemical entity such as an antimicrobial agent, might, on the one hand, permit early warning and subsequent chemical modification of antibiotics to permit them to elude potential resistance mechanisms and, on the other, lead to more prudent use of antibiotics under circumstances where the presence of specific resistance determinants can be predicted.

The development of resistance to antimicrobial agents is inevitable, in response to the strong selective pressure and extensive use of antibiotics. Resistance may develop as the result of mutation or acquisition, or a combination of the two. The use of antibiotics should be such as to delay the inevitable, and knowledge gained over the past 50 years if correctly interpreted and used to modify current practices appropriately should permit this. Unfortunately, for all of the antimicrobial agents in current use, we have reached a state of no return. The American Society for Microbiology Task Force on Antibiotic Resistance has made a number of recommendations to deal with the current crisis of antibiotic resistant microbes (see Table VIII; ASM, 1995). It has been suggested that such measures can, at the least, maintain and improve the status quo.

Two of the options that would permit continued success of antibiotic therapy in the face of increasing resistance are (1) the discovery of new antibiotics (by "new" this implies novel chemical structures) and (2) the development of agents, that might be used in combination with existing antibiotics, to interfere with the biochemical resistance mechanisms. Such a strategy has already been partially successful in the development of inhibitors of β-lactamases to permit "old" antibiotics to be used. However, there appears to be little success (or effort) in applying this approach to other antibiotic classes. In our discussion of biochemical mechanisms (Section II) we have noted a number of cases where this approach could be taken (e.g., fluorinated analogs of chloramphenicol).

With respect to novel antibiotics, several valid approaches exist: (1) natural product screening, especially directed at products of the 99.9% of microbes that cannot be grown in the laboratory; (2) combinatorial chemistry that can be used as a means of discovery of new active molecules or to new substitutions on

TABLE VIII   Recommendations from the American Society for Microbiology Task Force on Antibiotic Resistance

---

Establish a national surveillance system immediately
  Lead agency should be the National Center for Infectious Diseases of the U.S. Centers for
    Disease Control (CDC) and should involve the National Institute of Allergy and Infectious
    Disease of the National Institutes of Health (NIH), the Environmental Protection Agency,
    and the Food and Drug Administration

Strengthen professional and public education in the area of infectious disease and antibiotics to
  reduce inappropriate usage of antibiotics
  The curriculum for health professionals should include the appropriate handling, diagnosis,
    and treatment of infectious disease and antibiotic resistance
  Reduce the spread of infectious agents and antibiotic resistance in hospitals, nursing homes,
    day care facilities, and food production industries
  Educate patients and food producers
  Improve antimicrobial use for cost-effective treatment and preservation of effectiveness

Increase basic research directed toward development of new antimicrobial compounds, effective
  vaccines, and other prevention measures
  Fund areas directly related to new and emerging infections and antibiotic resistance
  Fund basic research in bacterial genetics and metabolic pathways
  Establish a culture collection containing representative antibiotic-resistance biotypes of
    pathogens
  Sequence genomes of microbial pathogens
  Develop better diagnostic techniques
  Develop vaccines and other preventative measures

---

known ring structures; and (3) rational chemical design based on identification of specific biochemical targets. The success of these methods will depend on the availability of cell-based and biochemical assays that will detect low concentrations of active molecules by high flux screening methods. It seems redundant to insist on the requirement for early identification of natural resistance mechanisms for any compounds of interest (thus permitting the design of analogs) and the study of structure–toxicity relationships at the earliset stage possible.

Last, but not least, the introduction of a novel antimicrobial agent into clinical practice must be accompanied by strict limitations on its use. No novel therapeutic should be used for other than human use under prescription, and no structural analog should be employed for "other" purposes.

ACKNOWLEDGMENTS

We thank the Canadian Bacterial Diseases Network and the Natural Sciences and Engineering Council of Canada for support.

REFERENCES

Aminoglycoside Resistance Study Group. (1994). Resistance to aminoglycosides in *Pseudomonas.*
  Trends Microbiol. 2(10), 347–353.

Arca, P., Hardisson, C., and Suárez, J. E. (1990). Purification of a glutathione S-transferase that mediates fosfomycin resistance in bacteria. *Antimicrob. Agents Chemother.* **34**, 844–848.

Arthur, M., Brisson-Noël, A., and Courvalin, P. (1987). Origin and evolution of genes specifying resistance to macrolide, lincosamide and streptogramin antibiotics: Data and hypothesis. *J. Antimicrob. Chemother.* **20**, 783–802.

Arthur, M., Reynolds, P., and Courvalin, P. (1996). Glycopeptide resistance in gram-positive bacteria. *Trends Microbiol.* **4**, 401–407.

ASM. (1995). Report of the ASM Task Force on antibiotic resistance. *Antimicrob. Agents Chemother.* (Suppl.), 1–23.

Bannam, T. L., and Rood, J. I. (1991). Relationship between the *Clostridium perfringens catQ* gene product and chloramphenicol acetyltransferases from other bacteria. *Antimicrob. Agents Chemother.* **35**, 471–476.

Bennett, P. M., and Chopra, I. (1993). Molecular basis of β-lactamase induction in bacteria. *Antimicrob. Agents Chemother.* **37**, 153–158.

Benveniste, R., and Davies, J. (1973). Aminoglycoside antibiotic-inactivating enzymes in actinomycetes similar to those present in clinical isolates of antibiotic-resistant bacteria. *Proc. Natl. Acad. Sci. U.S.A.* **70**, 2276–2280.

Berg, D. E. (1989). Transposable elements in prokaryotes. *In* "Gene Transfer in the Environment" (S. B. Levy and R. V. Miller, eds.), pp. 99–137. McGraw-Hill, New York.

Bergh, O., Borsheim, K. Y., Bratbak, G., and Heldal, M. (1989). High abundance of viruses found in aquatic environments. *Nature (London)* **340**, 467–468.

Bissonnette, L., and Roy, P. H. (1992). Characterization of In0 of *Pseudomonas aeruginosa* plasmid pVS1, an ancestor of integrons of multiresistance plasmids and transposons of gram-negative bacteria. *J. Bacteriol.* **174**, 1248–1257.

Bissonnette, L., Champetier, S., Buisson, J.-P., and Roy, P. H. (1991). Characterization of the nonenzymatic chloramphenicol resistance (*cmlA*) gene of the In4 integron of Tn1696: Similarity of the product to transmembrane transport proteins. *J. Bacteriol.* **173**, 4493–4502.

Blazquez, J., Baquero, M.-R., Canton, R., Alos, I., and Baquero, F. (1993). Characterization of a new TEM-type β-lactamase resistant to clavulanate, sulbactam, and tazobactam in a clinical isolate of *Escherichia coli*. *Antimicrob. Agents Chemother.* **37**, 2059–2063.

Brisson-Noël, A., Delrieu, P., Samain, D., and Courvalin, P. (1988). Inactivation of lincosaminide antibiotics in *Staphylococcus*. *J. Biol. Chem.* **263**, 15880–15887.

Bush, K., Jacoby, G. A., and Medeiros, A. A. (1995). A functional classification scheme for β-lactamases and its correlation with molecular structure. *Antimicrob. Agents Chemother.* **39**, 1211–1233.

Cannon, M., Harford, S., and Davies, J. (1990). A comparative study on the inhibitory actions of chloramphenicol, thiamphenicol and some fluorinated derivatives. *J. Antimicrob. Chemother.* **26**, 307–317.

Chao, L., Vargas, C., Spear, B. B., and Cox, E. C. (1983). Transposable elements as mutator genes in evolution. *Nature (London)* **303**, 633–635.

Chen, S.-T., and Clowes, R. C. (1987). Variations between the nucleotide sequences of Tn1, Tn2, and Tn3 and expression of β-lactamase in *Pseudomonas aeruginosa* and *Escherichia coli*. *J. Bacteriol.* **169**, 913–916.

Cole, S. T. (1994). *Mycobacterium tuberculosis*: Drug-resistance mechanisms. *Trends Microbiol.* **10**(2), 411–415.

Collis, C. M., and Hall, R. M. (1992). Site-specific deletion and rearrangement of integron insert genes catalysed by the integron DNA integrase. *J. Bacteriol.* **174**, 1574–1585.

Collis, C. M., Grammaticopoulos, G., Briton, J., Stokes, H. W., and Hall, R. M. (1993). Site-specific insertion of gene cassettes into integrons. *Mol. Microbiol.* **9**, 41–52.

Cookson, B. (1995). Aspects of the epidemiology of MRSA in Europe. *J. Chemother.* **7**(Suppl. 3), 93–98.

Couture, F., Lachapelle, J., and Levesque, R. C. (1992). Phylogeny of LCR-1 and OXA-5 with class A and class D β-lactamases. *Mol. Microbiol.* **6**, 1693–1705.

Dabbs, E. R., Yazawa, K., Mikami, Y., Miyaji, M., Morisaki, N., Iwasaki, S., and Furihata, K. (1995). Ribosylation by mycobacterial strains as a new mechanism of rifampin inactivation. *Antimicrob. Agents Chemother.* **39**, 1007–1009.

Datta, N., and Hughes, V. M. (1983). Plasmids of the same Inc groups in enterobacteria before and after the medical use of antibiotics. *Nature (London)* **306**, 616–617.

Davies, J. (1992). Another look at antibiotic resistance. *J. Gen. Microbiol.* **138**, 1553–1559.

Day, P. J., and Shaw, W. V. (1992). Acetyl coenzyme A binding by chloramphenicol acetyltransferase: Hydrophobic determinants of recognition and catalysis. *J. Biol. Chem.* **267**, 5122–5127.

Debets-Ossenkopp, Y., Sparrius, M., Kusters, J., Kolkman, J., and Vandenbroucke-Grauls, C. (1996). Mechanism of clarithromycin resistance in clinical isolates of *Helicobacter pylori. FEMS Microbiol. Lett.* **142**, 37–42.

Doyle, D., McDowall, K. J., Butler, M. J., and Hunter, I. S. (1991). Characterization of an oxytetracycline-resistance gene, *otrA*, of *Streptomyces rimosus. Mol. Microbiol.* **5**, 2923–2933.

Ferretti, J. J., Gilmore, K. S., and Courvalin, P. (1986). Nucleotide sequence analysis of the gene specifying the bifunctional 6′-aminoglycoside acetyltransferase 2″-aminoglycoside phosphotransferase enzyme in *Streptococcus faecalis* and identification and cloning of gene regions specifying the two activities. *J. Bacteriol.* **167**, 631–638.

Finken, M., Kirschner, P., Meier, A., Wrede, A., and Böttger, E. C. (1993). Molecular basis of streptomycin resistance in *Mycobacterium tuberculosis:* Alterations of the ribosomal protein S12 gene and point mutations within a functional 16S ribosomal RNA pseudoknot. *Mol. Microbiol.* **9**, 1239–1246.

Fourmy, D., Recht, M. I., Blanchard, S. C., and Puglisi, J. D. (1996). Structure of the A site of *Escherichia coli* 16S ribosomal RNA complexed with an aminoglycoside. *Science* **274**, 1367–1371.

Francia, M. V., de la Cruz, F., and García Lobo, J. M. (1993). Secondary sites for integration mediated by the Tn*21* integrase. *Mol. Microbiol.* **10**, 823–828.

George, A and Levy, S. B. (1983) Amplifiable resistance to tetracycline, chloramphenicol, and other antibiotics in *Escherichia coli:* involvement of a non-plasmid-determined efflux of tetracyciine. *J. Bacteriol.* **155**, 531–540.

Goldman, J. D., White, D. G., and Levy, S. B. (1996). Multiple antibiotic resistance (*mar*) locus protects *Escherichia coli* from rapid cell killing by fluoroquinolones. *Antimicrob. Agents Chemother.* **40**, 1266–1269.

Grinsted, J., de la Cruz, F., and Schmitt, R. (1990). The Tn*21* subgroup of bacterial transposable elements. *Plasmid* **24**, 163–189.

Hartl, D. L., Dykhuizen, D. E., Miller, R. D., Green, L., and de Framond, J. (1983). Transposable element IS*50* improves growth rate of *E. coli* cells without transposition. *Cell (Cambridge, Mass.)* **35**, 503–510.

Hillen, W., and Berens, C. (1994). Mechanisms underlying expression of Tn*10* encoded tetracycline resistance. *Annu. Rev. Microbiol.* **48**, 345–369.

Hinton, M., Linton, A. H., and Hedges, A. J. (1985). The ecology of *Escherichia coli* in calves reared as dairy-cow replacements. *J. Appl. Bacteriol.* **85**, 131–138.

Honoré, N., Nicolas, M.-H., and Cole, S. T. (1986). Inducible cephalosporinase production in clinical isolates of *Enterobacter cloacae* is controlled by a regulatory gene that has been deleted from *Escherichia coli. EMBO J.* **5**, 3709–3714.

Hughes, V. M., and Datta, N. (1983). Conjugative plasmids in bacteria of the 'pre-antibiotic' era. *Nature (London)* **302**, 725–726.

Huovinen, P., Sundström, L., Swedberg, G., and Sköld, O. (1995). Trimethoprim and sulfonamide resistance. *Antimicrob. Agents Chemother.* **39**, 279–289.

Jacoby, G. A., and Archer, G. L. (1991). New mechanisms of bacterial resistance to antimicrobial agents. *N. Engl. J. Med.* **324**, 601–612.

Jarlier, V., and Nikaido, H. (1994). Mycobacterial cell wall: Structure and role in natural resistance to antibiotics. *FEMS Microbiol. Lett.* **123**, 11–18.

Kahn, M., and Smith, H. O. (1986). Role of transformazomes in *Haemophilus influenzae* Rd transformation. *In* "Antibiotic Resistance Genes: Ecology, Transfer, and Expression" (S. B. Levy and R. P. Novick, eds.), pp. 143–152. Cold Spring Harbor Laboratory, Cold Spring Harbor, New York.

Kocabiyik, S., and Perlin, M. H. (1992). Altered substrate specificity by substitutions at Tyr218 in bacterial aminoglycoside 3′-phosphotransferase-II. *FEMS Microbiol. Lett.* **93**, 199–202.

Kokjohn, T. A. (1989). Transduction: Mechanism and potential for gene transfer in the environment. *In* "Gene Transfer in the Environment" (S. B. Levy and R. V. Miller, eds.), pp. 73–97. McGraw-Hill, New York.

Krause, R. M. (1994). Dynamics of emergence. *J. Infect. Dis.* **170**, 265–271.

Langlois, B. E., Dawson, K. A., Cromwell, G. L., and Stahly, T. S. (1986). Antibiotic resistance in pigs following a 13 year ban. *J. Anim. Sci.* **62**, 18–32.

LeClerc, J. E., Li, B., Payne, W. L., and Cebula, T. A. (1996). High mutation frequencies among *Escherichia coli* and *Salmonella* pathogens. *Science* **274**, 1208–1211.

Leclercq, R., and Courvalin, P. (1991). Bacterial resistance to macrolide, lincosamide, and streptogramin antibiotics by target modification. *Antimicrob. Agents Chemother.* **35**, 1267–1272.

Lévesque, C., Brassard, S., Lapointe, J., and Roy, P. H. (1994). Diversity and relative strength of tandem promoters for the antibiotic-resistance genes of several integrons. *Gene* **142**, 49–54.

Levy, S. B., and Miller, R. V., eds. (1989). Gene transfer in the environment. *In* "Environmental Biotechnology." McGraw-Hill, New York.

Levy, S. B., and Novick, R. P., eds. (1986). "Antibiotic Resistance Genes: Ecology, Transfer, and Expression." Banbury Report 24. Cold Spring Harbor Laboratory, Cold Spring Harbor, New York.

Lewis, K. (1994). Multidrug resistance pumps in bacteria: Variations on a theme. *Trends Biochem. Sci.* **19**, 119–123.

Mabilat, C., Goussard, S., Sougakoff, W., Spencer, R. C., and Courvalin, P. (1990). Direct sequencing of the amplified structural gene and promoter for the extended-broad-spectrum β-lactamase TEM-9 (RHH-1) of *Klebsiella pneumoniae*. *Plasmid* **23**, 27–34.

Marger, M. D., and Saier, M. H. (1993). A major superfamily of transmembrane facilitators that catalyse uniport, symport and antiport. *Trends Biochem. Sci.* **18**, 13–20.

Martinez, E., and de la Cruz, F. (1990). Genetic elements involved in Tn21 site-specific integration, a novel mechanism for the dissemination of antibiotic resistance genes. *EMBO J.* **9**, 1275–1281.

Miller, P. F., and Sulavik, M. C. (1996). Overlaps and parallels in the regulation of intrinsic multiple-antibiotic resistance in *Escherichia coli*. *Mol. Microbiol.* **21**, 441–448.

Monod, M., Mohan, S., and Dubnau, D. (1987). Cloning and analysis of *ermG*, a new macrolidelincosamide-streptogramin B resistance element from *Bacillus sphaericus*. *J. Bacteriol.* **169**, 340–350.

Murray, I. A., and Shaw, W. V. (1997). O-Acetyltransferases for chloramphenicol and other natural products. *Antimicrob. Agents Chemother.* **41**, 1–6.

Needham, C., Rahman, M., Dyke, K. G. H., and C., N. W., Noble, W. C. (1994). An investigation of plasmids from *Staphylococcus aureus* that mediate resistance to mupirocin and tetracycline. *Microbiology* **140**, 2577–2583.

Neu, H. C. (1992). The crisis in antibiotic resistance. *Science* **257**, 1064–1073.

Nikaido, H. (1994). Prevention of drug access to bacterial targets: Permeability barriers and active efflux. *Science* **264**, 382–387.

Nikaido, H. (1996). Multidrug efflux pumps of gram-negative bacteria. *J. Bacteriol.* **178**, 5853–5859.

Nikolich, M. P., Hong, G., Shoemaker, N. B., and Salyers, A. A. (1994). Evidence for natural horizontal transfer of *tetQ* between bacteria that normally colonize humans and bacteria that normally colonize livestock. *Appl. Environ. Microbiol.* **60**, 3255–3260.

Noble, W. C., Rahman, M., and Cookson, B. (1988). Transferable mupirocin resistance. *J. Antimicrob. Chemother.* **22**, 771.

Noble, W. C., Virani, Z., and Cree, R. G. A. (1992). Co-transfer of vancomycin and other resistance genes from *Enterococcus faecalis* NCTC 12201 to *Staphylococcus aureus*. *FEMS Microbiol. Lett.* **93**, 195–198.

O'Hara, K., Kanda, T., Ohmiya, K., Ebisu, T., and Kono, M. (1989). Purification and characterization of macrolide 2′-phosphotransferase from a strain of *Escherichia coli* that is highly resistant to erythromycin. *Antimicrob. Agents Chemother.* **33**, 1354–1357.

Olsen, R. H., and Shipley, P. (1973). Host range and properties of the *Pseudomonas aeruginosa* R factor R1822. *J. Bacteriol.* **113**, 772–780.

Olsen, R. H., Siak, J., and Gray, R. H. (1974). Characteristics of PrD1, a plasmid dependent broad host range DNA bacteriophage. *J. Virol.* **14**, 689–699.

Pang, Y., Brown, B. A., Steingrube, V. A., Wallace, R. J., Jr., and Roberts, M. C. (1994). Tetracycline resistance determinants in *Mycobacterium* and *Streptomyces* species. *Antimicrob. Agents Chemother.* **38**, 1408–1412.

Parent, R., and Roy, P. H. (1992). The chloramphenicol acetyltransferase gene of Tn*2424:* A new breed of *cat. Journal of Bacteriol.* **174,** 2891–2897.

Paulsen, I. T., Brown, M. H., and Skurray, R. A. (1996). Proton-dependent multidrug efflux systems. *Microbiol. Rev.* **60,** 575–608.

Payie, K. G., Rather, P. N., and Clarke, A. J. (1995). Contribution of gentamicin 2'-*N*-acetyltransferase to the O-acetylation of peptidoglycan in *Providencia stuartii. J. Bacteriol.* **177,** 4303–4310.

Rahman, M., Noble, W. C., and Dyke, K. G. H. (1993). Probes for the study of mupirocin resistance in staphylococci. *J. Med. Microbiol.* **39,** 446–449.

Rather, P. N., Munayyer, H., Mann, P. A., Hare, R. S., Miller, G. H., and Shaw, K. J. (1992). Genetic analysis of bacterial acetyltransferases: Identification of amino acids determining the specificities of the aminoglycoside 6'-*N*-acetyltransferase Ib and IIa proteins. *J. Bacteriol.* **174,** 3196–3203.

Rather, P. N., Orosz, E., Shaw, K. J., Hare, R., and Miller, G. (1993). Characterization and transcriptional regulation of the 2'-*N*-acetyltransferase gene from *Providencia stuartii. J. Bacteriol.* **175,** 6492–6498.

Recchia, G. D., and Hall, R. M. (1995). Gene cassettes: A new class of mobile element. *Microbiology* **141,** 3015–3027.

Richmond, M. H., and John, M. (1964). Co-transduction by a staphylococcal phage of the genes responsible for penicillinase synthesis and resistance to mercury salts. *Nature (London)* **202,** 1360–1361.

Roberts, M. C. (1994). Epidemiology of tetracycline-resistance determinants. *Trends Microbiol.* **2**(10), 353–357.

Roberts, M. C., Pang, Y. J., Spencer, R. C., Winstanley, T. G., Brown, B. A., and Wallance, R. J. (1991). Tetracycline resistance in *Moraxella (Branhamella) catarrhalis:* Demonstration of two clonal out breaks by using pulsed-field gel electrophoresis. *Antimicrob. Agents Chemother.* **35,** 2453–2455.

Rouch, D. A., Byrne, M. E., Kong, Y. C., and Skurray, R. A. (1987). The *aacA–aphD* gentamicin and kanamycin resistance determinant of Tn*4001* from *Staphylococcus aureus:* Expression and nucleotide sequence analysis. *J. Gen. Microbiol.* **133,** 3039–3052.

Salyers, A. A. (1993). Gene transfer in the mammalian intestinal tract. *Curr. Opin. Biotechnol.* **4,** 294–298.

Sanchez-Pescador, R., Brown, J. T., Roberts, M., and Urdea, M. S. (1988). Homology of the TetM with translational elongation factors: Implications for potential modes of tetM conferred tetracycline resistance. *Nucleic Acids. Res.* **16,** 1218.

Schottel, J., Mandal, A., Clark, D., Silver, S., and Hedges, R. W. (1974). Volatilisation of mercury and organomercurials determined by inducible R-factor systems in enteric bacteria. *Nature (London)* **251,** 335–337.

Scott, J. R. (1992). Sex and the single circle: Conjugative transposition. *J. Bacteriol.* **174,** 6005–6010.

Shaw, K. J., Rather, P. N., Sabatelli, F. J., Mann, P., Munayyer, H., Mierzwa, R., Petrikkos, G. L., Hare, R. S., Miller, G. H., Bennett, P., and Downey, P. (1992). Characterization of the chromosomal *aac(6')-Ic* gene from *Serratia marcescens. Antimicrob. Agents Chemother.* **36,** 1447–1455.

Shaw, K. J., Rather, P. N., Hare, R. S., and Miller, G. H. (1993). Molecular genetics of aminoglycoside resistance genes and familial relationships of the aminoglycoside-modifying enzymes. *Microbiol. Rev.* **57,** 138–163.

Shaw, W. V. (1984). Bacterial resistance to chloramphenicol. *Br. Med. Bull.* **40,** 36–41.

Sheridan, R. P., and Chopra, I. (1991). Origin of tetracycline efflux proteins: Conclusions from nucleotide sequence analysis. *Mol. Microbiol.* **5,** 895–900.

Shoemaker, N. B., Smith, M. D., and Guild, W. R. (1980). DNase resistant transfer of chromosomal *cat* and *tet* insertions by filter mating in pneumococcus. *Plasmid* **3,** 80–87.

Stewart, G. J. (1989). The mechanism of natural transformation. *In* "Gene Transfer in the Environment" (S. B. Levy and R. V. Miller, eds.), pp. 139–163. McGraw-Hill, New York.

Stokes, H. W., and Hall, R. M. (1989). A novel family of potentially mobile DNA elements encoding site-specific gene-integration functions: Integrons. *Mol. Microbiol.* **3,** 1669–1683.

Stotzky, G. (1989). Gene transfer among bacteria in soil. *In* "Gene Transfer in the Environment" (S. B. Levy and R. V. Miller, eds.), pp. 165–221. McGraw-Hill, New York.

Suárez, J. E., and Mendoza, M. C. (1991). Plasmid-encoded fosfomycin resistance. *Antimicrob. Agents Chemother.* **35**, 791–795.

Summers, A. O., Wireman, J., Vimy, M. J., Lorscheider, F. L., Marshall, B., Levy, S. B., Bennet, S., and Billard, L. (1993). Mercury released from dental "silver" fillings provokes an increase in mercury- and antibiotic-resistant bacteria in oral and intestinal floras of primates. *Antimicrob. Agents Chemother.* **37**, 825–834.

Taylor, D. E., and Chau, A. (1996). Tetracycline resistance mediated by ribosomal protection. *Antimicrob. Agents Chemother.* **40**, 1–5.

Timmermann, K. P. (1988). *Physiol. Plant.* **77**, 465–471.

Udo, E. E., Pearman, J. W., and Grubb, W. B. (1994). Emergence of high-level mupirocin resistance in methicillin-resistant *Staphylococcus aureus* in western Australia. *J. Hospital Infect.* **26**, 157–165.

Udou, T., Mizuguchi, Y., and Wallace, R. J., Jr. (1989). Does aminoglycoside-acetyltransferase in rapidly growing mycobacteria have a metabolic function in addition to aminoglycoside inactivation? *FEMS Microbiol. Lett.* **57**, 227–230.

Walker, J. B., and Skorvaga, M. (1973). Phosphorylation of streptomycin and dihydrostreptomycin by *Streptomyces*. Enzymatic synthesis of different diphosphorylated derivatives. *J. Biol. Chem.* **248**, 2435–2440.

Walsh, C. T., Fisher, S. L., Park, I.-S., Prahalad, M., and Wu, Z. (1996). Bacterial resistance to vancomycin: Five genes and one missing hydrogen bond tell the story. *Curr. Biol.* **3**, 21–28.

Weisblum, B. (1995). Insights into erythromycin action from studies of its activity as inducer of resistance. *Antimicrob. Agents Chemother.* **39**, 797–805.

Zakrzewska-Bode, A., Muytjens, H. L., Liem, K. D., and Hoogkamp-Korstanje, J. A. (1995). Mupirocin resistance in coagulase-negative staphylococci, after topical prophylaxis for the reduction of colonization of central venous catherters. *J. Hospital Infect.* **31**, 189–93.

Zilhao, R., and Courvalin, P. (1990). Nucleotide sequence of the *fosB* gene conferring fosfomycin resistance in *Staphylococcus epidermidis. FEMS Microbiol. Lett.* **68**, 267–272.

9

INFLUENZA: AN EMERGING MICROBIAL PATHOGEN

ROBERT G. WEBSTER
Department of Virology and Molecular Biology
St. Jude Children's Research Hospital
Memphis, Tennessee

I. INTRODUCTION

Given the existence in the aquatic bird reservoir of all known influenza A subtypes, we must accept the fact that influenza is not an eradicable disease. Prevention and control are the only realistic goals. If we assume that people, pigs, and aquatic birds are the principal variables associated with the interspecies transfer of influenza virus and the emergence of new human pandemic strains, it is important to maintain surveillance in these species. Live-bird markets that house a wide variety of avian species together (chickens, ducks, turkeys, pheasants, guinea fowl, and occasionally pigs) for sale directly to the public provide outstanding conditions for genetic mixing and spreading of influenza viruses; monitoring of the birds in these markets for influenza viruses will provide information that is relevant for both agricultural and human health. If pigs are the mixing vessel for influenza viruses, surveillance of influenza in this population may provide an early warning system for humans.

Why is influenza considered an emerging pathogen? Influenza occurs each year in human populations, so why should it be considered as an emerging pathogen? It is because the virus continues to evolve and new antigenic variants (drift strains) are emerging constantly giving rise to the yearly epidemics. In addition, strains to which the majority of the human population have no immunity appear suddenly, and the resulting pandemics vary from serious to catastrophic. The purpose of this chapter is to understand the property of the virus that permits this variation to occur, the mechanisms involved, and the reservoirs

Emerging Infections

of influenza viruses in the world and to attempt to understand the emergence of the next human pandemic.

II. HISTORY

The highly contagious, acute respiratory illness known as influenza appears to have afflicted humans since ancient times. The sudden appearance of respiratory diseases that persist for a few weeks and equally suddenly disappear are sufficiently characteristic to permit identification of a number of major epidemics in the distant past. One such epidemic was recorded by Hippocrates in 412 B.C., and numerous episodes were described in the Middle Ages. Historical data from A.D. 1500 until 1800 collected by Webster (1800) and Hirsch (1883) were reviewed by Noble (1982), and some features emerged: (a) epidemics occurred relatively frequently but at irregular intervals, and occasionally the disease seemed to disappear for periods of time; (b) epidemics varied in severity but usually caused mortality in the elderly; and (c) some epidemics such as those in 1781 and 1830 appeared to spread across Russia from Asia. There are some similarities with our current knowledge of the epidemiology of influenza; recent epidemics have appeared first in China and usually have had the highest mortality in the elderly. There is as yet no indication that influenza disappears from the human population.

Although influenza has killed untold millions throughout the centuries, the pandemic of 1918 to 1919 was particularly severe. Not only did Spanish influenza, as it was called, kill between 20 to 40 million people worldwide, it also altered the course of history (Crosby, 1976). General Erich F. N. Von Ludenddorff, Chief of Staff of the Imperial German Army, blamed the failure of the Marne Offensive not on the influx of fresh American troops, but rather on the diminished strength of the German Army, which he attributed at least in part to influenza. All the armies in Europe were hit hard by this outbreak; in fact, 80% of U.S. Army war deaths were due to influenza.

The severity of the pandemic of 1918 to 1919 greatly accelerated the search for the causative agent of influenza. In the late 1920s, Richard E. Shope showed that swine influenza could be transmitted with filtered mucus, suggesting that the causative agent was a virus (Shope, 1931). In 1933, a virus was isolated from humans by Wilson Smith, Sir Christopher Andrewes, and Sir Patrick Laidlaw of the National Institute for Medical Research in London, England (Smith *et al.*, 1933).

Retrospective seroepidemiological analysis, also known as seroarcheology, suggests that the influenza epidemic in humans in 1889 and 1890 was caused by a virus antigenically similar to more contemporary "Asian" strains (H2N8) (Mulder and Masurel, 1958). The epidemic of 1900 may have been caused by a "Hong Kong"-like strain, but with a neuraminidase like that of the A/Equine/Miami/1/63 strain. Swine-like influenza virus (H1N1) is believed to have caused the catastrophic pandemic in 1918 to 1919 (Masurel, 1968). The seroepidemiological information suggested that human strains recirculate.

Until the first human influenza virus was isolated in 1933, it was impossible to determine with certainty which pandemics were caused by influenza viruses. Since 1933, major antigenic shifts have occurred: in 1957 when the H2N2 subtype (Asian influenza) replaced the H1N1 subtype; in 1968 when the Hong Kong (H3N2) virus appeared; and in 1977 when the H1N1 virus reappeared.

III. NOMENCLATURE

Influenza viruses are divided into types A, B, and C based on the antigenic differences between their nucleoprotein (NP) and matrix (M) protein antigens. Influenza A viruses are further subdivided into subtypes, and the current nomenclature system (WHO Memorandum, 1980) includes the host of origin, geographical location of first isolation, strain number, and year of isolation. The antigenic description of the hemagglutinin (HA) and neuraminidase (NA) is given parenthetically, for example, A/Swine/Iowa/15/30 (H1N1). By convention, the host of origin of human strains is not included, for example, A/Puerto Rico/8/34 (H1N1) [PR8].

There are 15 subtypes of influenza A HA and nine NA subtypes, and all are found in aquatic birds. Three subtypes (H1, H2, H3) have so far been found in humans, two subtypes in pigs (H1, H3), and two in horses (H3, H7). Influenza A viruses have occasionally been isolated from sea mammals, mink, and other species, but do not appear to be maintained in these species (see below). Extensive surveillance of humans, animals, and birds has revealed only two additional subtypes in the past 20 years, H14 and H15 (Kawaoka *et al.*, 1990; Röhm *et al.*, 1996), suggesting that there may be a finite number of distinct influenza A viruses in nature and that representatives of each subtype continue to circulate, primarily in aquatic birds.

IV. STRUCTURE AND FUNCTION OF THE VIRUSES

The structure of influenza viruses of types A and B is similar, whereas influenza C reveals a different pattern of surface projections. When examined by electron microscopy, negatively stained preparations of viruses derived from infected eggs or tissue culture fluids reveal irregularly shaped spherical particles approximately 120 nm in diameter (Fig. 1). In contrast, most influenza virus isolates from humans and other species, after a single passage in culture, show a much greater shape variation, including the presence of greatly elongated forms (Choppin *et al.*, 1960). Morphological examination of primary influenza virus isolates from avian species before passage in culture systems revealed that some strains are predominantly spherical while others are extremely heterogeneous in size with long, filamentous, and bizarrely shaped particles. The morphological characteristics of influenza viruses are a genetic trait (Kilbourne, 1963), but spherical morphology appears to be dominant on passage in chicken embryos or tissue culture systems. The M gene seems to be a major determinant of this morphological difference, although the HA and NP genes may also contribute (Smirnov *et al.*, 1991).

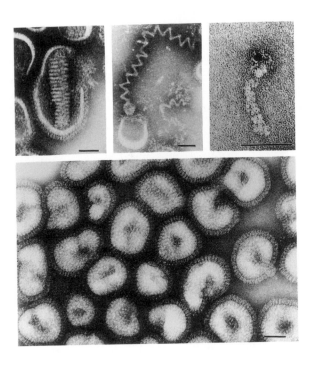

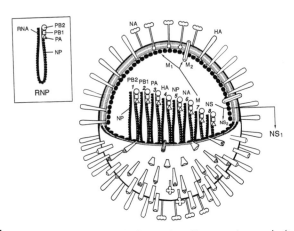

FIGURE 1 Structure of influenza virus. Electron micrographs (*top*) and diagram (*bottom*) of influenza virus. A to C illustrate the structure of the internal components, and D shows the external view. A substantial fraction (up to 50%) of influenza virions contain large helical internal components (A, B) which may contain individual ribonucleoprotein (RNP) segments (C) linked together. The individual RNPs each contain a binding site for the viral polymerase as seen by immunogold labeling of the end of the RNP segment (C). The external view of the virions (D) illustrate the pleomorphic appearance and the surface spikes. In all figures, bar = 50 nm. Courtesy of G. Murti, St. Jude Children's Research Hospital, Memphis, Tennessee. The diagram illustrates the main structural features of the virion. The surface of the particle contains three kinds of spike proteins—the hemagglutinin (HA), neuraminidase (NA), and matrix (M2) protein—embedded in a lipid bilayer derived from the host cell and covers and matrix (M1) protein that surrounds the viral core. The ribonucleoprotein complex making up the core consists of at least one of each of the eight single-stranded RNA segments associated with the nucleoprotein (NP) and the three polymerase proteins (PB2, PB1, PA). RNA segments have base pairing between their 3′ and 5′ ends, forming a panhandle. Their organization and the role of the NS2 in the virion remain unresolved. (From Murphy and Webster, 1996, Orthomyxoviruses. In "Fields Virology." 3rd edition. B. N. Field, et al. (eds.). Lippincott-Raven Publishers. Philadelphia, p 1401. with permission.)

The most striking feature of influenza virions is a layer of spikes projecting radially outward over the surface. These surface spikes on influenza A viruses are of three distinct types, corresponding to the HA, NA, and M2 components of the virus; the latter component is present in small numbers, and in influenza B strains N8 replaces M2 (Betakova et al., 1996; Brassard et al., 1996). Influenza C viruses possess only one surface spike with both HA and esterase activity. On influenza A and B viruses, the HA spike seen by electron microscopy appears rod-shaped, and the NA spike is mushroom-shaped. These differences are seldom, if ever, apparent on intact particles but are revealed after isolation of the proteins following detergent disruption of the virion envelope (Wrigley, 1979). The HA and NA glycoproteins are attached by short sequences of hydrophobic amino acids to a lipid envelope derived from the plasma membrane of the host cell. Within the lipid envelope lies the M1 protein, which is believed to be structural in function. Inside the M shell are eight single-stranded RNA molecules of negative sense (i.e., the virion RNA is complementary to the messenger RNA) associated with the nucleocapsid protein (NP) and three large proteins (PB1, PB2, and PA) responsible for RNA replication and transcription. The two virus-coded nonstructural proteins (NS1 and NS2) are found in infected cells; NS2 is also found in the virion (Lamb, 1989).

The organization of the eight single-stranded RNA segments within the virion is still poorly understood. Biochemical and biological studies support the idea that each RNA segment exists as a distinct ribonucleoprotein (RNP) complex (Compans et al., 1972). Genomic RNAs of influenza virus are held in a circular conformation in virions and in infected cells in a terminal panhandle form that might play an important role in replication (Hsu et al., 1987). Up to 50% of virions partially disrupted with Triton X-100 revealed large helical structures (Murti et al., 1980). It is possible that the helices represent the native organization of RNP and that multiple RNPs arise from the degradation of helices. Whether particles with helices are infectious remains to be determined, as does the question as to whether there is a mechanism by which the eight individual RNA segments are packaged or whether this is a random process.

V. VARIATION: ANTIGENIC DRIFT AND SHIFT

The influenza viruses are unique among the respiratory tract viruses in that they undergo significant antigenic variation. Both of the surface antigens of the influenza A viruses undergo two types of antigenic variation: antigenic drift and antigenic shift (Murphy and Webster, 1996). Antigenic drift involves minor antigenic changes in the HA and NA, whereas shift involves major antigenic changes in these molecules resulting from replacement of the gene segment.

A. Antigenic Drift

The HA is the major surface antigen of influenza virus and is subtype-specific. The HA from H3 human influenza viruses has been crystallized and the three-dimensional structure determined. Five antigenic domains (A to E) have been defined on the HA1 by comparative sequence analysis (Wiley *et al.*, 1981). Antigenic mapping with mouse monoclonal antibodies and sequence analysis of escape mutants confirm the number and location of the antigenic sites in H3 and H1 influenza viruses (Gerhard *et al.*, 1981; Webster and Laver, 1980). Crystallographic studies of escape mutants have confirmed that antibodies bind to those regions of the molecule where amino acid substitutions permitting escape from neutralization have been found (Knossow *et al.*, 1984).

After the appearance of a new subtype, antigenic differences between isolates can be detected within a few years using ferret antisera for analysis. Analysis with monoclonal antibodies indicates that major antigenic heterogeneity is detectable among different influenza virus isolates at any time (Stevens *et al.*, 1987). Antigenic drift occurs by accumulation of a series of point mutations resulting in amino acid substitutions in antigenic sites A to E at the membrane distal region of the HA. These substitutions prevent binding of antibodies induced by the previous infection, and the virus can infect the host. In influenza A viruses of humans and other mammals, antibodies play a role in selection of mutants, whereas in influenza A viruses in avian species and in influenza B and C viruses, antibody selection does not appear to be involved in selection (Air *et al.*, 1990).

Although antigenic drift can be mimicked in the laboratory by growth of the influenza viruses in the presence of monoclonal antibodies to a single antigenic site (e.g., to site A), it is not possible to select antigenic variants with mixtures of monoclonal antibodies to different antigenic sites (e.g., to sites A, B, C, etc.). The frequency of selection of variants by monoclonal antibodies to a single site *in vitro* is 10^{-5} (Yewdell *et al.*, 1979); with a mixture of monoclonal antibodies to two different antigenic sites the frequency would be 10^{-10}. How then does selection of drift variants occur in humans that have polyclonal responses and a mixture of antibodies to each antigenic site? Analysis of postinfection human sera indicate a limited antibody repertoire (Wang *et al.*, 1986). The restricted ability of human sera to neutralize monoclonal antibody-selected escape mutants (Natali *et al.*, 1981) suggests that the selection of variants may be sequential, involving several different human hosts.

B. Antigenic Shift

Since the first human influenza virus was isolated in 1933, antigenic shifts in type A influenza viruses have occurred in 1957 when the H2N2 subtype (Asian influenza) replaced the H1N1 subtype, in 1968 when the Hong Kong (H3N2) virus appeared, and in 1977 when the H1N1 virus reappeared (Fig. 2). Each of these major antigenic shifts have several characteristics in common: (a) their appearance was sudden, (b) they first occurred in China, (c) they were antigenically distinct from the influenza viruses then circulating in humans, and (d) they were confined to the H1, H2, and H3 subtypes.

On the basis of phylogenetic evidence, the most likely explanation for the appearance of new pandemic strains in humans is that they were derived from

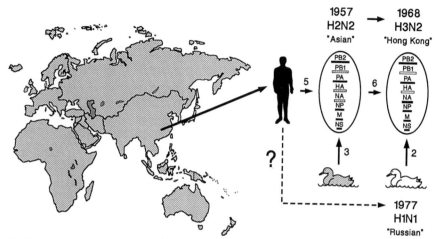

FIGURE 2 Emergence of human influenza pandemics in the twentieth century. Phylogenetic evidence suggests that an influenza virus possessing eight gene segments from avian influenza reservoirs was transmitted to humans and pigs before 1918 and replaced earlier strains. The site of origin of this strain is unresolved, but historical evidence suggests that this virus mutated, was carried from North America to Europe by American troops, and caused the catastrophic Spanish influenza pandemic of 1918. In 1957, the Asian pandemic virus acquired three genes (PB1, HA, and NA) from the avian influenza gene pool in wild ducks by genetic reassortment and kept five other genes from the circulating human strain. After the Asian strain appeared, the H1N1 strains disappeared from humans. In 1968, the Hong Kong pandemic virus acquired two genes (PB1 and HA) from the duck reservoir by reassortment and kept six genes from the virus circulating in humans. After the appearance of the Hong Kong strain, the H2N2 Asian strains were no longer detectable in humans. In 1977, the Russian H1N1 influenza virus that had circulated in humans in 1950 reappeared and spread in children and young adults. This virus probably escaped from a laboratory and has continued to cocirculate with the H3N2 influenza viruses in the human population.

avian influenza viruses either after reassortment with the currently circulating human strain or by direct transfer (Fig. 2). There is ample evidence for genetic reassortment between human and animal influenza A viruses *in vivo* (Webster *et al.*, 1971), and genetic reassortment has also been detected in humans (Cox *et al.*, 1983). Genetic and biochemical studies have concluded that the 1957 and 1968 strains arose by genetic reassortment. The 1957 Asian H2N2 strain obtained its HA, NA, and PB1 genes from an avian virus and the remaining five genes from the preceding human H1N1 strain (Kawaoka *et al.*, 1989; Scholtissek *et al.*, 1978a).

VI. RESERVOIRS OF INFLUENZA A VIRUSES

Influenza A viruses infect swine, horses, seals, and a large variety of birds as well as humans (Webster *et al.*, 1992). Influenza B viruses, on the other hand, infect only humans. Influenza C virus infects humans and swine. Phylogenetic studies of influenza A viruses have revealed species-specific lineages of viral genes and

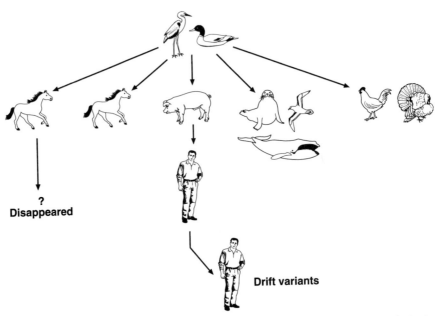

?
Disappeared

Drift variants

FIGURE 3 Reservoirs of influenza A viruses. Phylogenetic analysis supports the hypothesis that aquatic birds are the reservoirs of all influenza A subtypes. These viruses are in evolutionary stasis and accumulate mutational changes in their proteins at a surprisingly lower rate. There are five host-specific lineages, but periodically there are transmissions between lineages. The most ancient influenza A lineage in mammals is in horses (A/Equine/Prague/1/56 [H7N7]) and has not been detected in horses or other species in the past 10 years. After transmission to mammalian hosts, especially to humans, the viruses evolve rapidly, giving the near annual epidemics of influenza.

have demonstrated that the prevalence of interspecies transmission depends on the animal species. It has also been revealed that aquatic birds are the source of all influenza viruses in other species (Fig. 3).

A. Influenza Viruses in Birds

Representatives of each of the known subtypes of influenza A viruses have been isolated from aquatic birds. These viruses have been isolated worldwide from both domestic and wild species. The largest numbers of viruses have been isolated from feral water birds including ducks, geese, terns, shearwaters, and gulls, as well as from a wide range of domestic avian species such as turkeys, chickens, quail, pheasants, geese, and ducks. The disease signs associated with influenza A virus infections in avian species vary considerably with the strain of virus. Infection with most strains of influenza virus are completely asymptomatic. However, a few strains produce systemic infection accompanied by central nervous system (CNS) involvement, with death occurring within 1 week. The viruses that fall into this last category are members of the H5 and H7 subtypes such as A/FPV/Dutch/27 (H7N7) and A/Tern/South Africa/1/61 (H5N3) (Rott *et al.*, 1979). Besides these subtypes (H5, H7), the most serious problems are associated with influenza A subtypes (H1 through H13) in turkeys that cause chronic respiratory infections. For example, economic losses in a 2-year period in turkeys in Minnesota in the 1980s exceeded $5 million. Infection in turkeys is thought to be caused by virus strains that are periodically introduced from migrating ducks. Influenza occurs much less frequently in domestic chickens mainly because they are housed indoors. In 1983, however, an H5N2 influenza virus gained access to chicken houses in Pennsylvania, became virulent, and cost $61 million to eradicate.

In ducks, the majority of avian strains of influenza virus replicate predominantly in the cells lining the intestinal tract and also in the lungs and upper respiratory tract. The viruses gain access by passage through the digestive tract of the duck, despite the low pH of the gizzard, and are shed in high concentration in the feces. Influenza A viruses of each of the major subtypes that infect humans (H1N1, H2N2, H3N2) will also replicate in ducks; however, following intranasal administration these viruses replicate only in the upper respiratory tract and not in the intestinal tract. Some human strains will replicate to a limited extent in the cloaca if administered rectally.

Studies on wild ducks in Canada from 1975 to 1996 have shown the following: (a) a high percentage (up to 20%) of juvenile birds are infected with influenza virus when the birds congregate prior to migration; (b) none of the birds show any symptoms of infection; and (c) many subtypes of influenza virus are enzootic (Hinshaw *et al.*, 1980). In addition, influenza A viruses have been isolated from water samples collected from lakes in Canada at the time when wild ducks were marshalling before migration. Fecal samples collected from the shores of these lakes have also revealed influenza A viruses. The high concentration of duck influenza viruses in fecal material and the isolation of influenza A viruses from lake water offers a logical mechanism for the transmission of avian influenza viruses among feral ducks and to domestic avian and mammalian species.

The mechansim by which viruses persist in wild ducks from year to year is not clear. Some ducks may shed virus for as long as 30 days, and thus many passages would not be required to maintain the viruses in the population. The isolation of viruses from migrating ducks supports this possibility. It is conceivable that viruses remain viable in the frozen lakes over the winter until the birds return; however, viruses have not been detected when the birds were absent. Persistence of influenza in individual animals has not been reported and is unlikely. The available information supports the continuous circulation of influenza in migrating waterfowl with a low incidence of virus in birds in their overwintering sites in the subtropics.

The avirulent nature of avian influenza infection in ducks may be the result of virus adaptation to this host over many centuries, creating a reservoir that ensures perpetuation of the virus. It is suggested that ducks occupy a unique and very important position in the natural history of influenza viruses. Influenza viruses of avian origin have been implicated in outbreaks of influenza in mammals such as seals (Geraci *et al.*,1982), whales (Lvov *et al.*, 1978; Hinshaw *et al.*, 1986), pigs (Scholtissek *et al.*, 1983), and horses (Guo *et al.*, 1992), as well as in domestic poultry, especially in turkeys.

The HA plays a central role in the pathogenicity of avian influenza viruses. Infection with certain avian influenza viruses possessing the H5 or H7 HA is associated with high mortality in domestic poultry but very rarely in wild birds

(Alexander, 1986). The HA of these viruses is readily cleaved in tissue culture and does not require an exogenous protease for plaque formation. The virulent H5 and H7 viruses differ from the HAs of other influenza A subtypes in that their HAs possess multiple basic amino acids at the carboxyl terminus of HA1. This permits cellular proteases that recognize multiple basic amino acids to cleave the HA and render the virus infectious and capable of systemic spread *in vivo* (Steineke-Gröber *et al.*, 1992; Walker *et al.*, 1992).

Thus, the HA plays a central role in virulence of avian influenza A viruses. An important question concerns the potential for viruses that possess readily cleavable HAs to cause systemic disease in mammals including humans and whether the "Spanish" influenza of 1918 possessed such a cleavage site in its HA.

B. Influenza Viruses in Pigs

Pigs are susceptible to infection with all subtypes of avian influenza A viruses (Kida *et al.*, 1994) including different variants of the H1N1 and H3N2 subtypes of influenza A viruses. These include the H1N1 viruses of classic swine influenza, the H1N1 viruses antigenically similar to viruses isolated from avian sources, as well as the H1N1 viruses similar to strains from humans. The H3N2 variants of human influenza viruses also replicate in swine. Influenza in swine was first observed in the United States during the catastrophic 1918 to 1919 human influenza pandemic. The disease signs in pigs, as in humans, are characterized by nasal discharge, coughing, fever, labored breathing, and conjunctivitis (Shope, 1958). The isolation of swine influenza virus [A/Swine/Iowa/15/30 (H1N1)] by Shope and retrospective serological studies on humans indicated that the swine virus was antigenically similar to the type A influenza virus responsible for the human pandemic. Since then, swine influenza virus has remained in the swine population, circulating year-round and being responsible for one of the most prevalent respiratory diseases in pigs in North America.

In 1976, swine influenza virus (H1N1) was isolated from military recruits at Fort Dix. Antigenically and genetically indistinguishable isolates were subsequently obtained from a man and a pig on the same farm in Wisconsin. These studies confirmed the earlier serological and virus isolation studies that implicated swine viruses in human disease (Hinshaw *et al.*, 1978). Serological studies on slaughterhouse workers indicate that transmission of swine influenza viruses to humans occurs quite frequently (up to 20% of workers in 1977 had antibodies to swine influenza virus), but in the recent past none of these incidents has resulted in an epidemic of disease in humans. However, the swine virus is still occasionally isolated from humans with respiratory illness and occasionally is lethal (Rota *et al.*, 1989; McKinney *et al.*, 1990). Outbreaks of influenza in swine in Europe since 1980 have been associated with influenza A viruses that are antigenically and genetically distinguishable from classic swine (H1N1) viruses isolated from pigs in North America. These viruses are of avian origin and cause disease in pigs similar to classic swine influenza with high fever, 100% morbidity, and low mortality (Scholtissek *et al.*, 1993).

Influenza viruses antigenically identical to human H3N2 strains also infect swine and can cause clinical signs of disease (Kundin, 1970). The available evidence suggests that some of the subsequent variants of human H3N2 viruses since 1968 has been transmitted to pigs. There is evidence that H3N2 variants can persist in pigs after they have disappeared from the human population; thus, the A/Port Chalmers/1/73 variant of H3N2 has continued to circulate and cause disease in pigs in Europe through 1997 (Campitelli *et al.*, 1997). The role of the pig as an intermediate host will be considered later in Section XI.

C. Influenza Viruses in Horses

Although influenza virus was not isolataed from horses until 1956, there are historical records indicating that influenza virus probably infected horses centuries ago. Two different subtypes of influenza A viruses infect and cause disease in horses, A/Equine/Prague/1/56 (H7N7) and A/Equine/Miami/1/63 (H3N8). The former is commonly known as equine 1 and the latter, equine 2. Both viruses produce similar disease signs in horses, but equine 2 infections are usually more severe. The signs of disease include a dry hacking cough, fever, loss of appetite, muscular soreness, and tracheobronchitis. Secondary bacterial pneumonia almost always accompanies equine influenza. The recommendation to horse owners is that all equines with clinical influenza or those recovering from influenza must be rested for 3 weeks following infection (Bryans, 1975) or permanent disability can occur. A high incidence of inflammation of the heart muscle (interstitial myocarditis) has been recorded in horses during and after A/Equine/Miami/1/63 (H3N8) (equine 2) infections (Gerber, 1970). Other sequelae of equine influenza include purulent pharyngitis, purulent conjunctivitis, sinusitis, chronic laryngitis, and chronic bronchopneumonia.

Cost analysis of an outbreak of influenza in horses from a thoroughbred stable was $668 per horse in 1975 (Bryans, 1975) and is probably 10 times this cost in 1997, thereby generating interest in equine influenza vaccines. Formalin-inactivated vaccines are available, but these vaccines induce immunity in horses which lasts only 3 months. The available evidence indicates that the short-lived immunity is a property of the immune response of horses rather than low vaccine potency or genetic drift (Mumford *et al.*, 1983). Antigenic drift does occur but at a slower rate than in humans, making periodic changes in vaccine strains necessary.

Equine 1 influenza viruses have not been isolated anywhere in the world for over 15 years, raising the possibility that this virus has disappeared. However, antibodies to equine 1 influenza virus have been detected sporadically in horses, either from vaccination or from subclinical infections.

In 1989 a new H3N8 influenza virus appeared in horses in Northern China that derived all of its gene segments from an avian influenza virus (Guo *et al.*, 1990, 1992). The emergence of a new H3N8 influenza virus in horses in 1989 with all gene segments from an avian strain illustrates that avian to mammalian transfers occur at irregular intervals.

D. Influenza Viruses in Seals, Mink, and Whales

In 1979 to 1980, approximately 20% of the harbor seal (*Phoca vitulina*) population of the northeast coast of the United States died of a severe respiratory infection with consolidation of the lungs typical of primary viral pneumonia

(Geraci *et al.*, 1982). Influenza virus particles were found in high concentrations in the lungs and brains of the dead seals. Antigenic and genetic analysis established that this virus derived all its genes from one or more avian influenza viruses and was associated with severe disease in mammals. Biologically, the virus behaved more like a mammalian strain of influenza, replicating to high titer in ferrets, cats, and pigs. In humans, the replication of the A/Seal/Massachusetts/1/80 (H7N7) influenza virus is confined to the eyes, causing conjunctivitis. Infected persons recovered without complications though antibodies to the virus did not develop in the serum of infected individuals (Webster *et al.*, 1981). In squirrel monkeys, the seal influenza virus replicates in the lungs and nasopharynx after intratracheal administration and in the conjunctiva after administration to the eye (Murphy *et al.*, 1983). In the study, one monkey died of pneumonia, and the seal influenza virus was recovered from the spleen, liver, muscles, and lung, indicating that the virus has the capability for systemic spread in primates.

Influenza viruses have subsequently been isolated from seals found dead on the New England coast of the United States, including H4N5 in 1983, H4N6 in 1991, and H3N2 in 1992 (Hinshaw *et al.*, 1992). Influenza A viruses (H13N2 and H13N9) have also been isolated from the lungs and hilar nodes of one stranded pilot whale, although the relationship between influenza virus infection and stranding remains unknown. Genetic analysis of the H13N9 whale virus has shown that the virus was introduced from birds (Hinshaw *et al.*, 1986). Influenza A viruses (H1N3) have also been isolated from the liver and lungs of whales (Balaenopteridae) in the South Pacific (Lvov *et al.*, 1978).

Influenza viruses have been isolated from mink raised on farms. These viruses, which were of avian origin (H10N4), caused systemic infection and disease in the mink and spread to contacts (Klingborn *et al.*, 1985) but has not become established in this species.

These findings demonstrate that interspecies transmission of influenza A viruses occurs relatively frequently, mainly from aquatic birds to mammalian species. The epidemics tend to be self-limiting, and the newly introduced viruses do not seem to be maintained in animal species such as seals, whales, and mink.

VII. EVOLUTIONARY PATHWAYS FOR INFLUENZA VIRUSES

Studies on the ecology of influenza viruses has led to the hypothesis that all mammalian influenza viruses derive from the avian influenza reservoir. Support for this theory comes from phylogenetic analyses of nucleic acid sequences of influenza A viruses from a variety of hosts, geographical regions, and virus subtypes. Phylogenetic analyses of the NP gene show that avian influenza viruses have evolved into five host-specific lineages (Fig. 4): an ancient equine lineage, which has not been isolated in over 15 years; a recent equine lineage; a lineage in gulls; one in swine; and one in humans. The human and classic swine viruses have a genetic "sister group" relationship, which shows that they evolved from a common origin. It appears that the ancestor of the human and classic swine virus was an intact avian virus that, like the influenza virus currently circulating in pigs in Europe, derived all of its genes from avian sources (Gorman *et al.*, 1990; Gammelin *et al.*, 1990).

sequence analyses along with phylogenetic evidence of the influenza viruses isolated from horses in northeast China in 1989 and 1990 suggest that these viruses were of avian origin. The viruses were antigenically most closely related to H3 viruses of avian origin. Direct comparison of all eight influenza A virus genes against a wide array of isolates showed that all of the eight genes of the A/Equine/Jilin/89 (H3N8) are similar to those of avian viruses. Phylogenetic analyses of the eight genes of Equine/Jilin/89 showed that they are not derived from the currently circulating equine 2 (H3N8) viruses. It is not known if this H3N8 influenza has become established in horses in China.

A second example of the emergence of an influenza virus from the aquatic birds reservoir was the appearance of a highly pathogenic H5N2 influenza virus in domestic chickens in Mexico (Horimoto *et al.*, 1995). In October of 1993, there was decreased egg production and increased mortality among Mexican chickens, in association with serological evidence of an H5N2 influenza virus (Fig. 6). First isolated from chickens in May of 1994, after spreading widely in the country, the virus caused a mild respiratory syndrome in specific-pathogen-free chickens. Since eradication of the virus by destruction of infected birds was beyond the resources of the poultry industry in Mexico, a "field experiment" took place to determine the fate of an avirulent influenza virus after repeated cycles of replication in millions of chickens. By the end of 1994, the virus had mutated to contain a highly cleavable hemagglutinin, but it remained only mildly pathogenic in chickens. Within months, however, it had become lethal in poultry. Nucleotide sequence analysis of the HA cleavage site of the original avirulent strain revealed the amino acid sequence RETR, typical of avirulent viruses and unlike the KKKR sequence characterizing viruses responsible for the 1983 outbreak in U.S. poultry. Both mildly and highly pathogenic isolates con-

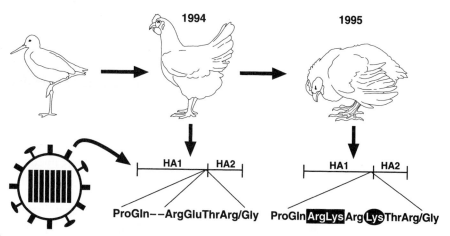

FIGURE 6 Molecular changes associated with emergence of a highly pathogenic H5N2 influenza virus in chickens in Mexico. In 1994, a nonpathogenic H5N2 influenza virus was detected in Mexican chickens that was related to an H5N2 virus isolated from shorebirds (ruddy turnstones) in Delaware Bay, United States, in 1991. The 1994 H5N2 isolates from chickens replicated mainly in the respiratory tract, spread rapidly among chickens, and were not highly pathogenic. Over the course of the next year the virus became highly pathogenic, and the HA acquired an insert of two basic amino acids (Arg-Lys) possibly by classic recombination and a mutation of Glu to Lys at position −3 from the cleavage site of HA1/HA2.

tained insertions and a substitution of basic residues in the HA connecting peptide, RKRKTR, which made the HA highly cleavable in trypsin-free chicken embryo fibroblasts. Phylogenetic analysis of the HA of H5 avian influenza viruses, including the Mexican isolates, indicated that the epidemic virus had originated from the introduction of a single virus of the North American lineage into Mexican chickens. This sequence of events demonstrates the stepwise acquisition of virulence by an avian influenza virus in nature.

The outcome of the "field experiment" will depend on an effective vaccination and biosecurity program; both nonpathogenic and pathogenic forms of the H5N2 influenza virus circulating in chickens in Mexico could be eradicated. This seems unlikely since the avirulent virus that is adapted to chickens spreads very effectively and causes minimal primary evidence of infection. A likely possibility is that the pathogenic variants will be self-limiting and will not be maintained. However, if the avirulent form is allowed to continue circulating, it is inevitable that there will be periodic outbreaks of highly pathogenic avian influenza. The lesson to be learned from this introduction of an avian influenza into domestic poultry is that the avirulent form of the viruses can circulate and accumulate multiple mutations and become a highly pathogenic strain that causes high mortality.

The third example is the introduction of avian influenza viruses into pigs in Europe and China and will be considered in the following section. These viruses may be of the greatest concern, for they may be en route to humans!

XI. IS THE PIG THE INTERMEDIATE HOST AND DOES IT SERVE AS AN EARLY WARNING SYSTEM FOR HUMAN PANDEMICS?

Current human influenza viruses are believed to have arisen by genetic reassortment between previous human influenza viruses and nonhuman viruses (Fig. 2). Where did the reassortment between genes of human and avian influenza viruses occur? Swine have been considered a logical intermediate for the reassortment of influenza viruses, for they can serve as hosts for viruses from either birds or humans (Scholtissek et al., 1993). Additionally, pigs have receptors for both avian and human influenza viruses (T. Ito, personal communication 1997), and pigs are susceptible to infection with all of the avian subtypes so far tested (H1–H13) (Kida et al., 1994).

To examine the roles of pigs as an intermediate host in the emergence of influenza viruses, we studied viruses in this species in the United States, Europe, and southern China. Analysis of the six genes coding for the internal protein of 73 classic swine H1N1 viruses from the United States by dot-blot hybridization revealed no exchange of genomes from different hosts (reassortment), nor were there viruses from hosts other than swine (Wright et al., 1992). All genes coding for internal proteins were characteristic of swine viruses.

Analysis of 104 swine influenza virus isolates from southern China including 32 H3N2 and 72 H1N1 isolates provided evidence for interspecies transmission and reassortment. The majority of H3N2 isolates were of human origin, all of the H1N1 isolates were of swine origin, and there were three reassortants between swine and human influenza viruses detected (Shu et al., 1996). There was no evidence for the transmission of avian influenza viruses to swine

(Lin *et al.*, 1994). In 1979, an avian influenza virus transmitted to pigs in Europe and become established in pigs, causing classic swine influenza disease signs (Pensaert *et al.*, 1981; Scholtissek *et al.*, 1983). This virus has reassorted with human influenza viruses maintained in pigs in Italy to produce "human–avian" reassortants containing the structure glycoproteins (HA and NA) of A/Port Chalmers/1/73 (H3N2)-like viruses and six gene segments from the avian-like virus circulating in European pigs (Castrucci *et al.*, 1993) (Fig. 7). Studies in children in The Netherlands have detected influenza viruses with this genetic composition in children with influenza symptoms (Claas *et al.*, 1994).

In contrast with earlier studies, antigenic and genetic analyses of influenza viruses isolated from pigs in Hong Kong in 1993 established that two different groups of H1N1 viruses were cocirculating among pigs that originated in sourthern China. One group belonged to the classic swine lineage, and the other belonged to the Eurasian avian lineage. These studies showed that an avian influenza virus spread from the avian reservoir to pigs in southern China (Guan *et al.*, 1996). Phylogenetic analysis indicates that the genes of the avian influenza H1N1 viruses form an Asian sublineage of the Eurasian avian lineage, suggesting that these viruses are an independent introduction into pigs in Asia.

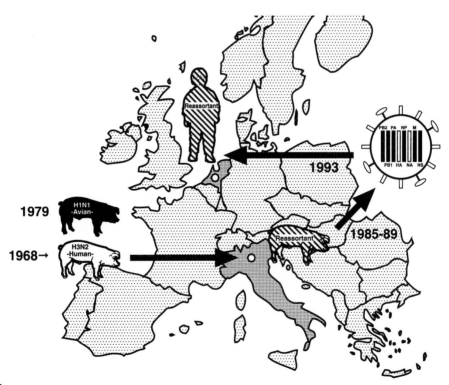

FIGURE 7 A model of emergence of an influenza A virus in mammals. In 1979, an avian H1N1 influenza virus transmitted to pigs in Europe and became established in European pigs. This virus reassorted with a human H3N2 influenza virus that continues to circulate in pigs in Italy, producing a human–avian virus containing the surface glycoproteins (HA and NA) from the human strain and six genes encoding the internal virion proteins from the avian H1N1 viruses. In 1993, human–avian reassortant viruses of this genetic composition were isolated from children in The Netherlands; the viruses did not become established in humans, but the studies demonstrate that pigs can be the "mixing vessel" for influenza and that such viruses can replicate and cause influenza in humans.

Earlier studies discussed above of influenza viruses from pigs in the geographical region of southern China, including Taiwan, from 1976 to 1982, revealed cocirculating H1N1 and H3N2 subtypes and detected a low incidence of reassortants (Lin *et al.*, 1994) but no evidence for avian influenza virus gene segments. During the 1993 study in pigs from China no H3N2 viruses were detected in pigs, suggesting that these viruses may be circulating at very low levels in pigs in China. Whether this avian H1N1 virus first detected in 1993 is a long-term resident of Chinese pigs or a recent introduction that will become established in the swine populations of China, as was the case in Europe after its introduction in 1979 (Scholtissek *et al.*, 1993), remains to be established.

The above studies establish that the pig may be the mixing vessel for reassortment of human and avian influenza viruses. The infrequent transmission of avian influenza viruses to pigs may be the initial limiting step in interspecies transmission. The establishment and spread of these viruses in pigs also appears to be limiting. A remaining question is whether subtypes other than H1N1 will transmit from the avian reservoir to pigs.

XII. IS CHINA AN EPICENTER FOR HUMAN PANDEMIC INFLUENZA VIRUSES?

Historical records and the appearance of the Asian, Hong Kong, and Russian pandemic strains of influenza virus in China suggest that the majority of pandemics of human influenza since about 1850 have originated in China. The exception seems to be Spanish influenza, which may have originated in military camps in Kansas and was taken to Europe by U.S. troops in 1918. This remains uncertain, however, as phylogenetic studies suggest that the Spanish influenza pandemic may have initially originated from the Eurasian continent.

The possibility has been raised that southern China is an influenza epicenter (Shortridge and Stuart-Harris, 1982). Unlike the temperate or subarctic regions of the world, where influenza in humans is a winter disease, in the tropical and subtropical regions of China influenza occurs year-round (Reichelderfer *et al.*, 1989). In China, influenza A viruses of all subtypes are prevalent in ducks and in water frequented by ducks, and the different subtypes are present year-round with a peak incidence in summer months. In China and other areas of southeast Asia, influenza viruses of the H1N1 and H3N2 subtypes are prevalent in pigs (Shortridge and Webster, 1979; Guan *et al.*, 1996). Farming practices (Scholtissek and Naylor, 1988), religious customs, and climate may all contribute to making southern China one place where influenza viruses originate; the high population of people, pigs, and ducks provides the opportunity for interspecies transmission and genetic exchange among influenza viruses.

Although the above considerations about the regions of the world where influenza pandemics originate are interesting, they are still merely speculation. Only circumstantial evidence exists for the appearance of pandemic influenza viruses in southern China. Epidemiological studies of the frequency of influenza virus transfer between species have not been done, and they merit attention. The molecular tools are at hand for answering these questions. There may be additional unknown ecological features, such as air pollution, genetic susceptibility, host range, and local customs, that influence the appearance of influenza viruses.

Over the past few years, an increasing number of strains from the People's

Republic of China have been included in the annual vaccine recommendations by the World Health Organization. This reflects increased surveillance of humans in China and will serve an important role in the early detection of new pandemic strains if they first occur in this part of the world.

XIII. PANDEMIC PLANNING

With the realization that a future pandemic of influenza in humans is inevitable and perhaps imminent, health authorities throughout the world have been preparing plans for dealing with this event. A federal interagency working group in the United States, together with consulting scientists, have developed a plan. Key items integrated into this plan are (i) influenza surveillance, (ii) vaccine development and production, (iii) vaccine utilization and coverage, (iv) chemoprophylaxis and therapy, and (v) influenza-related research. The detailed plan was not available at the time of writing in early 1997 but will be a vitally important document that will provide a comprehensive approach to dealing with influenza epidemics and pandemics.

A. Which Subtype?

Serological and virological evidence suggests that since 1889 there have been six instances of the introduction of a virus bearing an HA subtype that had been absent from the human population for some time. For the HA there has been a cyclical appearance of the three human subtypes with the sequential emergence of H2 viruses in 1889, H3 in 1900, H1 in 1918, H2 again in 1957, H3 again in 1968, and H1 again in 1977 (Fig. 2).

If recycling occurs then H2 should be the subtype that causes the next human pandemic. It is approximately 28 years since the H2 subtype last infected humans and, based on the reappearance of H1 after an absence of 27 years, there are enough susceptible people to support a pandemic. Influenza viruses of the H2 subtype have continued to circulate in the aquatic avian reservoir (Schäfer et al., 1993), the HA of which is essentially unchanged from the viruses that affected humans in 1957. In 1994–1995, forty-nine H2N2 virus isolates were made from live-bird markets and poultry in the United States (D. Senne, personal communication 1996). Do these viruses have the capacity to infect humans including immunosuppressed individuals or swine? There is no convincing evidence of the detection of H2 influenza viruses in pigs, but antibodies to H2 were detected in pigs in Guiyang in 1957 where the 1957 H2N2 strain was initially isolated. Since the Eurasian lineage of avian influenza viruses have been the source of gene segments that have reassorted with the circulating human viruses to produce the Asian/57 and Hong Kong/68 pandemic strains, we need to know if H2 viruses are currently prevalent in avian species in Eurasia and whether there is evidence for transmission to pigs.

If the next pandemic in humans is not caused by an H2 virus, which of the remaining subtypes (H4–H15) are possibilities? One would be H7, for it is known to infect mammals, with H7N7 being responsible for one of the two epidemic strains of influenza A in horses; it has also infected seals and caused conjunctivitis in humans (Webster et al., 1981). Preliminary results have been obtained for antibodies to H7 in humans in China (Shortridge, 1992; Zhou *et al.,* 1996). Other subtypes including H4 have been isolated from seals and H10 from mink, but these strains have not become established in these species. There is no theoretical reason why any subtype should not transmit to mammals for all subtypes replicate in pigs (Kida *et al.,* 1994). Those subtypes with high prevalence in aquatic birds throughout the world include H4, H3, and H6, which predominate in aquatic birds in China (Shortridge, 1982), Russia (Lvov, 1987), and North America, whereas H1, H2, and H4 predominate in Europe (Süss *et al.,* 1994).

B. When?

Prediction of when the next human pandemic will occur is not possible. Dr. Kennedy Shortridge remarks, "Put simply, each year brings us closer to the next pandemic" (Shortridge, 1995).

C. Severity?

Predictions of severity are not possible. Although we know some of the features determining pathogenicity in avian influenza viruses (e.g., insertion of basic amino acids at the cleavage site of the HA), we know very little about the molecular basis of pathogenicity of human influenza viruses and why one strain is more pathogenic than another.

Two significant events relevant to influenza pandemic planning occurred in mid-1997; the initial genetic characterization of the 1918 "Spanish" influenza virus was reported (Taubenberger et al., 1997) and a highly pathogenic avian H5N1 influenza virus caused a fatal case of influenza in a child in Hong Kong (de Jong et al., 1997).

Portions of four gene segments of the influenza genome from a paraffin block of a formalin fixed lung sample from a soldier that died of the 1918 "Spanish" influenza virus were sequenced. The sequence information indicated that the virus was a unique H1N1 strain most closely related to swine influenza virus and that the HA did not possess a series of basic amino acids at the cleavage site found in highly pathogenic avian strains. The avian H5N1 isolated from the tracheal aspirate of the deceased child contained eight basic amino acids at the cleavage site and was highly pathogenic in chickens. This virus is very closely related genetically to an H5N1 virus that caused high mortality in chickens in Hong Kong at the same time as the child died. The potential direct transfer of an avian influenza virus to humans and the realization that any of the influenza A subtypes may be the precursor of the next pandemic alerts us to the need to have reagents to all subtypes available for rapid diagnosis. Although there is currently no evidence for other human cases of influenza caused by this H5N1, it served to test the pandemic plan, point up some deficiencies, and has served to better prepare us for dealing with the next human pandemic virus.

ACKNOWLEDGMENTS

We thank Dayna Baker for manuscript preparation. This work was supported by U.S. Public Health Service Research Grants AI-29680 and AI-08831 from the National Institutes of Health, CORE Grant CA-21765, and American Lebanese Syrian Associated Charities (ALSAC).

10
THE EMERGENCE OF DENGUE: PAST, PRESENT, AND FUTURE

EDWARD C. HOLMES, LUCY M. BARTLEY,
AND GEOFFREY P. GARNETT

The Wellcome Trust Centre for the Epidemiology of Infectious Disease
Department of Zoology
University of Oxford
Oxford , United Kingdom

I. INTRODUCTION: THE RISE OF DENGUE

The mass movement of human populations has been an important factor in the history of the spread of infectious disease. The most striking examples have perhaps been the exploration and colonization of the Americas by the peoples of Europe and the later importation of slaves from Africa, which brought with it such infections as influenza, malaria, measles, smallpox, scarlet fever, and yellow fever. Because of the lack of indigenous immunity these diseases were able to spread rapidly, often with devastating effects.

To the list of Old World plagues that spread to the Americas with the advent of colonization, trading, and slavery can be added dengue, a febrile and sometimes hemorrhagic illness caused by a close relative of yellow fever virus. It is likely that dengue epidemics were first recorded at the end of the eighteenth century, with probable outbreaks in Cairo, Jakarta, and Philadelphia, although it is difficult to diagnose dengue from historical records because of its similarity to a number of other diseases, particularly yellow fever (hemorrhagic dengue and yellow fever are still mistaken today) and malaria. By the nineteenth century epidemics were commonplace in Asia, North America, and the Caribbean (reviewed in Gubler, 1988). Indeed, the term dengue itself, is based on the Swahili phrase "ki denga pepo," which describes a seizure caused by an evil spirit, first appeared in English during an epidemic which affected the Spanish West Indies in 1827–1828 (McSherry, 1993).

Today dengue is the most common vector-borne viral infection of humans, with up to 100 million cases reported annually and some 2 billion people at risk of infection in tropical and subtropical regions of Africa, Asia, and the Americas where the virus is often endemic (Monath, 1994; Monath and Heinz, 1996). The last 50 years have also seen the emergence of more serious forms of dengue disease, namely, dengue hemorrhagic fever (DHF) and dengue shock syndrome (DSS), which affect at least 500,000 people on an annual basis (although this is certainly an underestimate). Mortality rates are between 1 and 10% depending on the extent of hospital care, although they can reach 30% if patients go untreated (Gubler, 1988; Monath, 1994; Gubler and Clark, 1995; Monath and Heinz, 1996). Although DHF/DSS was reported sporadically early in the twentieth century, the first epidemics occurred in Southeast Asia during the 1950s, and the disease is particularly prevalent in this region today: in Thailand, for example, almost 900,000 cases of DHF/DSS were reported between 1958 and 1990, with a case–fatality rate of 1.57% (Monath and Heinz, 1996). The emergence of dengue, and particularly DHF/DSS, has been most dramatic in Latin America, with widespread epidemics since 1981 when DHF/DSS was first reported in Cuba (Gubler and Clark, 1995).

In addition to its impact on human health, dengue also entails an enormous economic burden through medical expenses, mosquito eradication procedures, and a decline in tourism. For example, the Cuban epidemic of dengue, which ran from July to October 1981, affected some 300,000 people with 34,000 cases of DHF and cost an estimated $100 million (Gubler, 1988). It is clear, therefore, that dengue represents one of the most important items on public health agendas and is a disease that appears to be increasing in prevalence. It is our goal in this chapter to review the factors that have led to the increased spread of dengue and consider its prospects for the future, particularly whether it poses a threat to countries with more temperate climates.

II. DENGUE: THE VIRUS AND THE DISEASE

A. Genetic Structure of Dengue

Dengue is a single-stranded, positive-sense RNA virus approximately 11 kb in length that encodes three structural and seven nonstructural genes. It is transmitted between primate hosts by mosquitoes of the genus *Aedes*. Although there are over 500 arthropod-borne viruses ("arboviruses"), of which some 100 cause disease in humans, dengue is a member of the genus *Flavivirus,* which share a common genome organization and antigenic determinants (Calisher *et al.,* 1989; Chambers *et al.,* 1990). There are 67 known flaviviruses, with those causing the most serious diseases in humans being dengue, Japanese encephalitis, and yellow fever. These viruses are all transmitted by mosquitoes and in their most serious forms cause hemorrhagic, encephalitic, and hemorrhagic/hepatic diseases, respectively (Monath and Heinz, 1996).

Like many RNA viruses, dengue exhibits extensive genetic diversity, a diversity which has important biological consequences, including hindering our attempts to develop effective vaccines. Specifically, dengue virus exists in four antigenically distinct serological types (designated DEN-1 to DEN-4) which were first identified by plaque reduction neutralization tests (Russell and Nisalak, 1967). These four serotypes are also distinct at the molecular level: for ex-

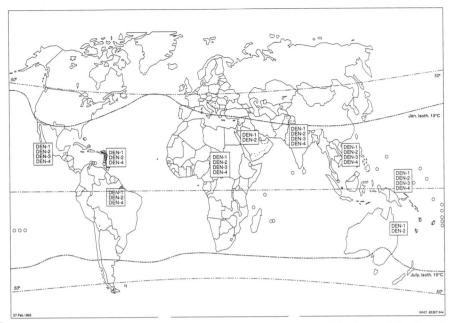

FIGURE I   Global distribution of the four dengue serotypes, DEN-1 to DEN-4, recorded at the end of 1995. Reprinted from the World Health Organization (WHO) with permission. The January (Northen Hemisphere) and July (Southern Hemisphere) 10°C winter isotherms, which mark the boundaries of dengue infection, are also indicated.

ample, viruses from different serotypes differ by up to 40% in amino acid sequence within the envelope (E) gene which encodes the major structural protein of the virion and is the target for neutralizing antibodies and T-cell responses. Phylogenetic analysis of E gene sequence data has revealed higher order relationships among the four serotypes: DEN-1 and DEN-3 are the most closely related (a relationship also found in neutralization assays), with DEN-2 joining next and DEN-4 being the most divergent (Blok *et al.,* 1992; Zanotto *et al.,* 1996). Whereas the dengue serotypes may have evolved in geographic isolation from one another, today all four cocirculate in many areas of Africa, Asia, and the Americas, although one serotype often dominates (Fig. 1).

B. Etiology of Severe Disease

Infection with dengue virus produces a range of clinical outcomes. The great majority of these are uncomplicated and after an incubation period of 2–7 days (although see Section III) lead to an acute debilitating, although self-limiting, febrile illness. This is characterized by a fever which may occur in one or two bouts, headache, lymphadenopathy, severe joint and muscle pains ("breakbone fever"), and a rash. This acute illness usually lasts for 8–10 days and is rarely fatal. Asymptomatic infections are also common. In some cases, however, dengue virus infections produce more serious clinical conditions, the most notable being dengue hemorrhagic fever (DHF) and dengue shock syndrome (DSS). Signs of DHF usually begin between the second and sixth day of the acute illness and may lead to spontaneous hemorrhaging which can be fatal if it occurs in the gastrointestinal tract or intracerebral regions. DSS occurs if the patient goes on to develop hypotension and profound shock due to plasma leakage and circulatory failure. This happens in about one-third of severe dengue cases (especially children) and is often associated with higher mortality.

Despite the prevalence and severity of DHF/DSS, its etiology remains uncertain. There are at least three possible explanations, the most cited being increased viral replication due to the enhanced infection of monocytes in the presence of preexisting dengue antibody at subneutralizing levels. This mechanism is known as antibody-dependent enhancement (ADE) (Halstead *et al.*, 1967, 1980; Halstead, 1988) and is illustrated schematically in Fig. 2. The initial dengue infection leads to a moderate viremia and serotype-specific antibodies which provide long-lasting (perhaps lifelong) immunity to the serotype of the infecting strain. For a short time (often less than 12 weeks) these antibodies are also able to inhibit subsequent infection with the other serotypes, but they soon decay to very low levels at which point they have little impact against heterologous strains. If the host is infected with a heterologous serotype before antibodies decay below a second threshold level then increased viral replication may result. Enhancement occurs on infection with a second serotype because infectious complexes of virions and immunoglobulin G (IgG) antibodies formed during the first infection gain access to peripheral blood monocytes and macrophage-like cells through the Fc γ and other receptors on the cell surface (Fanger and Erbe, 1992). The presence of enhancing antibodies therefore increases the num-

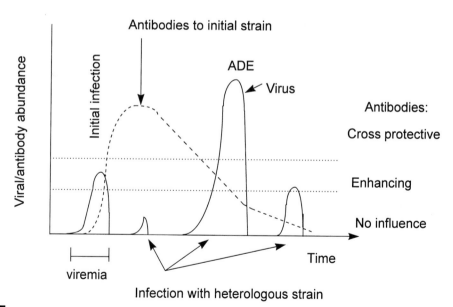

FIGURE 2   Schematic illustration of the relationship between the course of dengue infections and viremia in the individual host. The initial infection causes a moderate viremia and the production of serotype-specific antibodies. This protects against subsequent reinfection from the same serotype. The consequences of infection with another serotype depend on the titer of cross-reactive antibodies. Above one threshold value the growth in viral numbers is halted. However, below this threshold and above the next threshold the presence of antibodies enhances viral replication.

species, its association with humans in an urban setting makes it the main cause of large-scale epidemics. Concerted efforts have been made to eradicate. *A. ae-gypti* from certain parts of the world, and these schemes resulted in the removal of *A. aegypti* from most of Latin America during the 1950s and 1960s. Unfortunately, increased urbanization, economic development, and an expansion in the number of water-storage locations has allowed *A. aegypti* to reinvade this region since the 1970s, thereby increasing the incidence of both dengue and yellow fever (Monath, 1994). The possible implications of this increased incidence of dengue for the peoples of North America are discussed in Section V.

B. Transmission Dynamics of Dengue

The epidemiological pattern of dengue infections and disease is dependent on geographic location. The virus is hyperendemic in many tropical areas and particularly Southeast Asia, South Asia, and Africa, whereas other areas of the world only experience occasional dengue epidemics. Furthermore, there is concern that changing densities of mosquito populations may result in the occurrence of future epidemics in localities which currently have very low dengue incidence (Gubler, 1988; Monath, 1994; Rodriguez-Figueroa *et al.*, 1995). To understand the forces which control these epidemiological patterns it is necessary to describe the determinants of transmission of dengue virus, which are common to many other vector-borne infections.

The basic reproductive rate, R_0, of an infection is a measure of its potential reproductive success and therefore its ability to invade populations. Defined as the number of new infections caused by one infectious individual entering an entirely susceptible population, the basic reproductive rate for a vector-borne infection refers to the infection in the "definitive host," which are humans in the case of dengue. For indirectly transmitted microparasites like dengue there is a long history of using mathematical models to understand the transmission dynamics of infections, starting with the work of Ronald Ross (1911) on malaria. The addition of complexity and biological realism to these models, particularly by Macdonald (1957), is reviewed in detail by Anderson and May (1991).

Although developed with malaria in mind, the theoretical framework described here can also be applied to dengue as it was by Anderson and May (1991) and Newton and Reiter (1992). The simple model developed by Newton and Reiter is illustrated in Fig. 4. In this case the basic reproductive rate, R_0, is a product of the reproductive rate from human to mosquito R_h, and from mosquito to human R_v:

$$R_0 = R_h R_v \tag{1}$$

where

$$R_h = a\beta_h \left(\frac{1}{\nu}\right)\left(\frac{V}{H}\right) \tag{2}$$

and

$$R_v = a\beta_v p \left(\frac{1}{\mu_v}\right) \tag{3}$$

In Eqs. (2) and (3), a is the average biting rate per day of the vector species, β_h is the likelihood of transmission to the mosquito when it takes a blood meal

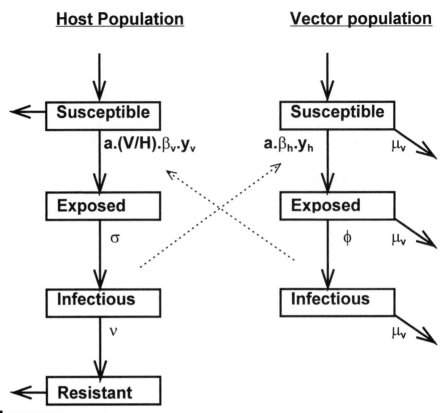

FIGURE 4   Flow diagram illustrating the infectious stages of human hosts and vector mosquitoes. People and vectors are assumed to be susceptible to dengue at birth. The rate at which they acquire infection (the "force of infection") depends on the prevalence of infection in the other species, the frequency of biting (the number of bites a vector takes is fixed, but the number of times a host is bitten depends on the vector per host ratio) and the transmission probability. On infection, people and vectors enter an incubating class from which they move to an infectious class. Humans recover rapidly from this infectious class, whereas vectors are assumed not to recover. y_v is the proportion of vectors infectious and y_h the proportion of humans infectious.

from an infectious human, and β_v is the likelihood of transmission to the human when it is fed on by an infectious mosquito. The recovery rate of humans from infectiousness, ν, is the inverse of the time during which a host is infectious. Likewise, μ_v, the mosquito death rate, is the inverse of the mean duration of infectiousness of mosquitoes. V is the number of vectors and H the number of hosts in the population, so that V/H is the vector density per host. Finally, p is the proportion of vectors which when infected become infectious.

One interesting aspect of these formulas is that there is a clear asymmetry in the terms R_h and R_v that in part reflects the obvious difference in mean life expectancy of humans and mosquitoes. This asymmetry has a number of consequences. For example, in the expression for the basic reproductive rate we are able to neglect the death rate of humans (as we can assume that this is much lower than the other processes) as well as any recovery of mosquitoes from infectiousness (which if it occurs at all is likely to be less common than the death of the vector). Furthermore, many vectors die before they actually become infectious (reflected in the term p). This is critical because if vectors both die and become infectious at a constant rate then the proportion becoming infectious would be

$$p = \frac{\phi}{\phi + \mu_v} \tag{4}$$

where ϕ is the rate at which vectors become infectious. Alternatively, if vectors become infectious after a fixed period τ and die at a constant rate, then the proportion becoming infectious is given instead by

$$p = e^{-\mu_v \tau} \tag{5}$$

Because the rate at which infected vectors become infectious is temperature dependent, the value of p is influenced by climate, as is the density of the vector population. As is discussed in Section V, this has important consequences for the possible spread of dengue into more temperate regions of the world.

A second and less intuitively obvious asymmetry in the basic model is that humans can be bitten by a virtually unlimited number of mosquitoes, whereas mosquitoes are limited in the number of blood meals they will take from humans. This asymmetry plays a role in determining the numbers of vectors which are on average likely to bite the initial infectious host and so spread the virus. Consequently, the mosquito density per host is included in the basic reproductive rate. However, this does not determine how many hosts an infectious vector will on average bite per day. This depends only on the frequency with which the vector takes blood meals.

Now that we have set out a basic model for the transmission dynamics of dengue, it is necessary to provide estimates for each of the parameters invoked. A review of parameters that are currently known for dengue is presented in Table I and illustrates our lack of knowledge about certain aspects of the epi-

TABLE I Parameters for a Model of Dengue Transmission

Variable	Name	Estimate (model value[a])	Source
$1/\sigma$	Intrinsic incubation period	4 to 7 days (not included)	47 experimental infections; Siler *et al.*, 1926
$1/\nu$	Infectious period of host	4 to 5 days (4)	Hospitalized patients, Jakarta; Gubler *et al.*, 1981
$1/\phi$	Extrinsic incubation period	7 days 90°F; 12 days 86°F (7, 12)	Watts *et al.*, 1987; McLean *et al.*, 1974
μ_v	Vector mortality rate	0.125 per day (0.125)	Sheppard *et al.*, 1969
β_v	Transmission probability, vector to host	1.0 per bite (1.0)	Gubler and Rosen, 1976; Putnam and Scott, 1995
β_h	Transmission probability, host to vector	No estimates (0.5)	
$a(V/H)$	Bites per host	1.5–4.8 per person per hour (0.25 per day)	Thailand over 12 months; Yasuno and Tonn, 1970
V/H	Vectors per host	20 per *room* (varied)	Yasuno and Tonn, 1970

[a] These are the values used to estimate the basic reproductive rate in Fig. 8b.

demiology of this infection, and particularly the probability of transmitting the virus from host to vector.

The course of dengue infection has been described in patients, volunteers, and animal models thereby providing information about the "intrinsic" incubation period ($1/\sigma$). It should be noted that although this period may play a role in transmission dynamics, and is included in Fig. 4, it is not important to the basic reproductive rate. In 47 experimental infections the symptoms occurred between 4 and 7 days in 89% of cases (Siler *et al.*, 1926). However, in an experimental inoculation with a trial DEN-3 vaccine, viremia occurred after 3 days and symptoms after 5 or 6 days in the three volunteers (Innis *et al.*, 1988). This suggests that the latent period from infection to infectiousness is shorter than the incubation period from infection to disease, which is typical of many viral infections (Anderson and May, 1991). The infectious period in human hosts is also short, with detectable viremia lasting for 4 to 5 days in hospitalized patients (Gubler *et al.*, 1981). Such patients may well have been atypical, but similar periods were observed in experimental infections of monkeys (Halstead *et al.*, 1973).

The period from infection to infectiousness in vectors (the "extrinsic" incubation period, $1/\phi$) has been shown to depend on temperature. McLean *et al.* (1974) observed a 6-day period in 4 mosquitoes at 32°C and a 13-day period in 12 mosquitoes at 27°C. Similarly, Watts *et al.* (1987) observed a period of 12 days in 5 mosquitoes at 30°C and a period as short as 7 days, but more often between 8 and 12 days, at 32° and 35°C. This is around the same duration as the life expectancy of the vector and reveals in part why high temperatures, by generating faster viral replication rates, are a requirement for large-scale dengue epidemics.

Vector population size, and hence vector to host density, is hard to measure accurately and is likely to vary greatly from place to place and on a seasonal basis. In a monastery in Bangkok, Yasuno and Tonn (1970) used a recapture method to estimate that there were 20 *A. aegypti* per room, whereas Newton and Reiter (1992) provided an estimate of 5 to 10 per room in Puerto Rico. Estimates of vector mortality rates were undertaken in Bangkok where Sheppard *et al.* (1969) used a recapture technique to estimate that the daily survival was 0.88 in *A. aegypti*. Similar estimates have been obtained in Kenya (McDonald, 1977; Trpis and Hausermann, 1986), and although the same is not true of all mosquito species, the death rate of *A. aegypti* appears to be constant with mosquito age (Clements and Paterson, 1981).

Estimating the biting rate of mosquitoes is also difficult. Most female mosquitoes take a number of blood meals before laying eggs, meals which on average weigh 109% of an unfed mosquito's body mass and take about 60 hr to digest (Lehane, 1991). *Aedes aegypti* may feed as many as three times during one gonotrophic cycle (Trpis and Hausermann, 1986), which may be increased by defensive host behavior (Kowden and O'Lea, 1978). After the vector ingests infected blood, the virus enters the mosquito midgut where it replicates and then disseminates into other organs, including the salivary glands where it replicates again. The virus in saliva can then be transmitted during the next blood meal (Gubler, 1988). Once infectious, mosquitoes remain so for life and suffer no pathogenic effects.

Temperature also influences the duration of the gonotropic cycle and the likelihood that mosquitoes will feed. Thus, the biting rate of the mosquitoes observed by Yasuno and Tonn (1970) varied from 1.48 to 4.77 bites per person per hour depending on the season. However, the figures are likely to be underestimates because both *A. aegypti* and *A. albopictus* are day biting species, with a biting peak just before sunrise and just after sunset, whereas Yasuno and Tonn (1970) measured biting for an hour in mid-morning. For 20 vectors per room to bite 5 times per hour consistently, they would each have to bite 6 times per day, which is clearly high when the digestion time of meals is considered. A figure of 0.25 bites per day as used by Newton and Reiter (1992) is probably more realistic and so is used in our model (see Section V). It is also worth noting that for the infection to be transmitted bites need to be on *different* people and that "biting" may constitute either a full blood meal or simply probing behavior, although this may also prove infectious. Finally, whether all bites of infectious human hosts transmit the virus to vectors is unclear. In the model developed in Section V, a host to vector transmission probability of 0.5 is assumed. Conversely, the transmission probability from vector to host is likely to be 100% for all contacts (Kraiselburd *et al.*, 1985).

Some of the parameter estimates presented above are likely to be independent of locality, whereas others, such as the density of vectors or the extrinsic incubation period, will be specific to location. In any locality the basic reproductive rate, R_0, determines how likely an epidemic is: an epidemic will not occur if the value is below 1.0 because an infection cannot replace itself, but it becomes increasingly likely as the reproductive rate increases.

The probability of an epidemic being caused by one infectious host, ρ, is given by

$$\rho = 1 - \frac{1}{R_0} \tag{6}$$

When stochastic effects are accounted for, the probability of extinction of a population is given by the death rate divided by the birth rate raised to the power N, the population size. R_0 is by definition the ratio of births to deaths for a population size of 1. Hence, the probability of one dengue case initiating an epidemic is 1 minus the probability of extinction. The importance of this expression and its corollary, the probability of epidemics of two different serotypes, is pointed out by Koopman *et al.* (1991): if the probability of one epidemic is $1 - (1/R_0)$ then the probability of two epidemics occurring with the introduction of two separate individuals with different virus serotypes is this probability squared. However, the more introductions of virus, the more likely it is that epidemics will occur.

Three attempts have been made to measure the basic reproductive rate of dengue, all in epidemic rather than hyperendemic situations and all using different methods. When Newton and Reiter (1992) attempted to estimate the parameters in the expression for R_0 presented above for the population of Puerto Rico, they obtained a value of around 1.9. In Brazil, Marques *et al.* (1994) used the initial growth rate of epidemics, Λ, as measured by the doubling time (t_d) of cases to estimate R_0. In this method,

$$t_d = \frac{\ln 2}{\Lambda} \tag{7}$$

when the infectious host growth rate (Λ) is constant, and where

$$\Lambda = \frac{R_0 - 1}{D} \qquad (8)$$

This is because the basic reproductive rate is the number of new infections that each infectious host initially causes before they are removed from the population (hence the minus 1). However, it takes the duration of infectiousness D to generate these new cases. This straightforward method led to estimates of R_0 in Brazilian cities ranging from 1.6 to 2.5. Similar figures, from 1.3 to 2.4, were derived for dengue-infected populations in Mexico (Koopman et al., 1991). In this case, R_0 was derived by estimating the proportion of the population that became infected during an epidemic, f, which is a transcendental function of the basic reproductive rate

$$f = 1 - e^{R_0 f} \qquad (9)$$

when the entire population is initially susceptible.

The most significant outcome of these similar estimates of R_0 of dengue is that they are all quite low, suggesting that even when an infectious host enters the population, only on about half the occasions is there likely to be an epidemic. The low R_0 is the result of a combination of parameter values that are not conducive to the spread of dengue in nonhyperendemic regions.

Finally, unlike many protozoan vector-borne infections, dengue induces acquired immunity. Therefore, the proportion of the population susceptible to infection is reduced following an epidemic. This in turn reduces the reproductive rate because infectious vector bites are wasted on immune hosts. In such circumstances, especially in small populations, the reproductive rate may fall below the critical value of 1.0, in which case the infection has to be maintained by reintroduction from other infected populations or host species, and through vertical transmission.

IV. THE MOLECULAR EVOLUTION OF DENGUE

Although dengue viruses are often difficult to isolate and propagate, the techniques of molecular virology and evolutionary biology have allowed us to conduct detailed studies of the causes and consequences of genetic variation in this virus. The first observation to be made is that a large amount of genetic variation exists within individual serotypes, where phylogenetic analysis has revealed distinct clusters, or genotypes, of viruses, some of which may differ in the likelihood that they cause DHF/DSS (Trent et al., 1989; Rico-Hesse, 1990; Deubel et al., 1993; Lewis et al., 1993; Lanciotti et al., 1994). Although the number of dengue strains from which sequence data are available is still sparse, especially from Africa, to date four genotypes have been found in DEN-2 and DEN-3 and two in DEN-1 and DEN-4, with a maximum intragenotype amino acid diversity of approximately 10% in the E gene.

The geographic distribution of dengue genotypes provides a number of clues as to the spread of the virus across the tropical world. An example is provided by DEN-2 for which most sequence data are available (Fig. 3; see Rico-

Hesse, 1990; Rico-Hesse *et al.*, 1997): genotype I is found in Asia, the Pacific, and most recently the Americas where it has been associated with severe epidemics of DHF (see Section II); genotype II is found in Asia and Africa; genotype III in Asia, the South Pacific, and the Americas; and genotype IV is found in a sylvatic cycle involving monkeys in West Africa. The fact that the same genotypes are found in diverse geographic locations, and that the sylvatic viruses constitute a distinct genotype, suggests that dengue epidemics are often caused by the importation of a new variant into a population that has no previous immunity to it, rather than by evolution from indigenous strains, some of which may be maintained in a sylvatic cycle.

An even more striking observation with respect to the genetic variation in dengue virus is that it appears to be increasing (Zanotto *et al.*, 1996). Although historical samples are not available in sufficient quantities to directly measure the change in viral diversity through time, it is possible to take a retrospective approach and reconstruct molecular phylogenies of dengue. These trees are distinctive in that most of the branching (lineage-splitting) activity has occurred in the recent past, near the tips of the tree. This suggests that there has been a massive increase in dengue genetic diversity (population size) in the recent past (Fig. 5). This change in branching pattern can be assessed more directly by plotting the number of lineages against the time at which they appear in the tree.

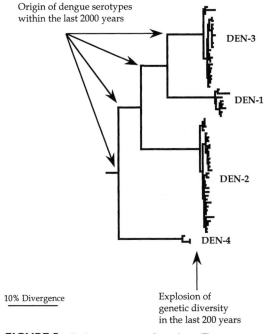

FIGURE 5   Phylogenetic tree of envelope (E) gene sequences from the four dengue serotypes. The tree was constructed using a maximum likelihood method (Felsenstein, 1993) on first and second codon positions (third positions are saturated with multiple substitutions) from 66 worldwide isolates of dengue and rooted by isolates from Japanese encephalitis virus. The points at which the serotypes diverged from each other and where the virus greatly increased in genetic diversity are also indicated. All horizontal branch lengths are drawn to scale.

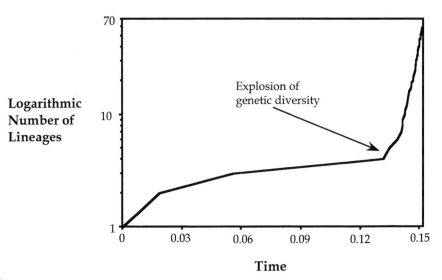

Explosion of genetic diversity

FIGURE 6   Plot of the logarithmic number of dengue lineages against the time at which they appear in a constant rate (KITSCH) phylogenetic tree. Time is scaled as the number of nucleotide substitutions from the root of the tree to the tips. The point at which dengue virus rapidly increases in genetic diversity is also indicated. The sequence data analyzed were also used to reconstruct the phylogenetic tree presented in Fig. 5.

Such an analysis reveals that dengue is characterized by two distinct periods of population history, in which a slow growth phase (where there is a low rate of linage splitting) is followed by the rapid growth phase near the present (Fig. 6). This explosion in the genetic diversity of dengue virus is especially striking when compared to other human pathogens: a similar analysis of the branching structure of trees of human immunodeficiency virus (HIV), for example, shows that this virus has been increasing in population size at an exponentially constant rate (Holmes *et al.*, 1995).

The simplest explanation for this change in transmission dynamics is that dengue was a low prevalence virus until the recent availability of new and susceptible human hosts, particularly within an urban environment, which initiated a massive viral population growth. This process was exacerbated by the increasing worldwide movement and mixing of human populations. Lineages of the virus which arose before this time had a lower chance of propagation because they would probably cause a localized epidemic in a small population and then die out because of a lack of susceptible hosts and vectors (Zanotto *et al.*, 1996).

As well as enabling us to reconstruct the population history of dengue virus, sequence analysis can also provide a time scale of its evolution. This is possible because dates of isolation are known for some dengue strains. This information can then be used to estimate rates of nucleotide substitution and to back-calculate the times of divergence of other strains. Applying this method to non-synonymous (amino acid changing) substitutions, Zanotto *et al.* (1996) estimated that the four serotypes of dengue arose within the last 2000 years and that the rapid increase in viral population size—the explosion in genetic diversity—occurred about 200 years ago, which corresponds well with what we know of the emergence of dengue from historical records. In other words, it was not until the last 200 years or so that lineages of dengue virus would be able to maintain a sustainable transmission network and so persist. Further evidence for this is that the dramatic expansion in dengue population size, reflected in the increased number of viral lineages, occurred at about the same time as an acceleration in the growth rate of the human population, thereby providing more susceptible hosts (Fig. 7).

Investigations into the molecular evolution of dengue have given us a new perspective into the origin and spread of this virus. Instead of thinking of a single point in time for the emergence of dengue we should in reality recognize three different stages. The first is the origin of the virus itself, that is, the point at which it diverged from the other flaviviruses. The molecular estimates of this divergence, which are the only source of evidence, place this at around 2000 years ago. The second period of emergence was when the virus first began to sustain itself in human populations, perhaps when it changed from being primarily a monkey (sylvan) disease to a human (urban) disease. Using both molecular methods and historical records we can date this to the end of the eighteenth

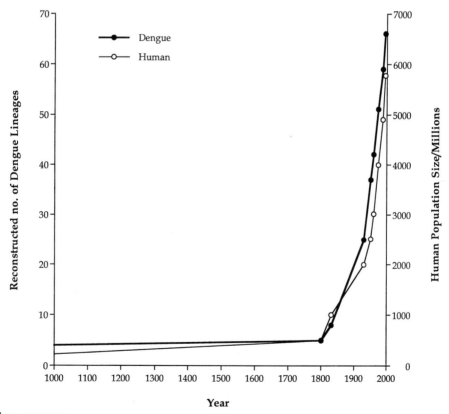

FIGURE 7 Association between the logarithmic number of lineages of dengue virus and the time they appear assuming a molecular clock (see Fig. 6) and the corresponding growth of the world's human population. From Zanotto et al. (1996) *Proc. Natl. Acad. Sci. USA* **93**, 548–553, with permission.

century, at which time human populations became both large and mobile. Finally, we can think of a third critical point in the emergence of dengue which corresponds to the recent rise of dengue hemorrhagic fever and dengue shock syndrome. These new diseases were first recognized on a large scale during the mid-1950s, again being caused by changes in human ecology and particularly increased urbanization and the widespread movement of populations.

V. THE FUTURE OF DENGUE

A. Will Dengue Spread to More Temperate Regions?

So far we have discussed the factors which have contributed to the emergence of dengue. With this knowledge in hand it is also possible to make some tentative predictions as to the future of this infectious disease. In particular, there has been concern that *Aedes albopictus,* which was introduced into the United States in truck tires imported from Asia in the mid-1980s, may become a major vector for dengue in the Americas. *Aedes albopictus* is a rural Asian mosquito which has adapted to urban environments and appears to be replacing *A. aegypti* in U.S. cities (Monath, 1993). More significantly, *A. albopictus* is able to tolerate colder temperatures than *A. aegypti* and may therefore be able to spread dengue into more temperate regions (Rosen *et al.,* 1983). Although some strains of *A. aegypti* are well adapted to temperate conditions, most prefer tropical and subtropical regions, and epidemic transmission of dengue by this species is uncommon in temperatures below 20°C. In the United States, for example, *A. aegypti*-transmitted dengue appears to be limited to a line corresponding to the 10°C winter isotherm (Fig. 1), found at a latitude of approximately 35° N (Shope, 1991). *Aedes albopictus,* however, has spread a further 7° N because it is able to suspend the hatching of eggs during cold weather through diapause, a physiological condition akin to hibernation in mammals. Fortunately, even though dengue virus has been isolated from *A. albopictus* in Brazil, this species has not yet been implicated in epidemics of disease in this region (Monath, 1994).

Whether dengue can spread through the United States, especially the southern states, because of the growing density of *A. albopictus* critically depends on whether $R_0 > 1$. With approximately one-eighth of all mosquitoes dying each day, only a fraction will become infectious. The lower temperatures found as one moves north will reduce this infectious fraction even further. In combination with lower vector densities and feeding rates, the reproductive rate of the virus will fall as it moves out of the hotter regions of the world.

The impact of mean temperature on dengue transmission has been shown in an epidemiological study in Mexico. In a logistic regression of variables influencing whether communities had experienced a dengue epidemic, Koopman *et al.* (1991) found that the two significant factors were average ambient temperature (related to the altitude at which the communities lived) and the number of standing water containers. Specifically, it was found that an increase in the average temperature of 3°–4°C may double the reproductive rate of the virus (Koopman *et al.,* 1991).

It is also possible to illustrate the impact of temperature on the likelihood of a dengue epidemic using our model of transmission dynamics described in Section III. Specifically, temperature influences the rate of progression from infection to infectiousness in vectors (the extrinsic reproductive rate). The proportion of mosquitoes surviving to become infectious under two models of progression is illustrated in Fig. 8a. Here we assume that mosquito survival is similar to that observed in Thailand (0.875 per day). One model assumes a constant rate of progression [given in Eq.(4)], while in the other mosquitoes become infectious after a fixed period [Eq. (5)]. The latter model is more biologically plausible and means that a smaller fraction will become infectious. These results can then be shown to affect the basic reproductive rate, R_0 (Fig. 8b), which is shown as a function of the vector per host ratio at two temperatures (30°C and 32° − 35°C) for both models of progression. It should be noted that the lowest reproductive rate still requires a high temperature, 30°C, and that it is assumed that humans are the main host, as may be the case for *A. aegypti*. If *A. albopictus* bites other host species, then R_0 will be greatly reduced. Finally, the basic reproductive rate can be translated into the likelihood of a dengue epidemic should an infectious human enter the population (Fig. 8c). Clearly, the greater the reproductive rate (which is a product of vector density) and the more introductions, then the greater is the chance of a dengue epidemic. Our assumptions about vector survival and association with humans are quite pessimistic, but under them it is clear that if vectors are not controlled epidemic dengue could be a problem. However, given the low reproductive rates observed in Brazil and Mexico, it seems unlikely to us that such high values of R_0 will be attained in the United States, so that the virus is unlikely to gain a foothold in this country given current climatic conditions, although a number of relevant parameter values are unknown.

The impact of temperature on the current epidemiology of dengue makes the historical record of dengue epidemics as far north as Philadelphia (presuming that they are indeed dengue) difficult to interpret. A number of hypotheses present themselves. First, the epidemics were due to infectious hosts entering the area and causing a declining but detectable number of cases. This explanation seems unlikely, however, given the short duration of infectiousness. The likelihood of very many infectious individuals entering the population simultaneously seems small even if the infection were perpetuated on vessels transporting slaves. Second, the dengue virus at this time was biologically different, enabling it to either replicate more rapidly in the vector, remain infectious in the human host for longer, or be more transmissible than the strains circulating today. Although this is possible, it begs the question of why such a strain of virus, which would be able to outcompete current dengue strains, is not found in tropical regions today. Third, there were high densities of mosquitoes in the summer when the rate of virus transmission was very high. Therefore, even though very few of the infected mosquitoes became infectious, the overall numbers were so large that this reduction did not matter and a dengue epidemic ensued. Given the infrastructure of New World cities of the period this last hypothesis presents itself as perhaps the most likely explanation.

As our study of the prospects of dengue in North America has taught us,

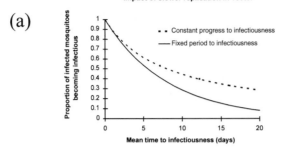

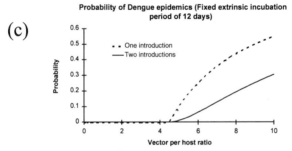

FIGURE 8   (a) Proportion of vectors becoming infectious as a function of the mean period from infection to infectiousness (the extrinsic incubation period), assuming a probability of survival of 0.875 per day. Two rates of progression were used: a constant rate of progression to infectiousness [Eq. (4)] and a fixed rate of progression [Eq. (5)]. (b) Relationship between the numbers of vectors per host and the basic reproductive rate of dengue [Eq. (3)] for two temperature-specific rates of progression from infection to infectiousness in the vector. The temperatures used are those for which empirical data on the progression to infectiousness are available, and even the lower value of 30°C is not common throughout the United States. In the fixed rate progression model, temperature has more influence in reducing the basic reproductive rate. Other parameter values are presented in Table I. The numbers of vectors per host required for a dengue epidemic are not high, but if lower temperatures reduce biting rates or the vector survival rates, then the reproductive rate will also be lowered. The biting rate incorporated, 0.25 bites per day, was that used by Newton and Reiter (1992) and may also be too high. (c) Assuming the slower, fixed rate progression to infectiousness, the probability of an epidemic with the introduction of one infectious host is illustrated for different vector to host densities [Eq. (6)]. This is squared to give the probability of two heterologous epidemics on the introduction of two hosts with different serotypes.

climatic temperature is a major factor determining the extent of dengue transmission. Another climatic factor which influences the extent of dengue infection is rainfall. Although dengue transmission occurs throughout the year in endemic tropical areas, there is a seasonal pattern in most countries. Outbreaks occur mainly in the wet season and after heavy rainfall, as this increases the number of breeding places for mosquitoes. Thus, median temperature during the rainy season has been shown to be a strong predictor of the extent of dengue infection (Koopman *et al.*, 1991). The association between temperature, rainfall, and the level of dengue infection has also be observed on a broader geographic scale through the Southern Oscillation Index (SOI) of El Niño, which has a major influence on the climate of the South Pacific (Hales *et al.*, 1996). Dengue epidemics appear to be correlated with positive values of SOI, when both temperature and rainfall are high. It is therefore possible that although the conditions for the transmission of dengue in the temperate world are not in place today, a future increase in the world's temperature due to global warming may allow this virus, along with other mosquito-borne infections, to greatly extend their geographic range.

B. What Are the Implications of the Increasing Genetic Diversity of Dengue Virus?

Whatever its cause, the expanding genetic variation of dengue virus is an important observation because it means that human populations are being exposed to increasingly diverse viral strains, some of which may evade immunity in previously exposed populations. In addition, viruses with an increased transmission potential or increased pathogenicity may eventually emerge, as may have already happened within DEN-2. The sylvatic strains of dengue, which currently appear to be genetically isolated from those implicated in human epidemics, may also serve as a reservoir for viruses with altered characteristics. Of most concern is that if the genetic diversity of dengue is indeed tied to the growth of the human host population, then genetic flexibility of dengue will only increase in the future. This is an evolutionary process which may yet produce a virus capable of causing large-scale epidemics in the temperate world.

Another outcome of the genetic diversity seen in dengue virus is that severe dengue disease might be caused by antibody-dependent enhancement of genetically divergent viruses. This also has a number of important epidemiological consequences. On the plus side, ADE may mean that two epidemics, rather than a single one, are necessary before serious disease occurs. In the arena of vaccination, however, ADE has more unwanted consequences. It may be necessary to produce a quadrivalent vaccine which is cross-protective against all four serotypes; otherwise, vaccination might only serve to increase the incidence of DHF/DSS. However, even a quadrivalent vaccine raises concerns about the nature of vaccine failure and whether mass vaccination would select for new dengue serotypes that could cause severe disease in those infected. An alternative approach has been to produce genetically engineered resistance to dengue infection in mosquitoes (Olson *et al.*, 1996). Despite progress in this area, the current lack of an effective vaccine or practical availability of immunized mosquitoes means that vector control measures, which had until the last 20 years eradicated *Aedes aegypti* from the Americas, are likely to remain the most efficient way to control the spread of dengue.

Finally, we can speculate as to the long-term evolutionary pressures on dengue. Normally in viral epidemics, previous experience in the population of cross-reactive viral antigens will select for dissimilar viral strains. For example, antigenic changes in influenza virus allow it to escape from herd immunity (Holmes and Garnett, 1994). However, because of the complicating effects of ADE, natural selection in the case of dengue could act *against* very similar types of virus, *for* moderately similar types, and be *neutral* for unrelated viral types. This would work if a raised viremia increased the probability of transmission from an infected human to a biting vector or if ADE increased the duration of viremia. Because DHF and DSS do not seem to involve a longer period when virus can be isolated (Gubler *et al.,* 1981), an enhanced transmission probability is a more likely mechanism, although its potential impact is limited as this probability cannot exceed unity in value. Unfortunately, the transmission probability from human to vector is the one parameter that has not been estimated empirically. In evolutionary terms, selection for moderately similar strains may be able to structure the viral population into distinct types which enhance viremia, with an optimal difference between them. Additionally, we may expect to see sequential epidemics of different strains of virus in hyperendemic areas where the earlier presence of one strain enhanced the reproductive rate of another. The seasonal forcing of epidemics by changes in temperature may structure such epidemics. Finally, the increasing genetic diversity of dengue may mean that viruses belonging to the same "serotype" will eventually be able to enhance infection with one another.

Whatever the future of dengue, the dynamics of the evolutionary process, particularly the potential for dengue virus and its vectors to adapt to new and changing environments, requires that we follow the progress of dengue epidemics with increasing vigilance.

ACKNOWLEDGMENTS

We thank Dr. E. A. Gould for helpful comments and The Wellcome Trust and The Royal Society for financial support.

REFERENCES

Anderson, R. M., and May, R. M. (1991). "Infectious Diseases of Humans: Dynamics and Control." Oxford Univ. Press, Oxford.

Blok, J., McWilliam, S. M., Butler, H. C., Gibbs, A. J., Weiller, G., Herring, B. L., Hemsley, A. C., Aaskov, J. G., Yoksan, S., and Bhamarapravati, N. (1992). Comparison of Dengue-2 virus and its candidate vaccine derivative: Sequence relationships with the flaviviruses and other viruses. *Virology* **187,** 573–590.

Burke, D. S., Nisalak, A., Johnson, D. E., and Scott, R. M. (1988). A prospective study of dengue infections in Bangkok. *Am. J. Trop. Med. Hyg.* **38,** 172–180.

Calisher, C. H., Karabatsos, N., Dalrymple, J. M., Shope, R. E., Porterfield, J., Westaway, E. G., and Brandt, W. E. (1989). Antigenic relationships between flaviviruses as determined by cross-neutralization tests with polyclonal antisera. *J. Gen. Virol.* **70**, 37–43.

Chambers, T. J., Hahn, C. S., Galler, R., and Rice, C. M. (1990). Flavivirus genome organization, expression, and replication. *Annu. Rev. Microbiol.* **44**, 649–688.

Chanas, A. C., Gould, E. A., Clegg, J. C. S., and Varma, M. G. R. (1982). Monoclonal antibodies to Sindbis virus glycoprotein E1 can neutralise, enhance infectivity and independently inhibit haemagglutination or haemolysis. *J. Gen. Virol.* **58**, 37–46.

Chiewslip, P., Scott, R. M., and Bhamarapravati, N. (1981). Histocompatibility antigens and dengue hemorrhagic fever. *Am. J. Trop. Med. Hyg.* **30**, 1100–1105.

Clements, A. N., and Paterson, G. D. (1981). The analysis of mortality and survival rates in wild populations of mosquitoes. *J. Appl. Ecol.* **18**, 373–399.

Deubel, V., Nogueira, R. M., Drouet, M. T., Zeller, H., Reynes, J. M., and Ha, D. Q. (1993). Direct sequencing of genomic cDNA fragments amplified by the polymerase chain reaction for molecular epidemiology of dengue-2 viruses. *Arch. Virol.* **129**, 197–210.

Fanger, M. W., and Erbe, D. V. (1992). Fc receptors in cancer and infectious disease. *Immunol. Res.* **11**, 203–216.

Felsenstein, J. (1993). PHYLIP (Phylogeny Inference Package) Version 3.5c. Distributed by the author. Department of Genetics, University of Washington, Seattle.

Gubler, D. J. (1988). Dengue. *In* "The Flaviviruses: Evolution and Ecology" (T. P. Monath, ed.), Vol. 2, pp. 223–260. CRC Press, Boca Raton, Florida.

Gubler, D. J., and Clark, G. G. (1995). Dengue/Dengue Hemorrhagic Fever: The emergence of a global health problem. *Emerging Infect. Dis.* **1**, 55–57.

Gubler, D. J., and Rosen, L. (1976). A simple technique for demonstrating transmission of dengue virus by mosquitoes without the use of vertebrate hosts. *Am. J. Trop. Med. Hyg.* **25**, 146–150.

Gubler, D. J., Suharyono, W., Lubis, I., Eram, S., and Sunarso, S. (1981). Epidemic dengue 3 in Central Java, associated with low viremia in man. *Am. J. Trop. Med. Hyg.* **30**, 1094–1099.

Guzman, M. G., Kouri, G. P., Bravo, J., Calunga, M., Soler, M., Vazquez, S., and Venereo, C. (1984). Dengue hemorrhagic fever in Cuba. I. Serological confirmation of clinical diagnosis. *Trans. R. Soc. Trop. Med. Hyg.* **78**, 235–238.

Hales, S., Weinstein, P., and Woodward, A. (1996). Dengue fever epidemics in the South Pacific: Driven by El Niño Southern Oscillation? *Lancet* **348**, 1664–1665.

Halstead, S. B. (1988). Pathogenesis of dengue: Challenges to molecular biology. *Science* **239**, 476–481.

Halstead, S. B., and O'Rourke, E. J. (1977). Dengue viruses and mononuclear phagocytes. I. Infection enhancement by non-neutralizing antibody. *J. Exp. Med.* **146**, 201–217.

Halstead, S. B., Nimmanitya, S., Yammat, C., and Russell, P. K. (1967). Hemorrhagic fever in Thailand, newer knowledge regarding etiology. *Jpn. J. Biol. Med.* **20**, 96–103.

Halstead, S. B., Shotwell, H., and Casals, J. (1973). Studies on the pathogenesis of dengue infections in monkeys. I. Clinical laboratory responses to primary infection. *J. Infect. Dis.* **128**, 7–16.

Halstead, S. B., Porterfield, J. S., and O'Rourke, E. J. (1980). Enhancement of dengue virus infection in monocytes by flavivirus antisera. *Am. J. Trop. Med. Hyg.* **29**, 638–642.

Holmes, E. C., and Garnett, G. P. (1994). Genes, trees and infections: Molecular evidence in epidemiology. *Trends Ecol. Evol.* **9**, 256–260.

Holmes, E. C., Nee, S., Rambaut, A., Garnett, G. P., and Harvey, P. H. (1995). Revealing the history of infectious disease epidemics using phylogenetic trees. *Philos. Trans. R. Soc. London Ser. B* **349**, 33–40.

Innis, B. L., Eckels, K. H., Kraiselburd, E., Dubois, D. R., Meadors, G. F., Gubler, D. J., Burke, D. S., and Bancroft, W. H. (1988). Virulence of a live dengue virus vaccine candidate: A possible new marker of dengue virus attenuation. *J. Infect. Dis.* **154**, 876–880.

Kliks, S. C., Nimmanitya, S., Nisalak, A., and Burke, D. S. (1988). Evidence that maternal dengue antibodies are important in the development of dengue hemorrhagic fever in infants. *Am. J. Trop. Med. Hyg.* **38**, 411–419.

Kliks, S. C., Nisalak, A., Brandt, W. E., Wahl, L., and Burke, D. S. (1989). Antibody-dependent enhancement of dengue virus growth in human monocytes as a risk factor for dengue hemorrhagic fever. *Am. J. Trop. Med. Hyg.* **40**, 444–451.

Klowden, M. J., and O'Lea, A. (1978). Blood meal size as a factor affecting continued host-seeking by *Aedes aegypti* (L.). *Am. J. Trop. Med. Hyg.* **27**, 827–831.

Koopman, J. S., Prevots, R. D., Marin, M. A. V., Dantes, H. G., Aquino, M. L. Z., Longini, I. M., and Amor, J. S. (1991). Determinants and predictors of dengue infections in Mexico. *Am. J. Epidemiol.* **133**, 1168–1178.

Kraiselburd, E. N., Gubler, D. J., and Kessler, M. J. (1985). Quantity of dengue virus required to infect rhesus monkeys. *Trans. R. Soc. Trop. Med. Hyg.* **79**, 248–251.

Kurane, I., and Ennis, F. A. (1992). Immunity and immunopathology in dengue virus infections. *Semin. Immunol.* **4**, 121–127.

Kurane, I., and Ennis, F. A. (1994). Cytokines in dengue virus infections: Role of cytokines in the pathogenesis of dengue hemorrhagic fever. *Semin. Virol.* **5**, 443–448.

Lanciotti, R. S., Lewis, J. G., Gubler, D. J., and Trent, D. W. (1994). Molecular evolution and epidemiology of dengue-3 viruses. *J. Virol.* **75**, 65–75.

Lehane, M. J. (1991). "Biology of Blood Sucking Insects." Harper Collins Academic, London.

Lewis, J. A., Chang, G.-J., Lanciotti, R. S., Kinney, R. M., Mayer, L. W., and Trent, D. W. (1993). Phylogenetic relationships of dengue-2 viruses. *Virology* **197**, 216–224.

Loke, H., Bethell, D., Day, N., White, N. J., and Hill, A. V. S. (1996). A TNF promoter variant, HLA and dengue in Vietman. *Hum. Immunol.* **47**, 680.

Macdonald, G. (1957). "The Epidemiology and Control of Malaria." Oxford Univ. Press, Oxford.

Mcdonald, P. T. (1977). Population characteristics of domestic *Aedes aegypti* (Diptera: Culicidae) in villages on the Kenya coast. II. Dispersal within and between villages. *J. Med. Entomol.* **14**, 49–53.

McLean, D. M., Clarke, A. M., Coleman, J. C., Montalbetti, C. A., Skidmore, A. G., Walters, T. E., and Wise, R. (1974). Vector capability of *Aedes aegypti* mosquitoes for California encephalitis and dengue viruses at various temperatures. *Can. J. Microbiol.* **20**, 255–262.

McSherry, J. (1993). Dengue. *In* "The Cambridge World History of Human Disease" (K. F. Kiple, ed.), pp. 660–664. Cambridge Univ. Press, Cambridge.

Marques, C. A., Forattini, O. P., and Massad, E. (1994). The basic reproductive number for dengue fever in São Paolo state, Brazil: 1990–1991 epidemic. *Trans. R. Soc. Trop. Med. Hyg.* **88**, 88–89.

Monath, T. P. (1993). Arthropod-borne viruses. *In* "Emerging Viruses" (S. S. Morse, ed.), pp. 138–148. Oxford Univ. Press, Oxford.

Monath, T. P. (1994). Dengue: The risk to developed and developing countries. *Proc. Natl. Acad. Sci. U.S.A.* **85**, 7627–7631.

Monath, T. P., and Heinz, F. X. (1996). Flaviviruses. *In* "Fields Virology 3rd Ed." (B. N. Fields, D. M. Knipe, P. M. Howley, R. M. Chanock, J. L. Melnick, T. P. Monath, B. Roizman, and S. E. Straus, eds.), pp. 961–1034. Lippincott-Raven, Philadelphia.

Newton, E. A. C., and Reiter, P. (1992). A model of the transmission of dengue fever with an evaluation of the impact of ultra-low volume (ULV) insecticide applications on dengue epidemics. *Am. J. Trop. Med. Hyg.* **47**, 709–720.

Olson, K. E., Higgs., S., Gaines, P. J., Powers, A. M., Davis, B. S., Kamrud, K. I., Carlson, J. O., Blair, C. D., and Beaty, B. J. (1996). Genetically engineered resistance to dengue-2 virus transmission in mosquitoes. *Science* **272**, 884–886.

Paradosa-Perez, M. L., Trujillo, Y., and Basanta, P. (1987). Association of dengue hemorrhagic fever with the HLA system. *Hematologia* **20**, 83–87.

Putnam, J. L., and Scott, T. W. (1995). The effect of multiple host contacts on the infectivity of dengue-2 virus infected *Aedes acypti. J. Parasitol.* **81**, 170–174.

Rico-Hesse, R. (1990). Molecular evolution and distribution of dengue viruses type 1 and type 2 in nature. *Virology* **174**, 479–493.

Rico-Hesse, R., Harrison, L. M., Salas, R. A., Torar, D., Nisalak, A., Ramos, C., Boshell, J., de Mesa, M. T. R., Nogueira, R. M. R., and da Rosa, A. T. (1997). Origins of dengue tupe 2 viruses associated with increased pathogenicity in the Americas. *Virology* **230**, 244–251.

Rodriguez-Figueroa, L., Rigau-Perez, J. G., Suarez, E. L., and Reiter, P. (1995). Risk factors for dengue infection during an outbreak in Yanes, Puerto Rico in 1991. *Am. J. Trop. Med. Hyg.* **52**, 496–502.

Rosen, L. (1977). The Emperor's New Clothes revisited, or reflections on the pathogenesis of dengue hemorrhagic fever. *Am. J. Trop. Med. Hyg.* **26**, 337–343.

Rosen, L., Schroyer, D. A., Tesh, R. B., Freier, J. E., and Lein, J. C. (1983). Transovarial transmission of dengue viruses by mosquitoes: *Aedes albopictus* and *Aedes aegypti. Am. J. Trop. Med. Hyg.* **32**, 1108–1119.

Ross, R. (1911). "The Prevention of Malaria," 2nd Ed. Murray, London.

Rudnik, A. (1965). Studies on the ecology of dengue in Malaysia: A preliminary report. *J. Med. Entomol.* **2**, 203.

Russell, P. K., and Nisalak, A. (1967). Dengue virus identification by the plaque reduction neutralization test. *J. Immunol.* **99**, 291–296.

Sheppard, P. M., MacDonald, W. W., Tonn, R. J., and Grab, B. (1969). The dynamics of an adult population of *Aedes aegypti* in relation to dengue hemorrhagic fever in Bangkok. *J. Am. Ecol.* **38**, 663–697.

Shope, R. E. (1991). Global climate change and infectious disease. *Environ. Health Perspect.* **96**, 171–174.

Siler, J. F., Hall, M. W., and Hitchens, A. P. (1926). Dengue: Its history, epidemiology, mechanisms of transmission, etiology, clinical manifestations, immunity and prevention. *Philipp. J. Sci.* **29**, 1–304.

Thein, S., Aung, M. M., Shwe, T. N., Aye, M., Zaw, A., Aye, K., Aye, K. M., and Aaskov, J. (1997). Risk factors in dengue shock syndrome. *Am. J. Trop. Med. Hyg.* **35**, 1263–1279.

Trent, D. W., Grant, J. A., Monath, T. P., Manske, C. L., Corina, M., and Fox, G. E. (1989). Genetic variation and microevolution of dengue 2 virus in Southeast Asia. *Virology* **172**, 523–535.

Trpis, M., and Hausermann, W. (1986). Dispersal and other population parameters of *Aedes aegypti* in an African village and their possible significance in epidemiology of vector-borne diseases. *Am. J. Trop. Med. Hyg.* **35**, 1263–1279.

Watts, D. M., Burke, D. S., Harrison, B. A., Whitmore, R. E., and Nisalak, A. (1987). Effect of temperature on the vector efficiency of *Aedes aegypti* for dengue 2 virus. *Am. J. Trop. Med. Hyg.* **36**, 143–152.

Yasuno, M., and Tonn, R. J. (1970). A study of biting habits of *Aedes aegypti* in Bangkok, Thailand. *Bull. WHO* **43**, 319–325.

Zanotto, P. M. de A., Gould, E. A., Gao, G. F., Harvey, P. H., and Holmes, E. C. (1996). Population dynamics of flaviviruses revealed by molecular phylogenies. *Proc. Natl. Acad. Sci. U.S.A.* **93**, 548–553.

Data from Uganda demonstrate a strong correlation between HIV infection and migration status. Among 5553 adults included in the study, HIV seroprevalence ranged from 5.5% among those who never moved to 12.4% among those who moved to a differenc village, to 16.3% among those who joined the cohort from another area during the study period (Decosas, 1992). A similar association between mobility and HIV status has been reported in Senegal. The findings suggest the hypothesis that in a situation of HIV endemicity, mobility is an independent risk factor for acquiring HIV regardless of point of origin or the destination. As discussed by Decosas *et al.* (1995), it is not necessarily the origin or the destination of migration, but the social disruption which characterizes certain types of migration, that determined vulnerability to HIV. The fact that population movements distribute HIV is secondary to the fact that certain types of migration cause HIV epidemics.

Once HIV moved from the rural area into the city, HIV spread regionally along highways, then by long-distance routes including air travel to more distant places (Quinn, 1994). This last step was critical for HIV and facilitated today's global pandemic. Social changes that allowed the virus to reach a larger population and to be transmitted were instrumental in the success of the various strains of the virus found in the human host. For HIV the long duration of infectivity (10 years or more) allowed the normally poorly transmissible virus many opportunities to be transmitted and to take advantage of cultural factors such as sexual transmission among multiple partners and spread via injecting drug use, as well as modern technology such as the early spread through blood transfusions and blood products. The concentrating effects that occur with the collection and distribution of blood products inadvertently disseminated the unrecognized pathogen HIV, before serological tests became available. Finally, microbes, like other living things, are constantly evolving (Morse, 1995). HIV has a high mutation rate and is rapidly evolving to new variants (Myers, 1994a,b). These genetic changes among HIV strains may result in differences in transmissibility, infectivity, or pathogenicity.

The early phases of emergence and dissemination influenced the next phase of escalation, which occurred during the 1980s and 1990s (Fig. 2). Transmission of HIV was amplified among many high-risk populations and spread to other populations at risk, including IDUs, heterosexual partners of infected individuals, female spouses of men who visit female sex workers, blood transfusion recipients, and eventually segments of the general population in both urban and rural settings (Quinn, 1996a). These phases of dissemination and escalation have occurred in nearly all countries to date, but they were most marked in sub-Saharan Africa and Asia. For example, this phase of escalation is currently illustrated in the densely populated region of southeast Asia, where the virus was only recently introduced and where nearly 5 million cumulative HIV infections have occurred within the last 5 years (Chin, 1995).

A fourth phase of the HIV pandemic has become more recently evident as HIV prevalence and reported AIDS cases appear to have stabilized in Australia, North America, and western Europe (Quinn, 1996a; Centers for Disease Control, 1995b,e) (Fig. 2). Although such changes might represent a positive development from a prevention perspective, they may also indicate a transition from epidemic HIV to endemic HIV infection. A stabilization in prevalence indicated that the number of deaths from AIDS equaled the number of new HIV infections

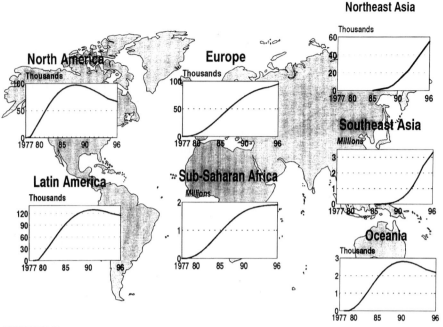

FIGURE 2   Annual rate of new HIV infections among adults by geographic region, 1977–1996. Sub-Saharan Africa and Southeast Asia figures are in millions; all others are in thousands. Data from Quinn (1996a) and Mann and Tarantola (1996).

(Centers for Disease Control, 1996a–e). In the United States in 1996 there were approximately 45,000 deaths due to AIDS and an estimated 45,000 new HIV infections. It is important to point out that stabilization may mask disproportionate increases in different modes of transmission, such as an increase in heterosexually transmitted HIV, or disproportionate increases in new HIV infections among young people as evidenced in the United States and Europe (National Research Council, 1990, 1993). Stabilization in HIV prevalence in reported cases is simply one more facet of the ever-changing global nature of emerging HIV pandemic.

Although there is a fifth phase to most epidemics, referred to as resolution, this phase has not occurred in any region of the world for HIV. While there is hope that the epidemic may be slowing in some regions and in some high-risk populations such as blood transfusion recipients, hemophiliacs, and homosexual men in developed countries, this decline is often counterbalanced by an increase in another segment of the population such as adolescents with sexually transmitted diseases (STDs), IDUs, and their sex partners (Haverkos and Quinn, 1995; National Research Council, 1993). With the current trends in transmission of HIV, it is unlikely that there will be any marked or significant declines in the next few years.

A. Dynamics of the Epidemic by Region

Within each region, each country, and each community, the HIV epidemic has established its own unique character, depending on the time of introduction of

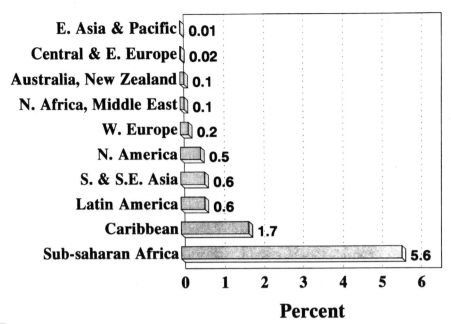

FIGURE 3 Estimated HIV prevalence rate by region, December 1996. HIV prevalence equals the number of persons living with HIV/AIDS in one specific region at a specific time divided by the total population of the region. Data from World Health Organization (1995a).

the virus, the social fabric of that community, its culture, its sexual networks, the mobility of the people, and the reaction of the government to mounting an AIDS control program. Countries in sub-Saharan Africa and the Caribbean currently have the highest national rates of adult HIV prevalence (Fig. 3). At the end of 1994 adult HIV prevalence ranged from 1 per 100,000 (0.001%) in Central Asian republics to more than 10% in five African countries (Botswana 18%; Zambia and Zimbabwe 17%; Uganda 15%; and Malawi 13%) (World Health Organization, 1995a). Part of this disparity can be attributed to the maturity of the epidemics in Africa and the more recent introduction of HIV into central and east Asia. Unfortunately, data regarding the incidence or current spread of HIV are scarce, and such information is urgently required to better estimate the future scope of the epidemic. The following sections review some of the more recent changes in the HIV epidemic for five regions of the world.

1. Sub-Saharan Africa

By the end of 1996, 18.5 million adults in sub-Saharan Africa had been infected with HIV (UNAIDS, 1996c). With nearly 5 million deaths due to AIDS, approximately 14 million adults and children are living with HIV in this region. Approximately 50 to 60% of HIV infections have been in east and central Africa, an area which accounts for only 15% of the total population of sub-Saharan Africa (UNAIDS, 1996c). In some urban populations, greater than 10% of the adult population is infected, and the annual incidence is estimated to be 3% (UNAIDS, 1996c). With high rates of heterosexual transmission, perinatal transmission, which can occur *in utero,* during delivery, or postnatally via breastfeeding, is common throughout sub-Saharan Africa. Serological surveys, particularly in urban centers of sub-Saharan Africa, have shown rates of 5 to 30% from HIV infection in pregnant women (U.S. Bureau of Census, 1994; UNAIDS, 1996c; National Research Council, 1996). In some countries the epidemic is more recent, with rates of infection increasing severalfold within only a few years, such as in Botswana and South Africa. For example, HIV prevalence in pregnant women in South Africa has increased from 4.3 to 11% in the province of Free State and from 9.6 to 18% in the province of Awazulu/Natal; overall, HIV prevalence rates on a national basis have increased from 2.4% in 1992 to 7.5% in 1994. Having estimated about 100,000 HIV infections in 1990, the country now estimates 1.2 million HIV infections in 1994 (UNAIDS, 1996c). Population mobility, patterns of sexual behavior, and societal factors are likely to influence the potential for such explosions.

One of the factors responsible for rapid spread of HIV in South Africa is the enormous migrant labor population (Quinn, 1994). Migrant laborers travel from neighboring countries to South Africa in search of work within mines and other growing industries. This migrant labor system also created a market for prostitution in mining towns and established geographic networks of relationships within and between urban and rural communities. The rapid growth of the mining industry with the migrant labor system it created led to an epidemic of STDs among these populations. Gonococcal infections and syphilis have been documented in 10 and 17% of mine workers, respectively. Women who provide migrant mine workers with sexual services also come from socially and economically marginalized groups in rural and urban areas, many of whom also have a high rate of HIV infection (>15% prevalence) and STDs, further potentiating the spread of both. These data illustrate one end of the spectrum of behavior where multiple partners and frequent partner change are common. This group may represent a "core" population involved in high-risk activity that acts as a major carrier of HIV (Anderson, 1992).

Within each country the HIV epidemics have progressed with different velocities in various population groups. Early in the evolution of the epidemic, urban populations and rural communities located along highways were more rapidly affected. Urban and trading centers continue to have a substantially higher prevalence of HIV than rural areas (Fig. 4). This pattern is by no means universal; population displacement, armed conflicts, and proximity to highways where intense migration of populations for economic reasons occurs have strongly inluenced the spread of HIV (Anderson, 1992; Decosas et al., 1995). A cluster sampling technique was used to document the difference in rural and urban HIV infection in countries of western and central Africa. The results of these studies have shown HIV prevalence rates below 1% in some rural areas with infection rates in urban centers considerably higher. In Rwanda the HIV seroprevalence was 17.8% in an urban sample of 1870 people and 1.3% in a rural sample of 742 people (Rwandan HIV Seroprevalence Study Group, 1989). However, more recent data suggest that significant rates of infection are now occurring in rural areas due to return migration and other factors. Some rural communities in Kenya, Tanzania, and Uganda now have infection rates similar to those observed in neighboring urban populations. In other countries, perhaps with poorer transport networks, this has not been the case.

In terms of further HIV spread, the consequences of contemporary civil and political unrest and subsequent population displacement, such as those cur-

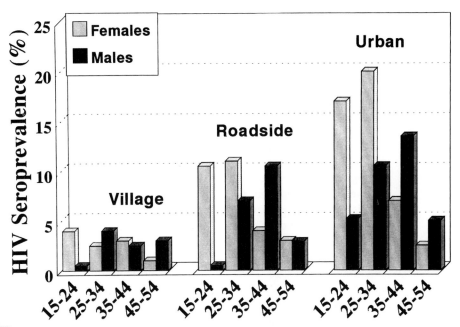

FIGURE 4 HIV seroprevalence for Mwanza, Tanzania, by age, sex, and residence, 1990–1991. Data from U.S. Bureau of Census (1994).

rently taking place in the Great Lakes region of Africa, are particularly disturbing (UNAIDS, 1996c). Rwanda and Burundi, for example, already have one of the oldest and most severe HIV epidemics in Africa. HIV seroprevalence rates of over 20% among pregnant women and 50% among STD clinic patients are common. With migration and displacement, it is expected that these rates may have only increased. In contrast, Zaire to date has experienced a relatively stable HIV prevalence. HIV prevalence in Kinshasa has continued to be 5% in pregnant women attending antenatal clinics. However, taking into account the current political events and the fact that Zaire with a population of 50 million is one of the largest African countries, its HIV situation may change rapidly with unpredictable and disastrous consequences (UNAIDS, 1996c; Institute of Medicine, 1992; Ryder and Hassig, 1998).

2. Asia

Because of the relatively recent introduction of HIV into the region, Asia is still in the early phases of an explosive HIV epidemic. As Asia has more than 60% of the worlds' population, the HIV epidemic could affect more people in this region than in any other area of the world (Brown and Xenos, 1994; Chin, 1995). Although HIV was introduced into this region in the late 1980s, there are already 4.7 million adults living with HIV, for a prevalence of 0.5% (World Health Organization, 1995) (Fig. 3). India and Thailand have 80% of the region's HIV infections, but HIV is rapidly spreading to neighboring countries including Vietnam, China, Cambodia, Indonesia, Malaysia, and Myanmar (Burma). Countries such as the Philippines, Singapore, and the Republic of Korea have had only limited spread to date, and the rate of growth appears to be substantially lower (UNAIDS, 1996a,b).

The pattern of HIV spread in Asia appears to be different from that described in other regions. HIV was initially introduced among IDUs in Thailand and in Manipur, northeastern India, where prevalence rose from 0 to greater than 55% in 4 years (Brown et al., 1994; Kitayaporn et al., 1994; Naik et al., 1991; Weniger and Brown, 1996). Data in the Yunnan province of China bordering Burma and Laos and considered part of the "Golden Triangle" of heroin exportation demonstrate an alarming HIV prevalence of 43 to 82% in IDUs (Cheng et al., 1994). Data from Malaysia and Vietnam show similar increases in HIV levels among IDUs. Following the rise of HIV infection among IDUs, HIV infection was noted among female sex workers. Although highly variable by region, HIV prevalences of 30 to 65% have been reported among female sex workers in various cities of Thailand and India (Bollinger et al., 1995). Transmission from these sex workers to their male clients, and subsequently to other sex partners including spouses, resulted in rapid spread of HIV to the general population. Among military recruits in Thailand in 1993, HIV prevalence was 4% overall and 12.4% in recruits from the northern province of Chiang Mai (Nelson et al., 1993; Celentano et al., 1996). Among pregnant women, HIV prevalence rose to 8% in Chiang Mai and Chiang Rai in northern Thailand, and the overall prevalence for pregnant women in the country was estimated at 2% (Thai Ministry of Public Health, 1995). In 1996 it was estimated that more than 1 million people in Thailand or 2% of the Thai population were HIV infected (UNAIDS, 1996a,b). If this rate of transmission continues, there will be 2 to 4 million cumulative HIV infections in Thailand by the year 2000. Some prevention efforts have, however, been taking effect. In a recent study HIV infection levels in military conscripts dropped from 3.6% in 1993 to 2.5% in 1995 (Nelson et al., 1996).

The HIV epidemic has been particularly explosive in India. With an estimated 3.5 million infected people, India is the country with the largest number of HIV-infected adults in the world, even though the adult prevalence has not reached 1%. In Bombay, the prevalence went from 2% in STD clinic attendees in 1990 to 36% in 1994 (UNAIDS, 1996a,b). HIV prevalence in sex workers rose from 1 to 51% in a 5-year period, and 2% of women attending antenatal clinics were HIV positive in 1994. In Pune the HIV seroprevalence in STD attendees was 23.4%, and the incidence was 26.1 per per 100 person-years of observation for female sex workers, 9.4% for men, and 8.4% for women who were spouses of men attending these clinics (Rodrigues et al., 1995; Mehendale et al., 1995). As in Africa, recurrent genital ulcer disease, urethritis, or cervicitis was independently associated with increased risk of seroconversion.

There is still great geographic variation of HIV infection in India. However, given the prevailing sex practices, a large population of HIV-infected female sex workers, the low social status of women, male patronage of sex workers, high rates of STDs, low rates of condom use, and the high frequency of IDUs in the north, it is likely that HIV will continue to spread in India, Thailand, and other neighboring countries (Quinn, 1996b; Chin, 1995). Similar to Africa the HIV epidemic is strongly associated with an increasing spread of tuberculosis (World Health Organization, 1995c; Harries, 1990; DeCock et al., 1992). In India, an estimated 1 to 2 million cases of tuberculosis occur every year. In Bombay, 10% of the patients presenting with tuberculosis are now HIV positive. In Thailand, public health authorities have also noted a marked increase in tuberculosis cases compared to a steady decline in previous years (Swasdisevi, 1994).

3. Oceania

Approximately 50,000 cumulative adult HIV infections have occurred in the Pacific region, with nearly half in Australia and New Zealand. The annual reported number of HIV infections in these latter two countries seems to have reached its peak, and since 1990 there has been an actual downward trend for both countries in the number of HIV infections reported each year (World Health Organization, 1995b). Most of these infections have occurred among homosexual men, and the gender ratio among infected individuals is 7:1, indicating a lower degree of heterosexual transmission than observed in other regions. The frequency of HIV infection among IDUs in Australia and New Zealand also remains lower than in western Europe and North America due to the early availability of sterile injection equipment (UNAIDS, 1996a,b).

4. The Americas

Approximately 3 million cumulative HIV infections have occurred in the countries of the Americas, with over 1 million in North America and 2 million in Latin America and the Caribbean. The United States has had the highest number of reported AIDS cases in the world, with over 560,000 cases and nearly 340,000 fatalities. AIDS became the leading cause of death in the United States among men aged 25 to 44 and the third leading cause of death among women in the same age grup, accounting for 23 and 11% of deaths, respectively (Centers for Disease Control, 1996a–e) (Figs. 5 and 6).

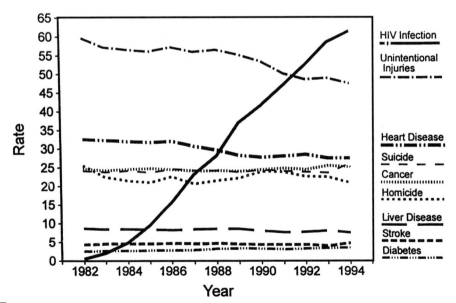

FIGURE 5 Death rates from leading causes of death among men aged 25 to 44 years, by year for the United States, 1982–1994. Data are expressed per 100,000 population. Data from Centers for Disease Control (1996b).

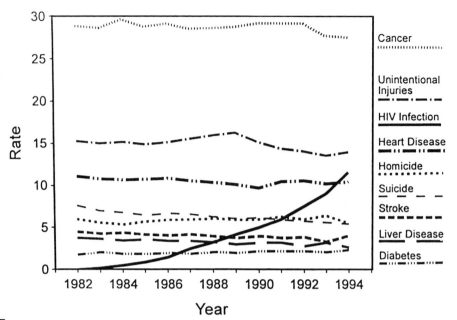

FIGURE 6   Death rates from leading causes of death among women aged 25 to 44 years, by year for the United States, 1982–1994. Data are expressed per 100,000 population. Data from Centers for Disease Control (1996a).

An evolving pattern in the epidemiology of HIV infection has been evident in the countries of the Americas. In the United States, the largest decline in the proportion of reported AIDS cases occurred among homosexual/bisexual men, while heterosexual contact had the greatest increase (Centers for Disease Control, 1995a,b; Rosenberg, 1995). Women, African Americans, Hispanics, and persons in the South and Northeast accounted for the highest percentage of increase in reported cases during 1996 (Centers for Disease Control, 1996a–e). Women represented 20% of the reported cases in 1996, nearly fourfold greater than the proportion reported in 1985. This increase was paralleled by an increase in HIV infection in pregnant women. In a survey of newborn infants, the average seroprevalence estimate for childbearing women was 1.7 per 1000 in 1992 (Davis *et al.*, 1995; Karon *et al.*, 1996). The highest prevalence was in the east and southeast regions of the United States, where gonorrhea and syphilis rates are also greater than in other areas of the country (Rogers *et al.*, 1994). This association between increasing HIV infection rates among women and association with STDs suggests further heterosexual transmission.

The epidemic in children under 13 years of age is closely associated with the epidemic in women. An estimated 7000 HIV-infected women delivered infants in the United States during 1993 (Centers for Disease Control, 1994). Assuming a perinatal transmission rate of 15 to 30%, approximately 1000 to 2000 infants were perinatally infected with HIV infection in 1993. HIV infection has become a leading cause of death for young children, currently ranked as seventh for children 1 to 4 years of age (National Center for Health Statistics,

1994). More recent trends in AIDS incidence among children have demonstrated a decline due to the current U.S. Public Health Service recommendations for routine counseling, voluntary prenatal testing for women, and the use of zidovudine to prevent perinatal transmission (Centers for Disease Control, 1996a–e).

For Latin America and the Caribbean the HIV pandemic is also continuing to evolve. Although the rate of spread of HIV has been slower there than in other developing regions of the world, the pandemic is well established and there is wide variation in the level of HIV infection. The dominant modes of transmission vary from one country to the next, ranging from epidemics that are predominantly homosexual and bisexual to epidemics associated with injecting drug use and others that are primarily associated with heterosexual transmission (Quinn *et al.*, 1989). Nevertheless, sexual transmission overall accounts for 80% of infections in the area, ranging from 64% in Brazil to as high as 93% in the Andean subregion (Bolivia, Colombia, Ecuador, Peru, Venezuela) (UNAIDS, 1996a,b; Pan American Health Organizaton, 1996).

Some countries are at particular risk for rapid dissemination of HIV from traditional at-risk groups to other vulnerable groups in the general population. Brazil and Mexico combined account for 70% of the region's infections outside of the United States (Pan American Health Organizaton, 1995, 1996). Although 18% of infections are in women, this proportion appears to be increasing as heterosexual transmission becomes more prominent. In Brazil the proportion of reported AIDS cases attributable to heterosexual transmission increased from 7.5% in 1987 to 26% in 1994 (UNAIDS, 1996a,b; Han, 1995). HIV infections among IDUs are also a growing problem. For example, in Argentina the prevalence of HIV infection increased from 30 to 50% in IDUs in the last 5 years. In Brazil a similar increase occurred from 20 to 60% (UNAIDS, 1996a,b).

In the Caribbean, there have been over 250,000 HIV infections, and the HIV prevalence rate (1.4%) is second only to that of sub-Saharan Africa (World Health Organization, 1995a). Haiti and the Dominican Republic account for 85% of the infections in the Caribbean (Pan American Health Organizaton, 1995). Because of social, political, and economic instability, HIV prevalence in Haiti rose from 2% in 1989 to an estimated 5% of the adult population in 1994 (Pape and Johnson, 1993). In urban areas the prevalence was estimated to be 10%. HIV prevalence is particularly high among sex workers, STD clinic attendees, and tuberculosis patients. High rates of HIV prevalence among pregnant women aged 14 to 24 illustrate the dissemination of HIV to the general population. In the Dominican Republic, HIV seroprevalence has been documented to be 11% among sex workers, 8% among STD patients, and 1.2% among women attending antenatal clinics (Pan American Health Organizaton, 1996). Barbados and Trinidad also have high infection rates (approaching 4%) among women attending antenatal clinics. The current male-to-female ratio of incident AIDS cases is 2:1. Women aged 15 to 19 now have higher annual incidence rates than men of the same age. Pediatric AIDS cases have been steadily rising and now account for 5% of all incident cases. In contrast, Cuba with two infections per 10,000 adults (0.02%) has the lowest prevalence in the region (UNAIDS, 1996a,b; Pan American Health Organizaton, 1996).

5. Europe

Approximately half a million individuals are estimated to be HIV infected in western and eastern Europe (UNAIDS, 1996a,b). Differences continue to exist in HIV transmission patterns between individual countries. The majority of AIDS cases in Scandinavia have occurred among homosexual/bisexual men, whereas IDUs constitute two-thirds or more of the cases reported in Italy and Spain. For all of Europe, the proportion of AIDS cases attributable to homosexual transmission fell from 62 to 36% between 1985 and 1992, and the proportion of transmission attributable to injecting drug use increased from 16 to 42% (UNAIDS, 1996a,b). Transmission through heterosexual intercourse has also increased, especially in urban populations with high rates of injecting drug use or STDs (Haverkos and Quinn, 1995; Prevots *et al.*, 1994; The European Study Group, 1993). Between 14 and 18% of HIV-infected people in Europe may have acquired their infection heterosexually. Seroprevalence studies demonstrate increasing HIV incidence among patients attending STD clinics, female sex workers and their clients, and pregnant women. In some countries such as Italy and Spain where IDU transmission has been common, heterosexual transmission is increasing rapidly.

Data regarding the frequency and transmission of HIV infection in eastern Europe are still limited, but they do provide some information. Homosexual and heterosexual transmission appear to be the predominant routes of infection in some countries, although several localized outbreaks of nosocomial transmission have been reported among infants and young children (UNAIDS, 1996a,b; Kozlov *et al.*, 1993). In the Czech and Slovak republics, about two-thirds of confirmed HIV infections are due to homosexual transmission. In Bulgaria 75% are linked heterosexual transmission. In Poland 70% of HIV-infected people are IDUs, and approximately 10% of IDUs surveyed are HIV infected. Following several years of low but slowly rising levels of HIV infection, an outbreak has occurred among IDUs in Ukraine (Kobyshcha *et al.*, 1996). In January 1995 1.4% of drug users were HIV positive. By August 1995 the prevalence had increased 10-fold to 13%. In Nikolayev, seroprevalence has reached 55% with rates in four other cities ranging from 4.2 to 12.1%. Similar increases are expected in other cities in neighboring countries. The Russian Federation may experience a similar epidemic. Although no HIV-positive individuals were found among 84,377 IDUs tested in 1994, 190 out of 45,507 tests in 1996 were positive (UNAIDS, 1996a,b). Since January 1996 the reported number of persons infected with HIV in Kaliningrad has increased 18-fold from 21 to 387, most of whom are IDUs (UNAIDS, 1996a,b). Following a trend observed in other countries, the male-to-female ratio of HIV-infected persons has begun to equalize. Infected men now outnumber infected women by 2:1 instead of 6:1. Similar risks of drug-related HIV spread in the near future hang over other eastern European countries such as Slovenia where drug use is on the rise and there is evidence of high-risk injecting behavior.

Sexual transmission of HIV in eastern Europe may be on the rise as evidenced by recent STD surveillance in the independent republics of the former Soviet Union. Between 1994 and 1995 syphilis incidence rates rose from 81.7 to 172 per 100,000 population in the Russian Federation, from 72.1 to 147.1

in Belarus, from 16.6 to 173 in Moldova, and from 32.6 to 123 in Kazakhstan (UNAIDS, 1996a,b; Gromyko, 1996). Consequently, eastern Europe has an explosive potential for HIV because transmission is occurring via all of the known risk behaviors, and little is being done to prevent its spread.

IV. POPULATION BIOLOGY AND THE SPREAD OF HIV

The character of the HIV pandemic in different regions has been largely influenced by the frequency of each of the three main modes of transmission—sexual, perinatal, and parenteral. From 75 to 80% of all HIV infections are transmitted through unprotected sexual intercourse (Mann and Tarantola, 1996). Heterosexual intercourse accounts for 70% of all adult HIV infections and homosexual intercourse for a further 5 to 10%. Transfusion of HIV-infected blood or blood products accounts for 3 to 5% of all adult HIV infections. In many parts of the world HIV transmission through transfusion of infected blood has been reduced by the use of voluntary blood donors, by routine screening of donated blood for HIV, and through a more rational use of blood aimed at reducing the number of transfusions (Greenberg *et al.*, 1988; Foster and Buve, 1995). The sharing of HIV-infected injection equipment by drug users accounts for 5 to 10% of all adult infections. Mother-to-child or vertical transmission accounts for more than 90% of all infections in infants and children, but only 10% of all HIV infections.

In developing countries heterosexual transmission has always been the predominant mode of spread (Krause, 1993). In Africa the male/female ratio is 1:1.4, and the peak of new infections occurs several years earlier in young women than in men (National Research Council, 1996). For example, the rate of newly acquired HIV infection is highest in 15- to 20-year-old women and 20- to 30-year-old men (Wawer *et al.*, 1991, 1994). In Mosaka, Uganda, HIV prevalence in 13- to 19-year-old females is 20 times higher than in males of the same age. However, the prevalence steadily increases in men, peaking 5 to 10 years later than in women. With increasing heterosexual transmission, the impact of the pandemic on women is rising sharply. As of late 1996, 11.3 million women have been infected with HIV (UNAIDS, 1996a,b) (Table 1). Worldwide the proportion of HIV-infected adult women rose from 25% in 1990 to 45% in 1995. By the year 2000 the number of new infections among women will be equal to that of men.

Variable rates of heterosexual transmission of HIV among sex partners of infected persons have been documented in studies in the United States, Europe, Africa, and Asia. Reported rates of infection for sexual partners of an infected person range from 15 to 50%. Biological factors that appear to be important in transmission efficiency include the degree of immunosuppression of the index case as measured by the level of CD4-positive lymphocytes, the level of plasma viremia, the level of virions present in semen or cervical secretions, lack of circumcision, and the presence of an ulcerative STD or an inflammatory nonulcerative STD such as gonorrhea or chlamydia (Quinn and Fauci, 1994; Quinn, 1996b; Holmberg *et al.*, 1989).

The strong association between HIV transmission and the presence of STDs has been termed "epidemiological synergy" because of the strong behavioral and biological association between HIV and STDs (Wasserheit, 1992). Numerous epidemiological studies from sub-Saharan Africa, Asia, Europe, and North America suggest that there is approximately a fourfold greater risk of becoming HIV infected in the presence of a genital ulcer caused by syphilis, chancroid, or herpes, and a two- to threefold greater risk in the presence of other STDs such as gonorrhea, chlamydia, and trichomoniasis (Laga *et al.*, 1993, 1994a; Cameron *et al.*, 1989; Plummer *et al.*, 1991; Pepin *et al.*, 1989). Because these latter diseases are more prevalent, they may be associated with greater attributable risk of HIV transmission. In a WHO report (World Health Organization, 1995d), the number of individuals infected with one or more of four curable STDs (i.e., gonorrhea, chlamydia, syphilis, and trichomoniasis) was estimated to be 333 million (Fig. 7). The greatest number of these STDs occurs in sub-Saharan Africa and Southeast Asia, the two regions with the highest rates of HIV infection (Heymann, 1995).

The observed links between STDs and HIV infection are compatible with three possible explanations (Institute of Medicine, 1996). First, STDs increase the infectivity of HIV. Persons who have both an STD and HIV infection may be more likely to transmit HIV to others due to the effects of STDs on HIV infectivity such as increased shedding of HIV. Second, STDs increase the susceptibility to HIV. Persons with an STD may be more susceptible to subsequent exposure to HIV because the STD may compromise the mucosal or cutaneous surfaces of the genital tract that normally act as a barrier against HIV. Third, the association between STDs and HIV remains confounded by sexual behavior and/or immune suppression in persons with sexually acquired HIV. HIV-infected persons may be more likely than uninfected persons to have another

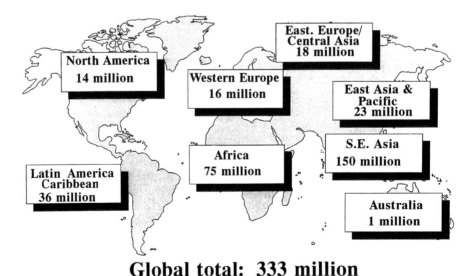

Global total: 333 million

FIGURE 7 Estimated new cases of curable STDs including gonorrhea, chlamydia, syphilis, and trichomoniasis among adults, and by region. Data from the World Health Organization (1995d).

STD due to high-risk sexual behavior or because HIV-related immune suppression predisposes to active STDs (e.g., by reactivating genital ulcers or by making an ulcerative disease harder to cure). Therefore, at any given level of sexual activity, persons exposed to an HIV-infected person may also be exposed to another STD.

The effect on HIV infectivity is strongly supported by the presence of HIV in genital ulcers (Plummer *et al.,* 1991; Kreiss *et al.,* 1989) as determined by the increased rate of detection of HIV DNA in cervical swab specimens among women with cervical inflammation (Clemetson *et al.,* 1993; Kreiss *et al.,* 1994), by the increased prevalence of HIV DNA in swab specimens in men with gonococcal urethritis (Moss *et al.,* 1995), and by increased concentrations of cell-free RNA in semen with gonococcal urethritis (Hoffman *et al.,* 1996). The impact of genital shedding on HIV on sexual or perinatal transmission has yet to be precisely measured, but it is likely to be a critical factor in transmission. An effect of other STDs on susceptibility to HIV is supported by studies of couples where one partner was HIV positive but the other was not. HIV-negative partners who develop genital infections were more likely to be subsequently infected with HIV than were those HIV-negative partners who did not have genital infections (de Vincenzi, 1994; Deschamps *et al.,* 1996).

The strong association between STDs and HIV infection among heterosexuals has been further documented by several population-based studies focused on the treatment of STDs. In a study by Hoffman *et al.* (1996), men with urethritis had a median HIV concentration of 125,000 copies/ml of HIV RNA in semen. After treatment this decreased to 37,000 copies/ml. The strongest decrease in viral load was seen among men who were treated for gonococcal urethritis. Similar data are available for female sex workers in Abidjan (Ghys *et al.,* 1996). The persistence of cervical vaginal HIV shedding was less frequent in female sex workers whose cervical vaginal ulcer had healed than in women whose ulcer had not healed as well as in sex workers who were cured of gonorrhea versus those who were not cured.

On a community level, investigators in Kinshasa, Zaire, found that STD diagnosis and treatment coupled with a condom distribution program for female sex workers successfully decreased the incidence and prevalence of both STDs and HIV (Laga *et al.,* 1994b). A community-based syndromic approach to the treatment of symptomatic STDs in Mwanza, Tanzania, led to a 42% reduction in HIV incidence compared to control villages (Grosskurth *et al.,* 1995). In the Rakai district of Uganda, a community-based trial of mass treatment of STDs was also associated with a marked decline in the prevalence and incidence of STDs, and information on HIV incidence should be forthcoming within the next year (Wawer *et al.,* 1996). The latter two studies are especially important since there was no clearly defined population of high-frequency transmitters to be specifically targeted, because in some regions the entire population is judged to be at risk (Laga, 1995).

The association of heterosexual transmission of HIV and STDs is relevant for many other areas of the world including the United States, where STDs continue to increase in incidence. Rates of curable STDs in the United States were the highest in the developed world and are higher than in some developing regions. Approximately 12 million new cases of STDs, 3 million among teenagers, occur annually in the United States (Centers for Disease Control, 1993). The reported incidence of gonorrhea in 1995 in the United States was 150 cases per 100,000 persons versus 3 cases per 100,000 in Sweden, and 18.6 per 100,000 in Canada (Centers for Disease Control, 1996a–e). The rate of primary, secondary, and early latent syphilis in the United States continues to increase, with 16.4 cases per 100,000 versus 0.5 cases per 100,000 in Canada. After sustaining a steady incidence rate of syphilis during the 1970s and early 1980s, the rate of syphilis increased sharply from 1987 through 1990 (Centers for Disease Control, 1995b; Quinn and Cates, 1992). The syphilis epidemic illustrates the ability of syphilis and other STDs to reemerge with alarming intensity in populations such as IDUs—particularly crack cocaine users—and their sex partners (Holmes, 1995; Centers for Disease Control, 1992). Approximately two-thirds of persons who acquire STDs are under age 25, now the leading age group at risk for HIV infection (Centers for Disease Control, 1993). Adolescents are the group at greatest risk for acquiring an STD including HIV for a number of reasons: they are more likely to have multiple sex partners, they may be more likely to engage in unprotected intercourse, and their partners may be at higher risk for being infected compared to most adults (Quinn, 1996b; Centers for Disease Control, 1993). Female adolescents are also more susceptible to cervical infections such as gonorrhea and chlamydia because the cervix of female adolescents and young women is especially sensitive to infection by certain STDs. In addition, adolescents and young people are at greater risk for substance abuse and other contributing factors that may increase risk for STDs than older persons (Alan Guttmacher Institute, 1994).

Estimating the transmission of HIV in a population is extremely complex. One needs to also consider partner selection patterns that lead to exposure of uninfected individuals to infected individuals in the population; the prevalence of HIV at the point where infected and uninfected individuals mix; and the existence of threshold levels of transmission required to sustain the infection in population subgroups (Anderson, 1991). These concepts are reflected in the basic biological and behavioral factors that determine transmission dynamics of STDs and HIV as reflected in the formula $R_0 = \beta Dc$ (Anderson, 1991; Holmes, 1995), where R_0 represents the reproductive rate of infection of a particular pathogen, that is, the number of new infections transmitted by one infected person in a susceptible population. In epidemiological terms this is known as the secondary infection rate. The overall magnitude of R_0 determines the size of the epidemic. The transmission coefficient, β, represents the infectivity rate or the probability that infection results after exposure. With regard to the dynamics of transmission, the magnitude of β during the acute primary infection basically determines the initial rate at which the epidemic spreads through a population. Infectiousness may be greatest during the acute primary infection and during advanced stages of the disease, when viral levels are known to be higher (Quinn and Fauci, 1994). D represents duration infectiousness.

Sustaining the epidemic is dependent on the next variable, c, which represents the average rate and variability of new sexual partner selection. This rate depends on many factors such as age, sex, and social variables. The structure of sexual mixing environment (i.e., sexual interaction among individuals) influences the shape of the HIV epidemic (Anderson *et al.*, 1991). Intensive sexual contact within a relatively circumscribed group partially explains the very rapid spread of HIV among homosexual men in the early phase of the epidemic. Sexual contact with partners from groups other their own results in further spread of the epidemic outside of the original high-risk group such as sexual transmission among homosexual men to IDUs and eventually to their heterosexual partners. Thus, an observed decline in the incidence of infection or disease within one group, as in homosexual men, may not necessarily indicate the worst of the epidemic is over if sexual mixing between groups is common (Anderson, 1991). The recent explosion of HIV infection among heterosexuals in Asia demonstrates how quickly HIV can spread within certain core groups such as female sex workers and their clients and IDUs, then subsequently, because of heterogeneous sexual mixing, to their spouses and the general population (Quinn, 1996b).

In the absence of prospective studies or formal trials of strengthened STD interventions to reduce sexual transmission of HIV in the United States, mathematical modeling may be essential to assess the potential impact of reducing STDs on HIV transmission. Such models are very complex and dependent on many assumptions related to sexual behaviors, natural history, the epidemiology of STDs, and the interactions between STDs and HIV (Institute of Medicine, 1996). Robinson *et al.* (1995) predicted that a 50% in reduction in duration of STDs could decrease HIV transmission by 43%, a prediction remarkably close to that observed in the intervention trial in Tanzania. Boily and Anderson (1996) developed a model which showed that HIV infection could not be established in the general U.S. heterosexual population in the absence of chlamydial infection or other STDs with comparable effects on HIV transmission. Over and Piot (1993) also supported the concept that reducing STDs could have a significant impact on sexually transmitted HIV infections. As mentioned previously, they estimated that successfully treating or preventing 100 cases of syphilis among high-risk groups for STDs would prevent 1200 HIV infections that would be linked to those 100 syphilis infections during a 10-year period.

Given the strong association between STDs and HIV transmission, prevention programs need to incorporate STD prevention strategies as a menas for preventing further HIV spread. New STDs appear as emerging and reemerging infections on a regular basis and are likely to continue to do so as long as rates of risky sexual behaviors remain high and global economic and demographic factors continue to promote emergence of new STDs.

A. Demographic Impact

The HIV pandemic has had severe effects on the population age structure, indenting the population pyramid in young adults, the main contributors to social and economic development (Quinn, 1994). The population of sub-Saharan Africa is predominantly young, in sharp contrast with the population structure in developed countries (Rossi-Espagnet *et al.*, 1991). Forty-five percent of the population in the region is under the age of 15 compared to one-third or less for other major geographic regions. A significant contributor to the elevated prevalence of infection in sub-Saharan Africa is the fact that behavioral factors associated with HIV transmission including multiple partners and transient relationships are generally more common among the young; furthermore, a high proportion of young adults are found in sub-Saharan African countries. Accordingly, the large number of young persons under age 15 who will soon enter their sexual and reproductive lives represent a priority group for AIDS and STD prevention.

In some urban centers of sub-Saharan Africa, western Europe, and the Americas, AIDS has already become the leading cause of death for both men and women, aged 15 to 49 years (DeCock *et al.,* 1990; Dondero and Curran, 1995; Centers for Disease Control, 1996a–e). AIDS kills people in their most productive years and ranks as the leading cause of potential healthy life-years lost in sub-Saharan Africa. In Abidjan, Côte d'Ivoire, it was estimated that 15% of adult male deaths and 17% of male years of potential life lost (YPLL) resulted from AIDS, whereas in women AIDS accounted for 13% of deaths and 12% of YPLL (DeCock *et al.,* 1990). In two community-based rural studies in the Mosaka and Rakai districts of Uganda, mortality among HIV-infected adults was over 100 per 1000 person-years of observation (PYO), an order of magnitude higher than among adults not infected with HIV (Mulder *et al.,* 1994; Sewankambo *et al.,* 1994) (Fig. 8). In both districts the adult HIV prevalence was 8% and 13%, respectively, and HIV was found to be the leading cause of death. Over 80% of deaths in the 20–29 age group occurred among those who were HIV infected. As a result, AIDS could double or triple the adult mortality rate in sub-Saharan African countries from levels that are already 8 times higher than in developed countries. In countries such as Uganda with an estimated 1.3 million infected persons out of a total population of 17 million, AIDS looms as a predominant health problem of the entire population.

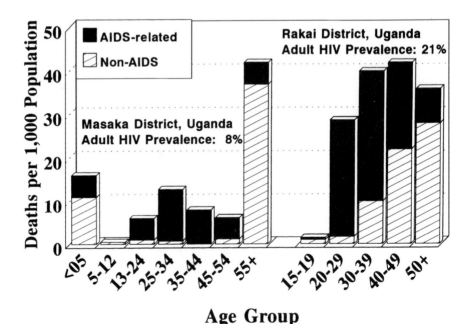

FIGURE 8 Evidence of AIDS impact on mortality in the Mosaka district and the Rakai district of Uganda, where adult HIV prevalence is 8 and 21%, respectively. Data are expressed as deaths per 1000 population per age group. Data from Mulder *et al.* (1994) and Sewankambo *et al.* (1994).

In addition to adult mortality, it is also estimated that the HIV pandemic has already resulted in the death of nearly 1 million children worldwide (Scarlatti, 1996). In a survey on child mortality in 10 central and eastern African countries, the death toll from AIDS in children under 5 is likely to rise from 159 per 1000 to 189 per 1000 by the year 2000 (Nicoll *et al.*, 1994). Even for those children who escape perinatal infection, survival rates will decrease because of the loss of one or both parents to AIDS. For each woman dying of AIDS in Africa, an average of two children will be orphaned, and by the year 2000 10 million children under 15 may be orphaned (Chin, 1990; Preble, 1990; Scarlatti, 1996). Overall infant and child mortality rates have increased as much as 30% more than previously projected as a direct consequence of perinatal HIV infection. Consequently, pediatric AIDS is now threatening much of the progress that has been made in child survival in developing countries during the past 20 years.

Because many AIDS deaths are concentrated among children and young adults, life expectancy has been reduced by more than 20 years in several countries, particularly in sub-Saharan Africa (Chin and Sato, 1994). It is projected that the population growth will decline more rapidly than expected and that the size of the African population in the year 2000 will be smaller than it would have been without AIDS (Chin and Sato, 1990; Way and Stanecki, 1991) (Fig. 9). The increasing HIV/AIDS cases have already overwhelmed the capacity of urban health systems in some countries, and demands for care will increasingly fall on poorly equipped and underfunded rural services, households, and individuals. It is estimated that 80% of hospital beds in an infectious disease hospital in Abidjan, Côte d'Ivoire, and 50% in a hospital in Kampala, Uganda, are occupied by people with HIV infection. In countries where 8% of the adult population is HIV-infected, surveys have measured a doubling of mortality due

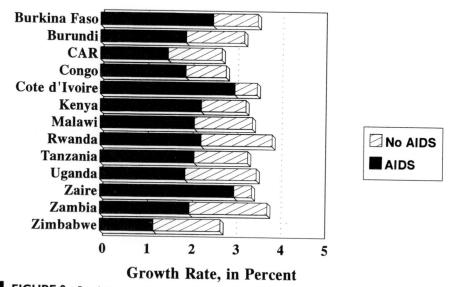

FIGURE 9 Population growth rates with and without AIDS for selected countries in Africa by the year 2000. Data from the U.S. Bureau of Census, International Program Center (U.S. Bureau of Census, 1994).

to HIV and a decrease of at least 5 years in life expectancy. Despite this pronounced effect on mortality and life expectancy, it is unlikely that the AIDS epidemic will decimate any population in the sub-Saharan region, largely because of current high population growth rates resulting from high fertility levels. In 1995 UNAIDS examined the demographic impact of HIV/AIDS in 15 sub-Saharan African countries with a prevalence of more tha 1% of the adult population. Below a prevalence of 1%, the impact of the epidemic on the national demographic picture is insignificant. In the 15 countries analyzed, the combined population was estimated to be 2 million people smaller than expected as a result of AIDS. By the year 2005 it will be 11.6 million smaller (291.8 million versus 303.4 million). As a result of the high fatality rate due to AIDS, life expectancy at birth decreased in these 15 countries from 52.8 to 49.6 years. By the year 2005 life expectancy without AIDS would have been 57.1 years, but with AIDS it may be reduced by more than 7 years to 49.6.

As a result of these demographic changes, HIV/AIDS poses a threat to economic growth in many countries already in distress. The World Bank estimates an annual slowing of growth of income per capita by an average of 0.6% per country in the 10 worst affected countries in sub-Saharan Africa (The World Bank, 1993). The powerful negative impact of AIDS on households, productive enterprises, and countries stems partly from the high cost of treatment, which diverts resources from productive investments, but mostly from the fact that AIDS affects people during their economically productive adult years when they are responsible for the support and care of others. In a 1993 World Development Report, HIV infections were one of the leading causes of disability-adjusted life years lost (DALYs), estimated at 14.7 million (Table 2). This measure attempts to engage the full loss of a healthy life as a result of death, disease, or injury. Currently, HIV infection constitutes about 2.8% of the world's global burden of disease or DALYs. However, these figures obscure the true magnitude of the epidemic since AIDS is already the greatest cause of disease burden in males and the fourth greatest cause in females in developing countries. In Africa, HIV accounts for 6.3% of the burden of disease, and in some African cities where the prevalence is much higher, the burden of disease is even greater. In Uganda over the next four years, AIDS is expected to cause half of the loss of years of productive life (The World Bank, 1993).

TABLE II   Top Five Causes for Disability-Adjusted Life Years (DALYs) Lost for Men and Women in Developing Countries, Ages 15–44

	Males		Females	
Rank	Cause	DALYs lost[a]	Cause	DALYs lost[a]
1	HIV	14.7	Maternal	27.9
2	Tuberculosis	13.3	STDs	13.8
3	Motor vehicle injuries	13.0	Tuberculosis	10.9
4	Homicide/violence	9.6	HIV	10.2
5	War	6.6	Depressive disorders	9.0

[a] Millions of DALYs lost. Data from The World Bank (1993).

This relationship between HIV and the economy is bidirectional. HIV results in higher morbidity and mortality among young adults, thereby reducing the quantity and quality of labor available to produce output or gross national product (GNP). A decline in GNP has a negative effect on other social indicators, such as life expectancy, access to education, future income, and capacity for the care of dependents. There are also structural changes within households brought about by the dissolution of families. Other variables include the depletion of the labor force, adverse impact on productive and social sectors, declining health services, shrinking revenues, and loss of military strength. The end result can be social and political unrest, destitution, and social and economic disintegration, which then leads to further spread of HIV infection. Thus, all sectors of the economy will feel the impact of the epidemic and will inescapably have to incur additional economic, social, and pyschological costs. Consequently, the epidemic, if unchecked, could transform the developmental performance of many countries. This conclusion appears to justify the implementation of activities that focus on behavioral and social change, which might reduce the spread of infection and likely generate enormous benefits in terms of avoidance of future costs.

Viral Evolution

HIV-1 evolved by rapid mutation and recombination, with both processes actively contributing to its genetic diversity (McCutchan et al., 1996). The diversity of the global AIDS pandemic is reflected in the heterogeneity of the viral subtypes of "clades" of HIV, which is based primarily on nucleotide changes in the genome of HIV. To date, two major types of HIV have been recognized: HIV-1 and HIV-2. While HIV-1 is the dominant type worldwide, HIV-2 is found principally in West Africa, with a small number of cases reported in Europe, Asia, Latin America, and North America. Both HIV-1 and HIV-2 are transmitted in similar ways, but it is apparent that HIV-2 is less efficiently transmitted and that the period between initial infection and illness may be longer for HIV-2 than for HIV-1 (Markovitz, 1993; Kanki et al., 1994; Marlink, 1994; Adjorlolo-Johnson et al., 1994; DeCock et al., 1993). Another large group of related retroviruses designated as simian immunodeficiency viruses (SIV) has also been discovered in nonhuman primates (Myers, 1994a). Currently, all known primate lentiviruses are grouped into five distinct phylogenetic lineages. One of these lineages includes all HIV-1 isolates as well as viruses from chimpanzees (SIV_{cpz}). The second lineage includes multiple strains of HIV-2 and viruses isolated from the sooty mangabes (SIV_{mn}) and from captive macaques (SIV_{mac}). The other three lineages include viruses isolated from different species of African monkeys.

Very little is known about the origin of HIV-1, but several theories abound. In one theory by Manfred Eigen (1993), using the mutational rate of HIV and the concept of quasi-species, estimates that HIV-1 is a descendant of an old viral family with an average lifetime of approximately 1000 years. With the large degree of genetic heterogeneity between HIV-1, HIV-2, and SIV sequences, it is likely that all three viral types evolutionarily diverged from a common ancestor between 600 and 1200 years ago. Thus, although HIV may not be a new virus, its pathogenicity or introduction into humans may have varied over the centuries, and it may have been introduced into humans more recently. Anderson *et al.* (1991) have further speculated on the basis of the extrapolation from yearly incidence data that the human infection may have occurred as far back as 150 years ago. Finally, epidemiological data and more recent genetic data of the different clades of HIV-1 suggest that the virus is evolving more rapidly than previously believed, and, with evidence of recombination, it has been estimated that human infection may have been as recent as within the last 50 years.

HIV-1 is subdivided into at least eight genetic clades (A–H) within the main group (M group) (Myers *et al.*, 1994). Each sequence clade is virtually equidistant from the others, representing a "star phylogeny." The distance separating the subtypes encompasses about 30% diversity in the *env* gene and about 14% in the *gag* gene. For comparison, viruses of different types such as HIV-1 and HIV-2 differ by at least 50% in both genes. Within clades the genetic distance can be half or greater than what is seen between clades. As HIV-1 sequences are determined and analyzed, they can usually be assigned to one of the eight clades (Myers *et al.*, 1994). Viral forms intermediate between clades have thus far not been found. This HIV variation may be accelerated in part by the dynamics of AIDS transmission, a phenomenon called "epidemic-driven variation" (Myers, 1994a). Analysis of a second group of HIV-1 subtypes called the O group (for "outlying") has provided a glimpse into the rise of HIV-1 in the world (Gurtler *et al.*, 1994). These HIV-1 strains are approximately 30% apart from one another in their *env* gene sequences—arms of a second-star phylogeny—and they are at least 50% apart in the *env* from the M group of the eight subtypes. While O forms may be minor contributors to the overall epidemic, they suggest a second, independent HIV-1 evolutionary event.

The evolution of HIV appears to reflect a geographic distribution of various genetic subtypes of HIV, undoubtedly reflecting "viral trafficking" patterns (Myers *et al.*, 1994). Molecular epidemiological studies of HIV subtypes have been conducted in several regions where HIV spread rapidly (Burke and McCutchan, 1996) (Fig. 10). HIV isolates in North America have been predominantly of the B clade. This suggests that of all viruses that could have been introduced, one form was introduced and was rapidly amplified through homosexual transmission in the mid-1970s, a phenomenon referred to as the "founder effect." The fulminating epidemic in Asia has been associated with B, C, and E clades, with E and C clades appearing to predominate. In Thailand such molecular studies delineated the separate introduction and largely independent spread of two different HIV subtypes, B and E, refuting earlier assumptions of a single epidemic (Ou *et al.*, 1993; Weniger *et al.*, 1994).

Molecular characterization has also been used to document several independent introductions of different HIV strains in such diverse places as eastern Europe, India, Brazil, and China (Myers *et al.*, 1994; Hu *et al.*, 1996; Burke and McCutchan, 1996). In South America and Europe, clade B viruses have also been predominant, yet clades A, C, D, F, and group O have also been recovered, but in smaller numbers (McCutchan *et al.*, 1996). In some countries in Africa which were either among the earliest centers of infection or had substantial population migration and transmission, at least five viral clades are known to be present. Similarly, diversity is now being documented in India and Brazil, and in several other countries there is evidence of recombinants such as A/E and

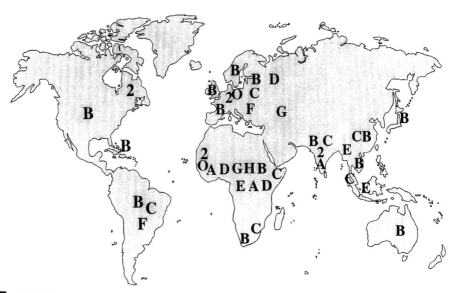

FIGURE 10　Geographic distribution of various clades of HIV infection. Points on the map represent the approximate locations where the predominant HIV strains have been reported and do not necessarily imply the actual distribution of all HIV strains, which is currently unknown. Modified from Hu *et al.* (1996).

B/F (Osmanov *et al.*, 1996; McCutchan *et al.*, 1996). Given the chance nature of HIV migrations and mutations, it is quite possible that one or more rare viral subtypes could be brought swiftly into ascendance: subtype F, a minor form in Brazil and Africa, has become a major form in Rumania, and subtype G, found rarely in central Africa, now appears to have been introduced into southern Russia (Janssens *et al.*, 1994; Bobkov *et al.*, 1994). The epidemiology of clade distribution worldwide is critical for vaccine development. In order for a vaccine to be useful internationally, it will need to induce sufficient broad (cross-clade) protection; otherwise, it will be of limited geographic usefulness, and "viral escape" can be anticipated (Levy, 1996; Osmanov *et al.*, 1996; World Health Organization Network for HIV Isolation and Characterization, 1994; Anderson *et al.*, 1996).

The extensive genetic variability of HIV-1 is primarily due to the high error rates of viral reverse transcriptase which result in approximately 10 genomic-based changes per replication cycle (Sharp *et al.*, 1994; Roberts *et al.*, 1988). In addition to the substitutions, deletions, and insertions, the frequency of these genetic errors is more difficult to estimate. Although viral reverse transcriptase can cause mutations throughout the viral RNA, many of these mutations are "synonymous," with changes that do not affect amino acid expression (Myers, 1994b). Only "nonsynonymous" mutations that cause amino acid changes would be selected during immune surveillance, and these alterations are most prominent in the *env* regions of HIV. The ability of HIV to mutate and establish new envelope subtypes over time may be due to attempts to escape host immune responses. HIV-1 undergoes continuous genetic variability within individual patients, a process leading to the evolution of quasi-species, with the heterogeneity usually not exceeding 2 to 5% in the *env* gene (Kuiken and Korber,

1994). Within a single geographic region, the range of genetic variability of the *env* gene is estimated to be 6 to 19%, although differences higher than 30% have been documented in sub-Saharan Africa (Delwart *et al.*, 1993). The extent of genetic variability in a given geographic location increases over time after introduction of a particular subtype in the population. Initially, the heterogeneity of the *env* gene can be as low as 3 to 5%, but this may further diversify at the rate of 1% per year (Osmanov *et al.*, 1996; Korber *et al.*, 1995).

Recent data have provided evidence that genetic recombination between two different HIV populations frequently occurs *in vivo*, resulting in biologically viable viruses with mosaic genomes, a phenomenon which results in additional HIV genetic variability and genetic shifts (Gao *et al.*, 1996; Hu and Temin, 1990; Robertson *et al.*, 1995). The identification of these recombinant viruses indicates that coinfection or superinfection with different genetic variants of HIV-1 does happen in nature (Artenstein *et al.*, 1995). This phenomenon suggests that active infection with one virus strain does not necessarily confer complete protection against infection with another strain. For example, available sequences of subtypes E and G are recombinant with subtype A (Gao *et al.*, 1996). Some recombinants possess the matrix and core of one subtype and the outer envelope of another, thus resembling pseudotypes. Subtype E viruses are apparently the progeny of an ancestral recombinant form that spread globally after the recombination event occurred (Nerurkar *et al.*, 1996). However, one study in Senegal showed that HIV-2-infected women appeared to have a lower incidence of HIV-1 than HIV seronegative women despite similar high-risk sexual behavior and the same frequency of STDs (Travers *et al.*, 1995). These data suggest that the protection observed may be a result of cross-reactive immunity to epitopes conserved between HIV-1 and HIV-2. However, this study also demonstrated for the first time sequential infection with two different HIV subtypes: first with HIV-2 and then with HIV-1 in a few individuals. Clearly, further studies are urgently needed to determine whether cross-reactive immunity can be induced to epitopes conserved within different strains of the virus.

Although several studies have definitely shown that HIV-2 is less infectious than HIV-1 in terms of sexual and perinatal transmission (Markovitz, 1993; Kanki *et al.*, 1994; Marlink, 1994; Adjorlolo-Johnson *et al.*, 1994; DeCock *et al.*, 1993), few data are available regarding the infectiousness or virulence of different clades of HIV-1. This issue of variable infectiousness between clades was recently raised from both *in vitro* and epidemiological studies. It was reported that subtypes C and E can infect and replicate more efficiently than subtype B *in vitro* in Langerhans cells, which are present in the vaginal mucosa, cervix, and foreskin of the penis, but not present in the walls of the rectum (Soto-Ramirez *et al.*, 1996). These data have been further supported by epidemiological studies suggesting that subtype E spread more easily than subtype B among female sex workers and their clients and among discordant couples in Thailand (Kunanusont *et al.*, 1995). It is important to note, however, that neither study was designed to fully control for multiple variables which may affect the risk of transmission, such as stage of HIV infection, frequency of exposure, condom use, or presence of STDs. The differential infectiousness of these clades may be solely based on epidemiological differences and not on any true biological replicative capabilities of different viral clades (Fransen *et al.*, 1996).

The ability to distinguish viruses by their molecular sequences can also be helpful in tracing transmission among individuals and the spread of viruses throughout various countries (Hu *et al.*, 1996; Anderson *et al.*, 1996). Molecular sequencing has already proven to be valuable for demonstrating the transfer of virus from mother to child, among sexual partners, and from a dentist to his patients (Ou *et al.*, 1992; Chant *et al.*, 1993; Myers, 1994a). Increased understanding of HIV heterogeneity may make it possible to preduct new clades that might arise through continual alterations in amino acid sequences in the envelope region. This knowledge will be important for vaccine development (World Health Organization Network for HIV Isolation and Characterization, 1994). Biological, serological, and molecular analysis can identify subtypes of HIV that may help further our understanding of HIV transmission and HIV strains and define the genetic diversity that may determine properties of viral replication and pathogenesis, as well as sensitivity to host immune responses.

Future Projections

The long-term course of the HIV pandemic cannot be predicted reliably; in the short term by the year 2000, there will be a cumulative total of 40 million infections in men, women, and children, more than 90% of which will be in developing countries. The projected cumulative total of adult AIDS cases will be approximately 15 million (UNAIDS, 1996a–c). By the year 2000 the cumulative number of HIV-related deaths in adults is predicted to rise from its current total of 5 million to more than 10 million individuals (d'Cruz-Grote, 1996). Unless interventions are more effective, between 15 and 20 million new HIV infections will occur in the next 5 years (UNAIDS, 1996a–c). In addition, more than 5 million children younger than 10 years will be orphaned as a result of AIDS-related fatalities (Chin, 1990; Preble, 1990; Nicoll *et al.*, 1994). The number of orphans will increase further in the early years of the twenty-first century as a result of deaths of mothers who were infected with HIV in the 1990s. As the epidemic matures in some parts of the world, large numbers of young people becoming sexually active will replenish the pool of susceptible individuals, especially in developing countries where the base of the age pyramid is quite broad. Evidence for a high incidence in young populations compared with older cohorts is already emerging for various countries. In the United States, the number of 13- to 21-year-olds who have become infected with HIV rose by 77% between 1991 and 1993 (Centers for Disease Control, 1996c). In sub-Saharan African countries, the highest HIV seroprevalence rates are in women between 15 and 25 years of age (National Research Council, 1996).

Heterosexual spread of HIV is causing an epidemiological shift of infection from high-risk populations such as homosexual men and IDUs to populations more reflective of the general populations, especially adolescents and young women of reproductive age. As a consequence, the number of persons with AIDS will continue to increase, causing unprecedented personal suffering, high direct costs for medical care, high indirect costs to society, and reduced economic output. The impact of HIV in many countries will be immense.

Nearly 75% of all new infections in the year 2000 will be in southeast Asia and sub-Saharan Africa (d'Cruz-Grote, 1996). By the turn of the twenty-first century, Latin America could also be confronted with a large-scale pandemic, which could emerge from the multiple, fragmented epidemics that have been documented in the region until now. Current trends observed in both Latin America and the Caribbean show that significant shifts are occurring in risk factors associated with reported cases of AIDS in the region, with an increasing frequency of heterosexual transmission. It is unlikely that the peak occurrence of AIDS estimated at 750,000 cases per year for sub-Saharan Africa will be reached until the middle of the next decade (Quinn, 1996a,b; UNAIDS, 1996c). In Asia, the annual number of AIDS cases will continue to increase steadily and should begin to level off at about 850,000 cases per year toward the end of the first decade of the twenty-first century. Although the absolute number of new HIV infections in Asia will be equal to or higher than those documented in Africa, the rate of infections or new cases per year as a percent of the adult population will still be lower in Asia than in sub-Saharan Africa (U.S. Bureau of Census, 1995; Chin, 1995). This is because the adult population of Asia is more than 5 times larger than that of sub-Saharan Africa. If unchecked, the rising incidence of AIDS in Asia will produce an unprecedented number of new cases in that region. On the basis of current data, cumulative HIV infections in Asia are conservatively projected to be more than 10 million (0.46% of the population) in the year 2000 and more than 18 million in 2010 (Chin, 1995). In sub-Saharan Africa the cumulative number of HIV infections is projected to reach about 15 million in the year 2000 (3%) and about 20 million in 2010 (1.7%) (UNAIDS, 1996c).

The effects of the epidemic in North America, western Europe, and Oceania will be far less than that described for Africa and Asia. The number of new AIDS cases in the year 2000 in developed countries will remain fairly constant, approximately 100,000 (Quinn, 1996a). This is because HIV-infected persons will eventually develop AIDS and the annual HIV incidence in these regions, projected to be close to 100,000, will nearly equal the number of fatalities. With improved and successful prevention campaigns such as zidovudine use during pregnancy, needle exchange programs, and improved STD treatment campaigns, the incidence of new HIV infections could potentially decline more.

CONCLUSION

The social, economic, demographic, and health impacts of the HIV epidemic are increasing in most countries. Especially dramatic is the spread of HIV among young adults, adolescents, and children in developing countries. In a number of industrialized countries, the spread of HIV is increasing rapidly in minority populations. There is also continuing spread of HIV to rural areas and to many members of the general population throughout the developing world. Globally, heterosexual transmission continues to rise, and in many countries the proportion of infected women is now equal to that of men. The extensive commercial sex industry, high prevalence of STDs, and injecting drug use continue to provide the potential for explosive epidemics in Indonesia, China, and several countries in western Africa and eastern Europe. In India, Thailand, Cambodia, and Myanmar, the explosion has already occurred (Quinn, 1996b).

Despite the setbacks and the enormous growth potential of the AIDS epidemic over the next few years, remarkable progress has been achieved in reducing the spread of HIV in some developing countries and in certain populations in industrialized countries (d'Cruz-Grote, 1996). Specifically, HIV incidence has declined in young men in Thailand due to a successful campaign on AIDS education and a condom distribution program (Hanenberg et al., 1994; Nelson et al., 1996). Impressive declines in HIV incidence have also been reported in homosexual/bisexual men in the United States, Australia, Canada, and western Europe, again due to a strong sex education campaign. HIV prevalence has also remained low in IDUs in a number of countries with needle exchange programs and drug rehabilitation programs (Institute of Medicine, 1995; Lurie and Drucker, 1996; Coutinho, 1995). In several developing countries in Africa, the aggressive treatment of STDs, coupled with condom distribution programs and educational programs, have been effective in decreasing HIV incidence rates in populations at high risk for HIV (Grosskurth et al., 1995; Laga et al., 1994b; Wawer et al., 1996; Han, 1995). As many as half of all future HIV infections might be prevented with adequately supported prevention programs. In Asia this could mean the prevention of several million AIDS deaths among young adults in their most productive years.

Until a safe and effective vaccine and/or inexpensive and effective therapeutic interventions are developed and made available globally, the primary prevention strategy must be focused on educational efforts to influence social, cultural, and behavioral factors. To control the AIDS epidemic, countries will need to not only promote individual behavioral change but also address the related problems of social disruption associated with mounting unemployment, accelerated urbanization, commercial sex, rapid decline in health services, and drug abuse. Fundamental social change such as improving the social status of women will be required if AIDS control efforts are to succeed.

With the magnitude of the current AIDS epidemic and the continued escalation and spread of HIV, it is evident that control and prevention of AIDS will require a sustained, long-term commitment. Prevention efforts should be specifically focused on women, young adults, adolescents, and marginalized communities. Special attention must be paid to the accelerating epidemics in India, Cambodia, and South Africa, as well as those countries and areas with potential for explosive epidemics such as other Asian countries, eastern Europe, and countries within sub-Saharan Africa and western Africa. Close linkages at the local, national, and global levels among epidemiologists, behavioral scientists, public health specialists, and nongovernmental and private voluntary organizations should be strengthened and maintained to improve prevention and care efforts and to monitor trends and evaluate the impact of such programs. The overall success of these programs in the regional and global AIDS control efforts will depend on a unifying international political and societal commitment.

REFERENCES

Adjorlolo-Johnson, G., DeCock, K. M., Ekpini, E., Vetter, K. M., Sibailly, T., Brattegaard, K., Yavo, D., Doorly, R., Whitaker, J. P., Kestens, L., Ou, C.-Y., George, J. R., and Gayle, H. D. (1994). Prospective comparison of mother-to-child transmission of HIV-1 and HIV-2 in Abidjan, Ivory Coast. *JAMA* **272,** 462–466.

Alan Guttmacher Institute. (1994). "Sex and America's Teenagers." New York.

Anderson, R. M. (1991). The transmission dynamics of sexually transmitted diseases: The behavioral component. *In* "Research Issues in Human Behavior and Sexually Transmitted Diseases in the AIDS Era" (J. N. Wasserheit, S. O. Aral, K. K. Holmes, and P. J. Hitchcock, eds.), pp. 38–45. American Society for Microbiology, Washington, D.C.

Anderson, R. M., May, R. M., Boily, M. C., Garnett, G. P., and Rowley, J. T. (1991). The spread of HIV-1 in Africa: Sexual contact patterns and the predicted demographic impact of AIDS. *Nature (London)* **352**, 581–589.

Anderson, R. M., Schwartlander, B., McCutchan, F., and Hu, D. (1996). Implications of genetic variability in HIV for epidemiology and public health. *Lancet* **347**, 1778–1779.

Artenstein, A. W., VanCott, T. C., Mascola, J. R., Carn, J. K., Hegerich, P. A., Gaywee, J., Sanders-Buell, E., Rebb, M. L., Dayhoff, D. E., Thitivichianlert, S. (1995). Dual infection with HIV-1 of distinct envelope (*env*) subtypes in humans. *J. Infect. Dis.* **171**, 805–810.

Barre-Sinoussi, F., Chermann, J. C., Rey, F., Nugeyre, M. T., Chamaret, S., Gruest, J., DAuguet, C., Axler-Blin, C., Vezinet-Brun, F., Rouzious, C., Rozenbaum, W., and Montagnier, L. (1993). Isolation of a T-lymphotropic retrovirus from a patient at risk for acquired immune deficiency syndrome (AIDS). *Science* **220**, 868–871.

Bobkov, A., Bachmann, M. H., Mullins, J. I., Louwagie, J., and Janssens, W. (1994). Identification of an HIV-1 *env* G sybtype and heterogeneity of HIV-1 in the Russian Federation and Belarus. *AIDS* **8**, 1649–1655.

Boily, M.-C., and Anderson, R. M. (1996). Human immunodeficiency virus transmission and the role of other sexually transmitted diseases: Measures of association and study design. *Sex Transm. Dis.* **23**, 312–330.

Bollinger, R. C., Tripathy, S. P., and Quinn, T. C. (1995). The human immunodeficiency virus epidemic in India. *Medicine* **74**, 97–106.

Brown, T., and Xenos, P. (1994). AIDS in Asia: The Gathering Storm. *In* "Asia–Pacific Issues," No. 16. East-West Center.

Brown, T., Sittirai, W., Vanichseni, S., and Thisyakom, U. (1994). The recent epidemiology of HIV and AIDS in Thailand. *AIDS*, **8** (Suppl. 2), S131–S141.

Burke, D. S., and McCutchan, F. E. (1996). Global distribution of human immunodeficiency virus type-1 clades. *In* "AIDS: Biology, Diagnostics, Treatment, and Prevention" (V. T. DeVita, Jr., S. Hellman, and S. A. Rosenberg, eds.), Lippincott-Raven, New York.

Cameron, D. W., Plummer, F. A., D'Costa, L. J., Ronald, A. R., Maitha, G. M., Gakinya, M. N., Cheong, M., Ndihya-Achola, J. O., Piot, P., Brunham, R. C. (1989). Female to male transmission of human immunodeficiency virus type 1: Risk factors for seroconversion in men. *Lancet* **2**, 403–407.

Celentano, D. D., Nelson, K. E., Suprasert, S., Eiumtrakul, S., Tulvatana, S., Kuntolbutra, S., Akarasewi, P., Matanasarawoot, D., Wright, N. H., Sirisopana, N., and Theetranont, C. (1996). Risk factors for HIV-1 seroconversion among young men in northern Thailand. *JAMA* **275**, 122–127.

Centers for Disease Control. (1981a). *Pneumocystis* pneumonia—Los Angeles. *Morbidity & Mortality Weekly Report* **30**, 250–252.

Centers for Disease Control. (1981b). Kaposi's sarcoma and *Pneumocystis* pneumonia among homosexual men—New York City and California. *Morbidity & Mortality Weekly Report* **30**, 305–308.

Centers for Disease Control. (1982). Acquired immune deficiency syndrome (AIDS): Precautions for clinical and laboratory staffs. *Morbidity & Mortality Weekly Report* **31**, 577–580.

Centers for Disease Control. (1983). Prevention of acquired immune deficiency syndrome (AIDS): Report of inter-agency recommendations. *Morbidity & Mortality Weekly Report* **32**, 101–103.

Centers for Disease Control. (1985a). Testing donors of organs, tissues, and semen for antibody to human T-lymphotropic virus type III/lymphadenopathy-associated virus. *Morbidity & Mortality Weekly Report* **34**, 294.

Centers for Disease Control. (1985b). Provisional Public Health Service interagency recommendations for screening donated blood. *Morbidity & Mortality Weekly Report* **34**, 1–5.

Centers for Disease Control. (1992). Syphilis cases, by 4-week period of report—United States, 1984–1992. *Morbidity & Mortality Weekly Report* **41**, 49.

Centers for Disease Control. (1993). "Division of STD/HIV Prevention. Annual Report 1992." Centers for Disease Control and Prevention, Atlanta, Georgia.

Centers for Disease Control. (1994). Recommendations of the U.S. Public Health Service Task Force on the use of zidovudine to reduce perinatal transmission of human immunodeficiency virus. *Morbidity & Mortality Weekly Report* **43** (No. RR-11).

Centers for Disease Control. (1995a). The First 500,000 AIDS Cases—United States, 1995. *Morbidity & Mortality Weekly Report* **44**, 849–853.

Centers for Disease Control. (1995b). "Division of STD/HIV Prevention. Annual Report 1994." Centers for Disease Control and Prevention, Atlanta, Georgia.

Centers for Disease Control. (1996a). HIV testing among women aged 18–44 years—United States, 1991 and 1993. *Morbidity & Mortality Weekly Report* **45**, 733–737.

Centers for Disease Control. (1996b). *HIV/AIDS Surveillance Report* **8**, 1–33.

Centers for Disease Control. (1996c). AIDS among children—United States, 1996. *Morbidity & Mortality Weekly Report* **45**, 1005–1010.

Centers for Disease Control. (1996d). "Division of STD/HIV Prevention. Sexually Transmitted Disease Surveillance 1995." Centers for Disease Control, Atlanta, Georgia.

Centers for Disease Control. (1996e). Update: Mortality attributable to HIV infection among persons aged 25–44 years—United States, 1994. *Morbidity & Mortality Weekly Report* **45**, 121–125.

Chant, K., Lowe, D., Rubin, G., Manning, W., O'Donoughue, R., Lyle, D., Levy, M., Morey, S., Kaldor, J., and Garsia, R. (1993). Patient-to-patient transmission of HIV in private surgical consulting rooms. *Lancet* **342**, 1548–1549.

Cheng, H., Zhang, J., Capizzi, J., Young, N. L., and Mastro, T. D. (1994). Introduction of HIV-1 subtype E into Yunnan, China. *Lancet* **344**, 953–954.

Chin, J. (1990). Current and future dimensions of the HIV/AIDS pandemic in women and children. *Lancet* **336**, 221–224.

Chin, J. (1995). Scenarios for the AIDS epidemic in Asia. *Asia-Pacific Population Research Reports* **No. 2** (Feb.).

Chin, J., and Sato, P. A. (1990). Projections of HIV infections and AIDS cases in the year 2000. *Bull. WHO.*, **68**, 1–11.

Chin, J., and Sato, P. A. (1994). Estimates and projections of the HIV/AIDS pandemic in sub-Saharan Africa. *In* "AIDS in Africa" (M. Essex, S. Mboup, P. J. Kanki, and M. R. Kalengayi, eds.), pp. 251–267. Raven, New York.

Clemetson, D. B. A., Moss, G. B., Willerford, D. M., Hensel, M., Emonyi, W., Holmes, K. K., Plummer, F., Ndinya-Achola, J., Roberts, P. L., Hillier, S., and Kreiss, J. K. (1993). Detection of HIV DNA in cervical and vaginal secretions. *JAMA* **269**, 2860–2864.

Coutinho, R. A. (1995). Annotation: Needle exchange programs—Do they work? *Am. J. Public Health* **85**, 1490–1491.

Curran, J. W., Jaffe, H. W., Hardy, A. M., Morgan, W. M., Selik, R. M., and Dondero, T. J. (1988). Epidemiology of HIV infection and AIDS in the United States. *Science* **239**, 610–616.

Davis, S. F., Byers, R. H., Lindegren, M. L., Caldwell, M. B., Karon, J. M., and Gwinn, M. (1995). Prevalence and incidence of vertically acquired HIV infection in the United States. *JAMA* **274**, 952–955.

d'Cruz-Grote, D. (1996). Prevention of HIV infection in developing countries. *Lancet* **348**, 1071–1074.

DeCock, K. M., Barrere, B., Diaby, L., Lafontaine, M.-F., Gnaore, E., Porter, A., Pantobe, D., Lafontant, G. C., Dago-Akribi, A., Ette, M., Odehouri, K., and Heyward, W. L. (1990). AIDS—The leading cause of adult death in the west African city of Abidjan, Ivory Coast. *Science* **249**, 793–796.

DeCock, K. M., Soro, B., Coulibaly, I.-M., and Lucas, S. B. (1992). Tuberculosis and HIV infection in sub-Saharan Africa. *JAMA* **2**, 4.

DeCock, K., Adjorlolo, G., Ekpini, E., Sibailly, T., Kouadio, J., Maran, M., Brattegaard, K., Vetter, K. M., Doorly, R., and Gayle, H. D. (1993). Epidemiology and transmission of HIV-2: Why there is no HIV-2 pandemic. *JAMA* **270**, 2083–2086.

Decosas, J, Kane, F., Anarfi, J. K., Sodji, K. D. R., and Wagner H. U. (1995). Migration and AIDS. *Lancet* **346**, 826–828.

Delwart, E. L., Shpaer, E. G., Louwagie, J., McCutchan, F. E., Grez, M., Rubsamen-Waigmann, H., and Mullins, J. J. (1993). Genetic relationships determined by a DNA heteroduplex mobility assay: Analysis of HIV-1 *env* genes. *Science* **262**, 1257–1259.

Deschamps, M. M., Pape, J. W., Hafner, A., and Johnson, W. D. (1996). Heterosexual transmission of HIV in Haiti. *Ann. Intern. Med.* **125**, 324–330.

de Vincenzi, I. (1994). A longitudinal study of human immunodeficiency virus transmission by heterosexual partners European Study Group on Heterosexual Transmission of HIV. *N. Engl. J. Med.* **331**, 341–346.

Dondero, T. J., and Curran, J. W. (1995). Excess deaths in Africa from HIV: Confirmed and quantified. *Lancet* **343**, 989–990.

Eigen, M. (1993). Viral quasispecies. *Sci. Am.* **July**, 42–49.

Foster, S., and Buve, A. (1995). Benefits of HIV screening of blood transfusions in Zambia. *Lancet* **346**, 225–227.

Fransen, K., Buve, A., Nkengasong, J. N., Laga, M., and van der Groen, G. (1996). Longstanding presence in Belgians of multiple non-B HIV-1 subtypes [letter]. *Lancet* **347**, 1403.

Gallo, R. C., Salahuddin, S. Z., Popovic, M., Shearer, G. M., Kaplan, M., Haynes, B. F., Palker, T. J., Redfield, R., Oleske, J., and Safai, B. (1984). Frequent detection and isolation of cytopathic retroviruses (HTLV-III) from patients with AIDS and at risk for AIDS. *Science* **224**, 500–503.

Gao, F., Yue, L., White, A. T., Pappas, P. G., Barchue, J., Hanson, A. P., Greene, B. M., Sharp, P. M., Shaw, G. M., and Hahn, B. H. (1992). Human infection by genetically diverse SIV$_{sm}$-related HIV-2 in West Africa. *Nature (London)* **358**, 495–499.

Gao, F., Robertson, D. L., Morrison, S. G., Hui, H., Craig, S., Decker, J., Fultz, P. N., Girard, M., Shaw, G. M., Hahn, B. H., and Sharp, P. M. (1996). The heterosexual HIV-1 epidemic in Thailand is caused by an intersubtype (A/E) recombinant of African origin. *J. Virol.* **70**, 7013–7029.

Ghys, P. D., Fransen, K., Diallo, M. O., Ettiegne-Traore, V., Maurice, C., Hoyi-Adonsou, Y. M., Kalish, M., Brown, T., Steketee, R., Coulibaly, I.-M., Greenberg, A. E., and Laga, M. (1996). The association between cervico-vaginal HIV-1 shedding and AIDS. *XI International Conference on AIDS, Vancouver* **WeC332** (abstract).

Greenberg, A. E., Nguyen-Dinh, P., Mann, J. M., and Kabote, N. (1988). The association between malaria, blood transfusion, and HIV seropositivity in a pediatric population in Kinshasa, Zaire. *JAMA* **259**, 545–549.

Gromyko, A. (1996). Epidemiological trends of AIDS and other sexually transmitted diseases in the eastern part of Europe. *XI International Conference on AIDS, Vancouver* **TuC205** (abstract).

Grosskurth, H., Mosha, F., Todd, J., Mwijarubi, E., Klokke, A., Senkoro, K., Mayaud, P., Changalucha, J., Nicoll, A., ka-Gina, G., Newell, J., Mugeye, K., Mabey, D., and Hayes, R. (1995). Impact of improved treatment of sexually transmitted diseases on HIV infection in rural Tanzania: Randomised controlled trial. *Lancet* **346**, 530–536.

Gurtler, L. G., Hauser, P. H., Eberle, J., von Brunn, A., Knapp, S., Zekeng, L., Tsague, J. M., and Kaptue, L. (1994). A new subtype of human immunodeficiency virus type 1 (MVP-5180) from Cameroon. *J. Virol.* **68**, 1581–1585.

Han, S. T. (1995). STD/AIDS—The need for a global response. *Venereology* **8**, 211–213.

Hanenberg, R. S., Rojanapithayakorn, W., Kunasol, P., and Sokal, D. C. (1994). Impact of Thailand's HIV-control programme as indicated by the decline of sexually transmitted diseases. *Lancet* **344**, 243–245.

Harries, A. D. (1990). Tuberculosis and human immunodeficiency virus infection in developing countries. *Lancet* 387–390.

Haverkos, H. W., and Quinn, T. C. (1995). The third wave: HIV infection among heterosexuals in the United States and Europe. *Int. J. STD & AIDS* **6**, 227–232.

Heymann, D. L. (1995). Sexually transmitted diseases and AIDS: Global and regional epidemiology. *Venereology* **8**, 206–210.

Hoffmann, I., Maida, M., Royce, R., Costello-Daly, C., Kazembe, P., Vernazza, P., Dyer, J., Zimba, D., Nkata, E., Kachenje, E., Bnda, T., Mughogho, G., Koller, C., Schock, J., and Gilliam, B. (1996). Concentration of HIV-1 in seminal plasma. *Eleventh International Conference on AIDS, Vancouver* **MoC903**, (abstract).

Holmberg, S. D., Horsburgh, C. R., Ward, J. W., and Jaffe, H. W. (1989). Biologic factors in the sexual transmission of human immunodeficiency viruses. *J. Infect. Dis.* **160**, 116.

Holmes, K. K. (1995). Human ecology and behavior and sexually transmitted bacterial infections. *In* "Infectious Diseases in an Age of Change: The Impact of Human Ecology and Behavior on Disease Transmission" (B. Roizman, ed.). National Academy Press, Washington, D.C.

Hu, D. J., Dondero, T. J., Rayfield, M. A., George, J. R., Schochetman, G., Jaffe, H. W., Luo, C. C., Kalish, M. L., Weniger, B. G., Pau, C. P., Schable, C. A., and Curran, J. W. (1996). The emerging genetic diversity of HIV. The importance of global surveillance for diagnostics, research, and prevention. *JAMA* **275**, 210–216.

Hu, W. S., and Temin, H. M. (1990). Retroviral recombination and reverse transcription. *Science* **250**, 1227–1233.

Institute of Medicine. (1992). "Emerging Infections: Microbial Threats to Health in the United States." National Academy Press, Washington, D.C.

Institute of Medicine. (1995). "Preventing HIV Transmission: The Role of Sterile Needles and Bleach." National Academy Press, Washington, D.C.

Institute of Medicine. (1996). "The Hidden Epidemic: Confronting Sexually Transmitted Diseases." (T. R. Eng and W. T. Butler, eds.). National Academy Press, Washington, D.C.

Janssens, W., Heyndrickx, L., Fransen, K., Motte, J., Peeters, M., Nkengasong, J. N., Ndumbe, P. M., Deleporte, E., Perret, J.-L., Atende, C., Piot, P., and van der Groen, G. (1994). Genetic and phylogenetic analysis of *env* subtypes G and H in central Africa. *AIDS Res. Hum. Retroviruses* **10**, 600–650.

Kanki, P. J., Travers, K. U., Mboup, S., Hsieh, C.-C., Marlink, R. G., Gueye-Ndiaye, A., Siby, T., Thior, I., Hernandez-Avila, M., Sankale, J.-L., Ndoye, I. (1994). *Lancet* **343**, 943–945.

Karon, J. M., Rosenberg, P. S., McQuillan, G., Khare, M., Gwinn, M., and Peterseh, L. R. (1996). Prevalence of HIV infection in the United States, 1984 to 1992. *JAMA* **276**, 126–131.

Kitayaporn, D., Uneklabh, C., Weniger, B. G., Lohsomboon, P., Kaewkungwal, J., Morgan, W. M., and Uneklabh, T. (1994). HIV-1 incidence determined retrospectively among drug users in Bangkok, Thailand. *AIDS* **8**, 1443–1450.

Kobyshcha, Y., Shcherbinskaya, A., Khodakevich, L., Andrushchak, L., and Kruglov, Y. (1996). HIV infection among drug users in Ukraine: Beginning of the epidemic. *XI International Conference on AIDS, Vancouver* **TuC204**, (abstract).

Korber, B. T. M., Allen, E. E., Farmer, A. D., and Myers, G. L. (1995). Heterogeneity of HIV-1 and HIV-2. *AIDS* **9** (Suppl. A), S5–S18.

Kozlov, A. P., Volkova, G. V., Malykh, A. G., Stepanova, G. S., and Glebov, A. V. (1993). Epidemiology of HIV infection in St. Petersburg, Russia. *J. Acquired Immune Defic. Syndr.* **6**, 208–212.

Krause, R. M. (1993). Emerging patterns of heterosexual HIV infection and transmission. *Int. J. STD. AIDS* **4**, 357–362.

Kreiss, J. K., Coombs, R., Plummer, F., Holmes, K. K., Nikora, B., Cameron, W., Ngugi, E., Ndinya-Achola, J. O., and Corey, L. (1989). Isolation of human immunodeficiency virus from genital ulcers in Nairobi prostitutes. *J. Infect. Dis.* **160**, 380–384.

Kreiss, J. K., Willerford, D. M., Hensel, M., Emonyi, W., Plummer, F., and Nkinya-Achola, J. (1994). Association between cervical inflammation and cervical shedding of human immunodeficiency virus DNA. *J. Infect. Dis.* **170**, 1597–1601.

Kuiken, C. L., and Korber, B. T. M. (1994). Epidemiological significance of intra- and inter-person variation of HIV-1. *AIDS* **8** (Suppl. 1), S73–S83.

Kunanusont, C., Foy, H. M., Kreiss, J. K., Rerks-Ngarm, S., Phanuphak, P., Raktham, S., Pau, C. P., and Young, N. L. (1995). HIV-1 subtypes and male-to-female transmission in Thailand. *Lancet* **345**, 1078–1083.

Laga, M. (1995). STD control for HIV prevention—It works. *Lancet* **346**, 518–519.

Laga, M., Manoka, A., Kivuvu, M., Malele, B., Tuliza, M., Nzila, N., Goeman, J., Behets, F., Batter, V., and Alary, M. (1993). Non-ulcerative sexually transmitted diseases as risk factors for HIV-1 transmission in women: Results from a cohort study. *AIDS* **7**, 95–102.

Laga, M., Diallo, M. O., and Buve, A. (1994a). Inter-relationship of sexually transmitted diseases and HIV: Where are we now? *AIDS* **8** (Suppl. 1), S119–S124.

Laga, M., Alary, M., Nzila, N., Manoka, A. T., Tuliza, M., Behets, F., Goeman, J., St. Louis, M., and Piot, P. (1994b). Condom promotion, sexually transmitted diseases treatment, and declining incidence of HIV-1 infection in female Zairian sex workers. *Lancet* **344**, 246–248.

Levy, J. A. (1996). HIV heterogeneity in transmission and pathogenesis. *In* "AIDS in the World II: Global Dimensions, Social Roots, and Responses" (J. M. Mann and D. J. M. Tarantola, eds.), p. 177. Oxford Univ. Press, New York.

Levy, J. A., Shimabukuro, J. M., and Oshiro, L. S. (1984). Isolation of lymphocytopathic retroviruses from San Francisco patients with AIDS. *Science* **225**, 840–842.

Lurie, P., and Drucker, E. (1996). An opportunity lost: Estimating the number of HIV infections due to the U.S. failure to adopt a national needle exchange policy. *XI International Conference on AIDS, Vancouver* **TuC324** (abstract).

McCutchan, F. E., and Myers, G. (1996). The global distribution of HIV-1 genotypes. *In* "AIDS in the World II: Global Dimensions, Social Roots, and Responses" (J. M. Mann and D. J. M. Tarantola, eds.), pp. 180–182. Oxford Univ. Press, New York.

McCutchan, F. E., Salminen, M. O., Carr, J. K., and Burke, D. S. (1996). HIV-1 genetic diversity. *AIDS* **10** (Suppl. 3), S13–S20.

Mann, J., and Tarantola, D. (eds.) (1996). "AIDS in the World II" Oxford Univ. Press, New York.

Mann, J. M., Chin, J., Piot, P., and Quinn, T. C. (1988). The international epidemiology of AIDS. *Sci. Am.* **10**, 82–89.

Markovitz, D. M. (1993). Infection with the human immunodeficiency virus type 2. *Ann. Intern. Med.* **118**, 211–218.

Marlink, R. (1994). Biology and epidemiology of HIV-2. *In* "AIDS in Africa" (M. Essex, M. Kalengay, P. Kanki, *et al.,* eds.), pp. 47–65. Raven, New York.

Mehendale, S. M., Rodrigues, J. J., Brookmeyer, R. S., Gangakhedkar, R. R., Divekar, A. D., Gokhale, M. R., Risbud, A .R., Paranjape, R. S., Shepherd, M. E., Rompalo, A. R., Sule, R. R., Tolat, S. N., Jadhav, V. D., Quinn, T. C., and Bollinger, R. C. (1995). HIV-1 incidence and predictors of seroconversion in patients attending sexually transmitted diseases clinics in India. *J. Infect. Dis.* **172**, 1486–1491.

Mertens, T. E., Belsey, E., Stoneburner, R., Low Beer, D., Sato, P., Burton, A., and Merson, M. H. (1995). Global estimates and epidemiology of HIV-1 infections and AIDS: Further heterogeneity in spread and impact. *AIDS* **8** (Suppl. 1), S259–S272.

Morse, S. S. (1995). Factors in the emergence of infectious diseases. *Emerging Infect. Dis.* **1**, 7–15.

Moss, G. B., Overbaugh, J., Welch, M., Reilly, M., Bwayo, J., and Plummer, F. A. (1995). Human immunodeficiency virus DNA in urethral secretions in men: Association with gonococcal urethritis and CD4 depletion. *J. Infect. Dis.* **172**, 1469–1474.

Mulder, D. W., Nunn, A. J., Kamali, A., Nakiyingi, J., Wagner, H. U., and Kengeya-Kayondo, J. F. (1994). Two-year HIV-1 associated mortality in a Ugandan rural population. *Lancet* **343**, 1021–1023.

Myers, G. (1994a). Molecular investigation of HIV transmission [editorial]. *Ann. Intern. Med.* **121**, 889–890.

Myers, G. (1994b). HIV: Between past and future. *AIDS Res. Hum. Retroviruses* **10**, 1317–1327.

Myers, G., and Korber, B. (1994). Future of human immunodeficiency virus. *In* "Evolutionary Biology of Viruses" (S. S. Morse, ed.). p. 36. Raven, New York.

Myers, G., Korber, B., Wain-Hobson, S., Jeang, K.-T., Henderson, L. E., and Pavlakis, G. N. (1994). "Human Retroviruses and AIDS 1994." Los Alamos National Laboratory, Los Alamos, New Mexico.

Naik, T. N., Sarkar, S., Singh, H. L., Bhunia, S. C., Singh, Y. I., and Pal, S. C. (1991). Intravenous drug users: A new high-risk group for HIV infection in India. *AIDS* **5**, 117–118.

National Center for Health Statistics. (1994). Annual summary of births, marriages, divorces, and deaths: United States, 1993. Hyattsville, Maryland: U.S. Dept. Health and Human Services, Public Health Service, CDC, **1994**, 18–20.

National Research Council. (1990). "AIDS: The Second Decade." National Academy Press, Washington, D.C.

National Research Council. (1993). "The Social Impact of AIDS in the United States." National Academy Press, Washington, D.C.

National Research Council. (1996). "Preventing and Mitigating AIDS in Sub-Saharan Africa: Research and Data Priorities for the Social and Behavioral Sciences" (B. Cohen and J. Trussel, eds.). National Academy Press, Washington, D.C.

Nelson, K. E., Celentano, D. D., Supraset, S., Wright, N., Eiumtrakul, S., Tulvatana, S., Matanasarawoot, A., Akarasewi, P., Kuntolbutra, S., and Romyen, S. (1993). Risk factors for HIV infection among young adult men in northern Thailand. *JAMA* **270**, 955–960.

Nelson, K. E., Celentano, D. D., Eiumtrakul, S., Hoover, D. R., Beyrer, C., Suprasert, S., Kuntolbutra, S., and Khamboonruang, C. (1996). Changes in sexual behavior and a decline in HIV infection among young men in Thailand. *N. Engl. J. Med.* **335**, 297–303.

Nerurkar, V. R., Nguyen, H. T., Dashwood, W.-M., Hoffmann, P. R., Yin, C., Morens, D. M., Kaplan, A. H., Detels, R., and Yanagiharak. (1996). HIV type 1 subtype E in commercial sex workers and injection drug users in southern Vietnam. *AIDS Res. Hum. Retroviruses* **12**, 841–843.

Nicoll, A., Timaeus, I., Kigadye, R. M., Walraven, G., and Killewo, J. (1994). The impact of HIV-1 infection and mortality in children under 5 years age in sub-Saharan Africa: A demographic and epidemiologic analysis. *AIDS* **8**, 995–1005.

Nzilambi, N., De Cock, K. M., Forthal, D. N., Francis, H., Ryder, R. W., Malebe, I., Getchell, J., Laga, M., Piot, P., and McCormick, J. B. (1988). The prevalence of infection with human immunodeficiency virus over a 10-year period in rural Zaire. *N. Engl. J. Med.* **318**, 276–279.

Osmanov, S., Heyward, W. L., and Esparza, J. (1996). HIV-1 genetic variability: Implications for the development of HIV vaccines. *In* "Development and Applications of Vaccines and Gene Therapy in AIDS" (G. Giraldo, D. P. Bolognesi, M. Salvatore, and E. Beth-Giraldo, eds.), pp. 30–38 (International Workshop, Naples, June 15–16, 1995). Karger, Basel.

Ou, C.-Y., Ciesielski, C. A., Myers, G., Bandea, C. I., Luo, C.-C., Korber, B. T. M., Mullins, J. I., Schochetman, G., Berkelman, R. L., Economou, A. N., Witte, J. J., Furman, L. J., Satten, G. A., MacIness, K. A., Curran, J. W., Jaffe, H. W., Laboratory Investigation Group, and Epidemiologic Investigation Group. (1992). Molecular epidemiology of HIV transmission in a dental practice. *Science* **256**, 1165–1171.

Ou, C.-Y., Takebe, Y., Weiniger, B. G., Luo, C.-C., Kalish, M. L., Auwanit, W., Yamazaki, S., Gayle, H. D., Young, N. L., and Schochetman, G. (1993). Independent introduction of two major HIV-1 genotypes into distinct high-risk populations in Thailand. *Lancet* **341**, 1171–1174.

Oucho, J. O., and Gould, W. T. S. (1993). Internal migration, urbanization, and population distribution. *In* "Demographic Change in Sub-Saharan Africa" (K. A. Foote, K. H. Hill, and L. G. Martin, eds.), pp. 256–296. National Academy Press, Washington, D.C.

Over, M., and Piot, P. (1993). HIV infection and sexually transmitted disease. *In* "WI priorities in developing countries" (D. T. Jamison, ed.), pp. 455–527. Oxford Univ. Press, New York.

Pan American Health Organization. (1995). "AIDS Surveillance in the Americas," Quarterly Report, 10 June 1995. PAHO/WHO, Washington, D.C.

Pan American Health Organization. (1996). "AIDS Surveillance in the Americas," Quarterly Report, 30 June 1995. PAHO/WHO, Washington, D.C.

Pape, J., and Johnson, W. D., Jr. (1993). AIDS in Haiti: 1982–1992. *Clin. Infect. Dis.* **17** (Suppl. 2), S341–S345.

Pepin, J., Plummer, F. A., and Brunham, R. C. (1989). The interaction of HIV infection and other sexually transmitted diseases: An opportunity for intervention. *AIDS* **3**, 3–9.

Piot, P., Plummer, F. A., Mhlau, F. S., Lamboray, J.-L., Chin, J., and Mann, J. M. (1988). AIDS: An international perspective. *Science* **239**, 573–579.

Plummer, F. A., Simonsen, J. N., and Cameron, D. W. (1991). Cofactors in male–female sexual transmission of human immunodeficiency virus type 1. *J. Infect. Dis.* **163**, 233–239.

Popovic, M., Sarngadharan, M. G., Read, E., and Gallo, R. C. (1984). Detection, isolation, and continuous production of cytopathic retroviruses (HLTV-III) from patients with AIDS and pre-AIDS. *Science* **224**, 497–500.

Preble, E. A. (1990). Impact of HIV/AIDS on African children. *Soc. Sci. Med.* **31**, 671–680.

Prevots, D. R., Ancelle-Park, R. A., Neal, J. J., and Remis, R. S. (1994). The epidemiology of heterosexually acquired HIV infection and AIDS in Western industrialized countries. *AIDS* **8**, (Suppl. 1), S109–S117.

Quinn, T. C. (1994). Population migration and the spread of types 1 and 2 human immunodeficiency viruses. *Proc. Natl. Acad. Sci. U.S.A.* **91**, 2407–2414.

Quinn, T. C. (1996a). The global burden of the HIV pandemic. *Lancet* **348**, 99–106.

Quinn, T. C. (1996b). Association of sexually transmitted diseases and infection with the human immunodeficiency virus: Biological cofactors and markers of behavioral interventions. *Int. J. STD AIDS* **7**, 17.

Quinn, T. C., Cannon, R. O., Glasser, D., Groseclose, S. L., Brathwaire, W. S., Fauci, A. S., Hook, E. W., 3rd. (1990). Association of syphilis with risk of human immunodeficiency virus infection in patients attending sexually transmitted disease clinics. *Arch. Intern. Med.* **150**, 1297–1302.

Quinn, T. C., and Cates, W. (1992). Epidemiology of sexually transmitted diseases in the 1990's. *In* "Advances in Host Defense Mechanisms" (T. C. Quinn, ed.), Vol. 8, pp. 1–37. Raven, New York.

Quinn, T. C., and Fauci, A. S. (1994). The changing demography of AIDS: Emergence of heterosexual transmission. *In* "Harrison's Principles of Internal Medicine" (K. J. Isselbacher, E. Braunwald, J. D. Wilson, *et al.,* eds.), Suppl. 9, pp. 1–9. McGraw-Hill, New York.

Quinn, T. C., Zacarias, F. R., and St. John, R. K. (1989). AIDS in the Americas: An emerging public health crisis [editorial]. *N. Engl. J. Med.* **320**, 1005–1007.

Roberts, J. D., Bebenek, K., and Kunkel, T. A. (1988). The accuracy of reverse transcriptase from HIV-1. *Science* **242**, 1171–1173.

Robertson, D. J., Hahn, B. H., and Sharp, P. M. (1995). Recombination in AIDS viruses. *J. Mol. Biol.* **40**, 249–259.

Robinson, N. J., Mulder, D. W., Auvert, B., and Hayes, R. J. (1995). Modeling the impact of alternative HIV intervention strategies in rural Uganda. *AIDS* **9**, 1263–1270.

Rodrigues, J. J., Mehendale, S. M., Shepherd, M. E., Divekar, A. D., Gangakhedkar, R. R., Quinn, T. C., Paranjape, R. S., Risbud, A. R., Brookmeyer, R. S., Gadkari, D. A., and Bollinger, R. C. (1995). The biological and behavioral risk factors for prevalent HIV infection in sexually transmitted disease clinics in India. *Br. Med. J.* **311**, 283–286.

Rogers, M. F., Caldwell, M. B., Gwinn, M. L., and Simonds, R. J. (1994). Epidemiology of pediatric human immunodeficiency virus infection in the United States. *Acta Paediatr. Suppl.* **400**, 5–7.

Rosenberg, P. S. (1995). Scope of the AIDS epidemic in the United States. *Science* **270**, 1372–1375.

Rossi-Espagnet, A., Goldstein, G. B., and Tabibzadeh, I. (1991). Survey of HIV-1 and other human retrovirus infections in a central African country. *Lancet* **1**, 941–943.

Rwandan HIV Seroprevalence Study Group. (1989). Nationwide community-based serological survey of HIV-1 and other human retrovirus infections. *Lancet* **1**: 941–943.

Ryder, R. W., and Hassig, S. E. (1988). The epidemiology of perinatal transmission of HIV. *AIDS* **2** (Suppl. 1), S83–S89.

Scarlatti, G. (1996). Paediatric HIV infection. *Lancet* **348**, 863–868.

Sewankambo, N. K., Wawer, M. J., Gray, R. H., Serwadda, D., Li, C., Stallings, R. Y., Musgrave, S. D., and Konde-Lule, J. K. (1994). Demographic impact of HIV infection in rural Rakai district, Uganda: Results of a population-based cohort study. *AIDS* **8**, 1707–1713.

Sharp, P. M., Robertson, D. L., Gao, F., and Hahn, B. (1994). Origins and diversity of human immunodeficiency viruses. *AIDS* **4** (Suppl. 1), S27–S42.

Soto-Ramirez, L. E., Renjifo, B., McLane, M. F., Marlink, R., O'Hara, C., Sutthent, R., Wasin, C., Vithagasain, P., Apichartpiyakul, C., Auewarakul, P., Pena Cruz, V., Choi, P. S., Osanthanondh, R., Muger, K., Lee, T. H., Essex, M. (1996). HIV-1 Langerhans cell tropism associated with heterosexual transmission of HIV. *Science* **271**, 1291–1293.

Swasdisevi, A. (1994). Clinical study of HIV disease in the lower area of northern Thailand in 1994. *J. Med. Assoc. Thai.* **77**, 440.

Thai Ministry of Public Health. (1995). "HIV/AIDS Situation in Thailand—Update: December 1994." Department of Communicable Disease Control, Ministry of Public Health, Bangkok.

The European Study Group. (1993). European community concerted action on HIV seroprevalence among sexually transmitted disease patients in 18 European sentinel networks. *AIDS* **7**, 393–400.

The World Bank. (1993). "The World Bank World Development Report 1993: Investing in Health." Oxford Univ. Press, London.

Travers, K., Mboup, S., Marlink, R., Gueye-Ndiaye, A., Siby, T., Thior, I., Traore, I., Dieng-Sarr, A., Sankale, J.-L., Mullins, K. C., Ndoye, I., Hsieh, C.-C., Essex, M., and Kanki, P. (1995). Natural protection against HIV-1 infection provided by HIV-2. *Science* **268**, 1612–1615.

UNAIDS. (1996a). "HIV/AIDS. The Global Epidemic." December. UNAIDS publication, Geneva.

UNAIDS. (1996b). "The Status and Trends of the Global HIV/AIDS Pandemic." July 5–6. UNAIDS publication, Geneva.

UNAIDS. (1996c). "Workshop on the Status and Trends of the HIV/AIDS Epidemics in Africa: Final Report." Harvard School of Public Health, Cambridge.

United Nations Economic Commission for Africa. (1991). "World Population Prospects 1990." United Nations, New York.

U.S. Bureau of Census. (1994). "Trends and Patterns of HIV/AIDS Infection in Selected Developing Countries: Country Profiles: June 1994," Research Note No. 14. U.S. Bureau of Census, Washington, D.C.

U.S. Bureau of Census. (1995). "HIV/AIDS in Asia," Research Note No. 18. U.S. Bureau of Census, Washington, D.C.

Wasserheit, J. N. (1992). Epidemiological synergy: Interrelationships between human immunodeficiency virus infection and other sexually transmitted diseases. *Sex. Transm. Dis.* **19**, 61–77.

Wawer, M., Serwadda, D., Musgrave, S., Konde-Lule, K., Musagara, M., and Sewankambo, N. (1991). Dynamics of spread of HIV-1 infection in a rural district of Uganda. *Br. Med. J.* **303**, 1303–1306.

Wawer, M., Sewankambo, N., Berkley, S., Serwadda, D., Musgrave, S. D., Gray, R. H., and Musagara, M. (1994). HIV prevalence in the Masaka District of Uganda. *Br. Med. J.* **308**, 171–173.

Wawer, M., Sewankambo, N. K., Gray, R. H., Serwadda, D., Paxton, L., Quinn, T. C., and Wabwire-Mangen, F. (1996). Community-based trial oif mas STD treatment for HIV control, Rakai, Uganda: Preliminary data on STD declines. *XI International Conference on AIDS, Vancouver*. **MoC433** (abstract).

Way, P. O., and Stanecki, K. (1991). "The Demographic Impact of the AIDS Epidemic on an African Country: Application of the Interagency Working Group on AIDS (IWGAIDS) Model," Staff Paper No. 58. U.S. Bureau of Census, Washington, D.C.

Weniger, B. G., and Brown, T. (1996). The March of AIDS through Asia [editorial]. *N. Engl. J. Med.* **335**, 343–344.

Weniger, B. G., Takebe, Y., Ou, C.-Y., and Yamazaki, S. (1994). The molecular epidemiology of HIV in Asia. *AIDS* **8** (Suppl. 2), S13–S28.

World Health Organization Network for HIV Isolation and Characterization. (1994). HIV-1 variation in WHO-sponsored vaccine evaluation sites: Genetic screening, sequence analysis and preliminary biological characterization of selected viral strains. *AIDS Res. Hum. Retroviruses* **10**, 1325–1341.

World Health Organization. (1995a). AIDS—Global situation of the HIV/AIDS pandemic. *Weekly Epidemiol. Rec.* **70**, 193–196.

World Health Organization. (1995b). HIV and AIDS in the western Pacific region. *AIDS Surveillance Report: Western Pacific Region* **5**, 1–8.

World Health Organization. (1995c). "WHO Report on the Tuberculosis Epidemic, 1995" (WHO/TB/95.183). WHO, Geneva.

World Health Organization. (1995d). An overview of selected curable sexually transmitted diseases. *Global Programme on AIDS*.

12

KOREAN HEMORRHAGIC FEVER AND HANTAVIRUS PULMONARY SYNDROME: TWO EXAMPLES OF EMERGING HANTAVIRAL DISEASES

NEAL NATHANSON

Department of Microbiology
University of Pennsylvania Medical Center
Philadelphia, Pennsylvania

STUART NICHOL

Division of Viral and Rickettsial Diseases
Center for Infectious Diseases
Center for Disease Control
Atlanta, Georgia

I. INTRODUCTION

Two important examples of emerging viral diseases are presented by Korean hemorrhagic fever (KHF) and hantavirus pulmonary syndrome (HPS), first identified in 1951 and 1993, respectively. Hantaan virus (HTN) is the cause of KHF, and Sin Nombre virus (SN) is the most common cause of HPS. Although viruses causing these diseases occur in totally different parts of the world, they have many features in common. Both cause enzootic infections of rodents, in which a single rodent species (*Apodemus agrarius* for HTN virus and *Peromyscus maniculatus* for SN virus) undergoes lifelong persistent infection, with shedding of the agent in the urine. It is presumed that human infections generally result from exposure to aerosolized urine, and they most frequently occur in persons who live or work under circumstances where they are very intimately exposed to infected rodents. Such exposure occurred in the early 1950s in military personnel occupying foxholes at the border of North and South Korea, and in the early 1990s in the southwestern area of the United States in a small number of persons living in close contact with infected rodents. In both instances, the diseases came to attention as outbreaks of serious human illness with distinctive clinical features and mortality ranging from 5% to 50%. Both diseases are caused by zoonotic viruses that may be transmitted to humans under special circumstances but cause infections that rarely spread from person to person.

The initial occurrences of KHF and HPS provoked massive efforts to isolate the causal agent. However, because hantaviruses are often fastidious in their ability to replicate in cell culture, it took over 25 years (from 1951 to 1978) before HTN virus, the etiologic agent of KHF, was identified, and its continued study over the period 1978–1990 gradually permitted the characterization of its biological properties and the elucidation of its genomic organization. The knowledge generated by these studies, together with advances in molecular genetics, made it possible to identify SN virus, the etiologic agent of HPS, within 1 month after the report of the first cases, with complete characterization of its genome and many of its biological and ecological properties being achieved within the next few years.

The history of the emergence of KHF and of HPS offers a fascinating and instructive exercise in comparative virology. In recounting these stories, we shall emphasize that history, indeed, does repeat itself, but that the rapid progress made in methods for investigation of viruses has telescoped the 30 years required to elucidate the cause of KHF into less than 3 years in the instance of HPS. Furthermore, it will be clear that the thorough study of the molecular, pathogenic, and ecological aspects of one viral infection provided information that was essential for the rapid and efficient unraveling of another unforeseen infection which came to light 40 years later on a different continent. Several publications summarize current information on the hantaviruses in considerable detail (Elliott, 1996; Gonzalez-Scarano and Nathanson, 1996).

II. KOREAN HEMORRHAGIC FEVER AND HANTAAN VIRUS

During the years 1951–1953, the United Nations conducted a military operation in Korea to stabilize the contested border between North and South Korea. During this period, troops deployed along the disputed border, near the Hantaan River, were subject to an acute febrile disease that carried a substantial mortality; it was estimated that more than 3000 cases occurred among UN troops during the war. The disease was termed Korean hemorrhagic fever (KHF), epidemic hemorrhagic fever, or hemorrhagic fever with renal syndrome (HFRS) (Lee, 1996; Smadel, 1959; Earle, 1954). KHF could take a variety of courses; about two-thirds of patients developed an influenza-like illness, about one-third exhibited hemorrhagic manifestations, and the overall mortality was 5–10%. In the typical hemorrhagic case, there was an acute onset of high fever, malaise, headache, and prostration. After a febrile phase lasting up to 1 week, there was a hypotensive phase of hours to days, during which thrombocytopenia and petechial hemorrhages often occurred; during this phase 10–15% of patients showed some degree of shock (about one-third of all deaths were caused by hypovolemic shock). Blood pressure then returned to normal or became hypertensive, with oliguria, and about half of deaths followed renal shutdown. A further phase of diuresis signaled the beginning of recovery, which was typically slow and might require at least 6 months.

It was recognized that KHF was similar or identical to diseases that had been described previously in the eastern USSR, Manchuria, and China under the name hemorrhagic nephrosonephritis, Song-go fever, or HFRS (Gonzalez-Scarano and Nathanson, 1996; Smadel, 1959). In fact, KHF continues to occur today; there are about 300–700 cases reported annually in South Korea, about 100,000 cases are notified annually in eastern China, and an unknown number occur in far eastern Russia (Lee, 1996; Gonzalez-Scarano and Nathanson, 1996). Careful descriptive study of outbreaks in UN troops established that (a) there was no evidence of person-to-person transmission, (b) infections were not spread through food and water, and (c) all cases in each minioutbreak appeared to have acquired infection while exposed on a small piece of terrain, usually abandoned farmland. On the basis of ecological observations before and during the Korean War, there was strong circumstantial evidence that this infection was acquired as an environmental exposure of troops or other persons to a zoonotic infection (Smadel, 1959). It had been postulated that wild rodents were the hosts, and that the infection was acquired directly or via an ectoparasite that served as a vector. Dramatic evidence for this mode of transmission was provided by illnesses among technicians (exposed in the laboratory only) who were attempting to isolate the causal agent from rodents captured in endemic areas (Kulagin et al., 1962; Gonzalez-Scarano and Nathanson, 1996).

In spite of repeated efforts, the agent could not be isolated, and there were no agent-specific tests for either antigen or antibody. However, it was established that the disease could be transmitted to humans by serum or urine obtained during the first few days of illness, and that the agent would pass through a Berkefeld N filter, indicating that it was a virus; furthermore, the agent was neutralized by convalescent serum (Smadel, 1959). About 25 years later, due to the tenacious efforts of a Korean investigator, Ho Wang Lee (Lee et al., 1978), an antigenic identification was achieved, using the convalescent sera of patients to stain the lungs of wild-caught *Apodemus agrarius* (field mice) which were naturally and persistently infected. It was shown that *A. agrarius* appeared to be the sole natural host for the virus in Korea and that other indigenous rodents were not infected. On the basis of this breakthrough, interest in the newly identified virus was rekindled. George French, another committed investigator, exhaustively tested large numbers of cell lines and finally found one that supported the replication of the virus (Vero E-6 cells are now the cell culture of choice) (French et al., 1981). The virus was named Hantaan (after the river dividing North and South Korea), characterized morphologically, biochemically, and genetically, and found to be the prototype of a new genus (*Hantavirus*) in the family *Bunyaviridae* (Lee, 1982, 1996; Gonzalez-Scarano and Nathanson, 1996).

A. Seoul Virus

In the course of studies in Korea, it was recognized that cases of hemorrhagic fever were occurring in urban areas outside the range of *A. agrarius*. Surveys of urban rodents showed that rats (*Rattus norvegicus* and *Rattus rattus*) carried antibodies against Hantaan virus and were the probable source of human infection (Lee et al., 1982). Beginning in 1979, cases of hemorrhagic fever were identified in Japan and Europe in research workers who were exposed to laboratory rats (Lee and Johnson, 1982). Initially, it was assumed that these rat viruses were identical to Hantaan virus. However, when isolates from captured rats were made in Vero E-6 cells and compared with prototype strains of Hantaan virus, the rat isolates could be distinguished from Hantaan virus. The rat virus was named Seoul virus after the site of the initial studies (Sugiyama et al., 1987;

Leduc *et al.*, 1986; Gonzalez-Scarano and Nathanson, 1996). Rodent infestation of shipping lead to the worldwide dissemination of Seoul virus, which may be found in animals captured in seaports around the globe (Leduc *et al.*, 1986). However, there have been relatively few reported cases of hemorrhagic fever reported, at least among humans living in coastal cities in the United States (Glass *et al.*, 1993; Yanagihara, 1990). Sporadic outbreaks of HFRS among research workers exposed to laboratory rats continue to occur, and the frequent transport of laboratory rats between research institutes constitutes an ongoing risk (Lee, 1996).

B. Puumala Virus

Nephropathia epidemica, an acute febrile disease with renal involvement, has been recognized for more than 50 years in several Scandinavian countries (Niklasson and Leduc, 1987; Gonzalez-Scarano and Nathanson, 1996). In Sweden, the epidemiological patterns were striking, with the highest rates in northern Sweden, in rural areas, and among men. These observations suggested that the infection was acquired from a zoonotic source, perhaps a rodent, and it was noted that *Clethrionomys glareolus*, the bank vole, was prevalent in areas with high rates of nephropathia epidemica. Attempts to isolate the agent had been consistently unsuccessful, but the isolation of Hantaan virus (which caused a similar but more severe disease) led to serological studies showing that convalescent human sera reacted with Hantaan virus (Niklasson and Leduc, 1987). Since *A. agrarius* are not found in Scandinavia, a survey was made for an alternate rodent host. Antigen was demonstrated in the lungs of wild-caught bank voles, and virus isolates were made from voles (using Vero E-6 cells) and were shown to be antigenically distinct from Hantaan virus. The agent was named Puumala virus after a site in Finland where some of the original isolates were obtained.

III. HANTAVIRUS PULMONARY SYNDROME AND SIN NOMBRE VIRUS

In May, 1993, a few cases of an acute pulmonary disease with high mortality were reported by the Indian Health Service to the New Mexico Department of Health, leading to an intensive collaborative investigation by several state health departments and the Centers for Disease Control (Hughes *et al.*, 1993; Marshall, 1993; Anonymous, 1993). By mid June, a total of 12 cases had been confirmed, of which 9 (75%) were fatal, occurring in New Mexico (9 cases), Arizona (2 cases), and Colorado (1 case). On the basis of an analysis of the first 100 cases, typical illnesses were characterized by abrupt onset of fever, myalgia, headache, and cough, followed by the rapid development of acute progressive pulmonary edema, leading to respiratory failure and hypovolemic hypotension, with death occurring from 2 to 10 days after onset and an overall mortality of 50% (Khan *et al.*, 1996). Most patients were healthy young adults (median 32 years, range 11–69 years), 35% were Native Americans and 63% were Caucasian, and there were approximately equal numbers of males and females. Pathologically, the disease was characterized by interstitial pneumonitis with a mononuclear cell infiltrate, edema, and focal formation of hyaline membranes (Zaki *et al.*, 1995). There was severe involvement of the capillary endothelium with loss of the functional integrity of the vascular–tissue barrier, accounting for the life-threatening clinical manifestations of infection, namely, pulmonary edema and hypotension.

When reports first came in of a "mystery" respiratory disease outbreak in late May 1993, an intensive effort to identify the causal agent was initiated, and testing of acute and convalescent sera against Hantaan, Seoul, and Puumala virus antigens showed the presence of hantavirus cross-reactive antibodies, with the strongest reaction being found against Puumala virus (Anonymous, 1993). This finding was unexpected because pathogenic hantaviruses had not previously been reported in the United States, and because the clinical syndrome differed from HFRS, the characteristic disease associated with known pathogenic hantaviruses. On the basis of this unforeseen serological finding, reverse transcriptase polymerase chain reaction (RT-PCR) was performed, using primers that represented conserved regions of the G2 glycoprotein of Prospect Hill/Puumala viruses or of Hantaan/Seoul viruses (Nichol *et al.*, 1993). Using the Prospect Hill virus primers, a DNA band of expected length was obtained, but the amplified sequence was 30% different from Prospect Hill, the most closely related hantavirus. This indicated that a new hantavirus, now designated Sin Nombre virus, was the cause of the illness, which is usually called hantavirus pulmonary syndrome (HPS). Remarkably, the tentative genetic identification of the causal agent was made about 30 days after the initial report of the first cases. As with other hantaviruses, it proved difficult to isolate Sin Nombre virus in cell culture, but repeated efforts involving many specimens and passage strategies led to the isolation of a strain that can be grown in Vero E-6 cells, permitting the sequencing of the entire genome (Chizhikov *et al.*, 1995; Schmaljohn *et al.*, 1995; Elliott *et al.*, 1994).

The identification of the causal agent as a hantavirus immediately led to a systematic trapping effort, in the areas of case concentrations, to identify a possible rodent reservoir. The deer mouse, *Peromyscus maniculatus*, was the dominant rodent captured, and a substantial proportion of animals were seropositive for Sin Nombre virus (Nichol *et al.*, 1993). Among seropositive deer mice, a high proportion also harbored the viral genome as detected by RT-PCR; the coexistence of virus and humoral antibody is typical of hantavirus persistence, and further studies have confirmed that the deer mouse is the primary reservoir of this virus (Childs *et al.*, 1994).

Using newly available diagnostic tools to confirm suspected cases, it was possible to construct an epidemic curve, which is shown in Fig. 1 for the first 100 cases reported through December, 1994 (Khan *et al.*, 1996). The graph indicates that cases of HPS occurred prior to the outbreak but that there was a sharp rise in the spring of 1993, leading to the recognition of the disease entity. In many instances, the emergence of a "new" infectious disease represents the recognition of a long-established entity, which comes to light because of the occurrence of an unusual cluster of illnesses at one place and time, and HPS represents a classic example of this phenomenon. By happenstance, data collected in central New Mexico indicated that there had been a 10-fold increase in the deer mouse population from May, 1992, to May, 1993 (Stone, 1993).

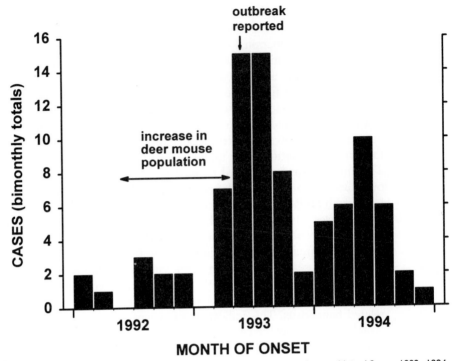

FIGURE I   Cases of hantavirus pulmonary syndrome by month of onset, United States, 1992–1994, omitting 16 cases with onsets prior to 1992. Redrawn from Khan *et al.* (1996) with permission.

This population surge was attributed to an increase in available food and coverage for the mice, which was associated with heavy snow and rain during the same period. It is postulated that this increase in mice, coupled with their high frequency of infection, increased the risk of exposure of humans who shared potential habitat with mice in highly rural areas. HPS case studies indicated that a high proportion (70–90%) of patients reported exposure to rodents, often associated with rodent infestations of their homes (Khan *et al.*, 1996).

As with other hantaviruses, Sin Nombre virus appears to be present throughout the range of the deer mouse, which extends over much of North America, including all of the United States except the southeastern region (Khan *et al.*, 1996). Cases are concentrated in the southwestern United States, however, which may reflect either a higher mouse density, a higher rate of mouse infection, or more sharing of habitats by humans and deer mice. The intense investigation of rodent hantaviruses has also revealed several other hantaviruses, particularly Black Creek Canal (BCC) virus, indigenous to the cotton rat *Sigmodon hispidus*, and Bayou (BAY) virus, indigenous to the rice rat *Oryzumus palustris* (Nichol *et al.*, 1996). BCC virus has been associated with a single HPS case close to Miami, Florida, although the virus has been found in cotton rats throughout the state. BAY virus has been associated with three HPS cases, one in Louisiana and two in Texas, and the virus has been detected in rice rats throughout the southeastern United States, from South Carolina to Texas (Nichol *et al.*, 1996; Khan *et al.*, 1996). More recent investigations have clearly demonstrated that numerous additional novel hantaviruses are present in a variety of rodent species in South America, several of which are associated with HPS cases. Clearly, this is not a phenomenon restricted to North America. Interestingly, although the vast majority of studies indicate that person-to-person transmission of hantaviruses does not occur, convincing data are emerging that such transmission has taken place during an HPS outbreak in Argentina (Kaiser, 1997). Ongoing studies will investigate this further and will also likely define additional hantaviruses and their rodent reservoirs throughout the Americas (Rowe *et al.*, 1995; Nichol *et al.*, 1996).

IV. HANTAVIRUSES AS A PARADIGM FOR EMERGING VIRAL DISEASES

The viruses of the genus *Hantavirus* of the family *Bunyaviridae* provide a paradigm for one class of emerging viral diseases. Table I summarizes the major ecological features of the salient members of this genus. Each virus has adapted to a single or a small group of rodent hosts with which it has a commensal relationship, being transmitted from adults to young animals and causing lifelong persistent infections with little or no acute disease consequence, although there may be late pathological abnormalities perhaps associated with immune complex deposition (Lyubsky *et al.*, 1996). Transmission to humans occurs under circumstances where people invade the microenvironment of the rodent

TABLE I Ecological Features of Different Virus Complexes within the Genus *Hantavirus*[a]

Virus complex	Wildlife host (rodent subfamily)	Geographic distribution	Human disease (severity)
Hantaan	*Apodemus agrarius,* striped field mouse (Murinae)	Eastern Asia, Eastern Europe	Hemorrhagic fever (severe)
Seoul	*Rattus* species, wild/laboratory rat (Murinae)	Eastern Asia, seaports worldwide	Hemorrhagic fever (moderate/mild)
Puumala	*Clethrionomys glareolus,* bank vole (Arvicolinae)	Europe, European Russia	Nephropathia epidemica (mild)
Prospect Hill	*Microtus pennsylvanicus,* meadow vole (Arvicolinae)	United States	Unknown
Sin Nombre	*Peromyscus maniculatus,* deer mouse (Sigmodontinae)	Western United States	Hantavirus pulmonary syndrome (HPS) (severe)

[a]Hemorrhagic fever is also known as hemorrhagic fever with renal syndrome (HFRS). In the United States, two other hantaviruses have been associated with a few cases of HPS (Black Creek Canyon virus, carried by *S. hispidus,* and Bayou virus, carried by *O. palustris*), and other new hantaviruses will probably be defined (Nichol *et al.*, 1996). After Gonzalez-Scarano and Nathanson (1996) with permission.

host and are inadvertently exposed to contaminated excreta, probably aerosolized urine. Although the virus can cross the species barrier, its lack of adaptation to the new human host is signaled by the nature of the infection, which is simultaneously much more pathogenic (than in its natural host) but much less transmissible (failing to spread from person to person). As long as societal pressures continue to produce novel ecological discontinuities, we can expect that zoonotic viruses will continue to "emerge" as one cause of new infectious diseases of humans.

It is striking that hantaviruses cause relatively benign persistent infections in their natural hosts but acute infections with severe disease in humans. Observations of this kind have suggested to some commentators that, during long-term coexistence, virus and host coevolve to a steady state in which infections are nearly harmless, due to reduced virulence of the virus and increased resistance of the host, and that introduction into a new species results in a much more virulent infection. However, a review of this theme points out that, even after coevolution, many viruses continue to cause relatively high mortality in their natural hosts (Nathanson *et al.*, 1997). An alternative viewpoint is that evolutionary pressure for survival can lead to virus–host relationships of different types, depending on the patterns of viral transmission and perpetuation (Yorke *et al.*, 1979; Nathanson, 1996). In some instances, such as hantavirus infections of rodents, lifelong persistence and continual viral shedding ensure widespread infection and perpetuation within small isolated populations. Other viruses are perpetuated by repeated cycles of acute infections in large populations, and such acute infections may result in either lifelong immunity or death.

Another feature of hantavirus biology is that once transmitted across the species barrier, the agents do not spread from person to person, in spite of the severity of disease (which implies replication to high titer in selected target tissues), although a recent outbreak in Argentina implies that there may be occasional exceptions (Kaiser, 1997). The failure of hantaviruses to spread from human to human is similar to a considerable number of zoonotic viruses that are occasionally transmitted to humans, without further transmission within the human population (Nathanson, 1997a). There are very few documented instances (human immunodeficiency virus in humans and canine parvovirus in dogs) in which a virus crosses the species barrier and becomes established within the new host species. It appears likely that adaptation to a new host species requires mutations which increase the ability of the evolved virus to replicate in key cell types in the new host species (Parrish, 1994; Nathanson, 1997b).

ACKNOWLEDGMENTS

This review summarizes work done by a large number of investigators whose combined efforts have identified and characterized viruses in the newly recognized genus *Hantavirus*. A more complete bibliography can be found in Elliott (1996) and in Gonzalez-Scarano and Nathanson (1996).

REFERENCES

Anonymous. (1993). Outbreak of acute illness—Southwestern United States. *Morbid. Mortal. Weekly Rep.* **42**, 421–424.

Childs, J., Ksiazek, T. G., Spiropoulou, C. F., Krebs, J. W., Morzunov, S. P., Maupin, G., Gage, K.,

Rollin, P. E., Sarisky, J., Enscore, R. E., Frey, J. K., Peters, C. J., and Nichol, S. T. (1994). Serologic and genetic identification of *Peromyscus maniculatus* as the primary rodent reservoir for a new virus in the Southwestern United States. *J. Infect. Dis.* **169,** 1271–1280.

Chizhikov, V. E., Spiropoulou, C. F., Morzunov, S. P., Monroe, M. C., Peters, C. J., and Nichol, S. T. (1995). Complete genetic characterization and analysis of isolation of Sin Nombre virus. *J. Virol.* **69,** 8132–8136.

Earle, D. (1954). Symposium on epidemic hemorrhagic fever. *Am. J. Med.* **16,** 619–709.

Elliott, L. T., Ksiazek, T. G., Spiropoulou, C. F., Morzunov, S. P., Monroe, M. C., Goldsmith, C. S., Humphrey, C. D., Zaki, S. R., Krebs, J. W., Maupin, G., Gage, K., Childs, J., Nichol, S. T., and Peters, C. J. (1994). Isolation of the causative agent of hantavirus pulmonary syndrome. *Am. J. Trop. Med. Hyg.* **51,** 102–108.

Elliott, R. M., ed. (1996). "The Bunyaviridae." Plenum, New York.

French, G., Foulke, R. S., Brand, O. A., Eddy, G. A., Lee, H. W., and Lee, P. W. (1981). Korean hemorrhagic fever: Propagation of the etiologic agent in a cell line of human origin. *Science* **211,** 1046–1048.

Glass, G. E., Watson, A. J., Leduc, J., Kelen, G. D., Quinn, T. C., and Childs, J. (1993). Infection with a ratborne hantavirus in US residents is consistently associated with hypertensive renal disease. *J. Infect. Dis.* **167,** 614–620.

Gonzalez-Scarano, F., and Nathanson, N. (1996). Bunyaviridae. *In* "Virology" (B. N. Fields, D. M. Knipe, and P. M. Howley, eds.), pp. 1473–1504. Lippincott-Raven, Philadelphia.

Hughes, J. M., Peters, C. J., Cohen, M. L., and Mahy, B. W. J. (1993). Hantavirus pulmonary syndrome: An emerging infectious disease. *Science* **262,** 850–851.

Kaiser, J. (1997). Human-to-human spread of hantavirus? *Science* **275,** 605.

Khan, A. S., Khabbaz, R. F., Armstrong, L. R., Holman, R. C., Bauer, S. P., Graber, J., Strine, T., Miller, G., Reef, S., Tappero, J., Rollin, P. E., Nichol, S. T., Zaki, S. R., Bryant, R. T., Chapman, L. E., Peters, C. J., and Ksiazek, T. G. (1996). Hantavirus pulmonary syndrome: The first 100 cases. *J. Infect. Dis.* **173,** 1297–1303.

Kulagin, S., Federova, N. I., and Ketiladze, E. S. (1962). A laboratory outbreak of hemorrhagic fever with renal syndrome (clinical–epidemiological characteristics) [in Russian]. *Zh. Microbiol. Epidemiol. Immunobiol.* **33,** 121–126.

Leduc, J., Smith, G. A., and Childs, J. E. (1986). Global survey of antibody to Hantaan-related viruses among peridomestic rodents. *Bull. WHO* **64,** 139–144.

Lee, H. (1982). Korean hemorrhagic fever. *Prog. Med. Virol.* **36,** 96–113.

Lee, H. W. (1996). Epidemiology and pathogenesis of hemorrhagic fever with renal syndrome. *In* "The Bunyaviridae" (R. M. Elliott, ed.), pp. 253–269. Plenum, New York.

Lee, H., and Johnson, K. J. (1982). Laboratory-acquired infections with Hantaan virus, the etiologic agent of Korean hemorrhagic fever. *J. Infect. Dis.* **146,** 645–651.

Lee, H., Lee, P. W., and Johnson, K. J. (1978). Isolation of the etiologic agent of Korean hemorrhagic fever. *J. Infect. Dis.* **137,** 298–308.

Lee, H., Baek, L. J., and Johnson, K. J. (1982). Isolation of Hantaan virus, the etiologic agent of Korean hemorrhagic fever, from wild urban rats. *J. Infect. Dis.* **146,** 638–644.

Lyubsky, S., Gavrilovskaya, I., Luft, B., and Mackow, E. (1996). Histopathology of *Peromyscus leucopus* naturally infected with pathogenic NY-1 hantaviruses: Pathologic markers of HPS viral infection in mice. *Lab. Invest.* **74,** 627–633.

Marshall, E. (1993). Hantavirus outbreak yields to PCR. *Science* **262,** 832–836.

Nathanson, N. (1996). Epidemiology. *In* "Virology" (B. N. Fields, D. M. Knipe, and P. M. Howley, eds.), pp. 251–271. Lippincott-Raven, Philadelphia.

Nathanson, N. (1997a). Emerging viral diseases: Societal causes and consequences. *ASM News* **63,** 1–6.

Nathanson, N. (1997b). Emergence of new viral infections: Implications for the blood supply. *Dev. Biol. Stand.* **80,** in press.

Nathanson, N., and Murphy, F. A. (1997). Evolution of viral diseases. *In* "Viral Pathogenesis" (N. Nathanson, ed.), pp. 353–370. Lippincott-Raven, Philadelphia.

Nichol, S. T., Spiropoulou, C. F., Morzunov, S. P., Rollin, P. E., Ksiazek, T. G., Feldman, H., Sanchez, A., Childs, J., Zaki, S. R., and Peters, C. J. (1993). Genetic identification of a hantavirus associated with an outbreak of acute respiratory illness. *Science* **262,** 914–917.

Nichol, S. T., Ksiazek, T. G., Rollin, P. E., and Peters, C. J. (1996). Hantavirus pulmonary syndrome and newly described hantaviruses in the United States. *In* "The Bunyaviridae" (R. M. Elliott, ed.), pp. 269–280. Plenum, New York.

Niklasson, B., and Leduc, J. (1987). Epidemiology of nephropathia epidemica in Sweden. *J. Infect. Dis.* **155**, 269–276.

Parrish, C. R. (1994). The emergence and evolution of canine parvovirus—An example of recent host range mutation. *Semin. Virol.* **5**, 121–132.

Rowe, J. E., St. Jeor, S. C., Riolo, J., Otteson, E. W., Monroe, M. C., Henderson, W. W., Ksiazek, T. G., Rollin, P. E., and Nichol, S. T. (1995). Coexistence of several novel hantaviruses in rodents indigenous to North America. *Virology* **213**, 122–130.

Schmaljohn, A. L., Li, D., Negley, D. L., Bressler, D. S., Turell, M. J., Korch, G. W., Ascher, M. S., and Schmaljohn, C. S. (1995). Isolation and initial characterization of a newfound hantavirus from California. *Virology* **206**, 963–972.

Smadel, J. E. (1959). Hemorrhagic fever. *In* "Viral and Rickettsial Diseases of Man" (T. M. Rivers and F. L. Horsfall, Jr., eds.), pp. 400–404. Lippincott, Philadelphia.

Stone, R. (1993). The mouse–pinon nut connection. *Science* **262**, 833

Sugiyama, K., Morikawa, S., Matsuura, Y., Tkachenko, E. A., Morita, C., Komatsu, T., Akao, Y., and Kitamura, T. (1987). Four serotypes of hemorrhagic fever with renal syndrome viruses identified by polyclonal and monclonal antibodies. *J. Gen. Virol.* **68**, 979–987.

Yanagihara, R. (1990). Hantavirus infection in the United States: Epizootiology and epidemiology. *J. Infect. Dis.* **12**, 449–457.

Yorke, J. A., Nathanson, N., Pianigiani, G., and Martin, J. (1979). Seasonality and the requirements for perpetuation and eradication of viruses in populations. *Am. J Epidemiol.* **109**, 193–123.

Zaki, S. R., Greer, P. W., Coffield, L. M., Goldsmith, C. S., Nolte, K. B., Foucar, K., Fedderson, R. M., Zumwalt, R. E., Miller, G. L., Khan, A. S., Rollin, P. E., Ksiazek, T. G., Nichol, S. T., Mahy, B. W. J., and Peters, C. J. (1995). Hantavirus pulmonary syndrome: Pathogenesis of an emerging infectious disease. *Am. J. Pathol.* **146**, 552–579.

13

EBOLA VIRUS: WHERE DOES IT COME FROM AND WHERE IS IT GOING?

FREDERICK A. MURPHY
School of Veterinary Medicine
University of California, Davis
Davis, California

C. J. PETERS
Special Pathogens Branch
Division of Viral and Rickettsial Diseases
National Center for Infectious Diseases
Centers for Disease Control and Prevention
Atlanta, Georgia

I. INTRODUCTION

Much has been written in the past five years about the natural history of the filoviruses, but most of this has been highly speculative, only faintly grounded in fact. The mystery about the reservoir host(s) and the means by which index human cases have become infected has been played up by the press/media into a metaphor for "The Ultimate Pathogen" (Kilbourne, 1983, 1996) or "The Andromeda Strain" (Crichton, 1969). Behind this press/media metaphor, however, is the fact that the filoviruses are dangerous on a global scale; we would like to explain this assertion and try to move discussion from the world of science fiction to the world of scientific inquiry. The questions of where the filoviruses came from, that is, where they exist in nature, and where they may be going, where they might emerge again as a public health problem, are the same questions that were once asked about yellow fever virus, measles virus, and other viruses. The difference is that, even now at the end of the twentieth century, the family *Filoviridae* is the only virus family containing human pathogens for which we do not have even an approximate answer.

This chapter is concerned primarily with the point in the natural history of the filoviruses where they are transmitted from their unknown reservoir host(s), perhaps even through an equally unknown intermediary host(s), to the index human host. To explore this point of zoonotic transmission, however, it is also necessary to consider events one step back and one step forward. Since we do not know the reservoir host(s), consideration here must be speculative, but since we do know a great deal about the circumstances of human-to-human transmission and nosocomial and community-acquired disease, consideration here will be based on data and experience. It is our feeling that focus on this point in the natural history of the filoviruses will help demystify other aspects of their natural history, such as those that have to do with disease pathogenesis, epidemiology, and clinical medicine.

We begin with a brief examination of the known filovirus disease outbreaks, searching in the facts surrounding these outbreaks for clues about reservoir host(s) and viral natural history. We proceed by asking the same questions in regard to episodes in hospitals caring for filovirus hemorrhagic fever patients. Then we beg the same questions in regard to what we know about the viruses and the infections they cause (in cell cultures, in experimental animals, in humans). We extend the perspective to the ecological and epidemiological characteristics of filovirus infections, and further extend this to recent field studies aimed at finding the reservoir host(s). Then we move on to a synthesis of the principles underpinning the natural history of zoonotic viruses and how they may apply to the filoviruses specifically. Finally, from an international public health perspective, we restate why we regard the filoviruses with such suspicion and why we think certain research and resources are needed to meet present international public health needs.

II. FILOVIRUS HEMORRHAGIC FEVER OUTBREAKS AND EPIDEMICS

A. Lessons from the Point of Zoonotic Transmission to Humans

The virus family *Filoviridae* only became known to medical science in 1967, although there has been speculation that the viruses have been around for at least as long as to have caused one of the plagues of Athens (Olson *et al.*, 1996). In 1967, African green monkeys (*Cercopithecus aethiops*) brought Marburg virus from Africa to Europe, resulting in 31 human cases (including six secondary cases) and seven deaths among workers handling the monkeys or their tissues (Siegert *et al.*, 1967); Smith *et al.*, 1967; Kissling *et al.*, 1968, 1970; Simpson, 1970; Martini and Siegert, 1971). Zoonotic transmission occurred in circumstances where there was very close contact between the monkeys and the humans. Monkeys were handled without substantial biocontainment equipment or practices; removal of kidneys and handling of cell cultures prepared from them was done with only rudimentary protocols to prevent bacterial contamination. In reviewing events before the point of zoonotic transmission, it was realized that many of the monkeys recently shipped from Uganda had died of a hemorrhagic disease. However, no specific antibodies were found in sera from monkeys subsequently captured in Uganda in the area where the monkeys had originated, and the source of the virus remains unknown. Because in later studies, all African green monkeys experimentally inoculated with the virus died acutely, it was postulated that monkeys might not be involved in the Marburg virus reservoir host cycle in nature. It was also apparent that the capacity of the virus for human-to-human spread was likely limited; this was evidenced by the low secondary attack rate among the many people exposed and the absence of tertiary cases. In fact, the human disease disappeared once the monkeys were eliminated. Since 1967, Marburg virus has reappeared only a few times, in Africa, in limited circumstances (Gear *et al.,* 1975; Conrad *et al.,* 1978; Smith *et al.,* 1982; Johnson *et al.,* 1996) (Table I).

Ebola virus did not appear on the scene until 1976, at which time two epidemics occurred, one in Zaire (now Democratic Republic of Congo), the other in the Sudan, together involving more than 550 cases and more than 430 deaths [Johnson *et al.,* 1977, 1978; Bowen *et al.,* 1977; World Health Organization (WHO), 1978a,b] As of the time of writing in early 1997, approximately 18 Ebola virus disease episodes have been identified, caused by four genetically distinct viruses, presenting an ever increasing geographic range within Africa (Heymann *et al.,* 1980; Teepe *et al.,* 1983; Baron *et al.,* 1983; WHO, 1979, 1982, 1995a–d, 1996, 1997) (Table I). There have also been two episodes of infection of scientists who were working on filoviruses in the laboratory, one involving Marburg virus and one the Sudan subtype of Ebola virus (Emond *et al.,* 1977; Nikiforov *et al.,* 1994); although no further transmission occurred in either episode, disease spread is certainly possible if this were to occur in circumstances where the diagnosis was missed.

The zoonotic source of the major epidemics of Ebola hemorrhagic fever, in northern Zaire and southern Sudan in 1976 and in Kikwit, Zaire, in 1995, has never been determined. Index human cases were in close contact with tropical forest ecosystems, but despite organized nonhuman specimen collecting expeditions and state-of-the-art laboratory technology for virus isolation (supplemented with viral antigen and viral RNA detection methods from 1995 onward), no trace of the zoonotic source of the virus has yet been found. The identification of the primary human index cases has been determined with high probability; however, attempts to backtrack the virus further, to the zoonotic contact point, have failed.

In the case of (a) the original Marburg hemorrhagic fever outbreak in Europe, (b) the disease outbreaks in monkeys in primate import quarantine facilities in the United States in 1989–1990 and 1996 caused by the Reston subtype of Ebola virus, (c) the single human case in the Côte d'Ivoire in 1994 caused by the Côte d'Ivoire subtype of Ebola virus, and (d) the outbreaks of disease in Gabon in 1994 and 1996 caused by the Zaire subtype of Ebola virus, the point of zoonotic transmission involved close association between nonhuman primates and primary human index cases [Siegert *et al.,* 1967; Jahrling *et al.,* 1990; Centers for Disease Control (CDC), 1990a,b, 1996; Peters *et al.,* 1991a; LeGuenno *et al.,* 1995; Simpson, 1995; WHO, 1996, 1997; LeGuenno, 1996; Georges *et al.,* 1996; Tukei, 1996; Amblard *et al.,* 1997]. However, attempts to backtrack from the implicated primates in these episodes have also failed to reveal a true reservoir host. In these episodes, most investigators have surmised that the nonhuman primates involved likely became infected from the same still mysterious reservoir host(s) that in other episodes seem to have exposed humans directly. This notion follows also from some evidence that transmission of filoviruses from monkey to monkey is not very efficient and that close contact among monkeys is needed to accomplish transmission just as it is in humans (e.g., antibody was not found in free-living chimpanzees in contact with those that died in the Taï Forest, Côte d'Ivoire, in 1994). Further, this notion is

TABLE I Filovirus Hemorrhagic Fever Cases, Outbreaks, and Epidemics

Virus	Year	Location	Cases (% mortality)	Circumstances of human infection
Marburg	1967	Germany and Yugoslavia	31 (23%)	Imported monkeys from Uganda were the source of the virus; humans became infected by contact with the monkeys or kidney cell cultures prepared from them; there were six secondary cases and no tertiary cases.
Marburg	1975	Zimbabwe/South Africa	3 (33%)	Unknown origin. A traveler was infected in Zimbabwe and died in Johannesburg, South Africa; secondary cases occurred in a companion and a nurse who were infected while providing patient care.
Ebola (subtype Zaire)	1976	Zaire	318 (88%)	Unknown origin. Yambuku and surrounding area. Spread occurred by close contact and by use of contaminated needles and syringes in hospitals and clinics.
Ebola (subtype Sudan)	1976	Sudan	284 (53%)	Unknown origin. Nzara, Maridi, and surrounding area. Spread was thought to be mainly by close contact. Nosocomial transmission and infection of medical care personnel were prominent.
Ebola (subtype Sudan)	1976	England	1 (0%)	Laboratory infection by needle-stick; no secondary spread.
Ebola (subtype Zaire)	1977	Zaire	1 (100%)	Unknown origin. Tandala. Single case in missionary hospital. Other sporadic cases may have occurred nearby at the same time.
Ebola (subtype Sudan)	1979	Sudan	34 (65%)	Unknown origin. Nzara. Recurrent outbreak at the same site as the 1976 Sudan epidemic.
Marburg	1980	Kenya	2 (50%)	Unknown origin. Index case infected in western Kenya, but died in Nairobi; a physician who was secondarily infected survived.
Marburg	1987	Kenya	1 (100%)	Unknown origin. Expatriate traveling in western Kenya.
Ebola (subtype Reston)	1989	United States	4 (0%)	Introduction of virus into quarantine facilities in Virginia, Texas, and Pennsylvania, by imported monkeys from the Philippines. Four humans were asymptomatically infected.
Marburg	1990	Russia	1 (0%)	Laboratory infection; no secondary spread.
Ebola (subtype Reston)	1990	United States	0 (0%)	Introduction of virus into same quarantine facilities in Virginia and Texas, from monkeys derived from the same export facility in the Philippines as in 1989. No human infections were identified.

Virus (subtype)	Year	Country	Cases (%)	Description
Ebola (subtype Reston)	1992	Italy	0 (0%)	Introduction of virus into quarantine facilities in Siena, by monkeys derived from the same export facility in the Philippines as involved in the U.S. episodes in 1989. No human infections were identified.
Ebola (subtype presumably Zaire)	1994	Gabon	44 (63%)	Unknown origin. Minkebe, Makokou area. Outbreak in gold-mining camps, in deep rain forest, previously thought to be yellow fever, identified retrospectively by serology in 1996.
Ebola (subtype Côte d'Ivoire)	1994	Côte d'Ivoire	1 (0%)	Ethologist became ill after conducting an autopsy on a wild chimpanzee in the Taï Forest; repatriated to Switzerland, recovered. Diagnosis by virus isolation and serology. The troop of chimpanzees had been decimated by a hemorrhagic disease in 1992 and 1994.
Ebola (subtype presumably Côte d'Ivoire)	1995	Liberia/Côte d'Ivoire	1 (0%)	Liberian refugee became ill, was hospitalized and recovered; serological diagnosis only (IgM, IgG antibodies).
Ebola (subtype Zaire)	1995	Zaire	317 (78%)	Unknown origin. Epidemic in Kikwit, traced to an index case who worked in the forest adjoining the city. The epidemic spread through families and hospital, and was terminated with the help of an international team.
Ebola (subtype Zaire)	1996	Gabon	37 (57%)	Mayibout area. A chimpanzee found dead in the forest was butchered and eaten; 19 primary cases occurred in the people who did the butchering, secondary cases occurred in family members.
Ebola (subtype Zaire)	1996	Gabon	60 (75%)	Unknown source. Booué area, with transport of patients to Libreville. Index case was a hunter who lived in a forest camp. Community spread occurred mostly by close contact with patients. A dead chimpanzee found in the forest at the time was implicated.
Ebola (subtype Zaire)	1996	South Africa	2 (50%)	A medical professional traveled from Gabon to Johannesburg, South Africa, after having treated Ebola-infected patients; he was hospitalized and a nurse became infected and died.
Ebola (subtype Reston)	1996	United States	0 (0%)	Introduction of virus into quarantine facility in Texas, from monkeys derived from the same export facility in the Philippines as in 1989. No human infections were identified.
Ebola (subtype Reston)	1996	Philippines	0 (0%)	Unknown origin. Same export facility in the Philippines was involved as in 1989. No human infections were identified.

consistent with the fact that the monkey species which have been studied experimentally have been far too susceptible to the lethal consequences of infection, and too likely to quickly burn out as host populations, to be able to perpetuate the viruses in nature over the long course. At a minimum, if a particular species of monkey were the reservoir host of Ebola virus, one would expect evidence of mortality such as seen with yellow fever in howler monkeys in the rain forests of Central and South America.

Although we really do not know the time or circumstances favoring Ebola transmission in its natural reservoir host niche, it is of interest that in the rainy seasons of 1976–1977 and then again in 1994–1997, circumstances have favored at least enough viral activity to have led to spread to humans living or working in the rain forest (Monath, 1996). These circumstances, of course, remind one of typical arthropod-borne and rodent-borne virus transmission cycles, with their annual periodicity and longer secular trends. In turn, this reminder might lead one to speculate that the four subtypes of Ebola virus represent adaptations to related reservoir hosts in similar habitats, much as we see with the hantaviruses and their rodent hosts.

The fact that the Ebola virus subtypes (i.e., serotypes, genotypes) which have caused human disease episodes have been different from one another (Zaire, Sudan, and Côte d'Ivoire subtypes and Gabon isolates which are not quite identical to Zaire isolates) makes it clear that a common source human-to-human transmission chain extending across sub-Saharan Africa is not the case; rather, virus subtypes lodged at or near each site of the recent human disease episodes have been responsible (Fisher-Hoch *et al.*, 1992a; Sanchez *et al.*, 1993; Georges Courbot *et al.*, 1997a,b). Indeed, Marburg and Ebola viruses are so very different from one another and the four recognized subtypes of Ebola virus are different enough from one another that we should consider each independently in regard to possible reservoirs and probable evolutionary pressures (Sanchez *et al.*, 1996). This being the case, it seems that it will be necessary to study each geographic site separately, being careful not to meld information too casually.

There seems to have been increasing incidence of Ebola hemorrhagic fever in western Africa over the past few years, evident as large and small outbreaks involving an ever-increasing geographic range. It has been said that this is just a matter of increasing recognition of human cases because of publicity, rather than a true emergence phenomenon. However, it seems more likely that epidemics such as that in Kikwit, Zaire, in 1995 would not have been overlooked earlier and that there really is an increasing incidence (CDC, 1995a–c; WHO, 1995a–d; Peters, 1996a). Because the reservoir host(s) is not known, ecological, environmental, and human behavioral changes that might have increased the opportunities for emergence in recent years are still matters of speculation, matters needing more study. Nevertheless, observations in these recent outbreak settings are all we have, so we must proceed from this point. One question in this regard is whether we should focus our attention on urban human-to-human transmission episodes (with particular focus on tracking back to index cases) or orient all research enterprise on the feral niche(s) where the viruses are perpetuated independently of human involvement.

B. Lessons from Disease Episodes in Hospitals Caring for Filovirus Hemorrhagic Fever Patients

It should be noted that the recent filovirus outbreaks, including the Ebola epidemic in Kikwit, Zaire, 1995, and the first epidemic in Mayibout, Gabon, 1996, were controlled with relatively simple measures (Muyembe-Tamfun and Kipasa, 1995; Georges et al., 1996; Calain, 1996; Amblard et al., 1997; Ivker, 1997; WHO, 1997). Indeed, the virus may never have spread within hospitals and communities, even communities such as Kikwit and Mayibout, if cultural attitudes and economic resources had favored the routine use of simple sanitary measures, strict barrier-nursing practices, and patient isolation. The terrible consequences that can follow such indifference are clear, yet this is a separate matter from the circumstance that brings the first human case, the index case, from the remote site of zoonotic exposure to the crowded sites of human-to-human transmission, starting in the caregiving family, proceeding to the hospital, and then extending to the community at large.

Within poorly equipped African hospitals and clinics, the larger Ebola hemorrhagic fever outbreaks have had a strong iatrogenic amplifier effect, whether this be from the reuse of syringes and needles as was the case in northern Zaire in 1976 or the lack of hygiene and the reality of close patient contact as was the case in Kikwit, Zaire, in 1995 (Johnson et al., 1977, 1978; WHO, 1978a,b; Pattyn, 1978; Baron et al., 1983; Khan et al., 1996). Indeed, it has been stated, "In Africa, hospitals cause Ebola."

Often, after initial outbreaks, hospitals have been closed, not by design but simply by fearful mortality among medical staff. In several instances, transmission has then declined and outbreaks have ended (Johnson et al., 1977, 1978; WHO, 1978a,b; Khan et al., 1996). Secondary attack rates outside the hospital have been about 5–15%, insufficient in most circumstances to sustain the outbreak. Factors contributing to the low secondary attack rate in villages include fundamental characteristics of filovirus infections per se, traditional shunning of the sick, modification of burial rituals once risk to family caregivers has been recognized, and other yet uncertain factors.

The hospital may also be a key to limiting the spread of the virus in the community. In the city of Kikwit, Zaire (population 300,000), where fewer traditional family-based caregiving practices may have been in place than in rural areas and traditional villages, the transport of patients to the hospital early in the course of their illness may have limited virus spread (Muyembe-Tamfun et al., 1996). The renovation of the Kikwit General Hospital (300 beds, plus a 60-bed maternity unit) in the years before the outbreaks in 1995 may have added to this trend of bringing patients to hospital early in their course of illness.

In regard to the question, "Where is Ebola going?" (i.e., "What will be the outcome of future Ebola episodes?"), it might be hoped that lessons learned in Kikwit, Zaire, 1995, would last. However, this remains to be seen. Some investigators have suggested that as early as 1 year after the epidemic, behaviors and practices had reverted to usual ways (Khan et al., 1996). This is especially distressing because experiences in Kikwit in 1995 provided such a good test of the concept that simple sanitary measures, strict barrier-nursing practices, and patient isolation could suffice to (a) terminate the viral transmission chain, (b) provide health care provider safety, and (c) provide adequate patient care.

The data from Kikwit, Zaire, are clear in this regard: whereas 76 medical staff were infected during the first weeks of the epidemic, after the institution of sanitary measures, barrier nursing, and patient isolation, although medical staff still entered the ward to care for patients, only one health care worker became infected (Calain, 1996). It should be noted that during the epidemic in Kikwit, strict barrier-nursing precautions included the use of a double gown (with impervious plastic lining), double gloves, boots, goggles, and mask (usually a HEPA-filtered mask) (Peters et al., 1991b; CDC, 1995d; CDC/NIH, 1993). The view that adequate hygiene, strict barrier-nursing practices, and patient isolation can limit the spread of Ebola virus to medical staff is also supported by the experience in Mayibout, Gabon, in 1996, where 32 patients were cared for without any infections in medical staff (LeGuenno, 1996).

III. THE FILOVIRUSES AND BASIC ASPECTS OF FILOVIRUS INFECTIONS

A. Lessons from Physical and Molecular Characteristics of the Filoviruses

Filovirus virions are enveloped and pleomorphic, appearing filamentous or bacilliform, or "∪"-shaped, or "6"-shaped, or circular. Particles have a uniform diameter of 80 nm but vary greatly in length (up to 14,000 nm; however, virions recovered from the peak-infectivity band after gradient centrifugation are more uniform—Ebola virions are about 1000 nm and Marburg virions about 800 nm in length) (Murphy et al., 1978; Regnery et al., 1981; Geisbert and Jahrling, 1995). Virions are covered by surface peplomers, about 7 nm in length, spaced at 10-nm intervals. Inside the virion envelope is a helical nucleocapsid about 50 nm in diameter with a helix periodicity of about 5 nm. Virus infectivity is rather stable at less than 20°C, but infectivity is rapidly destroyed at 60°C. Infectivity is sensitive to lipid solvents, β-propiolactone, formaldehyde, hypochlorite, quarternary ammonium and phenolic disinfectants, and UV- and γ-irradiation (Murphy et al., 1995; Feldmann and Klenk, 1996).

The genomes of Marburg and Ebola viruses are the largest thus far identified for any nonsegmented negative-strand RNA virus (19.1 kb); they contain seven linearly arranged genes that are organized in the same way as are the genes of rhabdoviruses and paramyxoviruses (gene order 3' NP-VP35-VP40-GP-VP30-VP24-L 5') (Sanchez et al., 1992, 1993; Feldmann et al., 1992, 1993, 1996a,b). An unusual characteristic of the genome organization of the filoviruses is the presence of gene overlaps, that is, short regions (17–20 bases) where the transcription start site of the downstream gene overlaps the transcription stop site of the upstream gene (Sanchez et al., 1993, 1996). The proteins of Ebola and Marburg viruses are quite similar, despite the level of divergence of their nucleotide sequences and their lack of serological cross-reactivity.

Thus, it would seem that although the basic physical and molecular characteristics of filovirus virions are interesting, and although their morphology does seem to add to the sense of mystery that has been fomented by the press/media, there is nothing that would really isolate these viruses from other human and animal viruses, nothing that would point to some unique habitat or reservoir host. On the other hand, some of the proteins of the filoviruses do lead us to wonder about the evolution of these viruses and their development of particular functional attributes that may have furthered their survival in their yet unknown habitat(s) and reservoir host(s).

For example, study of the Ebola glycoproteins has yielded several tantalizing findings, none of which has been fully evaluated (Feldmann *et al.*, 1994; Sanchez *et al.*, 1996). The glycoprotein genes of Marburg and the three Ebola subtypes differ significantly: the single Marburg glycoprotein is encoded in a single open reading frame, whereas the two virion glycoproteins of Ebola viruses are encoded in two reading frames and are expressed through transcriptional editing and translational frame-shifting. The Ebola *virion* glycoprotein (M_r 120,000–170,000) forms the surface peplomers, whereas the second glycoprotein (M_r 60,000), made in large amounts, is secreted extracellularly (Sanchez *et al.*, 1996). The participation of this soluble glycoprotein in the pathogenesis of Ebola disease in humans and experimental animals remains unknown; it may serve as some sort of immune decoy that minimizes the immune response to the virus. If so, does this point to any particular virus–host relationship, any particular reservoir host?

Filovirus glycoproteins are heaviliy glycosylated (carbohydrate may constitute one-third of the weight of the molecule; Feldmann *et al.*, 1993; Peters *et al.*, 1994), and it has been hypothesized over many years that viruses that exhibit this quality may better escape host defenses. Virions and virion budding sites on the plasma membranes of infected cells may not present T-cell recognition signals and may not present optimal targets for the efferent limb of the immune response once sensitization has occurred (i.e., they may not be targets for antibody action, macrophage action, and killer T-cell action).

One motif in the glycoproteins of Ebola and Marburg viruses has a high degree of sequence similarity with a putative immunosuppressive motif in the glycoproteins of oncogenic retroviruses. The retrovirus peptide has been shown to inhibit lymphocyte blastogenesis, decrease monocyte chemotaxis and macrophage infiltration, and inhibit the activity of natural killer cells (Cianciolo *et al.*, 1985; Kadota *et al.*, 1991; Burkreyev *et al.*, 1993; Volchkov *et al.*, 1995; Becker, 1995). It is not clear whether this conserved motif is important in the pathogenesis of filovirus hemorrhagic fever. Again, does this characteristic point to any particular virus–host relationship, any particular reservoir host?

B. Lessons from Genetic and Phylogenetic Properties of the Filoviruses

When Marburg and Ebola viruses were first discovered, it was thought that they might be rhabdoviruses (Pattyn, 1978). However, as characterization work proceeded and as genomic sequencing was initiated, it became clear that the viruses deserved the formation of a new family, the family *Filoviridae* (Murphy *et al.*, 1978; Kiley *et al.*, 1982, 1988). Later, when it was realized that the member viruses of the family *Filoviridae* exhibit phylogenetic characters (conserved domains in nucleoprotein and polymerase genes) in common with members of the families *Rhabdoviridae* and *Paramyxoviridae*, the three families were brought together in the order *Mononegavirales* (Muhlberger *et al.*, 1992; Murphy *et al.*,

1995). The similarities in these viruses that led to these taxonomic constructions suggest the possibility that more tractable, less hazardous substitute viruses might be chosen as stalking horses in searching for antiviral compounds and cytokines. Indeed, both vesicular stomatitis virus and respiratory synctial virus have been used to identify compounds active against Ebola virus (Huggins *et al.,* 1996). The finding of compounds effective against Ebola virus and respiratory synctial virus is particularly intriguing: the importance of the latter as a pediatric pathogen throughout the world might favor the commitment of the financial resources needed to bring development of candidate compounds to commercial reality. Given the minimal activity of the only three therapeutic modalities that have been tried so far, namely, hyperimmune horse serum, convalescent human plasma (and whole blood), and interferon-α, antiviral drug development is a high priority.

Phylogenetic analysis of the filoviruses shows clear separation of Marburg virus from the Ebola subtypes; however, each Ebola subtype shows a nearly equal difference from each of the others. The nucleotide sequence difference between the Ebola subtypes is about 47%, whereas the difference between Marburg and the Ebola viruses is about 72% (Sanchez *et al.,* 1992, 1993; Georges Courbot *et al.,* 1997a,b) (Fig. 1).

The degree of stability of filovirus sequences overall and the absence of genetic variability among Ebola virus isolates obtained within an outbreak match the character of other member viruses of the order *Mononegavirales.* Such genomic stability and the phenotypic (serotypic, pathotypic, geotypic) stability that follows on it, has typified wild-type measles virus isolates over the years and street rabies virus isolates from particular reservoir host niches (e.g., raccoons throughout the eastern United States). The Darwinian stabilizing influences operating here are not really understood but must involve constraints on multigenic viral replicative functions as well as constancy in environmental selective pressures (Holland, 1993). The degree of this genomic stability was made strikingly evident when isolates from the epidemic in Kikwit, Zaire, 1995, were compared with those from Yambuku, Zaire, 1976, and from several sites in Gabon, 1996 (sequences of glycoprotein genes of isolates from these disparate sources differed by less than 1.5%; Sanchez *et al.,* 1996; Georges Courbot *et al.,* 1997a,b; Amblard *et al.,* 1997; Nichol and Ksiazek, 1997).

In begging the question of the habitat(s) and reservoir host(s) of the filoviruses in nature, it is these genetic differences rather than the similarities that seem most important. As noted above, these differences are enough to suggest geographic isolation of each virus in or near the site where human (and nonhuman primate) disease occurred and where the virus was isolated. That is, these differences make it clear that a common source human-to-human transmission chain extending across sub-Saharan Africa is not the case. Only in epidemics does a single invariant viral genotype occur. Again, this observation suggests that we will have to investigate every Ebola virus emergence individually, building, we hope, on clues gathered in the first sites, such as the Taï Forest in Côte d'Ivoire, to be studied in comprehensive fashion (Formenty *et al.,* 1996; LeGuenno, 1996).

The genetic differences between the Ebola virus subtypes also raise the question of when and where we might encounter other variants: the progenitors

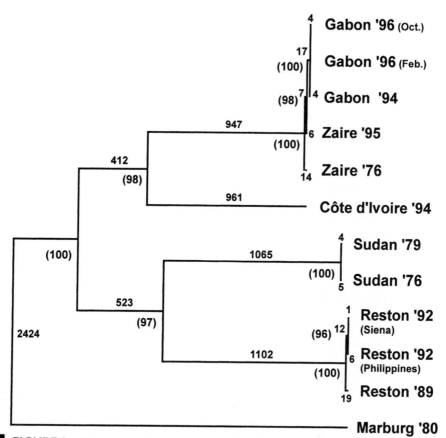

FIGURE I Phylogenetic tree showing relationships among the Ebola viruses. The entire coding region for the glycoprotein gene of the viruses shown was used in maximum parsimony analysis, and a single most parsimonious tree was obtained. Numbers in parentheses indicate bootstrap confidence values for branch points and were generated from 500 replicates (heuristic search). Branch length values are also shown. From Sanchez et al. (1996) and Georges Courbot et al. (1997), with permission.

representing the incremental steps in the evolution of this group of viruses, the evolutionary progeny of continuing selective pressures, the variants that evolve to invade different niches, different host species, and different organs. The emergence of the Reston subtype of Ebola virus from macaques in the Philippines, with its unusual and unexpected pathogenicity pattern (in nonhuman primates versus humans), certainly begs this question.

Most particularly, the question of mutability of the Ebola virus genome has led to public concern, fueled by the press/media, that the virus(es) will somehow acquire new means and patterns of transmission. The error-prone nature of RNA virus polymerases are usually mentioned when this notion is discussed. However, there is much more to acquiring a new phenotype than the incorporation of the occasional polymerase error. The probability that a random polymerase error will result in a new Ebola virus phenotype, say, a virus that is regularly transmitted by aerosol from severe respiratory tract infection, must be small given the amount of human disease seen to date without emergence of such a virus. It is difficult to assess whether it could happen at all, given our rudimentary understanding of the Ebola genome, the small amount of experimental work done on the whole question of aerosol transmissibility, and the early stages of our explanation of the significance of the quasi-species concept for pathogenesis and evolution of RNA viruses. We must "never say never," but at least changes in viral transmission patterns are not common, except when humans intervene.

C. Lessons from the Biological Nature of Filovirus Infections in Cell Culture

Is there relevant biological information from the virology laboratory that might help in pointing the way to the filovirus reservoir host(s)? The filoviruses are readily isolated in Vero cells (Vero is a continuous African green monkey cell line). The viruses can also be grown easily in several other mammalian cell cultures, with or without cytopathic effect; however, the viruses have not been successfully propagated in reptilian, amphibian, or mosquito cells (van der Groen et al., 1978; Swanepoel et al., 1996a,b). Although there are several dramatic and often-discussed correlations between viral growth in cell culture and in vivo, these have often been misleading. The one generalization that has most often proved true is that viruses which fail repeatedly to grow in cell cultures from certain hosts usually do not infect the same hosts in vivo. Thus, such studies of filoviruses must be done, must be interpreted from the perspective of searching for the reservoir host(s), and must not be prejudiced toward a mammalian host or to drive experimental laboratory and field investigations only toward mammalian host candidates.

The filoviruses, in cell culture systems (and in experimental animal host systems), have been difficult to neutralize with convalescent sera and have been resistant to the antiviral effects of interferon-α. These properties are by no means unique to the filoviruses, but they are somewhat unusual among human pathogens. Moreover, they are shared most prominently with viruses such as Lassa virus and lymphocytic choriomeningitis virus (arenaviruses), which are perpetuated in nature via persistent infection of specific rodent hosts.

There has even been speculation that the filoviruses may be plant viruses and/or that plants may play a role in their maintenance. Indeed, infection of plant cells in culture has been tried, but without success (Swanepoel et al., 1996a,b). Of course the idea of involvement of plants cannot be ruled out by such arbitrary experimentation or on theoretical grounds, but in this case consideration of known unique molecular and replicative properties of plant viruses makes us think that this is a very unlikely possibility.

Taken together, cell culture data might bias one toward a mammalian reservoir for the filoviruses. On further reflection and consideration of known properties of many of the viruses of birds, reptiles, amphibians, arthropods, etc., however, the data do not permit definitive conclusions or limiting predictions.

D. Lessons from Clinical and Pathological Characteristics of Filovirus Infections in Humans

An old truism states, "Understanding the nature of a disease in the individual patient is a key to understanding the nature of the disease in the population."

This truism must stem from diseases where there is one host species, one host population involved. However, with recognition of the extra complexity added by multispecies zoonotic transmission cycles, especially an unknown zoonotic transmission cycle, precise lessons seem hard to come by. Nevertheless, a review of the clinical and pathological nature of filovirus infections is warranted.

Marburg and Ebola virus subtypes Zaire, Sudan, and Côte d'Ivoire cause severe hemorrhagic fever in humans—"the evolution of disease often seems inexorable and invariable" (Piot *et al.*, 1978). Following an incubation period of usually 4 to 10 days (extreme range 2 to 21 days for infection by the Zaire subtype of Ebola virus), there is an abrupt onset of illness with initial nonspecific symptoms including fever, severe frontal headache, malaise, and myalgia. Early signs include bradycardia and conjunctivitis, and there may be a macropapular rash most readily evident on white skin (Pattyn, 1978; Peters *et al.*, 1994, 1996; Khan *et al.*, 1996). Deterioration over the following 2 to 3 days is marked by pharyngitis, nausea, and vomiting, progressing to hematemeses and melena. There is prostration and bleeding which is manifested as petechiae, ecchymoses, uncontrolled bleeding from venepuncture sites, and postmortem evidence of visceral hemorrhagic effusions. Death usually occurs 6 to 9 days after onset of clinical disease (range 1 to 21 days). Abortion is a common consequence of infection, and infants born to mothers dying of infection are fatally infected. Convalescence is slow and marked by prostration, weight loss, and often amnesia for the period of acute illness.

In filovirus infections of humans, there is infection of macrophages and endothelial cells throughout the body and infection of the parenchyma of multiple organs, especially the liver and spleen. The infection of these tissues is devastating, with swelling, hemorrhage, and focal necrosis (Murphy *et al.*, 1978; Dietrich *et al.*, 1978; Zaki *et al.*, 1996a,b). Disseminated intravascular coagulation is one of the mechanisms by which the patient is compromised. Destruction of lymphoreticular tissues may be partially responsible for the common absence of an effective immune response. Virus shedding from infected humans occurs from all body surfaces and orifices, including the skin and mucous membranes, and especially from hemorrhagic diatheses (Schnittler *et al.*, 1993; Zaki *et al.*, 1996a,b; Feldmann *et al.*, 1996a).

Of course, there is no way to extend these clinical and pathological observations to predict the nature of infection in the unknown reservoir host(s), but given the systemic nature of infection in humans and the similarity of this pattern of infection in susceptible experimental animals, especially nonhuman primates, it seems likely that if the reservoir host(s) is a mammal, its infection might also involve viral entry via the body surface (mucous membranes of the oro-naso-pharynx or eye, and/or breaks in the skin), and might also require hematogenous spread, systemic organ/tissue infection, and shedding via blood, mucosal surfaces, and the respiratory tract. This pattern of infection is common among other member viruses of the order *Mononegavirales,* such as measles virus, canine distemper virus, mumps virus, Newcastle disease virus in birds, and Sendai virus in mice. In other words, it might be more likely that filovirus infection of a mammalian reservoir host would not be superficial, involving primarily only the respiratory epithelium or intestinal epithelium. This hypothesis can be extended to avian, reptilian, and amphibian candidate reservoir hosts, so its predictive value is not too remarkable.

E. Lessons from the Pathogenetic Characteristics of Filovirus Infections in Experimentally Infected Animals

Another old truism might be stated, "Understanding the nature of a disease in an experimental animal model can hold the key to understanding its nature in its definitive host." In the case of the filoviruses, the same problem exists as in attempting to extend observations from human to laboratory animal infections; until we identify the reservoir host(s) of the filoviruses, we cannot know whether any of the experimental animals that have been studied bring us any closer to understanding the nature of infection in the reservoir host(s).

Nevertheless, many experimental animals have been studied. All of the usual laboratory mammalian species have been inoculated with Marburg and the Zaire subtype of Ebola virus (Kissling *et al.*, 1970; Murphy *et al.*, 1971; Murphy, 1978; Pokhodyaeu *et al.*, 1991; Pereboeva *et al.*, 1993; Ryabchikova *et al.*, 1993, 1996; Murphy and Nathanson, 1996; Peters, 1996b). There have been fewer studies of the Sudan and Reston subtypes and none yet reported of the Côte d'Ivoire (Fisher-Hoch *et al.*, 1992a; Geisbert *et al.*, 1992). Several species of monkeys, mice, guinea pigs, and hamsters are highly susceptible to Marburg virus and the Zaire subtype of Ebola virus, with infection usually ending in death. The Sudan subtype of Ebola virus often causes a self-limited infection in mice, guinea pigs, and the same species of monkeys.

In rhesus monkeys (*Macaca mulatta*), cynomolgus monkeys *(Macaca fascicularis)*, African green monkeys (*Cercopithecus aethiops*), and baboons (*Papio* spp.) inoculated with Marburg virus or the Zaire subtype of Ebola virus, the incubation period is 4 to 6 days, during which time virus replicates to high titer in the reticuloendothelial system (including lymph nodes and spleen), endothelium, liver, and lungs. With the onset of clinical disease, there is severe necrosis of these target organs, which is most evident in liver, and there is interstitial hemorrhage, which is most evident in the gastrointestinal tract (Murphy *et al.*, 1971; Murphy, 1978; Bazhutin *et al.*, 1992; P'yomkov *et al.*, 1995; Luchko *et al.*, 1995; Jaax *et al.*, 1996).

One observation in particular made during the episode of disease in monkeys in the quarantine facility at Reston, Virginia, in 1989, deserves further attention: infection of many macaques (*Macaca fascicularis*) by what turned out to be the Reston subtype of Ebola virus was characterized by the usual clinical signs and histopathological lesions, as noted above (Dalgard *et al.*, 1992; Hayes *et al.*, 1992; Jaax *et al.*, 1995; Jahrling *et al.*, 1996a). However, in addition there were inordinate amounts of respiratory and nasal secretions. These secretions contained over 10^6 plaque-forming units (pfu)/ml of Ebola/Reston virus, and no other viruses or bacteria were found. Given the concern over Ebola transmission by aerosol, this observation needs to be followed up (C. J. Peters, personal communication, 1997).

Thus, there is considerable similarity in the way filoviruses attack humans and certain nonhuman primates. As in other successful infections, the filoviruses have the capability to adapt to experimental animal species with which they have not likely had experience in nature. They are successful in gaining entry, escaping innate resistance factors, finding receptors on specific cells in several organs, finding routes and cellular substrates for systemic infection, overcoming acquired resistance factors such as the host immune response, and assuring shedding and continuation of the virus life cycle. As is the case in human infection, it must be that within this complex systemic pattern of infection, the filoviruses outmaneuver the specific host defense mechanisms of experimental animals by (a) their speed, as animals often die before it might be expected that an effective primary specific inflammatory/immune response would be elicited, and (b) their tropism(s), as the early reticuloendothelial and lymphoid tropisms likely minimize the response that might be elicited otherwise.

Thus, all in all, the same thing might be said from the lessons of filovirus infections in experimentally infected mammals as about infection in humans. Infection in a mammalian reservoir host might be systemic, peracute, and very productive of contagion, but such a hypothesis means little in the absence of direct knowledge.

Two species of insectivorous and one species of fruit-eating bats have been found to support the growth of Ebola virus very well (Swanepoel et al., 1996b): some bats were found to contain virus in their tissues and blood for as long as three weeks. Once again, there is the caveat that many viruses have been isolated from bats without evidence that these animals participate in the maintenance of the virus life cycle (e.g., St. Louis encephalitis virus, Japanese encephalitis virus, chikungunya virus, Rift Valley fever virus, Toscana virus) (American Committee on Arthropod-Borne Viruses, 1985). Nevertheless, bats are an attractive candidate to be the filovirus reservoir host. Most species of bats are migratory and so could account for the seasonality of Ebola virus appearances, and several other member viruses of the order *Mononegavirales* have bats as a primary or secondary reservoir host (e.g., rabies and other lyssaviruses such as the newly identified Australian bat lyssavirus; several other rhabdoviruses such as Mt. Elgon bat virus and Kern Canyon virus; Australian equine morbillivirus) (Murphy et al., 1995; Murray et al., 1995; Young et al., 1996; Fraser et al., 1996). Clearly, the possibility that certain bats may serve as the reservoir host of filoviruses will not be answered easily.

There is no evidence for latency or persistence in any filovirus infection in any experimental animals that have been studied (or in humans for that matter) (Fisher Hoch et al., 1992b; Khan et al., 1996). Occasional cases of subacute uveitis and orchitis have been observed in humans, but these have only reflected a short-term persistence of virus in tissues that are relatively protected from the acute inflammatory/immune response. Neither is there evidence that subclinical or silent productive infections play any important role in experimental animal models.

Evidence that there is a range of temperate virus strains in nature that might complicate our understanding of the pathogenesis and pathology of infections caused by the filoviruses is also lacking. We have no way of knowing whether the range in virulence from the "hottest" known Ebola virus, that is, the Zaire subtype, to the "coolest," that is, the Reston subtype, represents the full spectrum of biotypes/pathotypes in nature. The Reston subtype of Ebola virus may be temperate in humans, but it is quite virulent in nonhuman primates (Jahrling et al., 1996a). If there is a more complex interplay in nature between the filoviruses than we know about and if there are undiscovered less virulent subtypes, then we might easily go off on the wrong tangent in our speculations. Perhaps as long as we remind ourselves that "unnatural" virus–host pairings (e.g., Ebola virus Zaire subtype in humans or experimentally infected monkeys) may be more pathogenic than natural virus–host (reservoir host) pairings, and that for every generalization there is an exception, we can keep an open mind about the lessons from experimental pathology.

A limited number of nonmammalian species, such as pigeons, frogs, geckos, snakes, leafhoppers, spiders, and so forth, have been inoculated with Ebola virus, in every case with negative outcome (Swanepoel *et al.*, 1996a,b). Ebola virus replication has not been demonstrated after intrathoracic inoculation of several species of mosquitoes (Turrell *et al.*, 1996; Swanepoel *et al.*, 1996a,b); however, these negative results may have too casually dismissed all focus on the possible role of arthropods in filovirus transmission. The arthropod host specificity of most arboviruses is quite narrow, and there are many, many different arthropods that could be considered candidate filovirus hosts. Few exotic arthropods have been tested: the many species of biting flies, midges, mites, cimetid bugs, spiders, scorpions, etc., would each require particular wrinkles in experimental design. As in the case of any wild, exotic candidate animal host, it is important that adequate attention be given in such studies to the identification of specimens and the taxonomic system by which they are identified. Attention must also be paid to the archiving of data from such studies, via publication and public database development.

F. Lessons from the Characteristics of Various Other Viruses That Make Them Successful Pathogens

Over the years, certain viral characteristics have been judged in regard to their contribution to the overall "success" of particular viral pathogens. It might be of interest to judge the filoviruses in this regard, hoping thereby to find clues to their reservoir host(s). Characteristics of concern pertain partly to Darwinian forces favoring competitive survival (survival of the fittest) and partly to our sense of anthropocentric forces favoring shared survival [i.e., pertaining to the oft-stated notion that the successful pathogen should drift toward commensalism in its relationship with its natural host(s)] (Holland, 1993). However, every characteristic that may be judged as advantageous in defining a successful virus, perhaps even "The Ultimate Pathogen" (Kilbourne, 1983), also calls to mind examples where an opposite character is favored by other successful pathogens (Nathanson and Murphy, 1996). The following are some characteristics and a judgment of their importance to the success of the filoviruses:

1. *Capacity of the virus to grow rapidly:* Some of the most successful pathogens complete their life cycle in their reservoir hosts very quickly (e.g., Venezuelan equine encephalitis virus, Rift Valley fever virus, vesicular stomatitis virus, influenza virus, paramyxoviruses). The survival advantage here may involve the need for transmission via a fleeting intermediate host, for example, a mosquito that is active for only a short period seasonally, or the need for assuring transmission before host immunity intervenes. The filoviruses do grow rapidly, as indicated by their characteristic growth dynamics in cell cultures as well as their behavior in infected humans and monkeys. Does this characteristic carry over into the reservoir host(s)? Does this characteristic point to an arthro-

pod host or a host present in very large numbers, such as a rodent host, where rapid viral transmission favors staying ahead of host immunity and population immunity?

2. *Capacity of the virus to grow to high titer:* Capacity to grow to high titer is a corollary of the capacity of a virus to grow rapidly and is especially important in the life cycle of arboviruses. Vertebrate host viremia, dependent on productive viral growth in tissues, is necessary for the transmission of virus to an arthropod seeking a blood meal; because blood meal volumes are so small, high viremia likely represents a survival advantage to the virus. Enteric viruses also commonly grow to high titers, in this case so as to favor fecal contamination and the success of the fecal–oral transmission cycle. We know that the filoviruses grow to very high titers in humans and experimental animals; does this point to, as noted above, an arbovirus life cycle? Does this point to a fecal–oral transmission cycle?

3. *Capacity of the virus to be shed quickly:* Some successful pathogens grow quickly, as noted above, but have other mechanisms that increase shedding (e.g., rotaviruses, other enteric viruses, many respiratory viruses). Efficient shedding may be favored by short-lived clinical/physiological qualities of infection, such as diarrhea or productive coughing with catarrh. The filoviruses are shed quickly, but seemingly not in specifically produced body fluids (although in the setting of the primate quarantine facility where the Reston subtype of Ebola virus emerged, spread via respiratory secretions/excretions seems to have represented an exception). All in all, is it reasonable to predict that filoviruses do not employ in nature life cycles like those of the diarrhea viruses or the strict respiratory viruses?

4. *Capacity of the virus to replicate in certain key tissues that favor transmission:* Many successful pathogens employ specific tissues for shedding; for example, many poxviruses, although causing multiorgan systemic infection, are transmitted only after infecting skin epithelium, where they cause a virus-laden exanthem which is infectious by contact or by fomite (even in some cases by mechanical carriage by arthropods). Rabies virus, although neurotropic through most of its infection path, is transmitted in nature via virus shed from salivary gland epithelium. Human immunodeficiency virus (HIV), also systemic in its infection pattern, is quite lymphoreticular in its tropism, but transmission nearly always involves sexual contact or blood contact (sharing needles among intravenous drug users, blood contagion at birth, formerly blood transfusion and certain blood products). In humans and experimentally infected monkeys, the filoviruses are shed from the respiratory tract, skin, and mucous membranes, and especially from blood and blood-contaminated body fluids. Transmission has only occurred via close contact except in the unusual circumstances in monkey quarantine facilities where aerosol transmission has been evident (Peters *et al.,* 1991a,b; Jaax *et al.,* 1995). Do these comparisons point to filovirus transmission in nature only by close contact? Do they point away from transmission involving unusually restricted sites such as salivary glands? Do they point to a separation between major sites of virus replication and sites of virus shedding?

5. *Capacity of the virus to be shed even in the face of rising host immunity:* The capacity for viral shedding follows on the capacity of certain viruses to

evade host defenses and establish persistent infection. This characteristic is related to the capacity of certain viruses to be transmitted congenitally and others to be shed chronically. Often this pattern of infection involves a particular immunopathological interaction of the virus and the host immune system, and often it involves a sequestration of infection in immunologically privileged tissue sites, such as the kidney, salivary glands, and sexual organs. Often this pattern of infection is unique to one (usually the reservoir) but not all host species. The long-term shedding of arenaviruses and hantaviruses in reservoir host urine and saliva, as contrasted with the acute, self-limiting course of infection, with modest, short-lived shedding in humans, is exemplary. The recrudescent shedding of herpesviruses from ganglionic neurons and the long-term shedding of hepatitis B and C viruses by carriers are also models. The filoviruses do not seem to fit this category: in humans, nonhuman primates, and in all other experimental animals that have been studied, no persistent infection has been found. Thus, unless the behavior of the filoviruses in their reservoir host(s) is quite different, this would not seem to be a priority issue for immediate research.

6. *Capacity of the virus to survive after being shed*: Viral survival after shedding is usually an intrinsic quality of the virion, pertaining to its resistance to heat and other physical insults, solvents and other chemical insults, irradiation, etc. The range in environmental stability/instability among all human pathogenic viruses is very wide, indeed. Many but not all viruses that employ the fecal–oral transmission cycle are intrinsically "tough" (e.g., polioviruses, parvoviruses, and reoviruses are very resistant to environmental insults, whereas coronaviruses and toroviruses are not); many but not all viruses that are transmitted by the respiratory route or by other direct means are "fragile" (e.g., rhinoviruses, caliciviruses, and adenoviruses are rather resistant, whereas orthomyxoviruses, paramyxoviruses, and morbilliviruses are not). Further, most hepatitis viruses are rather "tough," having first to resist degradation in the intestine, and most arthropod-borne viruses are "fragile," never having to survive outside their vertebrate and arthropod hosts. Here, further information is needed about filovirus environmental stability. Anecdotes about Ebola virus surviving for months in blood at ambient temperature in the Kikwit hospital must be supported by controlled laboratory study. The evidence that we do have indicates that the filoviruses are rather average in stability (Cheprunov *et al.*, 1995; Belanov *et al.*, 1996). That is, the filoviruses are stable enough to represent the particular risk that has been evident in nosocomial contact transmission episodes and contaminated needle transmission episodes (even when syringe and needle have been held at room temperature for some time), but not enough to represent risk of remote environmental spread. The latter point is complemented by the knowledge that viruses which are most like the filoviruses (i.e., the member viruses of the order *Mononegavirales*) are not transmitted in nature via cycles involving long-term survival outside their host(s).

IV. ECOLOGICAL AND EPIDEMIOLOGICAL CHARACTERISTICS OF FILOVIRUS INFECTIONS

In regard to the issue of the natural history of the filoviruses, the perspective of disease ecology begs unique questions: What ecological and population processes account for the pattern and likelihood of disease emergence within a particular ecosystem? How can knowledge of spatial population dynamics increase the capacity for predicting the spread of disease? What are the ecological influences on evolutionary processes affecting a pathogen and its host(s) that may account for given patterns in disease and disease resistance? What long-term relationships between host and pathogen may be expected? How may the population dynamics and the reproductive ecology of the host, pathogen, and vector be modeled? How will the pathogen respond to global climate change and changing patterns of land and water use? What is the functional role of disease in ecosystem management? In sum, the perspective of disease ecology focuses on the overall *environment* and seeks to determine the influence of a pathogen, its host, and their interrelationship in the overall environmental situation. We shall attempt to apply this perspective to the question of the natural history of the filoviruses.

The epidemiological perspective is quite different; it begs questions relating directly to the determinants, dynamics, and distribution of the disease in the population at risk. Its focus is on the bases for risk of infection and disease in the *population,* as these are determined by characteristics of the virus, of individual hosts, and the host population. There is overlap in that there also may be emphasis on environmental and ecological factors that affect transmission from one host to another, but, as in the case of the field of disease ecology, the goal of infectious disease epidemiology is to meld understanding of all causative factors into a unified whole.

A. Lessons from the Ecological and Epidemiological Characteristics of Filovirus Infections

Filovirus index cases have often occurred in the tropical rainy season. Ebola hemorrhagic fever episodes in Zaire, Sudan, Côte d'Ivoire, and Gabon have all occurred in or near the end of the rainy season, and all have been associated with tropical forest or the marginal zone between tropical forest and savanna. On the one hand, this puts the search for the viruses in the most biologically diverse of all econiches, but, on the other hand, this at least narrows the search area somewhat. In this location, many reservoir host candidates exhibit seasonality in behavior (seasonal breeding, migration, contact with humans or non-human primates and other normal behaviors). Going further, this location may also point to reservoir host candidates with multiyear seasonal behavioral patterns. Does seasonality or multiyear seasonality of filovirus disease episodes point to arthropods? Are the most likely candidates arthropods that are capable of becoming infected and amplifying virus? True arboviruses exhibit marked seasonality in their distribution as evidenced by the distribution of virus in arthropod populations and by the distribution of disease in humans or domestic animals. Are the most likely reservoir candidates rodents, which exhibit marked seasonal breeding and feeding habits and which exhibit multiyear population and behavior differences because of varying food supplies? Are the most likely candidates bats or other seasonally migrating species? Or are the most likely candidates species that exhibit seasonal or multiyear seasonal variations that we do not even know about or recognize as significant in the context at hand?

Further, it must be recognized that within the ecosystems under consideration, the tropical rain forest and bordering savanna, there is great microniche isolation and ecological insularity; that is, there are many, many sites within larger geographical areas in which the filoviruses may invisibly coexist with their reservoir host(s). Many such econiches have never been examined in regard to any virological question, many may not even be known or defined at all. For example, when arbovirologists first studied the tropical rain forests of Africa and South America (Downs, 1973), it was not even understood that the forest canopy, the under-canopy, and each lower level down to ground level each represented a distinct, isolated econiche, each niche filled with different mosquitoes and with different mosquito-borne viruses. It took years of study just to begin to understand the complexity of these econiches. The real question here will be how to simplify the study of tropical rain forest and savanna ecosystems in a way practical enough to fit in with the limited global budget for filovirus field research.

Retrospective ecological studies have been performed at varying time intervals after most of the known filovirus outbreaks, but these studies have been limited when compared to studies of arbovirus ecosystems in tropical forests, as noted above. These classic studies, led in Brazil and Nigeria by O. R. and C. E. Causey and colleagues, as part of the Rockefeller Foundation Virus Program which ran from 1951 to 1970, depended on long-term staffing of field stations on site, with backup from a world-class reference laboratory, the Yale Arbovirus Research Unit, which later became the WHO World Reference Center for Arboviruses (Downs, 1973). The field programs in Brazil and Nigeria led to the identification of more than 100 new arboviruses and defined many of their reservoirs. In contrast, filovirus field studies have been carried out only through brief one-time expeditions to sites where human exposure had occurred, and even then such expeditions have usually been delayed until months or even years after human or nonhuman primate disease episodes. Given the very small number of filovirus field expeditions, their limited scope and scale, and their very narrow focus (i.e., to find one virus), it seems no wonder that there has been so little success. In this regard, the ongoing studies in the Taï Forest in Côte d'Ivoire become extremely important.

Would past arbovirus field programs, such as the Rockefeller Foundation Virus Program, as described above, or the long-running Institut Pasteur programs in several African countries, have recognized filoviruses if they had been present in arthropod or vertebrate specimens? All of these field studies carried out in the 1950s to 1970s were based on the inoculation of specimens (pooled, ground mosquitoes and far fewer animal and bird blood and tissue samples, etc.) into newborn mice. We know that Ebola virus, Zaire subtype, from human specimens, is lethal for this host (van der Groen et al., 1978), but it is not clear whether this virus or other filoviruses would have been identified in the serology-based system designed to identify and classify arboviruses. A few non-arthropod-borne zoonotic viruses were discovered through the Rockefeller Foundation and Institut Pasteur programs, but there is no way to know what the sensitivity of the programs was for such viruses. In any case, filoviruses were not identified in these programs, and perhaps, given the state of biocontainment in the field laboratories serving these programs at the time, it was fortunate that they did not appear (Casals, 1961, 1967; Downs, 1973).

We should examine the basis for the rarity of human infection from the ecological perspective. To do this, we have chosen to divide the subject into four premises, none of which are mutually exclusive and all of which pertain to the strategy that we would employ to search for the reservoir host(s) in the field.

1. *The reservoir host(s) is rare:* There are many species of animals and arthropods that exist in very small populations, often in very limited geographic areas and often in very restricted econiches. The adaptations necessary to assure the perpetuation of such species are many, and many remain unknown. We know very little about the viral flora of such species and even less about their overall zoonotic contribution to human viral diseases. The premise that the reservoir host(s) of the filoviruses is rare calls for a very difficult search approach, one based on the exhaustive examination of as many rare species as possible to find and trap. However, given the difficulty of specimen acquisition, testing this premise would put most demands on the field work and least on laboratory resources. Again, given the minimal level of our knowledge of the rare species in the tropical forests and adjoining savanna, there would seem to be minimal opportunity to focus this kind of search and there would be maximal dependence on good luck in finding the right niche.

2. *The reservoir host(s) is rarely infected:* There are many viral infections that seem to be very rare, occurring at a very low incidence in their host population. Most of the viruses of humans and animals that exist in this way do so through long-term or lifelong persistent infection, often with a long incubation period and/or intermittent low-level shedding. In some instances, such as with some of the agents of the transmissible spongiform encephalopathies, we have no idea how the infectious agent exists in nature. The point is that such infectious agents have adapted and have developed probably unique counters to their seemingly high risk of extinction. Considering the filovirus reservoir host(s), this premise also promises a most difficult search strategy. Here, we would focus on common animal and arthropod species, again also considering the possibility of focal distribution of virus within any overall species distribution and again considering the influence of subspeciation, genotypic variation, and topotype variation. Here, the search strategy would involve exhaustive collection of as large a number of candidate species as possible. Here, again, it would seem as if long-term field laboratory resources would be needed on site in Africa, but there would also be a large burden on the laboratory.

3. *The reservoir host(s) rarely comes in contact with humans:* There are many species of animals and arthropods that for many reasons do not come into contact with humans. Some such species are just ignored because they never have seemed valuable, interesting, or dangerous. Arboreal species, solitary species, camouflaged and reclusive species, as well as species not taken for food would be candidates. A new mind-set would be needed, focusing on the sorts of animals and arthropods that one might otherwise ignore, but systematic collection of moderate numbers of specimens from many such species would be a good start. Again, it would seem that a long-term field presence on site would be needed, but the laboratory burden would not be massive.

4. *The reservoir host(s) is not very infective because it rarely sheds virus, or rarely sheds virus in sites where humans are at risk of infection:* Many known

zoonotic viruses are rarely transmitted to humans (or to domestic animals), either because the reservoir host does not shed much virus (or sheds virus intermittently) or because its habits are not conducive to transmission to nonreservoir hosts. For example, many species represent nearly dead-end hosts for rabies virus. Hantaviruses, although often infecting a substantial proportion of their reservoir rodent host populations, cause few human infections because of the behavior of these rodents (e.g., *Peromyscus maniculatus*, the reservoir host of Sin Nombre virus, the etiologic agent of hantavirus pulmonary syndrome, presents risk of human infection when it enters houses seasonally, whereas *Clethrionomys* spp., the reservoir hosts of Puumala virus, the etiologic agent of nephropathia epidemica in Scandinavia, causes human infection rarely because of its reclusive habits). Eastern equine encephalitis virus, although common in its wild bird niche in parts of eastern United States, only rarely causes human disease because of the feeding habits of its mosquito hosts. Pursuing this premise in regard to the filovirus reservoir host(s) also leads to a search strategy rather like that for species that rarely come in contact with humans (or nonhuman primates). Pursuing this premise would also require the systematic collection of moderate numbers of specimens from many such species. Again, it would seem that a long-term field presence on site would be needed, but the laboratory burden would not be overwhelming.

5. *The virus requires a genetic adaptation before transmission can occur to humans:* Genetic adaptation is the premise underpinning viral "species jumping," the initial crossing of the species barrier. This is not so far from current thinking as one might suppose. The importance of rapid genomic mutation rates and adaptation in RNA viruses has been widely accepted, but it has not been well integrated into our thinking about the natural history of viruses. For example, Ebola virus, Sudan subtype, from human blood or from primary cell culture passage, is infective but not pathogenic for guinea pigs. It requires a few passages before becoming lethal for this new host. Pursuing this premise in regard to the filovirus reservoir host(s) might be the most difficult of all. It would involve much searching in the dark. It would require substantial field collection resources but additionally would require an exceptional scale of manipulative research, much of which would be difficult to tie back to candidate virus–host relationships in nature. Finally, it would dictate the extensive use of polymerase chain reaction (PCR), which would greatly increase overall costs.

On reflection, it is easy to see from the complexity of the above premises why there might continue to be a need for a broad range of candidate specimen collection and testing activities, especially in outbreak settings. The point where such searching becomes redundant or nonproductive, however, is a matter of judgment, not necessarily made clear to all investigators at the same time. In our view, this point has now been reached: we believe that it is time to move beyond episodic collecting in areas near human disease outbreaks to testing the above premises and to incorporating these premises into a comprehensive field/laboratory search enterprise. We believe that it is necessary to reinvent some of the long-term, on-site field strategies that guided the Rockefeller Foundation and Institut Pasteur arbovirus programs, adapting them to the problem at hand. We believe that such a program should also incorporate a basic virology research element, tying in viral molecular genetic approaches more tightly and bringing such approaches from the laboratory to the field.

B. Confounding Role of Serosurvey Data in Trying in Determine Ecological and Epidemiological Characteristics of Filovirus Prevalence

In several serosurveys, all performed with the indirect fluorescent antibody (IFA) technique, a high prevalence of Ebola antibodies has been found in apparently normal human populations. Given the absence of any confirmatory testing in these serosurveys, and the failure to find concordance when the IFA technique has been run comparatively with various confirmatory tests, it is remarkable that conclusions reached from such surveys still influence our ideas about the epidemiology and natural history of the filoviruses. The usual IFA technique employs acetone-fixed filovirus-infected cells (inactivated by γ-irradiation) as substrate for testing for the presence of antibodies in untreated human or animal serum (Elliott *et al.*, 1993).

As examples of the confusion that the use of the IFA test has caused, consider the fact that in surveys of humans from Africa, Alaska, and Panama, monkeys from Asia, and a variety of animal species obtained worldwide, in the absence of any recognized disease, a high prevalence of antibodies to filoviruses, particularly Ebola virus subtypes, has been reported (Pattyn, 1978; Johnson *et al.*, 1981; Stansfield *et al.*, 1982; van der Walls *et al.*, 1986; Meunier *et al.*, 1987a,b; Gonzalez *et al.*, 1989; CDC, 1990a,b; Peters *et al.*, 1991a; Johnson *et al.*, 1993a,b). This IFA reactivity is not a simple technical artifact; in fact, its cause remains unknown, although there are suspicions that it reflects cross-reactivity from infectious with extremely distantly related viruses such as other member viruses of the order *Mononegavirales*. More specifically, in one recent IFA-based study (admittedly, incorporating some confirmatory testing), it was reported that there is a high prevalence of antibodies in inhabitants of the Congolese basin of western Africa: up to 30% prevalence (Ebola virus, subtype Zaire) was reported in people living in the rain forest (with greater than 20% in the Pygmy ethnic group and 14% in Bantu people living in the same area), and up to 10% prevalence (Ebola virus, subtype Sudan) in people living in the savanna. Antibody to filoviruses was reported to also be present in domestic and wild animals: dogs, pigs, guinea pigs, and monkeys (*Cercopithecus aethiops* and *C. ascanius*) (Gonzalez, 1996).

Clearly, such studies can have an overwhelming influence on our thinking about the reservoir host(s) of the filoviruses. Such studies require independent confirmation, but moreover they should be conducted from their start with gold-standard techniques. What is needed is such a gold-standard test for filovirus antibodies in human and animal sera that would engender the confidence of all investigators in the field, a test that would have the same credibility as the virus neutralization test has had in assaying polio or measles or Japanese encephalitis virus antibodies. Until recently, unfortunately, candidate confirmatory tests have not proved particularly useful: (a) western blot test results have been ambiguous; (b) no viral hemagglutinin has been detected, so there can be no hemagglutination-inhibition test; and (c) very little or no neutralization by convalescent serum has been identified, so there can be no viral neutralization test.

As of 1997, the best bet for a gold-standard confirmatory test is a particular enzyme-linked immunosorbent assay (ELISA) (Ksiazek *et al.,* 1992). This ELISA is relatively simple to use; it is based on inactivated infected cell lysate as antigen and employs an essential negative control antigen test for every serum specimen tested. Serosurveys, employing this ELISA, carried out on patients, contacts of patients, and others in Kikwit in 1995 indicated an extremely low prevalence of antibodies, affirming that human infections are usually symptomatic. This test has been evaluated for specificity and sensitivity: it has been concordant with virus isolation results when used on sera from humans and monkeys known to have been infected with filoviruses, and it has been negative when used on sera from thousands of humans and monkeys from areas of North America where there never has been any evidence of the presence of a filovirus. Among the latter sera there were some that were reactive by the IFA test, but none exhibited ELISA reactivity. Seropositivity using this ELISA has also been shown to be maintained for a long period after infection: in a small number of human sera collected more than 10 years after infection, antibody has been detectable, whereas IFA results have been equivocal or negative.

V. ONGOING AND NEEDED RESEARCH ON FILOVIRUSES IN THE FIELD AND IN THE LABORATORY

A. Ongoing and Needed Field-Based Filovirus Research

Many filovirus investigators believe that a key feature of any search for the reservoir host(s) of the filoviruses must involve extensive examination of vertebrates near the places in Africa where human cases have occurred. It is argued that it is in such sites where the reservoir host(s) must participate in a transmission cycle, whether this involves direct contact transmission, fomite tranmission, arthropod transmission, or whatever. It is argued that vertebrates should be the primary focus of the search; vertebrates might at least serve as sentinels, providing evidence of past experience with the virus(es) by the presence of antibody. It is argued that vertebrates, specifically mammals, are the most likely reservoir hosts, based on results of infection of experimental animals and cultured mammalian cells. It is further argued that arboreal species should be considered leading candidate reservoir hosts, given the recent finding of a dead, Ebola-infected red colobus monkey in the Taï Forest, Côte d'Ivoire, in association with disease in chimpanzees (Formenty *et al.,* 1996; ProMED Internet news item, dated 16 November 1996). Finally, it is argued that migratory bats should be considered important candidate reservoir hosts. There have been anecdotes involving bats in several filovirus disease episodes, and, even though bats are common in tropical Africa, these clues must be followed up. Further, the involvement of bats in the natural history of some of the viruses most closely related to the filoviruses (i.e., some member viruses of the order *Mononegavirales*), along with the capacity of bats to sustain some viral infections for inordinately long periods, adds modest credibility to their candidacy. We believe that in pursuing such candidates, sound hypotheses must be established and explored. A sound work plan must be set up and followed so that when initial hypotheses are rejected there is a clear course of action to test the next in priority order. A proper shared information system is crucial in this regard, given the diversity of field projects to be undertaken and the geographic separation of scientists from different institutions from several different countries. The publication of all results will be essential if we are to be able to profit from experiences and findings, even if findings are negative.

Many of the above considerations have been employed in designing the collaborative search by investigators from several institutions for the Ebola virus reservoir host(s) in the forests around Kikwit, Zaire, and in the Taï Forest, Côte d'Ivoire (Ksiazek, 1996; Swanepoel *et al.*, 1996a,b; Formenty *et al.*, 1996). Investigators have started by examining sites that have been connected with the person identified as the index case in these outbreaks. These searches span critical ecological zones such as primary and secondary forest as well as savanna; however, there is a concentration on the rain forest zone because of the evidence mentioned above. As noted above, it has been decided that mammals should be the primary focus of the collections. However, not to be "too smart" about predicting a mammalian reservoir host(s), the investigators are testing any vertebrates that enter traps and are also testing some market animals obtained in nearby towns. The issue of an arthropod as reservoir host has been dealt with by collecting a large variety of species, including mosquitoes, ticks, sandflies, cimetids, spiders, and others.

In Kikwit, Zaire, 1995, a major question was whether to begin specimen collection in the dry season following the epidemic (which had begun in the rainy season) or to wait until the next rainy season. There was no clear choice. In favor of an immediate initiation of the studies was the well-known propensity for most zoonotic viruses to be transmitted intensively only at infrequent intervals; after all, the discovery and last known activity of Ebola virus in Africa had been in 1976–1979. There was always the possibility that a chronically infected reservoir host might still be present or some anomaly of animal species distribution might be noticed. Because most species in the rain forest live less than 1 year immediate action represented the best opportunity to detect the presence of antibodies in sentinel species. Against collecting at that time was the suggestion of rainy season proclivities of Ebola transmission and the likelihood of missing migratory species, intermittently active and short-lived arthropods, or other temporally unique opportunities. Logistics and politics also led to the decision to start immediately, in the dry season: the Zairian political climate was becoming more fragile, the cooperation of the local people was waning, the willingness of other institutions to collaborate was fleeting, the availability of funding was limited to the immediate accounting period, and, most importantly, the investigators hoped to do their work without the ensnaring mud and disruptive tropical downpours of the rainy season. The downside of this decision was addressed by more limited studies done during the next rainy season with a concentration at that time on migratory bats, unexpected species, and certain arthropods.

At the time of this writing in 1997, analysis of the samples collected during the field project is not yet finished. The complex and time-consuming process of identifying the animals collected occupied many experts for some time; for example, a new shrew species was identified in the course of this work. So far, no evidence of the presence of Ebola virus or antibodies to it have been found in any specimen (T. G. Ksiazek, R. Swanepoel, P. Jahrling, and colleagues, personal communication, 1997).

B. Ongoing and Needed Laboratory-Based Filovirus Research

There are many laboratory-based experiments that must be done, (a) to complement the above field-based research, (b) to support disease prevention and control activities, and (c) to bring the state of our basic knowledge about the filoviruses and the infections they cause to the same state that we have come to expect for all important pathogenic viruses and viral diseases. The following represent some research tacks of high priority; the listing should not be taken as all-inclusive.

1. *Molecular biology of the viruses:* Much progress has been made in characterizing the filoviruses, their genomes, and their proteins. However, we are just beginning to understand the function of the viral proteins, especially from the perspective of their role in the pathogenesis of disease. We know very little, indeed, about how viral gene expression and gene products contribute to the perpetuation of the viruses in nature. Because most molecular biology research is investigator-initiated, if funds and facilities [at biosafety level 4 (BSL 4)] were available, we feel that progress would be rapid in this area.

2. *Pathogenesis of viral infection:* Much progress has been made in regard to descriptive pathogenesis research, but now it is time for manipulative pathogenesis research approaches to be expanded. By manipulating the infectious processes themselves and the host responses that are engendered by the infection (innate inflammatory response, acquired immune response), clues regarding pathogenetic weak links would, it is hoped, be found. With other viral diseases, such clues have often been the keys to developing preventive and therapeutic regimes. In particular, attention must be given to understanding of the potential for filoviruses to employ aerosol transmission, especially in certain settings such as hospitals and experimental animal facilities.

3. *Immunology:* Given the poor neutralizing antibody response evoked by filovirus infections in naturally infected humans and in experimental animals, much more basic immunology research is warranted. Such research must focus on the details of filovirus antigen presentation and processing, means to overcome this hyporesponsiveness, and means to stimulate T-cell-based responses.

4. *Vaccinology:* At present, there is no justification for actual filovirus vaccine development: the number of people at risk is viewed as very small and the cost very large. However, given the long lead time involved and the refractory nature of the viruses in immunoprophylaxis experiments that have been done over the years, it seems prudent to extend basic immunology and molecular biology studies in ways that would accelerate vaccine development should it be necessary. We need to develop the means to be able to move quickly from principles to practice, should this become necessary (i.e., to move from an understanding of protective epitopes to vaccine candidates. It is not enough to wait

until molecular virology and pathogenesis research yields every bit of information that one would want for rational vaccine design—there is never enough information in this regard—but we do need basic information that can be translated into practical vaccine development on short notice. Should filovirus epidemics occur on a larger scale, we would then be in a position to protect the increased numbers of laboratory and field workers and medical care personnel that would be drawn into control programs.

5. *Immunoglobulin therapy:* Russian scientists first showed that very high titered anti-Ebola equine globulins may be valuable in post exposure prophylaxis (Mikhailov *et al.*, 1994; Borisevich *et al.*, 1995; Jahrling *et al.*, 1996b; Markin *et al.*, 1997). Anecdotal evidence has suggested that whole blood from convalescent patients may be protective when transfused into patients with Ebola hemorrhagic fever (Muyembe-Tamfun *et al.*, 1996). On the basis of these clues, work is underway at the Scripps Research Institute to develop highly avid neutralizing human monoclonal antibodies against Ebola virus. This work involves the use of mRNA immunoglobulin gene libraries constructed from bone marrow specimens obtained in Kikwit, Zaire. More of this kind of research must be supported.

6. *Therapeutic drug design and development:* The filoviruses are the only hemorrhagic fever agents for which we have no proven or investigational drug therapy. In the absence of success with vaccine development, there is a desperate need for drugs to protect laboratory workers in case of accident. There is also a desperate need for drugs to treat patients in hospital-based outbreaks in Africa. Indeed, the availability of drugs would also draw patients into hospitals during epidemics, thereby minimizing household and community transmission. The first favorable drug therapy results have finally been obtained (a S-adenosylhomocysteine hydrolase inhibitor) (Huggins *et al.*, 1996), but much more needs to be done. Here is a place where the combined resources of the U.S. Army Medical Research Institute of Infectious Diseases and the National Institute of Allergy and Infectious Diseases of the National Institutes of Health should be brought to bear.

VI. IMPLICATIONS CONCERNING GLOBAL PUBLIC HEALTH

Ebola virus must be dealt with in the context of its character as the etiologic agent of an emerging viral disease (Peters *et al.*, 1991a,b, 1994; 1996a; Murphy, 1993; Murphy and Nathanson, 1994; Sanchez *et al.*, 1995; Monath, 1996). To do this we should examine the emergence of human viral diseases in a historical context. Many of the most important viral diseases of history emerged only following the development of cities containing sufficient inhabitants to support their circulation. Measles is perhaps the best example: classic studies grounded in the work of Panum on the Faroe Islands in 1846 established that about 500,000 people are needed to support continuous transmission of measles virus (smaller populations are intermittently infected from outside) (Panum, 1940; Nathanson and Murphy, 1996). Such large population centers did not occur before the rise of irrigated agriculture in the Middle East around 5000 years ago. This development coincided with increasing domestication of sheep, goats, and cattle, which carry viruses considered to be the progenitors of human measles virus (sheep and goats: peste-des-petits-ruminants virus; cattle: rinderpest virus).

From this history lesson, we might speculate that future emergent viruses will come from new ecological niches, sites where new selective pressures favor the emergence of new variant viruses. The new megacities of Africa and other tropical zones of the world may provide such new niches for the emergence of new variant viruses and other infectious agents. Modern air transportation could deliver a new pathogen to any other megacity in the world in hours. Further, we might speculate that the most dangerous new, emerging viral disease would be one that is spread by the airborne route (Mims, 1991). This notion follows on the concept of "The Ultimate Pathogen" (Kilbourne, 1983, 1996). Diseases transmitted by the respiratory route, such as influenza, have proved to be very difficult to control, partly because of their rapid spread. This notion as it pertains to filovirus diseases has certainly been brought to the attention of the public by the press/media; even so, it must be dealt with by appropriate research. We know that Ebola virus has the capacity to invade the lung and to replicate very productively there. We suspect that invasion of lung comes late in course of infection in humans and evokes too little cough to generate an effective aerosol. However, we do not know whether this character of the infection might change in the future, with mutations favoring aerosol transmission being fixed through Darwinian selective forces.

Why should the average scientist or citizen be concerned about filoviruses? Is there a significant risk to Africa that compares with the everyday problems of malaria, yellow fever, pneumonia, diarrhea, and other more common causes of infectious disease mortality? Should there be a real concern in North America or Europe?

The danger from filoviruses is difficult to evaluate because of our limited knowledge base, and therefore these questions are difficult to answer objectively. There is a need to understand these viruses and the diseases they cause just because the risk they represent is unknown and the risk of future episodes is so unpredictable. This judgment takes nothing away from our need to understand the more common infectious agents and diseases of the tropics as well. We must not lose sight of the need to develop a knowledge base for all dangerous pathogens, particularly where the benefit from a small investment in epidemiological and laboratory research is likely to be so great. For example, we need to find the natural reservoir of the filoviruses and learn how their prevalence in the environment is regulated. We need to find out how transmission of these viruses to humans is regulated. In Africa, the emergence of Ebola virus could dramatically increase if the unknown reservoir increased in numbers, if it changed its behavior, or if ecological factors brought additional reservoir hosts into play. We need to know enough to anticipate such changes and to intervene rapidly should they occur.

Ecological changes that can contribute to disease emergence are common happenings in these times of rapid, uncontrolled exploitation of the tropical forests of the world and rapid, uncontrolled development of the cities of the tropics. Perhaps most important is the reality that across sub-Saharan Africa population centers lack the social organization that is needed for disease prevention and control. Present conditions of hygiene and sanitary management and the paucity of medical care and disease surveillance will continue, and they will continue to present risks of new infectious disease emergence. As western-style hospitals become more affordable for Africans, nosocomial Ebola amplification will increase. In this context, it is elementary to predict that outbreaks of filovirus disease will continue to occur in Africa, in all likelihood at an increasing frequency and in larger and larger epidemics.

Again, why should we in the developed world be concerned? Even if we say that we live in a global community and that there is a possibility that air travel could bring Ebola virus to our doorstep, quickly, what is the worst that might happen? If the worst that might happen is an occasional importation resulting in a small cluster of cases, possibly involving medical staff, should we be concerned? If such episodes are unpredictable in time and place, should we not just wait and react after the fact? Of course, the answer to such questions lies in past experiences: the same questions were asked when acquired immunodeficiency syndrome (AIDS) first appeared in Los Angeles and New York, and the wait-and-see answer did not serve our society well at all. One of the poorly understood findings from the Kikwit epidemic was that some Ebola patients were much more dangerous than others. Two individual patients were the cause of more than 50 contact cases (Khan *et al.*, 1996). We do not think that the concerned public would be satisfied if its public health leaders decided on a wait-and-see approach for dealing with Ebola or the other diseases with similar epidemic potential.

The over-arching global impact of emerging infectious diseases was begged by the U.S. Institute of Medicine study, published as *Emerging Microbial Threats* (Lederberg *et al.*, 1992), and answered by the Centers for Disease Control and Prevention Report, *Addressing Emerging Infectious Disease Threats, A Prevention Strategy for the United States* (CDC, 1994). The World Health Organization has answered similarly. The answer is based on the development of a global integrated enterprise, an early warning system, with new capacity for (a) disease surveillance, (b) diagnostics, (c) an integral research base, (d) a communications system, (e) a technology transfer system, (f) a global prevention/intervention and emergency response infrastructure, (g) a global training program, and (h) a stable funding base.

This enterprise need not be thought of as so expansive, so expensive, as to be unrealistic. For example, in regard to the filovirus diseases, surveillance need not be expensive and emergency response need only provide hospital hygiene and training and supplies for strict barrier-nursing practices and simple laboratory procedures to make diagnosis easier (Peters *et al.*, 1991b; CDC, 1995d; Lloyd *et al.*, 1996). In particular, this enterprise must be built on a more substantial research base, and this in turn requires adequate trained staff and laboratory facilities for work on the BSL 4 pathogens (CDC/NIH, 1993). Safe, productive research demands a core of trained, career scientists with knowledge of the pathogens and procedures to work with them, and these persons are not created in a short didactic course or readily carried over directly from other fields; indeed, there is underutilized BSL 4 research space in the United States. The nature and extent of present disease risks are such that present facilities around the world cannot support an appropriate scope and scale of urgently needed research work. Greater high containment laboratory capacity is urgently needed, along with funding to allow experts from academic institutions to collaborate with colleagues in government agencies in the needed work. This need must be met in all concerned developed countries, on behalf of the people of all less developed countries.

REFERENCES

Amblard, J., Obiang, P., Prehaud, C., Prehaud, C., Bouloy, M., and LeGuenno, B. (1997). Identification of the Ebola virus in Gabon in 1994. *Lancet* **349**, 181–182.

American Committee on Arthropod-Borne Viruses. (1985). "International Catalogue of Arboviruses" American Society of Tropical Medicine and Hygiene, San Antonio, Texas.

Baron, R. C., McCormick, J. B., and Zubeir, O. A. (1983). Ebola hemorrhagic fever in southern Sudan: hospital dissemination and intrafamilial spread. *Bull. WHO* **6**, 997–1003.

Bazhutin, N. B., Belanov, E. F., Spiridonov, V. A., Voitenko, A. V., Krivenchuk, N. A., Krotov, S. A., Omelchenko, N. I., Tereschenko, A. Y., and Khomichev, W. (1992). The influence of the methods of experimental infection with Marburg virus on the features of the disease process in green monkeys [in Russian]. *Vopr. Virusol.* **37**, 153–156.

Becker, Y. (1995). Retrovirus and filovirus immunosuppressive motif and the evolution of virus pathogenicity in HIV-1, HIV-2, and Ebola viruses. *Virus Genes* **11** (2–3), 191–195.

Belanov, Y. F., Muntyanov, V. P., Kryuk, V. D., Sokolov, A. V., Bormotov, N. I., P'yankov, O. V., and Sergeyev, A. N. (1996). Retention of Marburg virus infecting capability on contaminated surfaces and in aerosol particles [in Russian]. *Vopr. Virusol.* **41** (1), 32–34.

Borisevich, I. V., Mikhailov, V. V., Krasnianskii, V. P., Gradoboev, V. N., Lebedinskaia, E. V., Potryvaeva, N. V., and Timan'kova, G. D. (1995). Development and study of the properties of immunoglobulin against Ebola fever [in Russian]. *Vopr. Virusol.* **40** (6), 270–273.

Bowen, E. T. W., Lloyd, G., Harris, W. J., Platt, G. S., Baskerville, A., and Vella, E. E. (1977). Viral haemorrhagic fever in southern Sudan and northern Zaire. *Lancet* **1**, 571–573.

Burkreyev, A. A., Volchkov, V. E., Blinov, V. M., and Netesov, S. V. (1993). The G-P protein of Marburg virus contains a region similar to the immunosuppressive domain of oncogenic retrovirus P15E proteins. *FEBS Lett.* **323**, 183–187.

Calain, P. (1996). Protective measures and management of Ebola patients. *Proceedings of International Colloquium on Ebola Virus Research, Antwerp.*

Casals, J. (1961). Procedures for identification of arthropod-borne viruses. *Bull. WHO* **24**, 727–734.

Casals, J. (1967). Immunological techniques for animal viruses. *In* "Methods in Virology" (K. Maramorosch and H. Koprowski, eds.), Vol. 1, pp. 113–198. Academic Press, New York.

Centers for Disease Control and Prevention. (1990a). Update: Filovirus infection among persons with occupational exposure to nonhuman primates. *Morbid. Mortal. Weekly Rep.* **39**, 266–273.

Centers for Disease Control and Prevention. (1990b). Update: Filovirus infection in animal handlers. *Morbid. Mortal. Weekly Rep.* **39**, 221.

Centers for Disease Control and Prevention. (1994). "Addressing Emerging Infectious Disease Threats, A Prevention Strategy for the United States." Centers for Disease Control and Prevention, Atlanta, Georgia.

Centers for Disease Control and Prevention. (1995a). Outbreak of Ebola viral hemorrhagic fever—Zaire, 1995. *Morbid. Mortal. Weekly Rep.* **44**(19), 381–382.

Centers for Disease Control and Prevention. (1995b). Update: Outbreak of Ebola viral hemorrhagic fever—Zaire, 1995. *Morbid. Mortal. Weekly Rep.* **44**(20), 399.

Centers for Disease Control and Prevention. (1995c). Update: Outbreak of Ebola viral hemorrhagic fever—Zaire, 1995. *Morbid. Mortal. Weekly Rep.* **44**(25), 468–475.

Centers for Disease Control and Prevention. (1995d). Update: Management of patients with suspected viral hemorrhagic fever—United States. *Morbid. Mortal. Weekly Rep.* **44**(25), 475–479.

Centers for Disease Control and Prevention. (1996). Ebola–Reston virus infection among quarantined nonhuman primates—Texas, 1996. *Morbid. Mortal. Weekly Rep.* **44**(15), 314–316.

Centers for Disease Control and Prevention/National Institutes of Health. (1993). "Biosafety in Microbiological and Biomedical Laboratories," 3rd Ed. U.S. Department of Health and Human

Services, U.S. Government Printing Office, Washington, D.C., 1993-017-040-00523-17. HHS Publication No. (CDC) 93-8395.

Cheprunov, A. A., Chuev, Y. P., P'yankov, V., and Efimova, I. V. (1995). Effects of some physical and chemical factors on inactivation of Ebola virus [in Russian]. *Vopr. Virusol.* **40**(2), 40–43.

Cianciolo, G. J., Copeland, T. J., Oroszlan, S., and Snyderman, R. (1985). Inhibition of lymphocyte proliferation by a synthetic peptide homologous to retroviral envelope protein. *Science* **230**, 453–455.

Conrad, J. L., Isaacson, M., Smith, E. B., Wulff, H., Crees, M., Geldenhuys, P., and Johnston, J. (1978). Epidemiologic investigation of Marburg virus disease, Southern Africa, 1975. *Am. J. Trop. Med. Hyg.* **27**, 1210–1215.

Crichton, M. (1969). "The Andromeda Strain." Dell, New York.

Dalgard, D. W., Hardy, R. J., Pearson, S. L., Pucak, G. J., Quander, R. V., Zack, P. M., Peters, C. J., and Jahrling, P. B. (1992). Combined simian hemorrhagic fever and Ebola virus infection in cynomolgus monkeys. *Lab. Anim. Sci.* **42**, 152–157.

Dietrich, M., Schumacher, H. H., Peters, D., and Knobloch, J. (1978). Human pathology of Ebola virus infection in the Sudan. *In* "Ebola Virus Hemorrhagic Fever" (S. R. Pattyn, ed.), pp. 37–42. Elsevier/North Holland, Amsterdam.

Downs, W. G. (1973). "The Arthropod-Borne Viruses of Vertebrates. An Account of the Rockefeller Foundation Virus Program, 1951–1970." Yale Univ. Press, New Haven, Connecticut.

Elliott, L. H., Bauer, S. P., Perez-Oronoz, G., and Lloyd, E. S. (1993). Improved specificity of testing methods for filovirus antibodies. *J. Virol. Methods* **43**, 85–100.

Emond, R. T. D., Evans, B., Bowen, E. T. W., and Lloyd, G. (1977). A case of Ebola virus infection. *Br. Med. J.* **2**, 541–544.

Feldmann, H., and Klenk, H.-D. (1996). Marburg and Ebola viruses. *Adv. Virus Res.* **47**, 1–52.

Feldmann, H., Mühlberger, E., Randolf, A., Will, C., Kiley, M. P., Sanchez, A., and Klenk, H.-D. (1992). Marburg virus, a filovirus: Messenger RNAs, gene order, and regulatory elements of the replication cycle. *Virus Res.* **24**, 1–19.

Feldmann, H., Klenk, H.-D., and Sanchez, A. (1993). Molecular biology and evolution of filoviruses. *Arch. Virol. Suppl.* **7**, 81–100.

Feldmann, H., Nichol, S. T., Klenk, H.-D., Peters, C. J., and Sanchez, A. (1994). Characterization of filoviruses based on differences in structure and antigenicity of the virion glycoprotein. *Virology* **199**, 469–473.

Feldmann, H., Bugany, H., Mahner, F., Klenk, H.-D., Drenckhahn, D., and Scnittler, H. J. (1996a). Filovirus-induced endothelial leakage triggered by infected monocytes/macrophages. *J. Virol.* **70**, 2208–2214.

Feldmann, H., Slenczka, W., and Klenk, H.-D. (1996b). Emerging and reemerging of filoviruses. *Arch. Virol. Suppl.* **11**, 77–100.

Fisher-Hoch, S. P., Brammer, T. L., Trappier, S. G., Hutwagner, L. C., Farrar, B. B., Ruo, S. L., Brown, B. G., Hermann, L. M., Perez-Oronoz, G. I., Goldsmith, C. S., Hanes, M. A., and McCormick, J. B. (1992a). Pathogenic potential of filoviruses: Role of geographic origin of primate host and virus strain. *J. Infect. Dis.* **166**, 753–763.

Fisher-Hoch, S. P., Perez-Oronoz, G. I., Jackson, E. L., Hermann, L. M., and Brown, B. G. (1992b). Filovirus clearance in non-human primates. *Lancet* **340**, 451–453.

Formenty, P., Boesch, C., LeGuenno, B., Akoua-Koffi, C., and Diarra-Nama, J. (1996). Natural history of Ebola virus in the Tai Forest, Côte d'Ivoire. *Proceedings of International Colloquium on Ebola Virus Research, Antwerp.*

Fraser, G. C., Hooper, P. T., Lunt, R. A., Gould, A. R., Gleeson, L. J., Hyatt, A. D., Russell, G. M., and Kattenbelt, J. A. (1996). Encephalitis caused by a lyssavirus in fruit bats in Australia. *Emerging Infect. Dis.* **2**(4), 327–331.

Gear, J. S. S., Cassel, G. A., Gear, A. J., Trappler, B., Clansen, L., Meyers, A. M., Kew, M. C., Bothwell, T. H., Sher, R., Miller, G. B., Schneider, J., Koornhoff, H. J., Gomperts, E. D., Isaacson, M., and Gear, J. H. S. (1975). Outbreak of Marburg virus disease in Johannesburg. *Br. Med. J.* **4**, 489–493.

Geisbert, T. W., Jahrling, P. B., Hanes, M. A., and Zack, P. M. (1992). Association of Ebola-related Reston virus particles and antigen with tissue lesions of monkeys imported to the United States. *J. Comp. Pathol.* **106**, 137–152.

Geisbert, T. W., and Jahrling, P. B. (1995). Differentiation of filoviruses by electron microscopy. *Virus Res.* **39**, 129–150.

Georges, A. J., Renaut, A. A., Bertherat, E., Baize, S., Leroy, E., LeGuenno, B., Lepage, J., Amblard, J., Edzang, S., and Georges Courbot, M. C. (1996). Recent Ebola virus outbreaks in Gabon from 1994 to 1996: Epidemiologic and control issues. *Proceedings of International Colloquium on Ebola Virus Research, Antwerp.*

Georges Courbot, M. C., Lu, C. Y., Lansoud-Soukate, J., Leroy, E., and Baize, S. (1997a). Isolation and partial molecular characterisation of a strain of Ebola virus during a recent epidemic of viral haemorrhagic fever in Gabon. *Lancet* **349,** 181.

Georges Courbot, M. C., Sanchez, A., Lu, C. Y., Baize, S., Leroy, E., Lansout-Soukate, J., Tevi-Benissan, C., Georges, A. J., Trappier, S. G., Zaki, S. R., Swanepoel, R., Leman, P. A., Rollin, P. E., and Peters, C. J. (1997b). Isolation and phylogenetic characterization of Ebola viruses causing different outbreaks in Gabon. *Emerging Infect. Dis.* **3,** 59–62.

Gonzalez, J.-P., Josse, R., and Johnson, E. D. (1989). Antibody prevalence against hemorrhagic fever viruses in randomized representative Central Africa populations. *Res. Virol.* **140,** 319–331.

Gonzalez, J.-P. (1996). Human and animal filovirus surveillance in endemic areas of central Africa. *Proceedings of International Colloquium on Ebola Virus Research, Antwerp.*

Hayes, C. G., Buran, J. P., Ksiazek, T. G., Del Rosario, R. A., Miranda, M. E. G., Manaloto, C. R., Barrientos, A. B., Robles, C. G., Dayrit, M. M., and Peters, C. J. (1992). Outbreak of fatal illness among captive macaques in the Philippines caused by an Ebola-related filovirus. *Am. J. Trop. Med. Hyg.* **46,** 664–671.

Heymann, D. L., Weisfeld, J. S., Webb, P. A., Johnson, K. M., Cairns, T., and Berquist, H. (1980). Ebola hemorrhagic fever: Tandala, Zaire, 1977–78. *J. Infect. Dis.* **142,** 373–376.

Holland, J. J., ed. (1993). Genetic diversity of RNA viruses. *Curr. Top. Microbiol. Immunol.* **176,** 1–226.

Huggins, J., Tseng, C., Laughlin, C., and Bray, M. (1996). Antiviral drug therapy for filovirus infections. *Proceedings of International Colloquium on Ebola Virus Research, Antwerp.*

Ivker, R. (1997). Argument over Ebola in Gabon. *Lancet* **349,** 264.

Jaax, N., Jahrling, P., Geisbert, T., Geisbert, J., Steele, K., McKee, K., Nagley, D., Johnson, E., Jaax, G., and Peters, C. (1995). Transmission of Ebola virus (Zaire strain) to uninfected control monkeys in a biocontainment laboratory. *Lancet* **346,** 1669–1671.

Jaax, N. K., Davis, K. J., Geisbert, T. J., Vogel, P., Jaax, G. P., Topper, M., and Jahrling, P. B. (1996). Lethal experimental infection of rhesus monkeys with Ebola–Zaire (Mayinga) virus by the oral and conjunctival route of exposure. *Arch. Pathol. Lab. Med.* **120,** 140–155.

Jahrling, P. B., Geisbert, T. W., Dalgard, D. W., Johnson, E. D., Ksiazek, T. G., Hall, W. C., and Peters, C. J. (1990). Preliminary report: Isolation of Ebola virus from monkeys imported to USA. *Lancet* **335,** 502–505.

Jahrling, P. B., Geisbert, T. W., Jaax, N. K., Hanes, M. A., Ksiazek, T. G., and Peters, C. J. (1996a). Experimental infection of cynomolgus macaques with Ebola–Reston filoviruses from the 1989–1990 U.S. epizootic. *Arch. Virol. Suppl.* **11,** 115–134.

Jahrling, P. B., Geisbert, J., Swearengen, J. R., Jaax, G. P., Lewis, T., Huggins, J. W., Schmidt, J. J., LeDuc, J. W., and Peters, C. J. (1996b). Passive immunization of Ebola virus-infected cynomolgus monkeys with immunoglobulin from hyperimmune horses. *Arch. Virol. Suppl.* **11,** 135–140.

Johnson, E. D., Gonzales, J.-P., and Georges, A. (1993a). Haemorrhagic fever virus activity in equatorial Africa: Distribution and prevalence of filovirus reactive antibody in the Central African Republic. *Trans. R. Soc. Trop. Med. Hyg.* **87,** 530–535.

Johnson, E. D., Gonzales, J.-P., and Georges, A. (1993b). Filovirus activity among selected ethnic groups inhabiting the tropical forest of equatorial Africa. *Trans. R. Soc. Trop. Med. Hyg.* **87,** 536–538.

Johnson, E. D., Jaax, N., White, J., and Jahrling, P. (1995). Lethal experimental infections of rhesus monkeys by aeorsolized Ebola virus. *Int. J. Exp. Pathol.* **76,** 227–236.

Johnson, E. D., Johnson, B. K., Silverstein, D., Tukei, P., Geisbert, T. W., Sanchez, A. N., and Jahrling, P. B. (1996). Characterization of a new Marburg virus isolated from a 1987 fatal case in Kenya. *Arch. Virol. Suppl.* **11,** 101–114.

Johnson, K. M., Webb, P. A., Lange, J. V., and Murphy, F. A. (1977). Isolation and characterization of a new virus [Ebola virus] causing acute hemorrhagic fever in Zaire. *Lancet* **1,** 569–571.

Johnson, K. M., Webb, P. A., Justines, G., and Murphy, F. A. (1978). Ecology of hemorrhagic fever viruses: Arenavirus biology and the Marburg–Ebola riddle. *In* "Third Munich Symposium on

14

ARTHROPOD-BORNE PATHOGENS: ISSUES FOR UNDERSTANDING EMERGING INFECTIOUS DISEASES

WALTER J. TABACHNICK

Arthropod-borne Animal
Diseases Research Laboratory, USDA Agricultural Research Service
Laramie, Wyoming

I. INTRODUCTION [1]

The twenty-first century is nearly upon us and human health remains under assault from infectious diseases. The persistence of well-known infectious diseases and the emergence of new ones in new geographic areas has shown we are still unable to cope with emerging diseases. In this chapter "emerging" diseases are those that have newly appeared or have existed but are rapidly increasing in incidence or geographic range (Morse, 1993). There is a critical need for information to predict conditions contributing to "emergence" and to develop appropriate strategies that will reduce the chance of emergence or reduce the impact of emerging diseases on human and animal health. The arthropod-borne pathogens provide a unique set of issues that must be considered if we are to have this information.

A. Emerging Arthropod-borne Pathogens

Arthropod-borne pathogens figure prominently among the problems facing human health. The effects of many arthropod-borne pathogens on human and animal health are increasing and spreading. For example, the prevalence of malaria and Lyme disease have increased due to environmental changes associated with travel and urban reforestation, respectively (Morse, 1995). The focus of

---

[1] This chapter is dedicated to the memory of Dr. George B. Craig, Jr., a pioneer and inspirational leader in the field of vector biology.

Emerging Infections
Copyright © 1998 by Academic Press. All rights of reproduction in any form reserved.

this chapter will be on the arboviruses to illustrate emerging arthropod-borne disease issues, at the risk of overlooking other major parasitic diseases, such as malaria, leishmaniasis, filariasis, and onchocerciasis. Much of the following discussion on arboviruses is applicable to other arthropod-borne parasites. Several reviews describe emerging arboviral diseases (Monath, 1993; Murphy and Nathanson, 1994). The list of emerging arboviruses is impressive and includes human pathogens dengue, yellow fever, Rift Valley fever, Crimean–Congo hemorrhagic fever, Japanese encephalitis, Cache Valley, Venezuelan equine encephalitis, Oropouche; and animal pathogens African horse sickness, African swine fever, bluetongue, and vesicular stomatitis viruses.

The emergence of many arthropod-borne viruses is a result of changes in their arthropod vectors. The increase in incidence and spread of an arbovirus to a new region may be the result of an increase in the size of associated vector populations due to an environmental change, or due to a change in the distribution range of the vector. The recent emergence of several mosquito-borne arboviruses illustrates this. The surge of *Aedes aegypti* populations in the Caribbean basin coupled with intensive urbanization has resulted in an explosion of dengue unprecedented in the region. Rift Valley fever epizootics in sub-Saharan Africa have occurred after periods of high rainfall when huge numbers of adult *Aedes* vectors are present. The introduction of the Asian tiger mosquito, *Aedes albopictus,* into the United States is great cause for concern. The spread of this species to new regions of the world is a replay of an old story. The yellow fever mosquito *Ae. aegypti* most likely first spread from Africa to the New World probably aboard ships during the trans-Atlantic slave trade in the sixteenth through seventeenth centuries (Tabachnick, 1991a). *Aedes albopictus* probably spread recently to the New World as eggs in used tires on ships (Hawley, 1988). This mosquito is anthropophilic in blood feeding and is a capable vector of dengue viruses in Asia. *Aedes albopictus* infected with eastern equine encephalomyelitis virus have been found in the United States (Mitchell *et al.,* 1992). The potential emergence of an established pathogen in a new vector is paticularly worrisome in this case in view of the anthropophilism of this species and the deadly nature of the virus.

B. Arthropods, Arboviruses, and Evolution

There is a great deal of genetic variation in arthropod vectors (see Gooding, 1996), and there is great diversity in the arboviruses they vector (see Karabatsos, 1986; Monath, 1989; Calisher, 1994). Because diversity and genetic variation are the fuel of evolutionary change, it is interesting that the emergence of arthropod-borne pathogens and disease has been more the result of the spread of a vector to a new region and less the direct result of a change in virulence or of a switch to a completely different vector species. It is difficult to distinguish the evolution of a new pathogen from the discovery of a pathogen that has been present but simply not recognized. Whereas we have evidence that pathogens change virulence, and switch to new vectors, perhaps evolutionary time scales are such that changes in virulence or in host range occur rarely relative to what can be directly observed in a human lifetime. After all, there is ample evidence for the evolution of new biological plant and animal species (e.g., the formation of reproductive isolation). However, observing the actual formation of a new species has been rare. An example of an arbovirus switching to a new vector may have been the case in regard to o'nyong-nyong and chikungunya viruses (Monath, 1993). O'nyong-nyong virus appeared for the first time and, transmitted by *Anopheles* mosquitoes, caused an explosive epidemic in East Africa in 1959. It is believed that a change may have occurred in the closely related chikungunya virus, one that altered chikungunya's vector potential from *Aedes* to more abundant species of *Anopheles,* resulting in the appearance of o'nyong-nyong virus and the East African epidemic (Williams *et al.,* 1965). The mechanisms that allowed the appearance of o'nyong-nyong virus, the switch to the new vector, and the sequence of these events are unknown. A study of chikungunya, o'nyong-nyong, and the vector competence of *Anopheles* and *Aedes* might provide important information on the mechanisms that contributed to the evolution of these arboviruses and the circumstances of vector switching.

Although variation in virulence is well documented for many viruses, actual observations of evolutionary changes in virulence are not well documented, nor is an understanding of the causes of such changes. Arboviral diversity is widely recognized, and arbovirus evolutionary studies are developing evolutionary paradigms to explain virus phylogenies and evolution. Witness the growing numbers of evolutionary studies, for example, of the viruses causing dengue, vesicular stomatitis, eastern equine encephalitis, Venezuelan equine encephalitis, and bluetongue. Evolutionary studies of flaviviruses, based on RNA sequence comparisons of the gene controlling the E glycoprotein, show not only that the tick-borne and mosquito-borne flaviviruses are two phylogenetically distinct groups but also that the mosquito-borne dengue viruses have undergone an explosive radiation in the past 200 years (Zanotto *et al.,* 1996). Sequence analyses of viruses in the eastern equine encephalomyelitis complex (alphaviruses) suggested divergence of North and South American branches of the complex in the last approximately 1000 years, and that the North American groups diverged in the 1970s (Weaver *et al.,* 1994). Investigations of sequence diversity among several related arboviruses have shown geographic and regional differences. North and South American isolates of vesicular stomatitis virus show differences that have occurred over the past 20 or so years (Nichol *et al.,* 1993). Venezuelan equine encephalomyelitis virus in Florida diverged from a South American ancestor during the past 100–150 years (Weaver *et al.,* 1992). Divergence among various serotypes of bluetongue virus throughout the world show distinct geographic differences while sharing similarities due to their common African ancestor (Pritchard and Gould, 1995). However, despite an increasing understanding of viral diversity and viral evolution, our understanding of the specific viral genetic changes required for changes in virulence and host pathogenicity remain rudimentary.

C. Predictability and Arthropod-borne Emerging Diseases

A critical need for coping with emerging pathogens is information that will allow predictability and risk assessment. For vector-borne pathogens, we must have information that will allow us to predict the capacity of a pathogen to expand its range to a new vector, to predict the ability for geographic range expansion of vector species, and to predict changes in viral pathogenesis and virulence. We need greater understanding of vector–pathogen interactions so that we can develop new strategies to predict the danger from an emerging arthropod-borne pathogen. Only then will we increase our ability to conduct risk assessment for an emerging arthropod-borne disease for a specific geographic region or ecosystem. It may very well be that information providing predictability also will provide a basis for the design of appropriate control strategies to reduce the impact of these diseases.

Arthropod-borne disease control strategies now center on vertebrate host vaccination and vector control strategies. Either strategy carries costs and disadvantages that limit effectiveness. Vaccination schemes must include costs of vaccine development and of vaccine delivery to at-risk populations; both costs indirectly limit effectiveness. Vector control must include costs of implementation as well as the inevitable development of arthropod resistance. What is needed are new control strategies that will be more effective and efficient.

The necessity to provide predictability is obvious from current issues and concerns facing human and veterinary health authorities. Some concerns are as follows: (1) the risk of entry of yellow fever virus onto the Asian mainland, from which it historically has been absent, (2) the increasing risk of dengue hemorrhagic fever in the New World, as a result of the introduction of new dengue virus strains into that area, (3) the risk to the tropical world from the introduction or spread of malaria vectors beyond current distributions, (4) the risk from the introduction or spread of diverse strains of *Plasmodium,* (5) the potential for spread of Japanese encephalitis, Rift Valley fever, Oropouche, and many more, (6) the risk of alterations in pathogenicity as pathogens and nonpathogens invade new ecosystems, and (7) the need for information to assess the risk of emerging animal pathogens that could devastate livestock industries, animal trade, and agricultural economies.

Although bluetongue viruses are highly pathogenic in sheep, less than 1% of infected cattle have clinical signs. To prevent bluetongue viruses from spreading or emerging in bluetongue-free regions via viremic cattle, cattle from bluetongue-endemic countries, such as the United States, must be certified bluetongue-free for movement to bluetongue-free regions, such as Europe. The consequences to the United States in lost trade in livestock and livestock germplasm have been estimated at $120 million per year. The danger of introducing an exotic bluetongue virus and its being established in any country is unknown. A change in the pathogenesis of an endemic bluetongue virus or the entry of an exotic bluetongue virus and a concomitant increase in pathogenicity in U.S. cattle would be catastrophic. There are regions in the United States where 40–50% of cattle have antibodies to one or another of the five U.S. bluetongue virus serotypes. Any increase in the pathogenicity of an exotic bluetongue virus in U.S. cattle could devastate the U.S. cattle industry, and an introduction of a new bluetongue serotype could further restrict animal movement to countries lacking that particular serotype, bringing about even greater economic losses than currently exist. Among other exotic arboviruses that pose a similar danger to U.S. agriculture and animal trade are the vesicular stomatitis viruses, African horse sickness viruses, and Rift Valley fever viruses.

II. ROLE OF THE ARTHROPOD IN PATHOGEN TRANSMISSION

The distinctiveness of arthropod-borne pathogens in emerging infectious disease issues is due to the role of arthropod vectors. Despite more than 100 years of research, it is surprising how little appreciation there is for the complexity of this interaction. The arthropod has direct effects on the pathogen. More arbovirus genetic diversity is generated during association of the virus with the arthropod than during its association with its vertebrate host (Beaty *et al.*, 1997). For example, RNA viruses with segmented genomes undergo greater genetic reassortment in the vector than in the vertebrate host (e.g., the bluetongue viruses discussed by Beaty *et al.*, 1997). Current information suggests that the vector itself affects pathogenesis in vertebrates. The significance of including the arthropod in all studies of arthropod-borne pathogens is discussed elsewhere (Nuttall *et al.*, 1991). Although we generally appreciate that the arthropod is more than a "flying syringe," the flying syringe mentality may still result in a simplistic view of arthropod-borne diseases. The emergence of arthropod-borne pathogens may result from changes that have been addressed by others as "viral traffic" mechanisms, such as environmental changes brought about by human expansion into new regions or the creation of prime habitats for arthropod vector species (Morse, 1993). This chapter addresses the unique role of the arthropod vector, its direct effect on arbovirus biology, and the effects of vector–pathogen associations that must be considered to address emerging arthropod-borne disease issues.

A. Vector–Pathogen Interactions

I. Vector Competence

The general ability of an arthropod to transmit a pathogen involves a variety of traits described under the term "vector capacity." These include the intrinsic ability of the arthropod to become infected and transmit the pathogen, also called vector competence, and a variety of extrinsic traits, such as arthropod host preference for blood feeding, longevity, and population density (DeFoliart *et al.*, 1983). For purposes of understanding vector–pathogen interactions, understanding the complexity of events that occur within the vector, vector competence is important.

Once an infectious blood meal is obtained, the virus is subject to an array of events within the vector. It must pass through a variety of tissues, replicating in each until it reaches the salivary glands. At that juncture the virus can be transmitted to a susceptible vertebrate during subsequent vector blood feedings. The details of the mechanisms involved, almost entirely unknown, represent a fervent area of current research in vector biology (Tabachnick, 1994).

Despite the lack of knowledge of the underlying mechanisms, consider the effects of these events on an arbovirus. A small number of virions, perhaps at most a few thousand, are imbibed in the blood meal. These are amplified to many millions through rounds of replication until, at most, a few hundred to thousands are transmitted. The consequences of such bottlenecks and amplifications in both the vector and the vertebrate host on the genetic structure of virus populations need to be developed to understand the potential effects on pathogen diversity and population structure. The populational view seems to have been embraced by virologists. Most discussions appear to regard viruses as populations of related genotypes, populations that are continually undergoing dynamic changes in diversity and genetic variation, so that they consist of "quasispecies" (Domingo *et al.,* 1995). In particular, the arboviruses, most with RNA genomes, have the potential for tremendous population genetic diversity due to intramolecular and segment reassortment mechanisms (Beaty *et al.,* 1997). The effect of this diversity on the fitness of arboviruses and in particular on virulence in specific environments is only beginning to be explored. Changes in the fitness of vesicular stomatitis viruses accompany changes in population sizes or bottlenecks, such as I have described as occur in the vector, as well as fitness changes that accompany passage in vertebrate and invertebrate tissues. Vesicular stomatitis viruses passaged in insect cell culture have reduced virulence in vertebrate cells (Novella *et al.,* 1995). It is clear that factors that influence changes in virulence are only beginning to be understood.

The interaction of two evolutionary factors constitute a major theme in this discussion: chance effects and selection. Both play a role in developing vector–pathogen associations. The key to predictability will be to appreciate this, assess the nature of each factor, and then develop information on underlying mechanisms, information that will allow predictability and assessment of risk in specific circumstances. One might expect that predictability would become more difficult if chance effects are prominent. This is not necessarily so, as will be discussed later.

2. Salivary Gland Interactions

If there is a single feature of the arthropod that demonstrates its role as more than a flying syringe, it has been the identification of pharmacological factors in arthropod salivary glands that influence host hemostasis. Host hemostasis is the vertebrate reaction that interferes with the arthropod's ability to obtain a blood meal. Arthropods have an array of factors, secreted in the saliva, that inhibit vertebrate hemostasis (Titus and Ribeiro, 1990). These factors include inhibitors of platelet aggregation (apyrases), anticoagulants, vasodilators, and factors that modulate the vertebrate immune system. Sand flies, biting midges, and mosquitoes have salivary gland proteins that inhibit macrophage function. Collectively these factors permit the arthropod to obtain a blood meal, and it appears that many hematophagous arthropods have independently developed the ability to accomplish this using different proteins (Ribeiro, 1995). Whatever the factor, arthropod saliva provides an environment that is very conducive to pathogen transmission at the blood-feeding site. The density of hematophagous arthropods that are feeding on a single host at any given time may be important because the more blood feeding that occurs, the more salivary factors are introduced into the vertebrate host, and it is likely the more permissive the environment created for pathogen transmission. Increased biting fly activity with the associated immunosuppressive effects on animals may be the reason vesicular stomatitis viruses spread rapidly among cattle and horse populations during epizootics of the disease in the southwestern United States. That vesicular stomatitis epizootics in the southwest United States occur during seasons in which biting insect populations are high may be related to increased biting fly activity in general and the associated immunosuppressive effects on host animals. This hypothesis needs to be tested and verified. Arthropod salivary gland factors are only beginning to be introduced into our paradigm of arthropod-borne pathogens.

Several questions are relevant to the role of the arthropod salivary gland and the ability to predict emerging arthropod-borne pathogens. Do the salivary gland factors of particular vector species influence transmission differently depending on the pathogen (e.g., dengue viruses, vesicular stomatitis viruses, plasmodia, leishmania)? What is the impact of strains or genetic variants of each pathogen? Understanding these interactions is essential for predictability, risk assessment, and perhaps control. Specific salivary gland factors might be altered by genetically changing vectors or by manipulating the environment to prevent or reduce the chance of transmission. It may prove possible to use drugs or some other modification of the vertebrate environment to alter the effect of salivary gland proteins, thereby reducing transmission. If genetic transformation strategies prove successful in vector species, vectors could be genetically transformed so that vector salivary glands secrete a pathogen epitope that vaccinates susceptible hosts against the complete pathogen. Information on the mechanisms that add to predictability will also lead to new control strategies.

B. Vector Influence on Pathogenesis, Pathogen Evolution

If mechanisms for understanding vector–pathogen interactions remain largely unknown, and our understanding of pathogenesis is also limited, then how can we hope to arrive at some chance of predictability? In addition, is it necessary to understand every nuance of the complex mechanisms that may be involved in vector–pathogen interactions to achieve sufficient predictability? Several interesting opportunities may provide answers to solve this dilemma.

1. Coevolution

Vector–pathogen biologists have advocated the theory that arboviruses have coevolved with their arthropod vectors (Eldridge, 1990, 1993). Coevolution is a very attractive hypothesis for predictability. However, a coevolutionary argument implies a precise association between vector and pathogen that depends on a reciprocal interaction between them (Tabachnick, 1991b). Coevolution is an "association by descent" between species, and the extent they have influenced evolution of one another (Mitter and Brooks, 1983). A strict definition of coevolution states that specific changes in the pathogen were caused by parallel changes in the vector and, very critically, changes in the pathogen then caused further reciprocal changes in the vector. If reciprocity results in congruence between phylogeny of the pathogen and phylogeny of the vector, then there is parallel speciation of vector and pathogen.

There is a paucity of evidence in support of coevolution between arboviruses and their arthropod vectors in the strict sense (Tabachnick, 1991b). Indeed, there is little direct evidence for coevolution in the strict sense for many interspecies interactions (Futuyma, 1983; Mitter and Brooks, 1983), and therefore the term diffuse coevolution has been used to describe the evolution of a particular trait in one or more species in response to a trait or suite of traits in several other species (Futuyma and Slatkin, 1983). Futuyma (1986) cautioned that the term coevolution should be used only when alternative explanations that are not dependent on species interactions are exhausted. The introduction of myxoma virus into Australia to control rabbit populations illustrates coevolution between an arbovirus and a vertebrate. In the course of a decade after its introduction into Australia the virus evolved a lower degree of virulence while rabbits evolved greater resistance to the virus (Fenner, 1965).

It is likely that pathogen evolution has been influenced by the internal environment of the vector, and that other aspects of vertebrate biology have been influenced by pathogens. The arthropod host may indeed be the preferred site for arboviruses to evolve because persistent infection is commonly observed in competent arthropods and is conducive to intramolecular genomic changes and the accumulation of spontaneous mutations (Beaty *et al.*, 1997). It is the reciprocal effect for which evidence is lacking. There is no evidence that specific vector-borne pathogens have been prominent in the evolution of any specific vector traits, or in vector speciation. Diffuse coevolution is likely to have occurred. Vectors may have evolved some traits in response to arbovirus infection in general, but such a diffuse interaction may provide little information for predicting specific associations of viruses and vectors, information that is essential to address emerging disease problems.

Coevolution in the strict sense, however, provides a powerful model to predict vector–pathogen relationships. If coevolution in the strict sense proves to be prominent, then phylogenetic relationships offer a tool to predict potential vectors and to address emerging disease issues. We might predict that the potential for the spread or emergence of a pathogen is restricted to only those geographic regions where phylogenetically related vectors were present. Finding a newly discovered, related species of *Ae. aegypti,* or of *Anopheles gambiae,* would have to be assessed in terms of the expected effect of the new vector on the transmission and evolution of dengue viruses or *Plasmodium* spp., respectively, and on the reciprocal effect of the pathogens on the evolution of the new species. Suppose a new species were discovered that was very closely related to *Ae. aegypti.* A coevolutionary view would predict that this new species would influence dengue virus evolution. Would one predict that the newly identified vector must have or is an environment for the evolution of new dengue viruses associated with the newly identified species? A true believer in strict coevolution should search for this new virus because it must have played a role in the evolution of the new vector species.

The coevolutionary hypothesis is attractive. However, what vector–pathogen biologists are often referring to when they refer to coevolution is that pathogen evolution has occurred as a result of association with the vector. It has been argued that the punctuated evolution of vesicular stomatitis viruses in the New World is the direct result of changes in the associated vector species as the virus has spread north from Mexico into the United States (Nichol *et al.*, 1993; Fitch, 1996). Despite the absence of evidence for coevolution in the strict sense of reciprocal interactions, it is clear that both vectors and pathogens have distinct phylogenies and evolutionary relationships with other members of their respective taxa. There are probably constraints on vector–pathogen associations imposed by these separate phylogenies. It is this aspect that may offer opportunities for predictability.

2. Virulence of Arthropod-borne Pathogens

Evolutionary concepts and models figure prominently in attempts to understand expected changes in the virulence of pathogens under various situations. The classic view that the long-term effects of an association between a pathogen and host is always a reduction in virulence in the pathogen has been replaced by the more enlightened model that certain conditions related to the transmission efficiency of the pathogen can give rise to increasing pathogen virulence (May and Anderson, 1983; Levin, 1996). Evolutionary concepts are beginning to be applied to understand arthropod–virus associations (Scott *et al.*, 1994) and disease epidemiology (Ewald, 1994).

Ewald (1994) has offered several attractive evolutionary hypotheses concerning arthropod-borne pathogens and diseases. The analyses demonstrate that, in general, among regularly transmitted human pathogens, arthropod-borne pathogens are more virulent than non-arthropod-borne pathogens. He attributes this to the role of the vector in transmission. Pathogens that are transmitted by "the flying syringe" (the term is correctly used in this instance) have fewer constraints on their virulence compared to non-vector-borne pathogens, because they will be transmitted by the vector no matter how debilitating to the vertebrate host. Ewald (1994) calls this the "adaptive severity hypothesis." Vectors can indeed more readily obtain a blood meal from more debilitated than less debilitated or uninfected vertebrates (Day and Edman, 1983). Therefore, pathogen transmission may be facilitated despite greater virulence. In contrast, those pathogens that are transmitted by vertebrate to vertebrate direct contact will lose transmission capability if vertebrate hosts are so debilitated that contacts with other susceptible vertebrates are reduced. Ewald's is an attractive hypothesis that leads to his proposal that the use of bed nets for malaria control, by shielding infected individuals from mosquitoes, may tip the scales in the direction of the evolution of losing virulence—malaria should become more benign. We need more such testable hypotheses.

A second hypothesis developed to address the virulence of arthropod-borne pathogens is the "restricted adaptation hypothesis" (Ewald, 1994). This hypothesis argues that the reason for the generally greater virulence of arthropod-borne pathogens is that the pathogen is incapable of evolving to a benign equilibrium in both the vertebrate and invertebrate host. Therefore, evolution to becoming more benign is restricted to only one host. Because the flying syringe is more critical in transmission, the pathogen evolves benignness in the vector, enabling the pathogen to become virulent in the vertebrate host. If there is a chance for an arthropod-borne pathogen to evolve to becoming benign in a vertebrate host, it should be for those pathogens that passage the most between arthropods and that vertebrate. Comparisons of 13 closely related arthropod-borne pathogens that differ in the degree that they passage in humans showed that all the pathogens (except one) with the greater cycling in humans showed more severe clinical human disease (Ewald, 1983). Because the restricted adaptation hypothesis does not predict such an outcome, this observation using these selected pathogens supports the adaptive severity hypothesis.

C. Predictability

Earlier in Section II, B I raised the possibility that it may not be necessary to understand every aspect of vector–virus interactions to provide predictability and useful risk assessment. However, it is still essential to identify key mechanisms controlling these interactions, specifically to identify the molecular genetic mechanisms of vector–pathogen interactions. Identifying vector genes and genes of pathogens that are involved in vector–pathogen interactions will provide information on specific, important traits (Tabachnick, 1994). For example, identification of arthropod genes involved in vector competence variation will provide predictability of potential vectors and geographic regions at risk for disease by enabling us to determine the vector capacity of natural arthropod populations (Tabachnick, 1994). Which arthropod proteins, receptors, and enzymes are critical for virus infection and transmission? Further, information on how such genes, and the proteins they produce, respond in different environments will also allow us to predict and assess the danger of the entry of exotic arboviruses.

Similar factors that control the distributions and ranges of arthropod vector species need to be identified. Under what conditions will *An. gambiae, Ae. aegypti,* or *Ae. albopictus* spread to new regions? Certainly the consequences of global warming have to be considered, because increasing temperatures will affect the ranges of important arthropod vectors and will influence vector–pathogen interactions (Patz *et al.,* 1996). The temperature of incubation of an arbovirus in its vector is a critical factor in determining the length of time between infection of the arthropod and the ability of the vector to transmit the pathogen to a susceptible vertebrate. The effect of higher temperatures, those within the range of arthropod viability, will result in more infected arthropods that blood feed more frequently. However, the effect of higher temperatures on changing vector–pathogen associations, changing virulence, changing pathogenesis, and the emergence of new pathogens is unknown. It is possible that by being able to predict the consequences of environmental changes, such as global warming, we may be able to identify factors that will enable us to influence the environment. Changes in the environment might prevent the spread of vector populations, reduce vector population size, prevent new vector–pathogen associations, or reduce the chance of an increase in the virulence of a pathogen.

It is critical to establish the effects of the vector on the pathogenesis of arboviruses and parasites. The goal of efforts to provide gene maps of *An. gambiae, Ae. aegypti,* the North American bluetongue virus vector *Culicoides variipennis,* and other vectors are to use gene mapping strategies to identify vector genes that are critical to pathogen–vector interactions. Although it has been shown numerous times that vector susceptibility to infection with pathogens is under genetic control, there are very few examples of specific arthropod genes that control vector competence (Tabachnick, 1994). A single genetic locus has been identified that controls *C. variipennis* susceptibility to infection with the bluetongue viruses (Tabachnick, 1991c). However, the mechanism of action of this locus is unknown. Currently there is no vector–pathogen model for which the molecular details of the interaction are known. Achieving an understanding of the mechanisms controlling vector competence, coupled with structure–function studies of pathogens, will provide a basis to target specific functions and reduce vector ability. This will enable investigations of features of arthropod and pathogen interactions that affect virulence, pathogenesis, and the evolution of vector–pathogen relationships. There likely are aspects of vector–pathogen associations that may be predictable once key elements of the mechanisms controlling the interaction are identified.

III. APPROACHES FOR UNDERSTANDING THE COMPLEXITY OF VECTOR–PATHOGEN ASSOCIATIONS

A. Phyletic Approaches

Although there is little direct support for coevolutionary explanations for vector–pathogen associations, studies of phylogenetic relationships may provide boundaries that delimit phyletic constraints on these associations. Phylogenetic relationships of different vector species, or different virus groups, provide information about the genetic relationships among the members of each group. One may be able to draw inferences concerning limits on vector–pathogen associations using vector and virus phyletic relationships. Aspects of vector–pathogen associations are probably constrained by the phylogenies of both vector and pathogen. For example, *Ae. aegypti* and its African relatives do not become infected with human *Plasmodium* spp. and are not vectors of these pathogens. This is the case despite numerous opportunities for *Ae. aegypti* to feed on malaria-infected individuals. From this we may infer that it is unlikely that *Ae. aegypti* will soon become a vector of human malarial pathogens. The reverse is also true; *Anopheles* mosquitoes do not transmit the flaviviruses yellow fever and dengue. The phyletic approach offers the prospect of predictability with little understanding of the underlying mechanisms and nuances of *Plasmodium*–mosquito and yellow fever virus–mosquito interactions. Indeed, the identification of specific phyletic constraints that limit associations may provide opportunity to understand features controlling various interactions. The phylogenies of *Aedes* and *Anopheles* impose constraints, as do the phylogenies of *Plasmodium* and flaviviruses. Understanding how o'nyong-nyong, a relative of chikungunya virus, made the phyletic leap from *Aedes* to *Anopheles* in East Africa may provide important information on the phyletic constraints of these species.

The discussion of the use of phyletic relationships for predictive purposes is somewhat a straw man. One can choose specific traits that clearly are constrained due to the phylogeny of the groups in question. The anthropoid phylogeny imposes limitations on humans for developing the biological ability to fly, and insect phylogeny imposes limitations on the size of insects. What we require, if we are to advance predictability, is information on the underlying mechanisms controlling vector ability, competence, and pathogen interactions. As previously discussed, this knowledge is essential if we are to assess variation in vector–pathogen associations and determine the features of required traits that can be predicted using phyletic relationships of vectors and/or phyletic relationships of pathogens. Otherwise we are left with nice stories, with no testable hypotheses, and no means to separate features that are constrained by phylogeny from other, more subtle, controlling factors that are not limiting phyletic factors. For example, phylogenetically unrelated vectors that are infected and are competent vectors of the same virus may be an example of convergent evolution.

Data on arbovirus–vector associations are still far too incomplete to provide definitive tests of phyletic associations. One source for arbovirus–vector associations is the International Catalogue of Arboviruses (Karabatsos, 1986). To test for associations between arboviruses and mosquito vectors I surveyed the Catalogue for arboviruses of three genera, alphaviruses, bunyaviruses, and flaviviruses, and recorded the associations of viruses in each of the groups with species in each of three genera of mosquitoes, *Aedes, Anopheles,* and *Culex.* Figure 1 show results of this survey. More bunyaviruses were isolated from the three mosquito genera than were alphaviruses and flaviviruses combined. Caution is warranted in interpreting these results. They illustrate the difficulties using these data due to biases in the way the data in the Catalogue were collected. There are more species of *Aedes* and *Culex* listed in the Catalogue than of *Anopheles,* and there is probably a bias toward identifications from mosquitoes of arboviruses that are known to cause disease. There likely are more reports in the Catalogue for species and viruses from developed compared to underdeveloped countries, and from regions with field stations than from remote regions. Finally, some associations may have little to do with a direct vector–virus interaction but may be due to an arbovirus having a specific host association, and a host preference by the vector for that vertebrate.

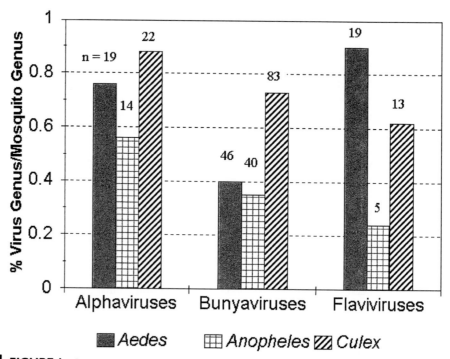

FIGURE I   Percentage of the total number of arthropod isolated alphaviruses, bunyaviruses, and flaviviruses isolated from *Aedes, Culex,* and *Anopheles* mosquitoes, where *n* is the total number of different viruses isolated from each virus–mosquito group. (From Karabatsos, 1986).

Pritchard, L. I., and Gould A. R. (1995). Phylogenetic comparison of the serotype-specific VP2 protein of bluetongue and related orbiviruses. *Virus Res.* **39**, 207–220.

Ribeiro, J. M. C. (1995). Blood-feeding arthropods: Live syringes or invertebrate pharmacologists? *Infect. Agents Dis.* **4**, 143–152.

Scott, T. W., Weaver, S.C., and Mallampalli, V. L. (1994). Evolution of mosquito-borne viruses. *In* "The Evolutionary Biology of Viruses" (S. S. Morse, ed.), pp. 293–324. Raven, New York.

Tabachnick, W. J. (1991a). Evolutionary genetics and arthropod-borne disease: The yellow fever mosquito. *Am. Entomol.* **37**, 14–24.

Tabachnick, W. J. (1991b). Reappraisal of the consequences of evolutionary relationships among California serogroup viruses (Bunyaviridae) and *Aedes* mosquitoes (Diptera: Culicidae). *J. Med. Entomol.* **28**, 297–298.

Tabachnick, W. J. (1991c). Genetic control of oral susceptibility to infection of *Culicoides variipennis* for bluetongue virus. *Am. J. Trop. Med. Hyg.* **45**, 666–671.

Tabachnick, W. J. (1994). Genetics of insect vector competence for arboviruses. *Adv. Dis. Vector Res.* **10**, 93–108.

Titus, R. G., and Ribeiro, J. M. C. (1990). The role of vector saliva in transmission of arthropod-borne disease. *Parasitol. Today* **6**, 157–160.

Weaver, S. C., Bellew, L. A., and Rico-Hesse, R. (1992). Phylogenetic analysis of alphaviruses in the Venezuelan equine encephalitis complex and identification of epizootic viruses. *Virology* **191**, 282–290.

Weaver, S. C., Hagenbaugh, A., Bellew, L. A., Gousset, L., Mallampalli, V., Holland, J. J., and Scott, T. W. (1994). Evolution of alphaviruses in the eastern equine encephalomyelitis complex. *J. Virol.* **68**, 158–169.

Williams, M. C., Woodall, J. P., Corbet, P. S., and Gillett, J. D. (1965). O'nyong-nyong fever: An epidemic virus disease in East Africa. VIII. Virus isolations from *Anopheles* mosquitoes. *Trans. R. Soc. Trop. Med. Hyg.* **59**, 300–306.

Zanotto, P. M. de A., Gould, E. A., Gao, G. F., Harvey P. H., and Holmes, E. C. (1996). Population dynamics of flaviviruses revealed by molecular phylogenies. *Proc. Natl. Acad. Sci. U.S.A.* **93**, 548–553.

15

"NEW" INTESTINAL PARASITIC PROTOZOA

ADEL A. F. MAHMOUD

Department of Medicine
Case Western Reserve University
University Hospitals
Cleveland, Ohio

I. INTRODUCTION

Since the 1960s, many so-called new intestinal parasitic protozoa have been identified as potential and actual human pathogens. Although most of these organisms were recognized over a century ago either in nature or in multiple animal and bird species, their role in causing human disease is new or was not appreciated for a long time. Most of these so-called new parasitic protozoa are intestinal pathogens such as *Giardia* and the spore-forming *Cryptosporidium,* microsporidia, *Isospora,* and *Cyclospora.* Other organisms that have been recently appreciated include *Babesia* infecting human erythrocytes and *Naegleria* causing a form of meningoencephalitis.

Central to understanding the biological relevance of these organisms as human pathogens is appreciating the reasons for their recent emergence. Whether this is due to acquisition of new virulence capabilities or to increased proximity of humans to reservoirs of infection in animal populations is now being debated and reexamined. Another significant factor that brought to prominence most of these organisms is the human immunodeficiency virus (HIV) epidemic and other immunosuppressive conditions. These host factors changed the immune capabilities of a significant segment of the population worldwide and provided a "new" cohort of susceptible hosts.

This chapter focuses on the group of intestinal protozoa which attracted clinical as well as scientific attention since the 1960s. *Giardia* infection became the most commonly identifiable intestinal pathogen associated with outbreaks of diarrheal disease in the United States in the 1970s. In addition, human infection with the spore-forming intestinal protozoa *Cryptosporidium*, microsporidia, *Isospora,* and *Cyclospora* have increasingly been reported not only in immunosuppressed individuals but as endemic pathogens in some communities and as causative organisms in water- and food-borne outbreaks of diarrheal disease. Whereas *Giardia* belong to the flagellate group of human pathogens, intestinal spore-forming protozoa share many biological features but differ in other aspects particularly as it applies to epidemiology and disease syndromes in humans.

II. *GIARDIA*

The protozoan now recognized as belonging to the genus *Giardia* was probably the first microorganism seen through a microscope by von Leeuwenhoek during the second half of the seventeenth century. The organism was rediscovered in 1885 by Vilein Lambl, but the recognition of its pathogenicity to humans was delayed for many decades (Table I). *Giardia* is now considered an intestinal pathogen endemic all over the world and is responsible for considerable human and animal disease (Farthing, 1993).

A. Biology

The characteristic morphological features of the genus *Giardia* are well established. The organisms exist as trophozoites and cysts (Fig. 1). The trophozoites inhabit the proximal small intestine of the host. They are pear-shaped (10–15 μm $\times$ 6–10 μm), binucleated, and possess eight flagella, two median bodies, and a characteristic ventral disc (Thompson *et al.*, 1993). The trophozoites contain several cytoplasmic organelles but lack mitochondria. *Giardia* cysts measure 8–12 μm by 7–10 μm, contain two to four nuclei, and are enclosed within a fibrous proteinaceous membrane. Detailed information on the internal structure of the trophozoites and cysts have been reported, particularly the ultrastructure of the ventral disc, which distinguishes *Giardia* from all other flagellates. A *Giardia*-specific protein termed giardin has been found only in the ventral disc (Peattie *et al.*, 1989) and consists mainly of α-giardin (33.8 kDa)

TABLE I   Approximate Dates of Description and Recognition of "New" Intestinal Protozoa as Causes of Human Disease

Organism	Description	First human cases proved
Giardia	1681	1960s
Cryptosporidium	1907	1976
Microsporidia	1857	1959
Isospora	?	1970s
Cyclospora	?	1977

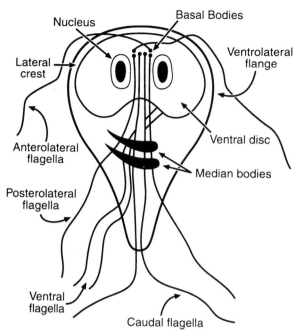

FIGURE I Schematic representation of the outer morphology and internal structures of a trophozoite of *Giardia duodenalis*.

and β-giardin (29 kDa). These molecules are thought to be central in the organization of the *Giardia*-unique ventral disc. Other characteristic structures include the median bodies (Feely *et al.*, 1990). Furthermore, several endosymbionts have been identified with *Giardia* trophozoites and cysts including bacteria and viruses. Among these, *Giardia* virus, a 32-μm double-stranded RNA virus, has received considerable attention (Wang and Wang, 1986) since it resembles viruses that infect yeast and fungi. The biological relevance of these viruses is unclear, but they may have an adverse effect on the organisms as demonstrated by the decrease in adherence or growth rate in infected parasites (Wang and Wang, 1991).

The life cycle of *Giardia* infection in humans or other animal hosts is simple (Fig. 2). Infection is established via ingestion of parasite cysts or in rare occasions the trophozoite stage. Excystation occurs in the stomach and may be aided by several factors including gastric acidity, pancreatic secretions, and carbon dioxide concentration (Feely *et al.*, 1991). The resulting trophozoites move to the duodenum and upper jejunum where they establish the active multiplicative phase of infection. Trophozoites multiply by binary fission, although new evidence suggests the existence of sexual multiplication which may have a significant impact on genetic diversity in these organisms (Tibayrene *et al.*, 1990).

Experimental infections demonstrate that giardiasis can be established with a very small inoculum as low as one trophozoite or 10 cysts. The next phase of the parasite life cycle following tyrophozoite multiplication and attachment involves transmission to another host directly or via environmental vehicles. Tro-

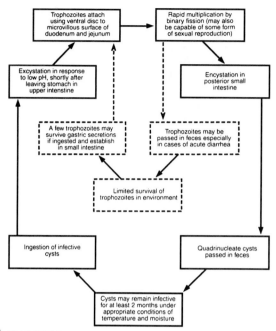

FIGURE 2   Life cycle of *Giardia duodenalis.*

phozoites may be seen in stool samples particularly in association with acute diarrhea. More commonly encystation occurs before expulsion of the organisms with stools. Encystation usually takes place in the distal small intestine and may be triggered by bile salts (Gillin *et al.,* 1988). The mature cyst contains four nuclei which result from one cycle of asexual multiplication of trophozoites. *Giardia* cysts may need a few days following shedding to become infective. The cysts are resistant to environmental conditions and can survive up to several months in soil. *Giardia* cysts also resist the usual chlorine concentration used in water purification but may be removed from drinking water by filtration.

The taxonomy of the genus *Giardia* has undergone several modifications. Currently, it is assumed that the more than 40 species (based on host specificity) actually belong to three morphologically distinct groups (Thompson *et al.,* 1990): *Giardia duodenalis* infecting mammals predominantly, *Giardia muris* of rodents, and *Giardia agilis* of amphibians. Two additional species have been proposed, but accurate taxonomic classification will have to wait the development of additional immunologic and molecular criteria. Furthermore, with the introduction of *in vitro* culture and cloning procedures, the basis for taxonomic classification of *Giardia* may be better established (Thompson *et al.,* 1990; Roberts-Thomson, 1993).

The 1980s and 1990s have witnessed considerable progress in understanding the immunology and molecular biology of *Giardia.* Of interest is the establishment of the concept of surface antigenic variation (Gottstein *et al.,* 1990). *Giardia* trophozoites are covered by several surface proteins including the variable surface proteins (Aley and Gillin, 1995). The proteins cover all exposed surfaces of the parasite and may be essential to its protection. One of these molecules, variable surface protein, is genetically controlled; trophozoites can change expression of this variable surface protein and therefore provide a different immunologic target (Gillin *et al.*, 1990; Pimenta *et al.*, 1991; Nash *et al.*, 1991). Organisms that have been isolated or cloned demonstrate several populations of trophozoites expressing variable surface protein. *Giardia* contains a repertoire of approximately 100 genes capable of producing different antigenic variants (Nash and Mowatt, 1992, 1993). The first complete variable surface protein gene has been cloned (Gillin *et al.*, 1990), and many features of the protein structure and attachment have been reported (Das *et al.*, 1991). The biological significance of variable surface proteins has been demonstrated in studies showing that antibodies against these molecules agglutinate and inhibit the attachment of *Giardia* trophozoites *in vitro* (Gillin *et al.*, 1990; McCaffery *et al.*, 1994). In addition, characterization of other *Giardia* molecular structures has provided a basis for understanding the parasite genome, although its size and ploidy are still being debated (Adam, 1991; Thompson *et al.*, 1993; Farthing, 1995).

B. Epidemiology

Human giardiasis is transmitted from infected humans or other animal species to susceptible individuals. Transmission occurs via fecal–oral or via waterborne routes. The sources of infection in the environment are either infected individuals or zoonotic infection in mammals such as dogs, cats, and beaver. Prevalence of infection in humans varies by age and sanitary standards. Infection is more common in children and particularly in developing countries, where prevalence rates of up to 100% have been reported. Infection may also be more common in individuals with low gastric acidity, after gastroectomy, and in nutritionally or immunologically deficient individuals (Farthing, 1993).

From the public health point of view, waterborne transmission of giardiasis is a complex challenge (Craun, 1990). While the possible sources of contamination of the water supply are multiple (Schantz, 1991), ensuring its safety is not easily achievable. The usual chlorine dose used in water purification does not kill *Giardia* cysts. A filtration step or shifting from surface to deep underground sources of drinking water is necessary; neither is economically feasible. Another critical mode of transmission occurs in day care centers where prevalence varies from 17 to 90% (Pickering and Engelkirk, 1988). Infection with *G. duodenalis* was found to be responsible for 4000 hospitalizations between 1979 and 1988 (Lengerich *et al.*, 1994). The rate of giardiasis was highest among children under 5 years of age and women of childbearing age. These data reflect the magnitude of endemic giardiasis in the United States and implicates day care centers as possible foci for dissemination of infection (Overturf, 1994).

C. Pathogenesis and Clinical Features

The exact pathogenetic mechanisms responsible for symptomatic giardiasis are not known. It appears that disease is closely related with the ability of the organism to attach to intestinal epithelium. The ventral disc plays a significant role in attachment, which may be mediated either by suction forces, by mechanical processes dependent on the contractile proteins in the organism, or through mannose-binding lectins (Farthing, 1993). The pathophysiological changes associated with symptomatic intestinal giardiasis focus on deranged intestinal epithelium morphology and function (Farthing, 1993). Among the more significant morphological changes reported is villous atrophy, and shortening and disruption of microvilli (Farthing, 1995). Functionally, reduction in disaccharidase activity is at its peak in association with maximal diarrhea and morphological evidence of epithelial damage. The physiological consequences of these changes include impairment of water, sodium, and chloride absorption in response to glucose (Buret *et al.*, 1992).

The parasite role in producing these morphological and functional abnormalities is evident from studies in experimental animals and infected humans (Farthing, 1993). The exact pathogenic mechanisms is, however, unclear. Suggestions include epithelial invasion, toxin production, and increased cell turnover. Host factors such as intraepithelial lymphocytes or bacterial overgrowth also have been proposed, but none has conclusively been confirmed.

Human infection with *G. duodenalis* results in a myriad of clinical features ranging from asymptomatic to acute or chronic intestinal disease and their metabolic consequences. The frequency of each clinical feature varies in different studies. More reliable information has been obtained from clinical observations in *Giardia* epidemics where the features of acute disease may be better described. Under such circumstances, diarrhea, abdominal pain, nausea, and steahorrhea are the most common clinical manifestations along with weakness and weight loss (Farthing, 1995); these features occur in 50–70% of infected individuals. Chronic giardiasis, on the other hand, may be associated with chronic diarrhea and steahorrhea. Malabsorption and weight loss may be significant under these circumstances.

D. Management

Giardia duodenalis infection in humans may be suspected because of epidemiological evidence or from the nature of the clinical presentation. Diagnosis is established by stool examination microscopically for identification of trophozoites or cysts or immunologically or molecularly (Isaac-Ranton, 1991; Farthing, 1995). The new nonparasitologically based procedures are much more sensitive and specific and do not necessitate microscopic expertise. Individuals with symptomatic infection should be treated with either quinacrine, metronidazole, or furazolidone (Medical letter, 1996). In addition, single-dose therapeutics such as tinidazole are becoming available outside the United States and have proved extremely effective (Farthing, 1995).

III. NEW "SPORE-FORMING" PROTOZOA

Spore-forming protozoa were first decribed in the late 1800s, but the recognition of their ability as human pathogens is a relatively new phenomenon (Table I). Because information on *Cryptosporidium* is more detailed than for the other three groups, it will be used to illustrate the biology and host–parasite relationship in these infections.

A. Biology

The four groups of organisms, namely, *Cryptosporidium*, microsporidia, *Isospora*, and *Cyclospora*, are classified as Protozoa. They all share the common characteristic of spore formation, which is necessary for dissemination in the environment and transmission to other hosts (Goodgame, 1996). The phylum Sporozoa contain species such as *Plasmodium* and *Toxoplasma* as well as *Cryptosporidium*, *Isospora*, and *Cyclospora*. In contrast, microsporidia belong to their own phylum. Speciation of each of these organisms is still based on morphological and ultrastructural characteristics as well as host specificity (Sleigh, 1992).

The life cycle of the four spore-forming protozoa follow the same pattern (Fig. 3). It is characterized by having all stages of development both sexual and asexual occur in the same host, a phenomenon known as monoxenous. Human infection is acquired by ingestion of viable spores which release the tissue-invading sporozoites in the small intestine. Excystation may result from environmental changes as the spores travel throughout the stomach and duodenum (O'Donoghue, 1995; Robertson *et al.*, 1993). Thereafter sporozoites adhere to enterocytes and become enclosed in parsitophorous vacuoles (Reduker *et al.*, 1985). This stage of the parasite, within the enterocytes, is identified as the vegetative forms or trophozoites; they undergo maturation and asexual reproduction through a process called merogony. The resulting merozoites are capable

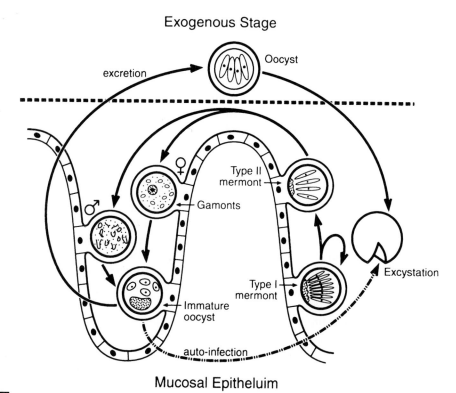

Exogenous Stage

Oocyst

excretion

♀ Type II mermont

♂ Gamonts

Immature oocyst

Type I mermont

Excystation

auto-infection

Mucosal Epitheluim

FIGURE 3 Representation of the life cycle of spore-forming protozoa. The example used applies to *Cryptosporidium*. Similar patterns occur in other organisms.

■ TABLE II   Morphological Features of the Oocyst
Stage of Intestinal Spore-Forming Protozoa

Organism	Oocyst size (μm)	Number of sporozoites
Cryptosporidium	4–6	4
Microsporidia	1–2	?
Isospora	20–30	4
Cyclospora	4–10	2

of infecting other enterocytes or proceed to differentiate into sexual forms responsible for spore formation (sporogomy) (Current *et al.,* 1986). The characteristic morphological features and size of the oocyst stage of these protozoa are
summarized in Table II. The life cycle of spore-forming protozoa may be completed in a couple of days; however, it varies from host to host and from species
to species. Most of the regulatory mechanisms responsible for proceeding from
one stage to another are unknown.

Studies have been initiated to characterize these pathogens antigenically
and molecularly. For example, anticryptosporidia polyclonal and monoclonal
antibodies have been produced (Nina *et al.,* 1992). Several immunodominant
parasite antigens have been identified; sera from infected individuals, however, recognize relatively limited antigens in parasite extracts (McDonald *et al.,*
1991). Some of the monoclonal antibodies against *Cryptosporidium parvum*
identify species-specific antigens (Ungar *et al.,* 1986), which may pave the way
toward better species definition and to immune-based fecal examination methods. The molecular biological understanding of spore-forming protozoa is at
its infancy. Expression libraries have been constructed, and the possibility of
isolating species-specific DNA sequences is being actively pursued (Webster
et al., 1993). Cloning and sequencing of several cryptosporidial structural genes
such as ATPase, microtubules, and actin are now available (Nelson *et al.,* 1991).
Similar studies for other spore-forming protozoa have been initiated (Adal
et al., 1995).

The issue of parasite–host specificity in intestinal Sporozoa and microsporidia is now the subject of intense debate. Original descriptions assigned separate species to parasites infecting separate animals. Epidemiological and clinical
observations, however, have supported the notion that *Cryptosporidium* isolated from several animal species such as sheep and cows are capable of infecting humans (Sterling and Arrowood, 1993). As the basis of identifying species
specificity is entirely morphological, these observations need to be confirmed
on a much firmer ground (e.g., biochemical or molecular). *Cryptosporidium*
have been identified in a wide range of vertebrate hosts. The genus *Cryptosporidium* has been classified in the order Eucoccidiorida (Table III), and a total of
21 species have been named (Levine, 1986). The term microsporidia is used to
describe a group of organisms belonging to the phylum Microspora. The phylum is classified into six clinically significant genera (Table IV) based on ultra-

TABLE III Taxonomic Classification[a] of *Cryptosporidium*

Classification	Characteristics
Phylum Apicomplexa	Apical complex present; all species parasitic
Class Sporozoasida	Reproduction asexual and sexual; oocysts produced
Subclass Coccidiasina	Life cycle generally involves merogony, gametogony, and sporogony; small gamonts
Order Eucoccidiorida	Merogony or "schizogony" present
Suborder Dimeriorina	Macro- and microgamonts develop independently; nonmotile zygote
Family Cryptosporidiidae	Oocysts contain four naked sporozoites (no sporocysts); endogenous stages with attachment organelle; monoxenous life cycle

[a] Based on classification system proposed by Levine (1986).

structural and chromosome cycles (Johnson *et al.,* 1990). The situation with *Isospora* and *Cyclospora* is simpler since there are only two species that are known to infect humans: *Isospora belli* and *Cyclospora cayetanensis.*

The next concern about the biology of these organisms relates to the definition of virulence. The question is central to all emerging infections and much more in the case of organisms that have been known to exist in nature and in several animal species for decades. What are the determinants of emergence of spore-forming intestinal protozoa? Is it a change in the immunocompetence of the host, ecological variables, or expansion of the host spectrum? Alternatively, is it a change in the virulence of these organisms? The limitations in attempting to answer the latter question include our current status of knowledge in regard to pathogenesis. We do not have even a simplistic explanation for how these organisms cause disease, and consequently our understanding of virulence is rudimentary. A redefinition of virulence from the simplistic but realistic concept of conventional wisdom as advanced by May and Anderson (1983) to alternative models for the evolution of microparasite virulence is needed. Levin (1996) proposed that there are two additional mechanisms that may lead to change in virulence, namely, coincidental evolution (Levin and Svanborg-Eden, 1990) and shortsighted within-host selection (Levin and Bull, 1994). The absence of a mo-

TABLE IV Taxonomy of Microsporidia That May Cause Human Infection and Disease

Class	Order	Family	Genus
Dihaplophasea	Dissociodihaplophasida	Nosematidae	*Nosema*
Haplophasea	Glugeida	Pleistophoridae	*Pleistophora*
		Encephalitozoonidae	*Encephalitozoon*
	Chytridiopsida	Enterocytozoonidae	*Enterocytozoon*
Not classified		Not classified	Not classified

lecular understanding of virulence and pathogenesis in intestinal spore-forming protozoa delays a better appreciation of the new phenomenon of their emergence as human pathogens.

B. Epidemiology

Intestinal spore-forming protozoa initiate human infection through ingestion of oocysts. These organisms are found in feces of infected individuals and multiple animal species. The oocysts are capable of surviving in soil for many months. Transmission to humans occurs via food, water, and in certain circumstances directly from infected to noninfected persons. Data on prevalence of these infections are not available in a uniform manner. The mean prevalence of *Cryptosporidium* infection in 56 surveys in developed countries excluding patients with HIV/AIDS (acquired immunodeficiency syndrome) was 4.9%, whereas the mean in 48 surveys in underdeveloped countries was 7.9% (Current and Garcia, 1991). These figures do not accurately represent the huge variation in prevalence from one country or region to another. Prevalence in asymptomatic individuals has been estimated at 0.3% in developed countries compared to 1.6% in the less developed regions of the world (Ungar, 1990). One also has to take into consideration the difficulties encountered in stool examination and variations in techniques. More recently serological surveys have indicated an increased percentage of positive results with age; the prevalence of anticryptosporidia antibodies in children is in the range of 5–10% and 50–70% in adults (Kuhls *et al.*, 1994). None of these surveys examined the correlation of antibodies to current or past clinical disease.

Prevalence data in HIV/AIDS patients relate to two major features: association with chronic diarrhea and CD4 counts (Goodgame *et al.*, 1993; Crowe *et al.*, 1991; Flanigan *et al.*, 1992; Blanshard *et al.*, 1992). *Cryptosporidium* has been demonstrated in the stools of 10–20% of AIDS patients with diarrhea; the prevalence in those with CD4 counts less than 100 cells/m^3 varies from 30 to 70%. Microsporidia infection has been reported from all continents except Antarctica. Most available information concerns patients with HIV/AIDS, with prevalences ranging from 7 to 50%; higher values are particularly reported in those with CD4 counts of <100 cells/mm (Weber and Bryan, 1994).

Infection in nonimmunocompromised hosts has been documented in several case reports. In 1995, 1996, and 1997 several outbreaks of *Cyclospora* infection were reported following drinking chlorinated water or consuming imported fruits. The situation in the summer of 1996 was particularly alarming. Outbreaks involving a considerable number of individuals were reported in 15 states (Colley, 1996). The source of infection was identified to be imported raspberries, although no conclusive evidence was obtained as to the source of infection.

The epidemiology of *Isospora* and *Cyclospora* is less well defined (Benator *et al.*, 1994; Wurtz, 1994). Both organisms have been detected worldwide. The higher occurrence of *Isospora* in developing countries may explain its higher prevalence in AIDS patients in areas such as Haiti. The more recent food-related outbreaks of *Cyclospora* in the United States add another dimension to its epidemiology (Colley, 1996; Ooi *et al.*, 1995; Chiodini, 1994).

Most alarming about the epidemiology of cryptosporidiosis is the new wave of waterborne epidemics (Soave, 1995). The organisms have been recovered from multiple sources of drinking water. The Milwaukee outbreak (more than 400,000 infections) was the largest recorded epidemic of waterborne disease in the United States since 1920. The outbreak represents a clear warning in regard to the inadequacy of water purification in many communities and the lack of epidemiological, parasitological, and public health expertise in many areas.

C. Pathogenesis and Clinical Features

Spore-forming intestinal protozoan infections changed the spectrum of organisms causing infectious diarrhea. Two settings brought these organisms to the forefront: acute or chronic diarrhea in AIDS patients and outbreaks of community-based diarrheal disease in immunocompetent individuals (Goodgame, 1996). The disease syndrome involves several possible mechanisms of diarrhea, but none has been firmly established. Pathologically, the small intestines of infected, symptomatic individuals demonstrate a range of changes including shortening of intestinal villi, cellular infiltrate, and often evidence of enterocyte destruction (Rabeneck *et al.*, 1993; Guarino *et al.*, 1995; Genta *et al.*, 1993; Goodgame *et al.*, 1996). The pathogenesis of diarrhea in cryptosporidiosis has received additional attention with the HIV/AIDS epidemic. The diarrhea is characteristically profuse and watery and may contain mucus but seldom blood or fecal leukocytes.

In the immunocompetent the course of infection is usually self-limited. Examination of the morphological changes in the gastrointestinal tract demonstrate mild to severe atrophy of the villous structures along with inflammatory cell infiltrate (Adal *et al.*, 1995). When the intestinal epithelium of infected symptomatic individuals was examined by electron microscopy, the infected enterocytes were demonstrated to lack microvilli, their cytoplasm showed multiple vacuolation, and their mitochondria were swollen and vacuolated as well. The functional consequences of these changes have been better studied in animal models. Malabsorption associated with bacterial overgrowth and osmotic pressure differentials across the intestinal border leads to watery diarrhea. Suggestions for the presence of enterotoxin-like activity in the parasite were made but not conclusively confirmed. Additional functional derrangements have been described such as reduction of lactase and alkaline phosphatase in the brush border. These abnormalities complete the spectrum of physiological disturbances associated with cryptosporidiosis. Disease due to the new intestinal protozoa may occur in other epidemiological settings such as traveler's diarrhea and acute diarrheas of children, particularly in less developed countries. The important message is the recognition of these organisms as possible etiologic agents of diarrheal disease occurring in specific epidemiological settings.

Extraintestinal disease is rare; colonic involvement may be seen in clinical presentations of several of these pathogens (Table V). Systemic dissemination occurs in microsporidiosis which may involve multiple organs such as liver, kidney, lungs, and brain (Pol *et al.*, 1993; Willson *et al.*, 1995; Molina *et al.*, 1995; Weber *et al.*, 1997).

TABLE V **Sites of Extraintestinal Disease Due to Spore-Forming Protozoa**

Organism	Colonic	Biliary	Systemic
Cryptosporidium	+	+	−
Microsporidia	+	+	+
Isospora		+	−
Cyclospora		−	−

D. Management

The critical challenge for practicing physicians is the recognition of the changing spectrum of etiologic agents of diarrheal disease in its multiple settings. Once any of the new intestinal parasitic protozoa is suspected, a close working relationship with the parasitology laboratory is essential. Although the diagnostic features of the different species are known, stool samples have to be processed using special concentration techniques and special stains (Arrowood and Sterling, 1989; Weber *et al.*, 1991; Beauvais *et al.*, 1993). The pace of developing immunologic diagnostics for examination of stool samples is accelerating, and reagents will soon be available (Grigoriew *et al.*, 1994). Serology, on the other hand, has not been helpful in diagnosis; rather, it is an important tool for epidemiological surveys (Grigoriew *et al.*, 1994).

Therapy for intestinal spore-forming protozoal infection is not satisfactory, particularly in AIDS patients with chronic diarrhea. Paromomycin is currently used for cryptosporidiosis with a response rate of 30–70%. Albendazole, though not approved by the U.S. Food and Drug Administration (FDA), may be used in microsporidiosis, while trimethoprim–sulfamethoxazole is effective against isosporiasis as well as in cyclosporiasis.

Control and prevention of these infections are complicated by their biology and epidemiology. Infection is transmitted via multiple vehicles, which necessitates reexamination of public health policies (Juranek, 1995) and the regulatory process for water and food safety.

REFERENCES

Adal, K. A., Sterling, R., and Guerrant, R. L. (1995). *In* "Infections of the Gastrointestinal Tract" (M. J. Blaser, P. D. Smith, J. I. Ravdin, H. B. Greenberg, and R. L. Guerrant, eds.), pp. 1107–1128. Raven, New York.

Adam, R. D. (1991). The biology of *Giardia* spp. *Microbiol. Rev.* 55, 706–732.

Aley S. B., and Gillin, F. S. (1995). Specialized surface adaptations of *Giardia lamblia. Infect. Agents Dis.* 4, 161–166.

Arrowood, M. J., and Sterling, C. R. (1989). Comparison of conventional staining methods and monoclonal antibody-based methods for *Cryptosporidium* oocyst detection. *J. Clin. Microbiol.* 27, 1490–1495.

Beauvais, B., Sarfati, C., Molina, J. M., Lesourd, A., Lariviere, M., and Derouin, F. (1993). Comparative evaluation of five diagnostic methods for demonstrating microsporidia in stool and intestinal biopsy specimens. *Ann. Trop. Med. Parasitol.* 87, 99–102.

Benator, D. A., French, A. L., Beaudet, L. M., Levy, C. S., and Orenstein, J. M. (1994). *Isospora belli* infection associated with acalculous cholecystitis in a patient with AIDS. *Ann. Intern. Med.* **121,** 663–664.

Blanshard, C., Jackson, A. M., Shanson, D. C., Francis, N., and Gazzard, B. G. (1992). "Cryptosporidiosis in HIV-seropositive patients. *Q. J. Med.* **85,** 813–823.

Buret, A., Hardin, J. A., Olson, M. E., and Gall, D. G. (1992). Pathophysiology of small intestinal malabsorption in gerbils infected with *Giardia lamblia. Gastroenterology* **103,** 506–513.

Chiodini, P. L. (1994). A "new" parasite: Human infection with *cyclospora cayetanensis. Trans. R. Soc. Trop. Med. Hyg.* **88,** 369–371.

Colley, D. G. (1996). Widespread, foodborne cyclosporiasis outbreaks present major challenges. *Emerging Infect. Dis.* **2**(4), 338–340.

Craun, G. F. (1990). Water-bourne Biardiasis *In* "Giardiasis" (E. A. Meyer, ed.), pp. 267–293. Elsevier, Amsterdam.

Crowe, S. M., Carlin, J. B., Steward, K. I., Lucas, C. R., and Hoy, J. F. (1991). Predictive value of CD4 lymphocyte numbers for the development of opportunistic infections and malignancies in HIV-infected persons. *J. Acquired Immune Defic. Syndr.* **4,** 770–776.

Current, W. L., and Garcia, L. S. (1991). Cryptosporidiosis. *Clin. Microbiol. Rev.* **4,** 325–358.

Current, W. L., Upton, S. J., and Haynes, T. B. (1986). The life cycle of *Cryptosporidium baileyi* n.sp. (Apicomplexa, Cryptosporidiidae) infecting chickens. *J. Protozool.* **33,** 289–296.

Das, S., Traynor-Kaplan, A., Reiner, D. S., Meng, T. C., and Gillin, F. D. (1991). A surface antigen of *Giardia lamblia* with a glycosylphosphatidylinositol anchor. *J. Biol. Chem.* **266,** 21318–21325.

Farthing, M. J. G. (1993). Diarrhoeal disease: Current concepts and future challenges—Pathogenesis of giardiasis. *Trans. R. Soc. Trop. Med. Hyg.* **87**(Suppl. 3), 17–21.

Farthing, M. J. G. (1995). *In* "Infections of the Gastrointestinal Tract" (M. J. Blaser, P. D. Smith, J. I. Ravdin, H. B. Greenberg, and R. L. Guerrant, eds.), pp. 1081–1105. Raven, New York.

Feely, D. E., Holberton, D. V., and Erlandsen, S. L. (1990). The biology of Giardia *In* "Giardiasis" (E. A. Meyer, ed.), pp. 11–49. Elsevier, Amsterdam.

Feely, D. E., Gardner, M. D., and Hardin, E. L. (1991). Excystation of *Giardia muris* induced by a phosphate–bicarbonate medium: Localisation of acid phosphatase. *J. Parasitol.* **77,** 441–448.

Flanigan, T., Whalen, C., Turner, J., Soave, R., Toerner, J., Havlir, D., Kotler, D. (1992). *Cryptosporidium* infection and CD4 counts. *Ann. Intern. Med.* **116,** 840–842.

Genta, R. M., Chappell, C. L., White, A. C., Jr., Kimball, K. T., and Goodgame, R. W. (1993). Duodenal morphology and intensity of infection in AIDS-related intestinal cryptosporidiosis. *Gastroenterology* **105,** 1769–1775.

Gillin, F. D., Reiner, D. S., and Boucher, S. E. (1988). Small intestinal factors promote encystation of *Giardia lamblia in vitro. Infect. Immun.* **56,** 705–707.

Gillin, F. D., Hagblom, P., Harwood, J., Aley, S. B., Reiner, D. S., McCaffery, M., So, M., Guiney, D. G. (1990). Isolation and expression of the gene for a major surface protein of *Gardia lamblia. Proc. Natl. Acad. Sci. U.S.A.* **87,** 4463–4467.

Goodgame, R. W. (1996). Understanding intestinal spore-forming protozoa: Cryptosporidia, Microsporidia, Isospora, and Cyclospora. *Ann. Intern. Med.* **124,** 429–441.

Goodgame, R. W., Genta, R. M., White, A. C., and Chappell, C. L. (1993). Intensity of infection in AIDS-associated cryptosporidiosis. *J. Infect. Dis.* **167,** 704–709.

Goodgame, R. W., Kimball, K., Ou, C. N., White, A. C., Jr., Genta, R. M., Lifschitz, C. H., Chappell, C. L. (1995). Intestinal function and injury in acquired immunodeficiency syndrome-related cryptosporidiosis. *Gastroenterology* **108,** 1075–1082.

Gottstein, B., Harriman, G. R., Conrad, J. T., and Nash, T. E. (1990). Antigenic variation in *Giardia lamblia:* Cellular and humoral immune response in a mouse model. *Parasite Immun.* **12,** 659–673.

Grigoriew, G. A., Walmsley, S., Law, L., Chee, S. L., Yang, J., Keystone, J., Krajden, M. (1994). Evaluation of the Merifluor immunofluorescent assay for the detection of *Cryptosporidium* and *Giardia* in sodium acetate formalin-fixed stools. *Diagn. Microbiol. Infec. Dis.* **19,** 89–91.

Guarino, A., Canani, R. B., Casola, A., Pozio, E., Russo, R., Bruzzese, E., Fontana, M., Rubino, A. A. (1995). Human intestinal cryptosporidiosis: Secretory diarrhea and enterotoxic activity in Caco-2 cells. *J. Infect. Dis.* **171,** 976–983.

Isaac-Ranton J. L. (1991). Laboratory diagnosis of giardiasis. *Clin. Lab. Med.* **11,** 811–827.

Johnson, A. M., Fielke, R., Lumb, R., and Baverstock, P. R. (1990). Phylogenetic relationships of *Cryptosporidium* determined by ribosomal RNA sequence comparison. *Int. J. Parasitol.* **20**, 141–147.

Juranek, D. D. (1995). Cryptosporidiosis: Sources of infection and guidelines for prevention. *Clin. Infect. Dis.* **21**(Suppl. 1), S57–S61.

Kuhls, T. L., Mosier, D. A., Crawford, D. L., and Griffis, J. (1994). Seroprevalence of cryptosporidial antibodies during infancy, childhood, and adolescence. *Clin. Infect. Dis.* **18**, 731–5.

Lengerich, E. J., Addiss, D., and Juranek, D. D. (1994). Severe giardiasis in the United States. *Clin. Infect. Dis.* **18**, 760–763.

Levin, B. R. (1996). The evolution and maintenance of virulence in microparasites. *Emerg. Infect. Dis.* **2**(2), 93–102.

Levin, B. R., and Bull, J. J. (1994). Short-sighted evolution and the virulence of pathogenic microorganisms. *Trends Microbiol.* **2**, 76–81.

Levin, B. R., and Svanborg-Eden, C. (1990). Selection and the evolution of virulence in bacteria: An ecumenical excursion and modest suggestion. *Parsitology* **100**, S103–S115.

Levine, N. D. (1986). The taxonomy of *Sarcocystis* (Protozoa, Apicomplexa) species. *J. Parasitol.* **72**, 372–382.

McCaffery, J. M., Faubert, G. M., and Gilin, F. D. (1994). *Giardia lamblia:* Traffic of a trophozoite variant surface protein and a major cyst wall epitope during growth, encystation, and antigenic switching. *Exp. Parasitol.* **79**, 236–249.

McDonald, V., Deer, R. M. A., Nina, J. M. S., Wright, S., Chiodini, P. L., and McAdam, K. P. W. J. (1991). Characteristics and specificity of hybridoma antibodies against oocyst antigens of *Cryptosporidium parvum* from man. *Parasite Immunol.* **13**, 251–259.

May, R. M., and Anderson, R. M. (1983). Parasite-Host Coevolution *In* "Coevolution" (D. J. Futuyama and M. Slatkin, eds.), pp. 186–206. Sinauer, Sunderland, Massachusetts.

Medical letter. (1995). *Drugs for Parasitic Infections* **37**, 99–108.

Molina, J. M., Oksenhendler, E., Beauvais, B., Sarfati, C., Jaccard, A., Deroin, F., Modai, J. *et al.* (1995). Disseminated microsporidiosis due to *Septata intestinalis* in patients with AIDS: Clinical features and response to albendazole therapy. *J. Infect. Dis.* **171**, 245–249.

Nash, T. E., and Mowatt, M. R. (1992). Characterization of a *Giardia lamblia* variant-specific surface protein (VSP) gene from isolate GS/M and estimation of the VSP gene repertoire size. *Mol. Biochem. Parasitol.* **51**, 219–227.

Nash, T. E., and Mowatt, M. R. (1993). Variant specific surface proteins of *Giardia lamblia* are zinc-binding proteins. *Proc. Natl. Acad. Sci. U.S.A.* **90**, 5489–5493.

Nash, T. E., Merritt, J. W., and Conrad, J. T. (1991). Isolate and epitope variability in susceptibility of *Giardia lamblia* to intestinal proteases. *Infect. Immun.* **59**, 1334–1340.

Nelson, R. G., Kim, K., Gooze, L., Peterson, C., and Gut, J. (1991). Identification and isolation of *Cryptosporidium parvum* genes encoding microtubule and microfilament proteins. *J. Protozool.* **38**(Suppl.), 52S–55S.

Nina, J. M. S., McDonald, V., Dyson, D. A., Catchpole, J., Uni, S., Iseki, M., Chiodini, P. L., and McAdam, K. P. W. J. (1992). Analysis of oocyst wall and sporozoite antigens from three *Cryptosporidium* species. *Infect Immun.* **60**, 1509–1513.

O'Donoghue, P. J. (1995). *Cryptosporidium* and cryptosporidiosis in man and animals. *Int. J. Parasitol.* **25**, 139–195.

Ooi, W. W., Zimmerman, S. K., and Needham, C. A. (1995). *Cyclospora* species as a gastrointestinal pathogen in immunocompetent hosts. *J. Clin. Microbiol.* **33**(5), 1267–1269.

Overturf (1994). Editorial response. (1994). Endemic giardiasis in the United States—Role of the day-care center. *Clin. Infect. Dis.* **18**, 764–765.

Peattie, D. A., Alonso, R. A., Hein, A., and Caulfield, J. P. (1989). Ultrastructural localization of giardins to the edges of disk microribbons of *Giardia lamblia* and the nucleotide and deduced protein sequence of alpha giardin. *J. Cell Biol.* **109**, 2323–2335.

Pickering, L. K., and Engelkirk, P. G. (1988). *Giardia lamblia. Pediatr. Clin. North Am.* **35**, 565–577.

Pimenta, P. F. P., Pinto da Silva, P., and Nash, T. (1991). Variant surface antigens of *Giardia lamblia* are associated with the presence of a thick cell coat: Thin section and label fracture immunocytochemistry survey. *Infect. Immun.* **59**, 3989–3996.

Pol, S., Romana, C. A., Richard, S., Amouyal, P., Desportes-Livage, I., Carnot, F., Pays, J. F., Berthelot, P. (1993). Microsporidia infection in patients with the human immunodeficiency virus and unexplained cholangitis. *N. Engl. J. Med.* **328**, 95–99.

Rabeneck, L., Gyorkey, F., Genta, R. M., Gyorkey, P., Foote, L. W., and Rosser, J. M. (1993). The role of microsporidia in the pathogeneisis of HIV-related chronic diarrhea. *Ann. Intern. Med.* **119**, 895–899.

Reduker, D. W., Speer, C. A., and Blixt, J. A. (1985). Ultrastructure of *Cryptosporidium parvum* oocysts and excysting sporozoites as revealed by high resolution scanning electron microscopy. *J. Protozool.* **32**, 708–711.

Robertson, L. J., Campbell, A. T., and Smith, H. V. (1993). *In vitro* excystation of *Cryptosporidium parvum. Parasitology* **106**, 13–19.

Roberts-Thomson, I. C. (1993). Genetic studies of human and murine giardiasis. *Clin. Infect. Dis.* **16**(Suppl. 2), S-98–S-104.

Schantz, P. M. (1991). Parasitic zoonoses in perspective. *Int. J. Parasitol.* **21**, 161–170.

Sleigh, M. A. (1991). The Nature of Protozoa *In* "Parasitic Protozoa" (J. P. Kreier and J. R. Baker, eds.), 2nd Ed., Vol. 1, pp. 1–53. Academic Press, San Diego.

Soave, R. (1995). Editorial response: Waterborne cryptosporidiosis—Setting the stage for control of an emerging pathogen. *Clin. Infect. Dis.* **21**, 63–64.

Sterling, C. R., and Arrowood, M. I. (1993). Cryptosporidia *In* "Parasitic Protozoa" (J. P. Kreier and J. R. Baker, eds.) 2nd Ed., Vol. 6, pp. 156–225. Academic Press, San Diego.

Thompson, R. C. A., Lymbery, A. J., and Meloni, B. P. (1990). Genetic variation in *Giardia* Kunstler, 1882: Taxonomic and epidemiological significance. *Protozool. Abstr.* **14**, 1–28.

Thompson, R. C. A., Reynoldson, J. A., and Mendis, A. H. W. (1993). Giardia and Giardiasis *Adv. Parasitol.* **32**, 71–160.

Tibayrenc, M., and Ayala, F. J. (1991). Towards a population genetics of microorganisms: The clonal theory of parasitic protozoa. *Parasitol. Today* **7**, 228–232.

Ungar, B. L. P. (1990). Cryptosporidiosis in humans (homosapiens) *In* "Cryptosporidiosis of Man and Animals" (J. P. Dubey, C. A. Speer, R. Fayer, eds.), pp. 59–82. CRC Press, Boston.

Ungar, B. L. P., Soave, R., Fayer, R., and Nash, T. E. (1986). Enzyme immunoassay detection of immunoglobulin M and G antibodies to *Cryptosporidium* in immunocompetent and immunocompromised persons. *J. Infect. Dis.* **153**, 570–578.

Wang, A. L., and Wang, C. C. (1986). Discovery of a specific double-stranded RNA virus in *Giardia lamblia. Mol. Biochem. Parasitol.* **21**, 269–276.

Wang, A. L., and Wang, C. C. (1991). Viruses of parasitic protozoa. *Parasitol. Today* **7**, 76–80.

Weber, R., and Bryan, R. T. (1994). Microsporidial infections in immunodeficient and immunocompetent patients. *Clin. Infect. Dis.* **19**, 517–21.

Weber, R., Bryan, R. T., Bishop, H. S., Wahlquist, S. P., Sullivan, J. J., and Juranek, D. D. (1991). Threshold of detection of *Cryptosporidium* oocysts in human stool specimens: Evidence for low sensitivity of current diagnostic methods. *J. Clin. Microbiol.* **29**, 1323–7.

Weber, R., Deplazes, P. Flepp, M. Mathis, A., Baumann, R., Sauer, B., Kuster, H., Luthy, R. (1997). Cerebral microsporidiosis due to *Encephalitozoan cuniculi* in a patient with human immunodeficiency virus infection. *N. Engl. J. Med.* **336**, 474–478.

Webster, K. A., Pow, J. D. E., Giles, M., Catchpole, J., and Woodward, M. J. (1993). Detection of *Cryptosporidium parvum* using a specific polymerase chain reaction. *Vet. Parasitol.* **50**, 35–44.

Willson, R., Harrington, R., Steward, B., and Fritsche, T. (1995). Human immunodeficiency virus 1-associated necrotizing cholangitis caused by infection with *Septata intestinalis. Gastroenterology* **108**, 247–51.

Wurtz, R. (1994). *Cyclospora:* A newly identified intestinal pathogen of humans. *Clin. Infect. Dis.* **18**, 620–623.

16

TRANSMISSIBLE SPONGIFORM ENCEPHALOPATHIES OF MAN AND ANIMALS

JAMES HOPE

BBSRC Institute for Animal Health
Compton Laboratory
Compton, Berkshire, England

I. INTRODUCTION

Scrapie, Creutzfeldt–Jakob disease (CJD), Gerstmann–Straussler–Sheinker (GSS) syndrome, and related diseases of mink (transmissible mink encephalopathy), mule deer, and elk (chronic wasting disease) are classified as the transmissible degenerative (or spongiform) encephalopathies (TSE). Since the 1980's, new species have been affected including cattle (bovine spongiform encephalopathy), cats (feline spongiform encephalopathy), and a variety of captive zoo felines and antelope, and a new form of CJD in humans has recently emerged. Iatrogenic transmission of CJD in humans occurs, and these diseases can be transmitted from affected to healthy animals by inoculation or by feeding diseased tissues. The transmissible factor or prion has yet to be fully characterized. This chapter tells of the TSEs of humans, sheep, and cattle and covers recent progress in the understanding of their etiology.

A. Scrapie

Scrapie of sheep has been known in Europe for centuries and has spread to most parts of the world, excluding Australasia and Argentina, with the migrations of humans and livestock. It is characterized by altered behavior, hypersensitivity to sound or touch, loss of condition, pruritus, and associated fleece loss and skin abrasions, and incoordination of the hind limbs. Diagnosis is confirmed postmortem by the examination of brain tissue for a triad of histopathological signs: vacuolation (Fig. 1), loss of neurons, and gliosis (Hadlow, 1995).

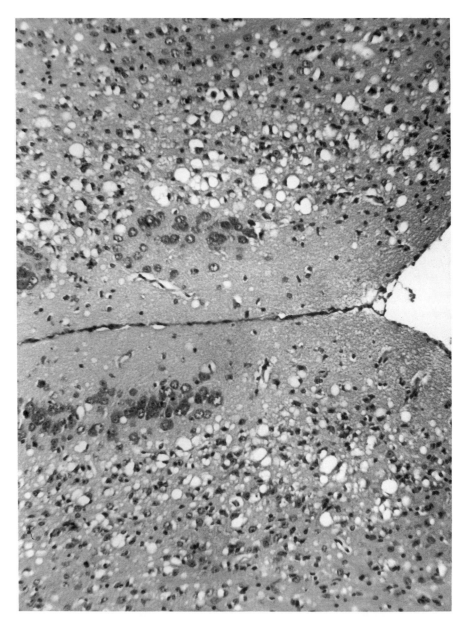

FIGURE I   Vacuolar degeneration in the cerebral cortex of a mouse infected with the ME7 strain of scrapie. This diagnostic pathology gives the appearance of a "sponge" at low magnification and gives rise to the generic name of these infectious diseases: the transmissible spongiform encephalopathies (magnification X 40). (Photograph courtesy of Dr. Moira E. Bruce.)

Scrapie has been reported in most breeds of sheep, and, within a flock, it appears to occur in related animals. The natural clinical disease has a median peak incidence in flock animals of 3.5 years, with a range of 2.5 to 4.5 years covering the vast majority of cases (Hunter *et al.*, 1992). For most of this period, the infected animal is clinically normal and indistinguishable from its uninfected flockmates. The within-flock incidence of clinical disease is usually 1–2 cases per 100 sheep per year, but there have been several instances of 40–50% of animals of a flock succumbing to the disease within a year. A number of genetic markers have been identified as risk factors, and the introduction of gene typing has greatly facilitated interpretation of field studies on the incidence of natural and experimental disease (Goldmann *et al.*, 1994).

II. CREUTZFELDT–JAKOB DISEASE

Creutzfeldt–Jakob disease is a progressive dementia with clinical signs suggesting dysfunction of the cerebellum, basal ganglia, and lower motor neurons. It is associated with gradual mental deterioration leading to dementia and confusion, and a progressive impairment of motor function. Most patients die within 6 months of onset of clinical signs, and there are no verified cases of recovery. Pathologically the lesions of the brain include variable vacuolation of the neuropil, astrocytosis, and, in about 10% of CJD cases, amyloid plaques. Gerstman–Straussler syndrome is a familial variant of CJD with an extended, clinical time course.

The incidence of CJD-related disease in humans is remarkably constant at 0.5–1 cases per million of population per year throughout the world and so is not linked to the incidence of any of the animal diseases. This low incidence casts doubt on the role of infection in its propagation within the population (but see below). About one in seven cases are familial and linked to mutations in the open reading frame (ORF) of the PrP gene. There has been a large amount of clinical and pathological studies on human cases of neurological disease which seem to be associated with these rare mutations of the PrP gene (for a review, see Brown *et al.*, 1994). In some families, there is complete penetrance of the phenotype and so the mutation is regarded as the cause of the disease. Apart from iatrogenic cases induced by transplantation of infected tissues or inoculation of contaminated pharmaceuticals of human origin, there is no epidemiological evidence for horizontal transmission of the disease. A stochastic event involving conversion of the PrP protein to its disease-associated isoform and the chance mutation of a benign, ubiquitous viral-like agent are two mechanisms that have been suggested to explain the incidence of sporadic cases (see section on the nature of the agent). There is no cure for the clinical condition although genetic counseling, where applicable, may effectively prevent transmission of disease from one generation to the next.

There is considerable clinical and pathological heterogeneity in the human prion diseases, and although genetic typing and nucleotide sequencing of the PrP ORF have provided some unifying concepts, mutation in the PrP protein does not appear to be the whole story. Other genetic factors including linkage to the E4 allele of ApoE gene have been implicated as risk factors for the occurrence of CJD (Amouyel *et al.*, 1994; but see also Salvatore *et al.*, 1995).

A. Emerging Human Transmissible Spongiform Encephalopathies

Fatal familial insomnia (FFI) has emerged as the newest member of the human transmissible encephalopathies. It is characterized by a dysfunction of the autonomic nervous system usually presenting with insomnia and problems of appetite, temperature, and blood pressure regulation. At postmortem examination, the pathology of the brain is mostly neuronal loss and degeneration of the thalamus with little or no vacuolation of the neuropil. Classification of FFI as a prion disease was originally based on its association with an asparagine (N) to aspartic acid (D) mutation at codon 178 of the PrP gene, a mutation which is also linked to a classic form of CJD. Which of the two pheontypes prevails appears to depend on the amino acid encoded by codon 129 of the same PrP allele: in FFI codon 129 encodes methionine, whereas in CJD codon 129 encodes valine. Homozygosity at codon 129 also appears to be risk factor in the development of sporadic CJD, but each polymorphism at this codon is fairly common and not thought to be pathogenic per se. The classification of FFI has been confirmed by transmission of disease to laboratory mice (Tateishi *et al.*, 1995).

In March 1996, the U.K. government announced their concern about a new variant of CJD, and details of these cases have subsequently appeared (Will *et al.*, 1996). To date, 15 cases of CJD have been identified in the United Kingdom with a new neuropathological and clinical profile of extensive PrP deposition, cerebellar amyloid plaques, spongiform change most evident in the basal ganglia and thalamus, prolonged duration (up to 2 years), atypical electroencephalogram (EEG), early ataxia, and behavioral and psychiatric disturbances. Similar cases have not been identified in archival patient files or elsewhere in Europe [apart from a single French case (Chazot *et al.*, 1996)], and there would appear to be a risk factor for this variant unique to the United Kingdom. Its coincidence with a novel bovine TSE (BSE) which is largely restricted to the United Kingdom has led to speculation, as yet unproved, that this new form of CJD represents a cross-species transmission of infection from cattle.

B. Bovine Spongiform Encephalopathy

Bovine spongiform encephalopathy (BSE) has devasted the U.K. cattle industry in the 1980s and 1990s (Bradley and Wilesmith, 1993). From isolated cases first reported in 1986 and some retrospectively identified in May 1985, a major epidemic was underway by 1988 which has to date claimed over 160,000 cattle within the British Isles. Some other countries have also confirmed cases: Switzerland (250+), Ireland (200+), Portugal (30), France, and Germany with one or two cases in Italy, Denmark, Canada, the Netherlands, Oman, and the Falkland Islands.

The disease produces a progressive degeneration of the central nervous system and was named because of the spongelike appearance of BSE brain tissue when seen under the light microscope (Wells *et al.*, 1987). Warning signs of the illness include changes in the behavior and temperament of the cattle. The affected animal becomes increasingly apprehensive and has problems of movement and posture, especially of its hind limbs. The cow (or bull) has increased sensitivity to touch and sound, displays loss of weight, and, as the disease takes hold of its nervous system, suffers a creeping paralysis. This clinical phase of BSE lasts from a fortnight to over 6 months. Although the majority of animals affected have been dairy cows, this neurological disease can occur in either sex with a modal age of onset of 4–4.5 years (range 1.8–18 years). Most cases of

BSE have occurred in cattle between the ages of 3 and 5 years, and for most of its development time the disease gives no telltale sign of its presence (Wilesmith *et al.*, 1988).

The neuroligical lesions in BSE-affected cow brains are virtually identical to those found in scrapie-affected sheep and include the spongiform change which gives BSE its name. From its clinical and neuropathological signs, BSE was immediately suspected to belong to the scrapie family of transmissible spongiform encephalopathies. This has been confirmed by biochemical studies (Hope *et al.*, 1988a) and by experimental transmission of BSE to mice (Fraser *et al.*, 1988), sheep, and goats (Foster *et al.*, 1993) among other species.

C. Origins of Bovine Spongiform Encephalopathy and Its Current Status

Epidemiological analyses of BSE-affected herds identified a protein feed supplement to be the most likely source of infection (Wilesmith *et al.*, 1988), and it was made illegal in the United Kingdom to feed ruminant-derived protein to ruminants in July 1988. During the late 1970s changes in the rendering process which salvage compounds of nutritional and commercial value from abattoir waste are thought to have led to a less efficient system for inactivating scrapie-affected sheep offal and, in turn, to a contaminated protein supplement. Subsequent recycling of BSE-infected cattle waste in this process may have contributed to the persistence of the disease.

The ruminant feed legislation was aimed at removing the source of infection from cattle born after 1988, and the epidemic is now showing signs of rapid decline; by April, 1996, the number of confirmed BSE cases had dropped to below 1200 a month following a peak incidence of over 1000 cases a week in 1993. There have been over 26,000 cases of BSE in cattle born after the feed ban; however, the very low within-herd incidence (about 2%) of the disease makes vertical or horizontal transmission within herds unlikely, and illegal or unknowing feeding of contaminated protein to calves is suspected as the reason for the "born after the ban" cases.

In parallel with the BSE epidemic, natural causes of transmissible spongiform encephalophaties have also emerged for the first time in cattle-related species: greater kudu, oryx, and in the cat family puma, cheetah, and domestic cats (Kirkwood and Cunningham, 1994; Pearson *et al.*, 1992). Apart from some cases in the greater kudu, contaminated feed is suspected but difficult to prove because of the absence of detailed feeding records.

To avoid human exposure to BSE, a ban on the use of specific cattle tissues for human consumption was introduced in November 1989 to January 1990 in the United Kingdom, and in September 1990 this ban was extended to their use as feed to any animal or bird. A comprehensive survey of the epidemiology of BSE has been published, and it is now acknowledged that almost 1 million BSE-infected carcasses may have entered the human food chain prior to 1995 (Anderson *et al.*, 1996). The specified bovine offals (SBOs) banned for use were brain, spinal cord, tonsil, thymus, spleen, and the lower intestine (duodenum to rectum) from cattle over the age of 6 months. This high-risk category of tissues was based on the levels of infectivity detected by mouse bioassay in natural cases of sheep scrapie. New data on the direct assay of infectivity by mouse bioassay in these cattle tissues indicates the levels of agent may not be so high nor so widespread in cattle as they are in sheep (Fraser and Foster, 1993). However, cattle-to-cattle transmission appears much more efficient than the equivalent cattle-to-mouse experiment, and there is no estimate of the relative efficiency of cattle-to-human transmission. Preliminary attempts to model infection of humans by injecting BSE into mice genetically engineered to express the human (in place of the murine) prion protein have given negative results so far, but the data are too incomplete to encourage complacency (Collinge et al., 1995; Hope, 1995).

III. EPIDEMIOLOGY

One of the main problems of understanding the epidemiology of human TSE disease is to explain its low incidence, an incidence which appears incompatible with a sustainable infection within the population. There are two main views on this dichotomy: the first proposes that the disease is not infectious but arises de novo in each individual as the result of a somatic or germline mutation in the prion protein gene; the other stresses our lack of knowledge of the prevalence of infection rather than the incidence of clinical disease and proposes that there is a widespread inapparent infection of the population by a benign agent and only in certain genetically susceptible individuals or by its mutation to a pathogenic form will this ubiquitous agent produce disease. In either case, the predominant form of natural transmission is predicted to be vertical in accord with field observation in humans.

In scrapie of sheep, the other common natural TSE, there is evidence for both vertical and horizontal transmission of disease. Maternal transmission of the infection from ewe to offspring either in utero or immediately after birth is thought to be the major route of propagation of the disease within a flock, but lateral transmission is also documented (Dickinson et al., 1974). Factors associated with the horizontal spread of infections such as host susceptibility, source, and route of infection (see pathogenesis, p. 453) have been investigated for many years in sheep and rodent models of disease. The low incidence of clinical disease in affected flocks is usually interpreted to mean the agent is not highly contagious, and this is supported by the low or zero infectivity in body fluids or secretions. Recently, hay mites cohabiting the pasture of flocks of sheep in Iceland have been implicated as an insect vector of disease (Wisniewski et al., 1996).

Medium levels of infectivity in placentas and amniotic fluids may contaminate pasture or pens and surrounds where lambing occurs and persist for long periods (Pattison et al., 1974). Dissemination of agent by other routes—in feces, urine, and milk—is less likely as little or no infectivity has ever been detected in these excretions. Of the common routes of entry (or reentry) into the body—ingestion, inhalation, contact, and coitus—the natural route is probably via the mouth or skin abrasions. Both are well-documented portals of entry for experimental transmission of scrapie and other spongiform encephalopathies to rodents and ruminants. Infectivity has been detected in the eyes and lungs of natural cases of disease, and conjunctival instillation of scrapie in mice can produce disease (Scott et al., 1993). Nevertheless, there have been no accounts of experimental, aerosol transmission, and infectivity in lungs may be due to secondary transport and infection; however, this emphasizes the need for adequate protection when handling tissues infected with TSE, and recommended safety precautions for laboratory workers include the use of face masks, avoidance of aerosols, and eye protection. Sexual intercourse does not seem to be a risk factor in the transmission of these diseases in humans or animal.

A. Spread of Pathogen within the Natural Host

Apart from the recognized familial incidence, verification of the oral route of transmission of natural scrapie and the pathogenesis of disease is based on painstaking work by many workers, notably Hadlow and colleagues. They measured infectivity by rodent bioassay in various neonatal and maternal sheep tissues and body fluids in flocks of sheep with a high incidence of natural scrapie or in experimentally infected animals. The early appearance of infectivity in tonsil, retropharyngeal and mesenteric–portal lymph nodes, and intestine suggested that primary infection was occurring by way of the alimentary tract, either prenatally from pathogen in amniotic fluid or postnatally from a contaminated environment (Hadlow *et al.*, 1982).

Exactly where in the alimentary tract the pathogen gains access to sites of replication and transport is a point of debate at the moment. Uptake by the oropharyngeal tract may circumvent normal protective processes against oral infection, and this route into the central nervous system has been implicated for scrapie and BSE. The acidic pH of the stomach is an effective barrier against this mode of transmission for a wide variety of organisms, but not for scrapie-like agents which are highly stable at low pH; similarly, other physicochemical inactivators such as cholate in bile salts which protect against enteric, enveloped viruses may simply disperse prions, enhancing their uptake and transport across the gut epithelium. For some viruses, there is uptake of nonreplicating particles by cells overlying Peyer's patches (in the ileum), and intact virus particles in smooth-surfaced cytoplasmic vesicles in these cells have been seen by electron microscopy. These cells appear to hand over virus particles to adjacent mononuclear cells for antigen processing and replication, and some of the particles then enter the local lymphatic system (Mimms and White, 1984). This paradigm matches the kinetics of infectivity levels in various tissues following oral infection of rodents with scrapie, but it is difficult to show in the natural disease.

After oral infection of mice, neural spread of the pathogen occurs from the gastrointestinal tract, perhaps via the enteric and sympathetic nervous systems, to the spinal cord. Neuroinvasion may be initiated either via infection of Peyer's patches or directly by infection of nerve endings in the gut wall, and the lymphoreticular system may be bypassed completely (Kimberlin and Walker, 1989a). However, with alternative routes of infection (intraperitoneal, subcutaneaous), the spleen and other lymphatic tissues are important sites where the pathogen appears to replicate to a threshold level before breakthrough into the peripheral nervous system and transport to the lower spinal cord (Fraser and Dickinson, 1978; Kimberlin and Walker, 1989b; Lasmezas *et al.*, 1996). It is the PrP-producing follicular dendritic cell which appears important for accumulation of infectivity in the spleen (Kitamoto *et al.*, 1991), although whether this cell type can also support agent replication is not known. Retrograde axonal transport to the central nervous system (CNS) then precedes its devastating effects on brain and brain stem.

Because the physiology and living conditions of humans, sheep, cattle, and laboratory rodents differ widely, it would be surprising if this model for the entry and spread of agent through the body were generally applicable, and much still needs to be done to understand pathogenesis in humans and ruminants. In early kuru studies, Asher and colleagues (1976) found epidemiological evidence that the agent may enter the body through breaks in the skin and mucous membranes; this agreed with contemporary data that, following subcutaneous injection of scrapie, rodents showed early replication in lymphoid tissue and later in other organs. Taylor and colleagues have shown that scrapie infection can be established readily through skin abrasions in immunocompetent but not immunodeficient (SCID) mice (Taylor *et al.*, 1996). However, in kuru and CJD studies (from clinically ill patients), brain was regularly found to be infected, but only occasionally was infectivity detected in other organs such as spleen and lymphoid tissue: paradoxically, as the titer increased in the CNS it decreased or disappeared from lymphoid tissue (Asher *et al.*, 1976). Current failure to detect infectivity (and PrP^{Sc}) in peripheral tissues of CJD/GSS patients and BSE-affected cattle cautions against making too many general statements about pathogenesis on the basis of sheep and rodent studies.

There is no evidence of maternal transmission of human TSEs although this has often been noted in field studies of natural scrapie in sheep. It suggests a possible means of scrapie eradication, but to date attempts to circumvent maternal transmisison in sheep by embryo transfer techniques have had mixed success, not least because of the "carrier status" problem and the lack of information on the range of natural strains of scrapie and their interaction with the host genome. Obviously the vertical and horizontal methods of transmission are not mutually exclusive and may change in relative importance with time. For example, the miniepidemic of spongiform encephalopathy in African kudu may have had a common source with BSE (i.e., feeding of contaminated protein), but in subsequent cases lateral transmission is implicated (Kirkwood *et al.*, 1994). This emphasizes that the initial wave of disease may be started by one mechanism but continued by another (e.g., maintenance at a low level by lateral or maternal transmission).

One of the major problems in predicting the incidence and consequences of the BSE epidemic or the emergence of TSEs in novel species lie in the unique properties of the spongiform encephalopathy infectious agent.

IV. NATURE OF THE TRANSMISSIBLE SPONGIFORM ENCEPHALOPATHY AGENT

Viruses, bacteria, bacteriophages, and all other conventional forms of life show phenotypic or strain variation which is encoded by their nucleic acid genomes. Strain variation is also a common feature of various scrapie isolates in mice, hamsters, sheep, and goats; however, although selection and mutation of murine scrapie strains is documented, a coding molecule has yet to be defined. The two main criteria used to distinguish strains of mouse-passaged scrapie are (1) the ranking of the incubation periods they produce in mice of the three *Sinc* genotypes, s7s7, s7p7, and p7p7, and (2) the severity and location of vacuolar degeneration induced in the brains of terminal cases of disease (Fraser, 1976). The occurrence of different strains of pathogens has implications for much of the epidemiology and genetics of these diseases (see below). For example, passage of BSE from seven unrelated cattle sources into a panel of *Sinc* s7s7, s7p7, and p7p7 mice have given a remarkably uniform pathology and ranking of incubation period, differing from over 20 other transmissions of sheep and goat scrapie. Transmissions to mice of spongiform encephalopathy from six species (including sheep and goats, kudu, and oryx) which have been experimentally or naturally infected with BSE have given similar results to direct BSE transmissions from cattle (Bruce *et al.*, 1994). Although the molecular basis of this uniformity is uncertain, similar transmissions from the "suspect-BSE" CJD cases described in the United Kingdom (Will *et al.*, 1996) may distinguish between a cattle or some other source of disease.

A. Biochemistry of Infectivity: Fibrils, Rods, and the Prion Protein

High titers of infectivity are recovered in preparations of membranes purified from TSE-affected brain and other tissue; this infectivity is not significantly reduced by disruption of membranes using deoxycholate (DOC), sarcosinate (Sarkosyl), and other mild detergents and can be concentrated and pelleted by differential centrifugation. This infection material is heterogeneous in size and physical properties and has yet to be isolated as a band by density-gradient centrifugation. This has hindered its biophysical characterization. Viewed by electron microscopy, these highly enriched fractions of infectivity contain fibrils of various shapes and sizes as well as ferritin particles and amorphous material. Surprisingly, these fractions are homogeneous biochemically. One isolate of hamster scrapie (263K) survives prolonged treatment with high concentrations of proteinase K (PK) and contains little else but a M_r 27,000–30,000 protein. This is the prion protein (PrP27-30), and, although different isolates from mouse and other species are more susceptible to proteases than the 263K protein and so are sometimes harder to detect, PK-resistant PrP (PrP-res or PrP^{Sc}) is found in all TSE isolates and has become a biochemical marker for disease and the infectious agent (Hope *et al.*, 1986, 1988a,b). Several groups have reported a stoichiometry of 100,000 molecules PrP^{Sc} per infectious particle using rodent models of these diseases (Scott *et al.*, 1991). The fibrils (scrapie-associated fibrils or rods) are aggregates of PrP^{Sc} and provide a morphological marker of infection/disease (Fig. 2).

PrP^C is a phosphoinositol–glycolipid-anchored membrane glycoprotein found in brain and, to a lesser extent, other tissues. The primary structure of the PrP is virtually constant in mammalian species. It is a glycoprotein of 33,000–35,000 Da which is anchored to the cell plasma membrane by a phosphatidylinositol–glycolipid attached to its carboxyl-terminal amino acid. The hamster protein (PrP^C and PrP^{Sc}) has 208 amino acids (PrP$^{23-231}$), and the normal isoform is completely degraded by proteases under conditions which leave a 27,000- to 30,000-Da, protease-resistant core of the PrP^{Sc} isoform intact (PrP27-30) Hope *et al.*, 1986. PrP$^{81-230}$ is equivalent to the PK-resistant core of mouse PrP^{Sc} (Hope *et al.*, 1988b), and its expression in transgenic mice has been shown to be suffi-

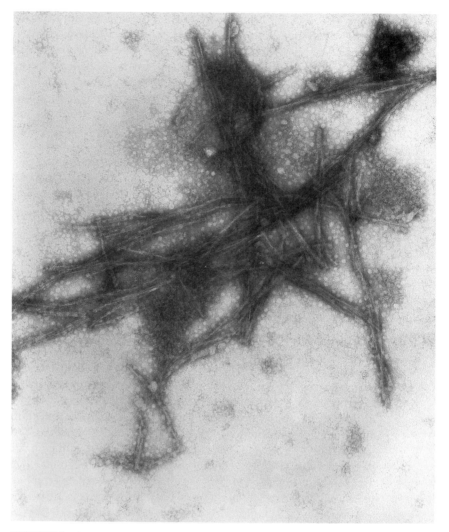

FIGURE 2 Scrapie-associated fibrils (SAF) from scrapie-affected mouse brain. Magnification: 100,000×. (Photomicrograph courtesy of Dr. Peter H. Gibson.)

cient to support replication of infectivity and the development of disease (Fischer *et al.*, 1996).

In vitro formation of PrP^{Sc} from PrP^C has been shown in infected cell cultures (see below) and in a cell-free system where the conversion is driven by addition of PrP^{Sc} template. The cell-free system mimics several aspects of the *in vivo* disease including species and strain specificities. From these test-tube studies, two distinct models for the formation of PrP^{Sc} have evolved: in both, exogenous PrP^{Sc} forms catalytic heterodimers with PrP^C which results in the formation of more PrP^{Sc}; in one, these "heterodimers" are real, whereas in the other they actually represent the growing face of a PrP^{Sc} fibril or aggregate. The latter, "seeded" polymerization model fits better with the kinetics of *in vitro* conversion PrP^C to protease-resistant PrP, although *de novo* "PrPSc" made in a test tube has yet to be shown to be infectious. This mechanism of conversion resembles a crystallization process in that it is rate-limited by nucleus formation and accelerated by seeding.

Although the structures and pathway of conversion between PrP^C and PrP^{Sc} have yet to be worked out, the atomic coordinates of a soluble, independent folding domain of the protein (residues 121–230) have been defined by nuclear magnetic resonance spectroscopy (Riek *et al.*, 1996). Knowledge of the full structure of PrP^C and PrP^{Sc} may help the design of chemicals engineered to prevent the conversion process and so help predict transmission between species and limit the effects of these diseases. To date, simple comparison of the primary sequences of PrP from different species has failed to aid the prediction of whether a particular source or new strain of TSE will transmit from one species to another.

B. *Sinc*, the Prion Protein Gene, and PrP-less Mice

Genetic linkage studies in various species have implicated PrP as a product of the *Sinc* locus (or homologous locus), and further evidence for this congruency has been provided using mice genetically engineered to lack one or both copies of the PrP gene. Mice lacking a PrP gene appear normal, and thew3 PrP^{null} mice neither replicate infectivity nor develop disease when challenged with doses of infectivity that would be lethal to their PrP$^{+/+}$ littermates. Levels of PrP^C appear to be crucial in determining the timing and duration of clinical disease; hemizygous PrP$^{+/-}$ mice, which express roughly half the normal (PrP$^{+/+}$) amount of PrP^C in the brain, have a significantly longer clinical phase and incubation period of disease compared with the survival characteristics of their PrP$^{0/0}$ littermates. These experiments have confirmed a key role for PrP^C as either substrate for a pathogenic isoform (PrP^{Sc}) or as a receptor molecule for a more conventional pathogen. Transgenic mice with human or bovine PrP genes in place of their own may be more susceptible than their wild-type littermates to inoculation with human or bovine infectious particles.

C. Cell Biology of Transmissible Spongiform Encephalopathy

The presence of PrP^C in a cell or organism appears to be a necessary if not sufficient condition for it to be able to replicate the TSE pathogen. Although the PrP gene is suppressed in many embryonic and adult mouse tissues (Manson *et al.*, 1992), its deletion from the mouse genome does not appear to affect normal development, behavior, and fertility. Animals homozygous (0/0) and heterozygous (0/+) for this mutation can be inbred to produce stable lines of mice and appear normal (Bueler *et al.*, 1992; Manson *et al.*, 1994a). Some workers (Colling *et al.*, 1996; Collinge *et al.*, 1994; Manson *et al.*, 1995) but not others (Herms *et al.*, 1995; Lledo *et al.*, 1996) have observed a subtle deficit in synaptic transmission in tissue slices, and altered circadian rhythms and sleep patterns have also been described (Tobler *et al.*, 1996) in these PrP-less (0/0) mice. Gross abnormalities such as loss of Purkinje cells and cerebellar ataxia are an inconsistent observation (Sakaguchi *et al.*, 1996) and need to be further substantiated as specific deficits of PrP gene deletion. However, 0/0 mice survive intracerebral inoculation of doses of scrapie agent lethal to wild-type mice, and 0/+ animals, which have decreased levels of PrP protein in the brain, have significantly extended survival times compared to +/+ mice (Bueler *et al.*, 1993; Manson *et al.*, 1994b). Titration of infectivity in brain and spleen of 0/0 mice following challenge with prions has failed to detect replication although infectivity persists in these animals long after infection (Sailer *et al.*, 1994): these experiments show that expression of the PrP gene is a prerequisite for replication and development of disease in the mouse.

Much further work needs to be done to understand these cellelar interactions in the brain and peripheral tissues, not least because their understanding should guide the development of therapeutics. Interestingly, polyanions such as dextran sulfate, Congo red, and pentosan sulfate which can inhibit PrP^C to PrP^{Sc} formation *in vitro* (Caughey and Raymond, 1993) may also extend the survival time of hamsters or mice coinfected with some strains of scrapie (Ehlers and Diringer, 1984; Farquhar and Dickinson, 1986; Ingrosso *et al.*, 1995; Kimberlin and Walker, 1986; Ladogana *et al.*, 1992); however, the large doses and extended therapy needed in these cases may preclude their use in practice. To date, no conventional antiviral agent or antibiotic has been effective in delaying the time course of scrapie or other natural TSEs.

REFERENCES

Amouyel, P., Vidal, O., Launay, J. M., and Laplanche, J. L. (1994). The apolipoprotein-E alleles as major susceptibility factors for Creutzfeldt–Jakob disease. *Lancet* **344**, 1315–1318.

Anderson, R. M., Donnelly, C. A., Ferguson, N. M., Woolhouse, M. E. J., Watt, C. J., Udy, H. J., Mawhinney, S., Dunstan, S. P., Southwood, T. R. E., Wilesmith, J. W., Ryan, J. B. M., Hoinville, L. J., Hillerton, J. E., Austin, A. R. and Wells, G. A. H. (1996). Transmission dynamics and epidemiology of bse in British cattle. *Nature (London)* **382**, 779–788.

Asher, D. M., Gibbs, C. J., Jr. and Gajdusek, D. C. (1976). Pathogenesis of subacute spongiform encephalopathies [review]. *Ann. Clin. Lab. Sci.* **6**, 84–103.

Bradley, R., and Wilesmith, J. W. (1993). Epidemiology and control of bovine spongiform encephalopathy (bse). *Br. Med. Bull.* **49**, 932–959.

Brown, P., Gibbs, C. J., Rodgersjohnson, P., Asher, D. M., Sulima, M. P., Bacote, A., Goldfarb, L. G., and Gajdusek, D. C. (1994). Human spongiform encephalopathy—The National-Institutes-of-Health series of 300 cases of experimentally transmitted disease. *Ann. Neurol.* **35**, 513–529.

Bruce, M., Chree, A., McConnell, I., Foster, J., Pearson, G., and Fraser, H. (1994). Transmission of bovine spongiform encephalopathy and scrapie to mice—Strain variation and the species barrier. *Philos. Trans. R. Soc. London Ser. B-Biol. Sci.* **343**, 405–411.

Bueler, H., Fischer, M., Lang, Y., Bluethmann, H., Lipp, H. P., Dearmond, S. J., Prusiner, S. B., Aguet, M., and Weissmann, C. (1992). Normal development and behavior of mice lacking the neuronal cell-surface prp protein. *Nature (London)* **356**, 577–582.

Bueler, H., Aguzzi, A., Sailer, A., Greiner, R. A., Autenried, P., Aguet, M., and Weissmann, C. (1993). Mice devoid of PrP are resistant to scrapie. *Cell (Cambridge, Mass.)* **73**, 1339–1347.

Caughey, B., and Raymond, G. J. (1993). Sulfated polyanion inhibition of scrapie-associated PrP accumulation in cultured cells. *J. Virol.* **67**, 643–650.

Chazot, G., Broussolle, E., Lapras, C., Blattler, T., Aguzzi, A., and Kopp, N. (1996). New variant of Cruetzfeldt–Jakob disease in a 26-year-old French man. *Lancet* **347**, 1181.

Colling, S. B., Collinge, J., and Jefferys, J. G. R. (1996). Hippocampal slices from prion protein null mice—Disrupted Ca^{2+}-activated K+ currents. *Neurosci. Lett.* **209**, 49–52.

Collinge, J., Whittington, M. A., Sidle, K. C. L., Smith, C. J., Palmer, M. S., Clarke, A. R., and

Jefferys, J. G. R. (1994). Prion protein is necessary for normal synaptic function. *Nature (London)* 370, 295–297.

Collinge, J., Palmer, M. S., Sidle, K. C. L., Hill, A. F., Gowland, I., Meads, J., Asante, E., Bradley, R., Doey, L. J., and Lantos, P. L. (1995). Unaltered susceptibility to bse in transgenic mice expressing human prion protein. *Nature (London)* 378, 779–783.

Dickinson, A. G., Stamp, J. T., and Renwick, C. C. (1974). Maternal and lateral transmission of scrapie in sheep. *J. Comp. Pathol.* 84, 19–25.

Ehlers, B., and Diringer, H. (1984). Dextran sulphate 500 delays and prevents mouse scrapie by impairment of agent replication in spleen. *J. Gen. Virol.* 65, 1325–1330.

Farquhar, C. F., and Dickinson, A. G. (1986). Prolongation of scrapie incubation period by an injection of dextran sulphate 500 within the month before or after infection. *J. Gen. Virol.* 67, 463–473.

Fischer, M., Rulicke, T., Raeber, A., Sailer, A., Moser, M., Oesch, B., Brandner, S., Aguzzi, A., and Weissmann, C. (1996). Prion protein (Prp) with amino-proximal deletions restoring susceptibility of Prp knockout mice to scrapie. *EMBO J.* 15, 1255–1264.

Foster, J. D., Hope, J., and Fraser, H. (1993). Transmission of bovine spongiform encephalopathy to sheep and goats. *Vet. Rec.* 133, 339–341.

Fraser, H. (1976). The pathology of a natural and experimental scrapie [review]. *Front. Biol.* 44, 267–305.

Fraser, H., and Dickinson, A. G. (1978). Studies of the lymphoreticular system in the pathogenesis of scrapie: The role of spleen and thymus. *J. Comp. Pathol.* 88, 563–573.

Fraser, H., and Foster, J. D. (1993). Transmission to mice, sheep and goats and bioassay of bovine tissues. *In* "Transmissible Spongiform Encephalopathies" (R. Bradley and B. Marchant, eds.), pp. 145–159. Commission of the European Communities, Brussels, Belgium.

Fraser, H., McConnell, I., Wells, G. A. H., and Dawson, M. (1988). Transmission of bovine spongiform encephalopathy to mice. *Vet. Rec.* 123, 472.

Goldmann, W., Hunter, N., Smith, G., Foster, J., and Hope, J. (1994). Prp genotype and agent effects in scrapie—Change in allelic interaction with different isolates of agent in sheep, a natural host of scrapie. *J. Gen. Virol.* 75, 989–995.

Hadlow, W. J. (1995). Neuropathology and the scrapie–kuru connection. *Brain Pathol.* 5, 27–31.

Hadlow, W. J., Kennedy, R. C., and Race, R. E. (1982). Natural infection of Suffolk sheep with scrapie virus. *J. Infect. Dis.* 146, 657–664.

Herms, J. W., Kretzschmar, H. A., Titz, S., and Keller, B. U. (1995). Patch-clamp analysis of synaptic transmission to cerebellar Purkinje-cells of prion protein knockout mice. *Eur. J. Neurosc.* 7, 2508–2512.

Hope, J. (1995) Mice and beef and brain diseases. *Nature (London)* 378, 761–762.

Hope, J., Morton, L. J. D., Farquhar, C. F., Multhaup, G., Beyreuther, K., and Kimberlin, R. H. (1986). The major polypeptide of scrapie-associated fibrils (SAF) has the same size, charge-distribution and N-terminal protein-sequence as predicted for the normal brain protein (PrP). *EMBO J.* 5, 2591–2597.

Hope, J., Reekie, L. J. D., Hunter, N., Multhaup, G., Beyreuther, K., White, H., Scott, A. C., Stack, M. J., Dawson, M., and Wells, G. A. H. (1988a). Fibrils from brains of cows with new cattle disease contain scrapie-associated protein. *Nature (London)* 336, 390–392.

Hope, J., Multhaup, G., Reekie, L. J., Kimberlin, R. H., and Beyreuther, K. (1988b). Molecular pathology of scrapie-associated fibril protein (PrP) in mouse brain affected by the ME7 strain of scrapie. *Eur. J. Biochem.* 172, 271–277.

Hunter, N., Foster, J. D., and Hope, J. (1992). Natural scrapie in British sheep: Breeds, ages and PrP gene polymorphisms. *Vet. Rec.* 130, 389–392.

Ingrosso, L., Ladogana, A., and Pocchiari, M. (1995). Congo red prolongs the incubation period in scrapie-infected hamsters. *J. Virol.* 69, 506–508.

Kimberlin, R. H., and Walker, C. A. (1986). Suppression of scrapie infection in mice by heteropolyanion 23, dextran sulfate, and some other polyanions. *Antimicrob. Agents Chemother.* 30, 409–413.

Kimberlin, R. H., and Walker, C. A. (1989a). Pathogenesis of scrapie in mice after intragastric infection. *Virus Res.* 12, 213–220.

Kimberlin, R. H., and Walker, C. A. (1989b). The role of the spleen in the neuroinvasion of scrapie in mice. *Virus Res.* 12, 201–211.

Kirkwood, J. K., and Cunningham, A. A. (1994). Epidemiologic observations on spongiform encephalopathies in captive wild animals in the British-Isles. *Vet. Rec.* **135**, 296–303.

Kirkwood, J. K., Cunningham, A. A., Austin, A. R., Wells, G. A. H., and Sainsbury, A. W. (1994). Spongiform encephalopathy in a greater kudu (*Tragelaphus strepsiceros*) introduced into an affected group. *Vet. Rec.* **134**, 167–168.

Kitamoto, T., Muramoto, T., Mohri, S., Dohura, K., and Tateishi, J. (1991). Abnormal isoform of prion protein accumulates in follicular dendritic cells in mice with Creutzfeldt-Jakob disease. *J. Virol.* **65**, 6292–6295.

Ladogana, A., Casaccia, P., Ingrosso, L., Cibati, M., Salvatore, M., Xi, Y. G., Masullo, C., and Pocchiari, M. (1992). Sulphate polyanions prolong the incubation period of scrapie-infected hamsters. *J. Gen. Virol.* **73**, 661–665.

Lasmezas, C. I., Cesbron, J. Y., Deslys, J. P., Demaimay, R., Anjou, K. T., Rioux, R., Lemaire, C., Locht, C., and Dormont, D. (1996). Immune system-dependent and system-independent replication of the scrapie agent. *J. Virol.* **70**, 1292–1295.

Lledo, P. M., Tremblay, P., Dearmond, S. J., Prusiner, S. B., and Nicoll, R. A. (1996). Mice deficient for prion protein exhibit normal neuronal excitability and synaptic transmission in the hippocampus. *Proc. Nat. Acad. Sci. U.S.A.* **93**, 2403–2407.

Manson J., West, J. D., Thomson, V., McBride, P., Kaufman, M. H., and Hope, J. (1992). The prion protein gene—A role in mouse embryogenesis. *Development (Cambridge, UK)* **115**, 117–122.

Manson, J. C., Clarke, A. R., Hooper, M. L., Aitchison, L., McConnell, I., and Hope, J. (1994a). 129/ola mice carrying a null mutation in PrP that abolishes messenger-RNA production are developmentally normal. *Mol. Neurobiol.* **8**, 121–127.

Manson, J. C., Clarke, A. R., McBride, P. A., McConnell, I., and Hope, J. (1994b). Prp gene dosage determines the timing but not the final intensity or distribution of lesions in scrapie pathology. *Neurodegeneration* **3**, 331–340.

Manson, J., Hope, J., Clarke, A. R., Johnston, A., Black, C., and MacLeod, N. (1995). PrP gene dosage and long term potentiation. *Neurodegeneration* **4**, 113–115.

Mimms, C. A., and White, D. O. (1984). "Viral Pathogenesis and Immunology," 1st Ed., pp. 1–398. Blackwell, Oxford.

Pattison, I. H., Hoare, M. N., Jebbett, J. N., and Watson, W. A. (1974). Further observations on the production of scrapie in sheep by oral dosing with foetal membranes from scrapie-affected sheep. *Br. Vet. J.* **130**, 65–67.

Pearson, G. R., Wyatt, J. M., Gruffyddjones, T. J., Hope, J., Chong, A., Higgins, R. J., Scott, A. C., and Wells, G. A. H. (1992). Feline spongiform encephalopathy—Fibril and PrP studies. *Vet. Rec.* **131**, 307–310.

Riek, R., Homemann, S., Wider, G., Billeter, M., Glockshuber, R., and Wurthrich, K. (1996). NMR structure of the mouse prion protein domain PrP(121–231). *Nature (London)* **382**, 180–182.

Sailer, A., Bueler, H., Fischer, M., Aguzzi, A., and Weissmann, C. (1994). No propagation of prions in mice devoid of PrP. *Cell (Cambridge, Mass.)* **77**, 967–968.

Sakaguchi, S., Katamine, S., Nishida, N., Moriuchi, R., Shigematsu, K., Sugimoto, T., Nakatani, A., Kataoka, Y., Houtani, T., Shirabe, S., Okada, H., Hasegawa, S., Miyamoto, T., and Noda, T. (1996). Loss of cerebellar Purkinje-cells in aged mice homozygous for a disrupted Prp gene. *Nature (London)* **380**, 528–531.

Salvatore, M., Seeber, A. C., Nacmias, B., Petraroli, R., Dalessandro, M., Sorbi, S., and Pocchiari, M. (1995). Apolipoprotein E in sporadic and familial Creutzfeldt-Jakob disease. *Neurosci. Lett.* **199**, 95–98.

Scott, J. R., Reekie, L. J. D., and Hope, J. (1991). Evidence for intrinsic control of scrapie pathogenesis in the murine visual-system. *Neurosci. Lett.* **133**, 141–144.

Scott, J. R., Foster, J. D., and Fraser, H. (1993). Conjunctival instillation of scrapie in mice can produce disease. *Vet. Microbiol.* **34**, 305–309.

Tateishi, J., Brown, P., Kitamoto, T., Hoque, Z. M., Roos, R., Wollman, R., Cervenakova, L., and Gajdusek, D. C.(1995). First experimental transmission of fatal familial insomnia. *Nature (London)* **376**, 434–435.

Taylor, D. M., McConnell, I., and Fraser, H. (1996). Scrapie infection can be established readily through skin scarification in immunocompetent but not immunodeficient mice. *J. Gen. Virol.* **77**, 1595–1599.

Tobler, I., Gaus, S. E., Deboer, T., Achermann, P., Fischer, M., Rulicke, T., Moser, M., Oesch, B.,

McBride, P. A., and Manson, J. C. (1996). Altered circadian activity rhythms and sleep in mice devoid of prion protein. *Nature (London)* **380**, 639–642.

Wells, G. A. H., Scott, A. C., Johnson, C. T., Gunning, R. F., Hancock, R. D., Jeffrey, M., Dawson, M., and Bradley, R. (1987). A novel progressive spongiform encephalopathy in cattle. *Vet. Rec.* **121**, 419–420.

Wilesmith, J. W., Wells, G. A. H., Cranwell, M. P., and Ryan, J. B. M. (1988). Bovine spongiform encephalopathy—Epidemiological-studies. *Vet. Rec.* **123**, 638–644.

Will, R. G., Ironside, J. W., Zeidler, M., Cousens, S. N., Estibeiro, K., Alperovitch, A., Poser, S., Pocchiari, M., Hofman, A., and Smith, P. G. (1996). A new vaiant of Creutzfeldt–Jakob disease in the UK. *Lancet* **347**, 921–925.

Wisniewski, H. M., Sigurdarson, S., Rubenstein, R., Kascsak, R. J., and Carp, R. I. (1996). Mites as vectors for scrapie. *Lancet* **347**, 1114.

17

MALARIA: A GLOBAL THREAT

KAREN P. DAY

Department of Zoology
The Wellcome Trust Centre for the Epidemiology of Infectious Diseases
University of Oxford
Oxford, United Kingdom

Sir Ronald Ross met Dr. Patrick Manson on his second visit to England in 1894. This association of two great men is one of the most fascinating romances in the history of tropical medicine. After several years of painstaking work on the mosquito–malaria hypothesis, on 20th August 1897 Ronald Ross found oocysts on the stomach wall of dappled winged mosquitoes (probably Anopheles stephensi*). This epoch-making discovery was made while working in a small laboratory in Secunderabad, India. He then wrote the following sonnetelles to his wife:*

> *This day relenting God*
> *   Hath placed within my hand*
> *A wondrous thing: and God*
> *   be praised. At His command.*
>
> *Seeking His secret deeds*
> *   With tears and toiling breath*
> *I find thy cunning seeds.*
> *   O Million-murdering Death.*
>
> *I know this little thing*
> *   A myriad men will save.*
> *O Death, where is thy sting.*
> *   Thy victory. O grave!*

About this discovery Ross writes: "On turning to the stomach with an oil-immersion lens I was struck at once by the appearance of some cells which seemed to be slightly more substantial than the cells of the mosquito's stomach usually are, still very delicate and colourless. There were a dozen of them lying among cells of the upper half of the organ and though, somewhat more solid than these, contained granules of black pigment similar to that of Plasmodium *in finger blood! They varied from 12 to 16 μm in diameter and were full of stationary vacuoles."*

V. P. Sharma

I. INTRODUCTION

Malaria is caused by infection with *Plasmodium* spp. transmitted to humans by the bite of anopheline mosquitos. A recent report by the World Health Organization (WHO, 1996) has identified malaria as a major cause of morbidity and mortality in tropical and sub-tropical regions of the world. The disease has been classified as an "emerging infection" by many national and international health authorities (Lederberg *et al.*, 1992), due to the increased global incidence of the disease. Malaria is making a dramatic comeback in areas where it was once eliminated or suppressed. Large parts of the African subcontinent remain endemic for malaria with reduced prospects for health improvement. Social change and human migration are causing increased risk of malarial disease. International travel in the absence of safe and effective prophylaxis is creating additional health problems for nonimmune travelers. As the global malaria situation worsens it pays to learn the lessons of previous attempts to control this killer disease.

A. Malaria Until the 1960s

It is postulated that malaria originated in tropical Africa at the dawn of humanity and became endemic in communities since the time that agriculture began in settled groups. The infection later spread and established itself in the great centers of riverine civilization in Mesopotamia, India, North China, and the Nile valley, from which it invaded the Mediterranean shores. From these five main foci, malaria extended its hold over most of the tropical world as well as spread through temperate areas of the world (Bruce-Chwatt, 1965). The distribution of malaria in the nineteenth century is shown in Fig. 1. Transmis-

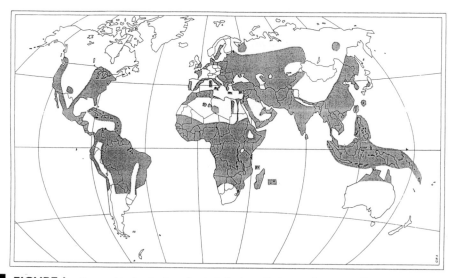

FIGURE 1   Geographic distribution of malaria in the mid-nineteenth century. (From Wernsdorfer, W. H. and McGregor, I. (1988). "Malaria Principles and Practice of Malariology." Churchill-Livingston. With permission.)

sion of the *Plasmodium* spp. has impacted significantly on human biology and history.

Human evolution has been markedly influenced by this disease as a consequence of the association between pathogenicity and parasite replication in the human erythrocyte. Polymorphisms in a number of erythrocyte genes appear to have been selected as they reduced risk of life-threatening malarial disease in human populations of the Mediterranean, Africa, Asia, and Melanesia (Hill, 1992). Before the advent of antimalarial drugs, many geographic areas in the tropics were uninhabitable by Europeans who lacked this innate resistance to infection.

Malarial disease has also influenced military campaigns throughout the history of civilization (Bruce-Chwatt, 1988). A few examples from the twentieth century illustrate what malaria can do to an army. The French, German, and British armies were paralyzed by an epidemic of malaria on the Macedonian front in the First World War. There were over 500,000 cases among U.S. soldiers during World War II. There were more than 80,000 cases of malaria diagnosed in American troops in Vietnam from 1965 to 1971. More casualities were reported due to malaria in the Vietnam War than any other cause. Consequently the disease is still considered a high priority for research by the military, who are actively supporting drug and vaccine development against malaria in at least the United States (Lederberg, *et al.*, 1992).

Although the parasites that cause malaria infection were only discovered by Laveran in 1880, drugs to treat malaria, known clinically as "intermittent fever," were found much earlier. The Jesuit missionaries working in Peru in the seventeenth century became familiar with the native remedy for treatment of intermittent fever made from the bark of a tree, later named the cinchona tree. This Peruvian bark was introduced into Europe in 1639 and was widely used for the treatment of intermittent fever during the seventeenth and eighteenth centuries in both Europe and her colonies. The prophylactic effect of the *Cinchona* bark was demonstrated in 1826 in British sailors visiting Sierra Leone. The active ingredient of the bark was identified as quinine in 1820 by two French pharmacists. This discovery allowed the standardization of the drug dosages to achieve more effective treatment and prophylaxis. Resistance to quinine was first observed as early as 1910 in Brazil (Neiva, 1910; Nocht and Werner, 1910). Chemical synthesis of antimalarial drugs began in earnest after the First World War by screening various compounds related to quinine against bird malarias. A number of synthetic antimalarials were discovered from the 1930s to the 1950s, but the most useful, inexpensive, and effective of these was chloroquine.

Large-scale interventions to control malaria were only initiated early in the twentieth century. Robert Koch, the great German bacteriologist, proposed the idea of mass chemotherapy and prophylaxis of endemic malaria by systematic distribution of quinine in 1900 (Koch, 1900). This method was introduced into New Guinea and elsewhere with some success but was abandoned in German colonies in Africa as unsustainable. Vector control was proposed after the discovery of the mosquito transmission of malaria in 1897 by Ross (1897). Ross (1911) also developed a quantitative epidemiological framework to describe the transmission of malaria. He showed that when the vector density was below a critical level the disease disappears. Elimination of anopheline breeding places by drainage and use of larvicides were the first methods of vector control implemented by Ross and others at the beginning of the twentieth century. They had variable success dependent on the level of transmission, local ecological conditions, as well as financial and logistic sustainability of these interventions. The growth of understanding of the mosquito transmission of malaria also led to enhanced personal protection. Mosquito nets were used while sleeping, and the screening of houses became standard practice for those financially able to afford protection.

As early as the 1920s vector control targeted at the adult mosquito was proposed as a more effective way to control malaria compared to larviciding. The proof of this proposal awaited the discovery of a long-lasting residual insecticide capable of killing adult anophelines. Such a compound named DDT, although discovered in 1874, was first screened for residual insecticide activity against anophelines in Rome in 1944. It was readily demonstrated that spraying this compound on house walls continued to kill adult anophelines for at least 10 months (Missiroli, 1948). This discovery revolutionized malaria control. Residual indoor spraying of DDT presented a sustainable and cost-effective method of vector control even for rural areas. After initial successes with this method in Europe and Asia, a global campaign to eradicate malaria in a framework of interaction between national health authorities sharing common borders was established in 1957 under the auspices of the World Health Organization. The global strategy involved an initial "attack phase" in which residual spraying played the major role and a "consolidation phase" of eliminating remaining foci of infection by case detection and distribution of antimalarial drugs. The success of this campaign in many parts of the world can be seen in Fig. 2. By 1956 many malarious areas of the world were in the maintenance and consolidation phase of the program. Europe, most of the United States, and

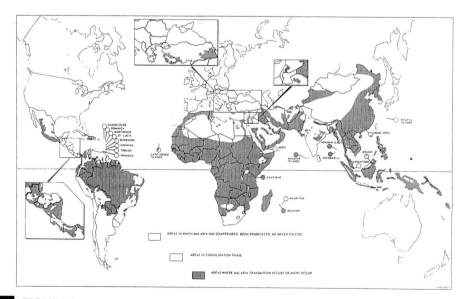

FIGURE 2   The global malaria situation in 1967. (From Wernsdorfer, W. H. and McGregor, I. (1988). "Malaria Principles and Practice of Malariology." Churchill-Livingston. With permission.)

Australia had eradicated malaria. The eradication campaign, however, failed to eliminate the malaria problem in much of tropical Africa, India, the Western Pacific, and Southeast Asia.

B. Malaria Since the 1960s

The brief period of optimism regarding the potential conquest of malaria disappeared from 1965 to 1969 due to changes in the biology of both the parasite and the anopheline vector as well as logistic problems regarding implementation and management of the eradication campaign. Evolution of antimalarial drug resistance to chloroquine was detected in *Plasmodium falciparum.* DDT spraying selected for mosquito behavioral changes such that a number of species became outdoor rather than indoor resting. Mosquito resistance to insecticides also developed. By 1970 nearly 20% of all malarious areas had either vector resistance to DDT or *P. falciparum* resistance to chloroquine, or both (Lepes, 1981). The optimism of the eradication era was replaced by the sober reality that the malaria situation in many endemic areas was now worse than ever. Malaria, which had been elminated or effectively suppressed in many parts of the world, has greatly increased in incidence since the mid-1970s. Over 2 billion people worldwide are at risk of malaria infection in tropical and subtropical regions of the world. Although the true extent of malaria morbidity and mortality is largely unknown, current estimates indicate that more than 300 million people are infected annually, resulting in about 120 million clinical cases worldwide and an estimated 0.5 to 1.2 million fatalities in Africa every year (WHO, 1996).

We face the new millennium both with the increased potential of the malaria parasite to kill due to the emergence of antimalarial drug resistance and with limited ways to control the mosquitos that transmit malaria. The breakdown of health services as well as limited financial resources for malaria control compound the problem faced by health professionals in many endemic areas. Social change is also contributing to changing patterns of global incidence of malaria. Migration to urban areas has increased, with 39% of the world's population living in cities in 1992 compared to only 29% in 1950. It appears that migrants to urban areas can contribute either infection or a new reservoir of susceptible hosts dependent on whether they come from an area where malaria is endemic or not. Adaptation of urban vectors to urban conditions has also occurred. Conversely, economic development has resulted in the migration of urban dwellers to rural areas with consequent increase in malaria morbidity and mortality in previously unexposed hosts. There has also been a massive increase in the number of refugees, many of whom are relocated to malarious areas. Movement of people from the developed to the developing world is also increasing with expansion of international travel for commerce and tourism with the consequent increase in exposure of nonimmune individuals to malaria.

This chapter reviews malaria as an emerging infection in the context of up-to-date considerations of the biology and epidemiology of malaria. Five key factors have been identified which have or are currently influencing the increased global threat of malaria. These are discussed in Section 4 after a general discussion of relevant biology and epidemiology in Sections 2 and 3. In Section 5, I briefly address the threat of global warming and how this may worsen the malaria situation. I conclude with a discussion of the prospects for the future (Section 6).

II. THE BIOLOGY OF *PLASMODIUM* SPP.

Malaria in humans is caused by infection with protozoan parasites of the genus *Plasmodium*. Four species of *Plasmodium* infect humans, namely, *Plasmodium falciparum*, *P. vivax*, *P. malariae* and *P. ovale*. The former species is considered to be the most virulent as it causes the condition known as cerebral malaria, which is often fatal. All four species are transmitted by anopheline mosquitos.

A. Life Cycle

Key features of the life cycle are the requirement of transmission between the anopheline and human host with expansion of the parasite population in both hosts by asexual replication in a range of tissues such as liver and bloodstream in the human host and the mosquito midgut. The basic malaria life cycle is shown in Fig. 3. One *P. falciparum* sporozoite transmitted by a mosquito bite can invade a liver cell, which after 8 to 10 days can release 10,000 merozoites capable of invading erythrocytes. Replication of parasites in erythrocytes and their subsequent release from these cells results in induction of fever and immune responses which cause the characteristic pathology of malaria. The production of transmission stages, known as gametocytes, results after commitment of infected cells to sexual development. Patterns of gametocyte production differ among the *Plasmodium* spp. (Carter and Graves, 1988). The life cycle continues after ingestion of male and female gametocytes by anopheline mosquitos as described in Fig. 3 and Section 2.3.

B. *Plasmodium* Diversity

Molecular genetic study of *P. falciparum* became possible after the pioneering work of Trager and Jensen in 1976 who reproduced the asexual life cycle of the parasite in *in vitro* culture. The availability of cloned lines of *P. falciparum* facilitated phenotypic and genetic characterization of individuals of this species. Considerable variability in isoenzymes, drug resistance, adhesion, and antigenic and genotypic characteristics of cloned lines of *P. falciparum* have been demonstrated (Kemp *et al.*, 1990). Inability to grow any other *Plasmodium* species of humans in culture has hindered progress toward detailed genetic characterization of these parasites. Limited studies of isoenzyme variability of isolates of *P. vivax* from patients have demonstrated extensive diversity of this parasite (Joshi *et al.*, 1989).

The existence of within-species diversity of *P. falciparum* must be considered in the context of the fact that sex is an obligatory part of the life cycle of this parasite. This aspect of the natural history of *P. falciparum* differs from that of most other microparasites of viral, bacterial, and protozoan origin which can undergo sexual recombination but generally do not do so. The obligatory sexual phase in the malaria life cycle means that the generation of novel geno-

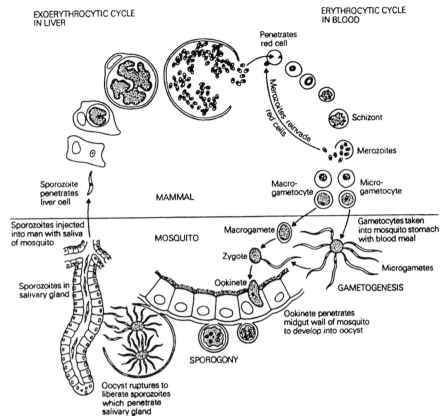

EXOERYTHROCYTIC CYCLE
IN LIVER

ERYTHROCYTIC CYCLE
IN BLOOD

Penetrates
red cell

Merozoites reinvade red cells

Schizont

Merozoites

Sporozoite
penetrates
liver cell

MAMMAL

Macro-
gametocyte

Micro-
gametocyte

Sporozoites injected
into man with saliva
of mosquito

MOSQUITO

Macrogamete

Gametocytes taken
into mosquito stomach
with blood meal

Zygote

Microgametes

Sporozoites in
salivary gland

Ookinete

GAMETOGENESIS

Ookinete penetrates
midgut wall of mosquito
to develop into oocyst

SPOROGONY

Oocyst ruptures to
liberate sporozoites
which penetrate
salivary gland

FIGURE 3   Life cycle of *Plasmodium* spp. Sporozoites are transmitted in the saliva of the female mosquito as she takes a blood meal. The first stage of development in the host, the exoerythrocytic cycle, occurs in the liver when sporozoites invade hepatocytes. They undergo massive asexual multiplication, resulting in the release of thousands of merozoites into the blood. The merozoites invade host erythrocytes, and another sequence of asexual replication occurs, with the parasite passing through recognizable stages of rings, trophozoites, and multinuclear schizonts. The schizont ruptures together with the erythrocyte, releasing up to 24 merozoites for further invasion and the cycle continues. Some merozoites can enter a sexual phase, developing into male and female gametocytes. The gametocytes are ingested by a mosquito and undergo sexual reproduction, allowing genetic recombination, within the midgut to produce ookinetes. These undergo a third phase of asexual multiplication, producing sporozoites that enter the salivary glands, for inoculation into a new host. (From Vickerman and Cox, 1967.)

types can occur during conventional meiosis when two genetically distinct clones of a species are cotransmitted from human host to anopheline vector. Thus, sex has the potential for generating considerable genomic diversity within each species. Sex creates the potential to increase fitness of individuals of the species in a changing environment.

It has been shown that the haploid genoe of *P. falciparum* has 14 chromosomes (Kemp. *et al.*, 1990). A number of polymorphic loci lie on different chromosomes and thus will undergo assortment independent of one another during meiosis. Coinfection with different genotypes is common in the human hosts resident in most endemic areas, thereby creating the possibility for outcrossing during the obligatory sexual phase in the mosquito host. Two studies have measured the rate at which cross-fertilization occurs in natural parasite populations of Papua New Guinea (PNG) and Tanzania (Babiker *et al.*, 1994a; Paul *et al.*, 1995; Hill *et al.*, 1995). Mating patterns, as assessed by heterozygosity of oocyst stages in the midgut wall of the mosquito, were found to differ in the two areas in relation to transmission intensity. The 10-fold higher transmission intensity observed in Tanzania compared to PNG resulted in higher levels of oocyst heterozygosity. This, it is possible that the evolution of multigenic phenotypes, such as drug and vaccine resistance, may occur at different rates in these two endemic areas in epidemiologically relevant time frames of the order of 5 to 10 years (Paul *et al.*, 1995).

P. falciparum has also been shown to undergo clonal antigenic variation; in other words, a single cloned trophozoite-infected erythrocyte has the capacity to switch its surface antigenic properties by intrinsic molecular mechanisms (Biggs *et al.*, 1991; Roberts *et al.*, 1992). Members of a multigene family designated the "*var* genes" have been shown to encode this variant surface antigen phenotype (Baruch *et al.*, 1995; Smith *et al.*, 1995; Su *et al.*, 1995). Each parasite genome contains approximately 50 different *var* genes. Antigenic switching involving differential expression of individual *var* genes at any point in time is believed to allow the parasite to evade variant-specific host immune responses. This immune evasion strategy is believed to prolong the survival of the parasite within the human host to ensure transmission to the mosquito vector in natural environments where vectors may appear seasonally or transiently.

It has been hypothesized that both clonal antigenic variation and allelic diversity of single-copy antigen genes of *P. falciparum* play an important role in both the survival of the parasite within the human host as well as the transmission success of the parasite between hosts within an endemic area (Anders and Smythe, 1989; Day and Marsh, 1991). These ideas, and others, have been formalized in a series of papers describing (Gupta and Day, 1994; Gupta *et al.*, 1994a) or criticizing (Saul, 1996; Tibayrenc and Lal, 1996) a "strain theory" of malaria transmission. Gupta and Day have proposed that the *var* genes represent strain determining loci. This remains to be proved. As yet we understand little of how parasite diversity impacts either on the epidemiology of malaria in a variety of transmission situations or on our ability to control malaria. Molecular epidemiology studies of parasite diversity will no doubt be a growing area of research over the coming years as this diversity represents a major obstacle to control by vaccination and drugs. The genetics of the parasite has previously been ignored in the development of theoretical frameworks for control (e.g., Dietz, MacDonald 1957 1988).

Geographic diversity of *P. falciparum* appears to exist in at least the distribution of alleles of a merozoite surface antigen (Conway *et al.*, 1992; Creasey *et al.*, 1990), suggesting that selection may operate (Conway, 1997). To date large-scale, global population genetic studies using neutral loci such as microsatellite markers have not been completed. Such studies will demonstrate whether *P. falciparum* represents one global population which is interbreeding rather than a series of discrete populations. This information will be vital to understand patterns of spread of multigenic drug and vaccine resistance in a world where human migration will increasingly play a significant role in the spread of infectious disease.

C. Malaria and the Anopheline Mosquito

Malaria can only be transmitted by female anopheline mosquitos when they take a human blood meal. The male *Anopheles* feeds on nectar and fruit juices while the female feeds primarily on blood. She takes a blood meal in order to lay eggs. This feeding occurs every 2 to 3 days, thereby allowing the transmission of malaria: initial ingestion of gametocytes, parasite development over 10 to 14 days, and subsequent release of sporozoites from the salivary gland occurs throughout repeated mosquito blood feeding known as the ovapositon cycle. Transmission of malaria can be interrupted by reducing the life span of the adult female so that parasite development (i.e., the sporogonic cycle) cannot be completed. Macdonald drew attention to this fact in his mathematical analysis of vector control of malaria transmission in 1957 (Macdonald, 1957).

The taxonomy of the genus *Anopheles* has been described by Service (1993). There are six subgenera which largely reflect the geographic origins of the mosquitos, that is, Old and New World, South and Central America, North America and Northern Mexico, Africa, Australasia and the Pacific, and Southeast Asia. There are 422 species of *Anopheles* mosquitos worldwide, and at present only 70 of these species are vectors of malaria under natural conditions. Behavioral differences in the feeding and resting habits of adult anophelines are readily observed. Some species feed in houses and rest there afterward, whereas others will feed indoors and rest outdoors. Other species only feed outside and never enter houses. The feeding habits of *Anopheles* spp. also vary greatly. Some species feed primarily on humans, whereas others prefer to feed on animals. Many endemic areas have more than one vector species, where each species can be defined as either a main or a subsidiary vector of malaria transmission. Each vector species may play more or less dominant roles in the transmission of malaria in different geographic regions. The flight range of anophelines is generally less than 2 to 3 km from their breeding places. Thus, migration of infected humans is more important in the dispersal of malaria than movement of infected mosquitos. Macdonald (1957) classified the natural distribution of the main vectors of malaria into 12 epidemiological zones.

Sibling species differing in behavior, morphology, genetic characteristics and ability to transmit *Plasmodium* spp. have been identified. Hence the concept of species complexes was introduced to *Anopheles* taxonomy. For example, *Anopheles gambiae*, the most important malaria vector in Africa, was shown to be a species complex of at least six sibling species, rather than a single species. Sibling species can interbreed, producing sterile males but fertile females.

The natural environment has profound effects on the biology of *Anopheles* spp. (Molineaux, 1988). Individual species have evolved as a result of adaptation to local ecological conditions. Larval stages of different vector species breed in surface water of varying depths, salinity, and level of oxygenation and are variably affected by the level of light, shade, and vegetation. Species also vary in the development of the aquatic larval stages and gonotrophic maturation of adult mosquitos in relation to ambient temperature. The longevity of adult vectors increases with the relative humidity of the air. The reproductive potential of vectors is enormous in favorable environmental conditions. Density-dependent constraints do, however, operate via competition and predation. Changes in climate can alter the type and distribution of vector species as can man-made changes to the environment.

The transmission of malaria from one human host to another requires the anopheline mosquito to take up infectious transmission stages (gametocytes) in the human blood meal. Appropriate development of the parasite in the midgut and salivary gland of the mosquito is then necessary to complete the sporogonic cycle. The biochemical basis of vector–parasite interactions is little understood but is currently under active investigation (Shahabuddin and Kaslow, 1993). It is clear from laboratory studies that polymorphisms in both parasite and vector molecules involved in transmission will occur in natural populations.

Molineaux (1988) identified four factors critical to the transmission of *Plasmodium* spp. by the adult anopheline vector. Any of these could be effected by demographic, climatic, natural, or man-made changes to the environment.

1. Density of vectors: Since human hosts are sporadically infectious, the density of vectors feeding on humans will clearly influence transmission.

2. Vector susceptibility: Variability in the capacity of different species and geographical strains of anophelines to transmit different *Plasmodium* spp. and geographical strains within a species has been observed. Molineaux summarizes experiments by a number of investigators defining the susceptibility of various anopheline species to *P. falciparum* and/or *P. vivax* from different geographic areas. These studies were largely conducted from the 1960s to the 1980s to define whether European, U.S., and Australian vectors of malaria could transmit African, Asian, or Melanesian strains of different *Plasmodium* spp. If such experimental infections were possible this would imply that infected migrants from endemic areas could reintroduce malaria to areas where malaria had been eradicated. European vectors were not susceptible to African or Indian strains of *P. falciparum*, whereas U.S. vectors showed variable susceptibility. This area of research has received little attention for the past 20 years. Given the changing global patterns of human migration and mosquito distribution, it may be timely to consider contemporary experiments of vector susceptibility.

3. Frequency with which the vector takes human blood meal: Feeding frequency will depend on temperature and host preference and will determine the potential of the mosquito become infected and transmit the infection from mosquito back to humans.

4. Duration of sporogony: The incubation period in the vector, that is, the time from infection to development of sporozoites in the salivary gland, is determined by both temperature and the genetics of the parasite. There is a minimum temperature around 15°C below which *Plasmodium* spp. will not develop.

III. EPIDEMIOLOGY OF MALARIA

Much has been written about the epidemiology of malaria during the twentieth century. It would be naive to attempt a summary of current knowledge. Instead, I will try to highlight aspects of epidemiology relevant to consideration of malaria as an emerging infection. The discussion primarily focuses on *P. falciparum* as the epidemiology of malaria caused by the parasite is the most studied.

A. Malarial Infection

Description of the patterns of infection and disease caused by *Plasmodium* spp. is a necessary part of planning, implementation, and evaluation of control. Standard methodologies (Gilles and Warrell, 1993) have been available for the quantitative description of parasite infection by microscopy since the end of the nineteenth century when the parasitological description of the life cycle was completed in humans and mosquitos. Diagnosis in most endemic areas still relies on detection of *Plasmodium* spp. in human blood or mosquito tissues by microscopic methods. Alternatively, species-specific, sensitive enzyme-linked immunosorbent assay (ELISA) tests are available for detection of sporozoites in salivary glands and trophozoites in human blood for large-scale epidemiological studies. Although the contribution of parasite diversity to the slow acquisition of immunity is assumed, molecular epidemiology data defining within-species diversity has not been incorporated into routine malariometric surveys to date. Basic research is underway to define field protocols using DNA amplification by polymerase chain reaction (PCR) to investigate the role of parasite diversity in generating the typical epidemiological patterns of malaria infection and disease (Babiker *et al.*, 1994; Felger *et al.*, 1994; Ntoumi *et al.*, 1995; Paul *et al.*, 1995; Snounou *et al.*, 1993).

In areas where malaria is endemic, an individual may experience many infections with different genotypes of *P. falciparum*. Crude annual incidence rates of up to 15 genetically distinct infections per annum for children aged 2 to 4 years have been calculated for the north coast of PNG by genotyping parasite DNA from finger prick blood samples (Carneiro *et al.*, 1997a). These rates most likely underrepresent the true incidence since longitudinal surveillance of semi-immune children living under conditions of stable malaria transmission in PNG show that these children may have as many as 11 genotypes present in their circulation over a period of 60 days using only a single polymorphic marker (Bruce *et al.*, 1996). There is rapid turnover of these infections at PCR-detectable levels (Daubersies *et al.*, 1996; Bruce *et al.*, 1996; Carneuro *et al.*, 1997). Generally, infections in semiimmune children are asymptomatic with occasional episodes of mild malarial disease and the rare occurrence of severe disease. The incidence of disease is usually associated with the incidence of a new parasite genotype (Carneiro *et al.*, 1997b; Contamin *et al.*, 1996). Descriptive results of the dynamics of infections in individuals have now been reported from a number of endemic areas revealing high levels of turnover. The outcome of infection depends on host as well as parasite factors. Molecular epidemiological studies are now showing that previous exposure to many distinct parasite genotypes is associated with the development of a nonsterilizing immunity reflected in the age-specific patterns of infection and disease.

There are "many epidemiologies" observed with respect to malaria transmission. These have been characterized by degrees of endemnicity ranging from high to low as well as epidemic (Molineaux, 1988; Gilles and Warrell, 1993). Descriptive studies of the epidemiology of malaria in areas of stable malaria transmission have revealed distinct age-specific patterns of parasite prevalence and density for trophozoites and gametocytes. Typical age-specific patterns of prevalence of infection for a highly endemic area are shown in Fig. 4. Both prevalence and density of trophozoites and gametocytes decrease with increas-

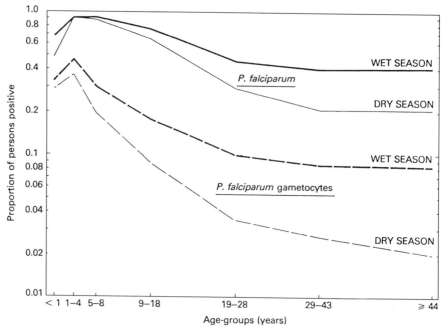

FIGURE 4 Prevalence of *P. falciparum* in the West African savanna. (From Wernsdorfer, W. H. and McGregor, I. (1988). "Malaria Principles and Practice of Malariology." Churchill-Livingston. With permission.)

ing age (Molineaux, 1988). The density of *P. falciparum* gametocytes is always less than the density of trophozoites. This feature of the epidemiology is a focus of current discussion. Analysis of *P. falciparum* infection prevalence in anophelines has demonstrated that infection levels rarely exceed 10–20% of the mosquito population in areas of high transmission. They are generally less than 1% in most endemic areas. This reflects the fact that more humans are infected than infectious.

During the era of vector control the epidemiology of malaria was considered in the context of transmission as measured by parasite prevalence and density in both the human and the mosquito. Measures of the impact of control were defined in relation to these malariometric parameters as well as changes in crude death rates (Molineaux, 1988). The change in policy leading to emphasis on case management rather than vector control stimulated research on malarial disease. Greenwood and colleagues working in the Gambia focused attention on the description of malarial disease (Greenwood *et al.*, 1987; 1991; Marsh, 1992) as well as measurement of the health impact of malaria interventions by clinical assessment of the study population (Greenwood *et al.*, 1987; Snow, 1988).

B. Malarial Disease

The 1980s and 1990s have seen intensive study of the epidemiology of malarial disease caused by *P. falciparum* (Greenwood *et al.*, 1991; Marsh, 1992; Trape *et al.*, 1994). This research has largely been done in areas of stable malaria transmission and predominantly in Africa. These studies have revealed new information relevant to the design of interventions to control disease. There has been little research done on the severity of disease caused by infection with other *Plasmodium* spp. Generally, they are considered to be less virulent.

The spectrum of malarial disease caused by *P. falciparum* infection can be broadly classified into mild and severe disease (Marsh, 1992). Mild malarial disease is characterized by fevers and rigors associated with the release of parasite toxins after rupture of erythrocytic schizonts (Bate *et al.*, 1992). These toxins induce release of tumor necrosis factor (TNFα) from macrophages which in turn induces fever and a general activation of macrophages to release nitric oxide. It is believed that both the fever and nonspecific immune mechanisms induced by TNFα regulate parasite density to protect the host (Kwiatkowski, 1995). In a minority of cases malaria infection progresses to cause life-threatening syndromes, the most common of which are severe malarial anemia (SMA) and cerebral malaria (CM). Severe malarial anemia is due to massive hemolysis caused by the rupture of erythrocytes at schizogony, destroying erythrocytes faster than the host is able to replace them. Cerebral malaria, although not fully understood, has been attributed to two phenomena, which may act independently or in concert. Blockage of brain microvasculature by cytoadherence of parasitized erythrocytes to brain endothelium is generally believed to cause CM, based on the observation of such adhesion in autopsy specimens. A second mechanism involves cytokine induction of secondary mediators such as nitric oxide which may cause aberrant neurotransmission and intracranial hypotension as a result of excessive vasodilation (Clark and Rockett, 1994). The two mechanisms may be linked by the involvement of cytokines such as TNFα, causing upregulation of adhesion receptors, as well as the effects of nitric oxide (Kwiatkowski, 1991).

Distinct age-specific patterns of infection and disease are seen in areas of stable malaria transmission where *P. falciparum* infection predominates (Brewster *et al.*, 1990; Marsh, 1992). The mean number of clinical attacks per child per year declines at a time when parasite prevalences are rising. Similarly, risk of death due to malaria also occurs at an age when parasite prevalences are increasing. The dysjunction between the patterns of infection and disease is best explained by proposing that the immunity that protects against disease develops in the first 5 to 6 years of exposure to malaria, whereas the nonsterilizing immunity that regulates infection occurs after 15 years of residence in an endemic area (Gupta and Day, 1995). The age-specific fever threshold, that is, the density of parasites which induce febrile illness, is observed to decrease with increasing age (Rogier *et al.*, 1996). This is believed to be due to the age-specific immune-mediatd mechanisms of parasite tolerance modifying the TNFα inducing activity of malaria toxins.

Clinical studies in the Gambia and Kenya have shown that the peak incidence of cerebral malaria occurs at an older age than the peak incidence of severe malarial anemia and mild malarial disease. This can be explained by invoking either the hypothesis that cerebral malaria only occurs after certain host developmental changes occur in the brain (Marsh, 1992) or the hypothesis that only "rare strains" of *P. falciparum* cause this disease whereas all "strains" of *P. falciparum* can cause severe malarial anemia in young children (Gupta *et al.*, 1994b). Descriptions of the epidemiology of malarial disease in different geographic areas is a subject of intense activity at present, as is the molecular basis of parasite virulence and of pathogensis.

Geographic differences in the incidence of severe malarial disease due to *P. falciparum* infection have been described. Severe malarial anemia appears to be a more common cause of severe disease in Tanzania compared to coastal Kenya and the Gambia where cerebral malaria is more prevalent. It has been known anecdotally that the incidence of cerebral malaria in Melanesia was lower than that seen in Africa. This has been formally documented in two studies in Papua New Guinea and Vanuatu (Maitland *et al.*, 1996). Transmission intensity, seasonality, host and parasite genetics, as well as health seeking behavior may be responsible for such geographic differences.

Snow *et al.* (1997) have compared the incidence of severe disease in several African sites after standardizing measurements of the force of infection and health seeking behavior. Paradoxically, they found the risks of severe disease were lowest among the population with the highest transmission. The highest risks were observed among the populations exposed to low or moderate transmission. They interpret these findings as indicating that intense exposure to malaraia in early life, coincident with the operation of other mechanisms, may reduce risk of disease. Lowering of parasite transmission, and thus immunity, in such populations may lead to a change in both the clinical spectrum of severe disease and the overall burden of severe malaria morbidity.

There has been much less research on malarial disease in areas of unstable malaria transmission. Boyd (1949) described the changing patterns of incidence of acute malarial disease with differing levels of transmission intensity (Fig. 5).

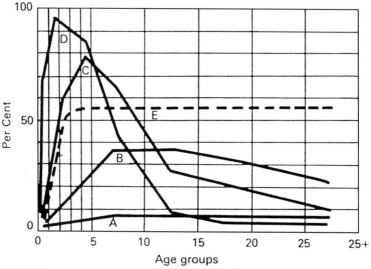

FIGURE 5 Incidence of acute malaria infections with transmission at different endemic levels. For low endemicity (curve A), a person may attain adolescence before infection is acquired and may escape altogether. For moderate endemicity (curve B), maximum incidence occurs in childhood and adolescence, though still not unusual for adult life to be attained before acquiring infection. For high endemicity (curve C), by late infancy or early childhood practically all are infected. Little acute illness is observed in adolescents and still less in adults. For hyperendemicity (curve D), most individuals acquire infection in early infancy, but acute manifestations are less frequent in childhood and are unusual in adults. The dashed line (curve E) shows that unless due to exotic parasites, epidemics can only occur in populations where malaria was either previously absent or persisted at low or moderate endemic levels. They are characterized by a high incidence at all age periods. (Boyd, 1949.) (From Wernsdorfer, W. H. and McGregor, I. (1988). "Malaria Principles and Practice of Malariology." Churchill-Livingston. With permission.)

As transmission intensity declines, the incidence of mild disease is not restricted to children but is also found in adults. There is some evidence that acquisition of malaria infection in the older age classes results in a different form of complicated or severe malaria compared to that seen in children is areas of stable transmission (Warrell, 1993).

C. Innate Human Resistance to Malaria

Haldane (1948) first suggested that the geographic distribution of β-thalassemia may be due to the heterozygous condition affording protection against malaria. Since this "malaria hypothesis" the α-thalassemias and a number of host erythrocyte polymorphisms have been geographically associated with malaria (reviewed by Hill, 1992; Marsh 1993). Case/control studies in the Gambia (Hill et al., 1991) and Papua New Guinea (Genton et al., 1995) have now shown that a number of polymorphisms are associated with reduced incidence of severe malarial disease. These include the heterozygous condition of the sickle cell trait (hemoglobin AS), glucose-6-phosphate dehydrogenase deficiency, and a deletion of band 3, causing Melanesian ovalocytosis and α-thalassemia. Polymorphisms in the promoter region of TNFα (McGuire et al., 1994) and in certain human leukocyte antigen (HLA) alleles (Hill et al., 1992) have also become associated with reduced risk of severe disease in the Gambia. This area of research is expanding as interest in genome studies and human evolution have been topical. These molecular epidemiology studies may give us insights into both human and parasite biology as well as identify risk factors for severe malarial disease.

The observation that host genetic factors can modify disease outcome is also of importance when we consider a world changing in global patterns of human migration (see Section 4.3). Morbidity and mortality due to malaria will be substantially greater on exposure of individuals who are innately susceptible to this disease. Such groups may be selectively targeted for interventions.

IV. FACTORS IN THE EMERGENCE OF MALARIA

As stated above there are numerous factors in the emergence of malaria. This section focuses on five factors I believe to be important in this process.

A. Failure of Malaria Eradication

A detailed history of the attempted eradication of malaria has been reported elsewhere (Gramiccia and Beales, 1988). I will draw on information collated by these authors to give a brief synopsis of information relevant to the subject of this review. The attempted eradication of malaria by residual insecticide spraying had freed 727 million people (i.e., 53% of the world's population of originally malarious areas excluding sub-Saharan Africa) of the risk of malaria by 1970. This progress had saved a great many lives and contributed to the economic development of many areas including Europe, Asia, and the Americas.

The goal of eradication also created a health infrastructure which later formed the backbone of general health services.

As stated above the goal of eradication was dropped in 1969 due to technical problems. These included the emergence of DDT resistance in anophelines; behavioral changes observed in anophelines such that indoor resting mosquitos became outdoor resting, thereby avoiding contact with residual insecticides; the development of resistance to chloroquine in *P. falciparum*; withdrawal of external resources; and manpower, training, and infrastructure problems. The absence of alternative inexpensive and effective control measures resulted in the "eradication" strategy being replaced by one of "control" in the period 1970 to 1978. Existing tools of vector control and case management were to be used within the socioeconomic constraints of national health budgets. During the period of conversion of eradication programs into malaria control programs, the malaria situation began to deteriorate. The number of reported cases doubled between 1974 and 1977 (WHO, 1992). An evaluation of the global situation at the end of 1989 (WHO, 1991) showed that out of a world population of about 5160 million people, 1400 million (27%) lived in areas where malaria never existed or disappeared without specific malaria interventions; 1650 million (32%) lived in areas where endemic malaria disappeared after implementation of control and the malaria free situation had been maintained; 1620 million (31%) lived in areas where endemic malaria had been considerably reduced after control measures had been introduced, but transmission had been reinstated and the situation was unstable or deteriorating; and 490 million (10%) lived in areas, mainly in tropical Africa, where endemic malaria remained basically unchanged and no national antimalaria program was ever implemented. A continuous upward trend in malaria incidence has been observed in parts of the Americas and Asia since 1989. Epidemic malaria associated with high morbidity and mortality has become a major health problem in semiarid areas where malaria control was once effective. It is assumed that residents of these endemic areas have gradually lost their immunity during control and are highly susceptible when transmission returns. Alternatively, they may be exposed to "new strains" of the parasite to which they had no preexisting immunity. A study from Madagascar supports the later hypothesis (Deloron and Chougnet, 1992). These authors describe long-lived immunity which protects against clinical malaria.

B. Antimalarial Drug Resistance

The late 1950s saw the emergence of resistance of *P. falciparum* to the antimalarial drug chloroquine. This first occurred in Indochina and South America and has subsequently spread to all areas where the parasite is endemic. The consequences for the control of malarial disease have been disastrous as chloroquine was an effective, inexpensive, and relatively safe drug not easily replaced by available antimalarials. The frequency and degree of chloroquine resistance are highest in the areas longest affected, with variable levels of resistance in areas more recently afflicted. The latter point can be well illustrated by examination of a data set from Tanzania which shows that in the early period of introduction of drug resistance both the frequency and level of resistance may fluctuate

(Koella *et al.*, 1990). Koella (1993) suggested that these data may be best explained by frequency-dependent selection of resistant strains occurring as a result of herd immunity to "strain-specific" antigens. The importance of the interaction between "strain-determining" loci and drug resistance loci is not well understood and warrants more research. It may be possible that vaccination which achieves even a nonsterilizing immunity may improve the efficacy of antimalarial drugs. A number of interesting epidemiological features of the spread of chloroquine resistance have been highlighted by Wernsdorfer (1991, 1994). In particular, the slow spread of chloroquine resistance into Africa from Asia is contrasted with the explosive spread of resistance once it had established in East Africa. Geographic patterns of both vector susceptibility and human migration will have played a role in this process.

The spread of chloroquine resistance has necessitated the use of alternative drugs (reviewed by Wernsdorfer, 1994) such as sulfonamide–pyrimethamine combinations, quinine/tetracyclines, mefloquine, halofantrine, and artemisinin derivatives. Resistance to some of these alternative drugs has now become a problem in several geographic locations. Resistance to sulfonamide–pyrimethamine combinations, which replaced chloroquine as a frontline treatment, has been reported throughout Southeast Asia, western Oceania, South America, and more recently East and West Africa. Multidrug resistance has been reported in parasite isolates from the Thai/Cambodian border and the Thai/Myanmar border, necessitating a shift to the last line drug, namely, the artemesinin derivatives. The lack of interest of the pharmaceutical industry to develop new antimalarial drugs makes the global malaria chemotherapeutic situation alarming.

Drug resistance has not been reported for *P. malariae* and *P. ovale*, whereas resistance of *P. vivax* to chloroqine was first reported in PNG in 1989 (Rieckmann *et al.*, 1989). Drug resistance in *P. vivax* is generally considered far less serious than in *P. falciparum* because it is significantly less virulent.

Current research activities aim to define the molecular mechanisms of chloroquine (Wellems, 1992) and pyrimethamine/sulfadoxine resistance (Plowe *et al.*, 1995; Wang *et al.*, 1997; Reeder *et al.*, 1996). Molecular correlates of drug resistance may help track the spread of drug resistance more efficiently as well as give insights into alternative drug design (Plowe *et al.*, 1995). This information can also be used in combination with measured inbreeding (Hill *et al.*, 1995; Paul *et al.*, 1995) to predict the time frame of spread of multigenic drug resistance when used in appropriate population genetic models (Curtis and Otoo, 1986; Dye and Williams, 1997; Hastings, 1997). Such predictions may help implement drug usage policy in endemic areas. Current debate in the malaria field is focused on the question of whether drugs should be used in combination or sequentially (Kremsner *et al.*, 1997; White and Olliaro, 1996). The same debate occurred in tuberculosis health policy in the 1960s, and the answer was clearly to use combinations.

C. Social Change and Malaria

The way human populations move and live is a dynamic process largely driven by economic opportunities and occasionally social unrest and war. Some examples of how such changes affect malaria transmission are discussed below.

I. Urbanization

The trend toward growing numbers of the human population living in urban areas is on the increase, with 56% of the world's population predicted to be living in urban areas by the year 2025 (Knudsen and Sloof, 1992). This trend is especially true in developing countries where malaria is endemic. For example, India had 2590 towns with a total combined population of 62 million in 1951 (Sharma, 1996). By 1991, 217 million people were living in 3768 urban areas. A study of urbanization and malaria in Africa documented the population increase of the town of Brazzaville from 92,520 in 1955 to 500,761 in 1983 (Trape and Zoulani, 1987). How does increased urbanization impact on malaria?

Migration from rural to urban areas can lead to the movement of infected people to the towns with consequent enhancement of malaria transmission within the town. The converse may also be true where nonimmune migrants arrive in an urban area where malaria transmission is occurring. In Sudan it has been found that, despite variation between districts, urbanization tends to lead to reduced human malaria transmission (Robert, 1986). Moreover, these results are similar to those of Trape and Zoulani (1987) in the Congo. This study showed variation between urban districts in number of bites per human host per night from 7.26 (in the wet season) to areas in which no *Anopheles* were collected in 42 nights. Even though entomological parameters (daily survival rate, life expectancy, infective life, and stability index) were the same in urban and rural areas, all the highest urban zones had less transmission than the surrounding rural areas, which varied between 35 and 95 bites per man per night. Transmission within towns is not uniform; both studies found considerable variation within the urban areas in the number of infective bites a person would receive in one night. In these studies it was the periurban districts, areas normally inhabited by poor migrants, that experienced the highest number of bites and displayed the highest level of malaria prevalence (Trape and Zoulani, 1987). These periurban areas account for a large proportion of the population of cities of developing countries; in India between 25 and 40% of the urban population lives in periurban areas with no proper water supply or drainage (Sharma, 1996). Even in Indian towns, however, piped water is provided for only a few hours per day or a few times per week. In such circumstances people must store water so that they can have constant access to it. Such water provides good mosquito breeding sites. For example, Cambay city (Gujarat State, India) was shown to have more than 10,000 breeding places for *Anopheles stephensi*, which has readily adapted to the periurban environment. Similarly, some towns in Andra Pradesh had 80% of their overhead water tanks positive for *An. stephensi* larvae even with weekly antilarval measures. Local governments are generally unwilling to clean up periurban slums as residents cannot afford to pay, or they move when they do have money. Thus, the migration of people to urban areas tends to increase malaria incidence in poor people living in periurban areas.

Urban migration can also increase malaria indices of urban areas in other ways apart from increasing transmission in periurban areas. People can move between rural areas and the city with consequent importation of new infections. This may increase the reservoir of parasite diversity and hence overall transmission. Repeated human migration between rural areas and towns has been identified as a significant factor in keeping malaria endemic in Delhi (Sethi *et al.*, 1990).

Living in a city can also be an important socioeconomic factor that determines the clinical consequences of malaria incidence. Whereas periurban slum areas may experience higher transmission rates, Trape *et al.* (1987) discovered that the per capita death rates were similar in both poor and affluent communities and less than the rural village death rates. This was due to the fact that city dwellers had better access to antimalarial drugs available on the open market while rural people had less access to drugs. Thus, urbanization can reduce the chances of dying of malaria rather than the risk of malaria per se for the poor.

2. Economic Development and Changes in Land Usage

The ecology of mosquito vectors and the epidemiology of malaria can change as a result of deforestation (Walsh *et al.*, 1993). Deforestation has reduced the 50 million hectares of forest present in India in 1950 to 22 million hectares. This deforestation has displaced people who, being homeless and poor, have tended to move to urban areas. Forests normally have high malaria incidences and so continual deforestation means that exforest dwellers provide a constant source of malaria for the rest of the country. Marshy land and poor drainage around irrigation zones provides breeding grounds for *Anopheles culicifacies,* and the slow running streams that feed irrigated fields allow *An. fluviatilis* to breed. The area under irrigation to India has increased from 23 to 90 million hectares since 1951, maintaining endemic malaria in 200 million people in these areas. Similar increases in malaria transmission in irrigated areas have been noticed in other countries (Amerasinghe *et al.*, 1992). Conversely, the Malnad foothills of the western Ghats were sprayed with DDT in the 1950s and 1960s while the region was extensively replanted with coffee plantations. Forests were cleared, ground cover of leaf litter was removed, and many of the small streams in the area were blocked with dams. This has led to an apparently permanent reduction in malaria; a 50,000-km^2 area is still free of malaria. Thus, poor planning when modifying environments can easily create large numbers of poor people susceptible to increased malaria incidence, while well-planned environmental modification that increases the wealth production in the area can decrease the risk of malaria.

Population migration and environmental migration can occur simultaneously, and the effect this has on malaria does depend on the socioeconomic status of the people and the region. Most examples of these interactions are complicated by vector resistance to pesticides, but a malariological history of Swaziland has shown that this is not the case in this area (Packard, 1986). Before World War II colonial farmers tended to displace native farmers to the low-veld areas of Swaziland. This increased the malaria incidence in these areas, but the colonial occupiers did not seem to notice or care. When sugar farming began in these areas in the late 1940s insecticide treatments reduced the parasite rates in children from 65 to 2% after only three years of spraying. Insecticide sprays were very effective in interrupting transmission, and to prevent DDT resistance developing in the vectors the frequency of spraying was reduced and had effectively stopped by 1959 when the proportion of people carrying parasites in their blood was found to be 0.11% (out of 15,682 tested, compared with 23% in 1950). Cessation of spraying did not increase malaria incidence instantly. By this time the lowveld areas had become relatively prosperous, and mining and heavy industry arrived. The population of these areas had increased since the start of sugar farming by 144% (compared with a national increase of 58%) but still was not large enough to fill all of the jobs available. The local work force could obtain more money and work in better conditions in mining and heavy industry compared with sugar farming, and so the sugar farmers imported labor from Mozambique. Mozambique had no effective malaria control strategy, and so the imported workers were often parasitemic. Malaria cases appeared in the sugar farms in 1960, spread throughout the whole lowveld area, and have been increasing in frequency ever since even though migrant laborers were no longer entering the country in large numbers by the late 1970s. Thus, the prosperity of the lowveld region protected it from malaria, but the arrival of poor immigrants seeking work reintroduced it. The relative affluence of the lowveld areas has protected the rest of Swaziland. People normally move around to find work in poor economic times (as the Mozambique laborers did in the 1960s and 1970s, but the continued demand for work in the lowveld areas and limited movement through the rest of the country prevented malaria from being reintroduced throughout Swaziland. There is now concern that with drops in sugar prices people will leave the lowveld and so spread the malaria to the rest of the country. A drop in sugar prices also makes local industry less likely to invest money in insect control measures, and so the threat of poverty reduces chances for malaria control.

Thus, interactions between social factors and malaria incidence are quite complex. Malaria transmission can be altered because of migration to urban areas by a variety of mechanisms, even if transmission is lower in these areas. Human modification of the environment can also result in altered incidence of malaria. These alterations in the incidence of malaria all result in an increase in malaria if associated with poverty.

Poverty can make people move to find work. Such movement can increase the incidence of malaria as explained above. Migration is also a major factor that contributes to the spread of drug resistance. Once drug resistance has evolved it can spread by transmission within a host population. If a vector is widespread then the resistance can rapidly spread through that transmission zone. For example, the distribution of chloroquine resistance became effectively identical to the distribution of *P. falciparum* in South America within 10 years of resistance being noticed because of the homogeneity of vector fauna in the Amazon basin (Wernsdorfer, 1994). When gaps between transmission zones exist, the migration of infected humans between these areas can transport drug-resistant *P. falciparum* strains. The Balcad area of Somalia had good vector control strategies that meant only a low level of malaria occurred throughout the year. Those that caught malaria were usually symptomatic but could be treated effectively with chloroquine (Warsame *et al.*, 1990). This effective treatment was extremely important in maintaining the low malaria incidence. When migrant laborers from areas with reduced chloroquine sensitivity entered the region, the subsequent failure of chloroquine treatment (once the resistant strain was established by 1988) caused an epidemic that upset the transmission dynamics in that area and reestablished malaria at pre-vector control levels. So the immigration of people seeking employment from areas of *P. falciparum* drug resistance not only introduced it to Balcad but also increased the malaria incidence by interfering with the stable transmission dynamics produced by the vector control strategies.

Human migration can also interact with natural features to establish drug resistance in new areas. Sudanese workers returning from Qatar in 1988 did so at the time of flooding and increased rainfall. This allowed increased vector reproduction and so increased transmission potential. Chloroquine resistance is present in Qatar, but was not noticed until late 1988 in Sudan. Many cases were in the families of workers who traveled from Qatar, strongly suggesting that the migrant workers were responsible for importing chloroquine resistance into Sudan (Novelli *et al.*, 1988).

Multidrug resistance can also be propagated by human migration. Perhaps the best example of this is found at the Thai/Cambodia border where *P. falciparum* is now resistant to all drugs but the artemisinin derivatives (Wernsdorfer, 1994). Very little treatment is available within Cambodia, but drugs are freely available inside Thailand. Mining work is available in Cambodia, so Thai workers tend to work in Cambodia but get malaria treatment in Thailand. Refugees also leave Cambodia, giving an average of 3000 people crossing the border per day. Pyrimethamine resistance developed on the border in the early 1950s because subclinical doses were used for presumptive treatment of suspected malaria. This acted as a strong selection pressure because the doses were not potent enough to wipe out all the parasites in a patient; the drug would merely act to kill those parasites most susceptible to the drug. Introducing chloroquine into salt supplies in the late 1950s led to this drug being useless by 1970; drug-free salt was obtained by many people, and even if drugged salt was obtained the doses received were subclinical. This was followed by sulfadoxine and pyrimethamine in combination until 1982 when quinine and tetracycline were briefly used before poor compliance made the combination of mefloquine, sulfadoxine, and pyrimethamine (MSP) the main set of antimalarials to be used. Mefloquine was used from 1985 to 1988 on the border by refugee agencies and the military in large amounts. This wholesale use of many types of drugs meant that at the border in 1989 quinine or pyrimethamine alone had a 90% failure rate. By 1991 the MSP combination had a 30% failure rate on the border and a 70% failure rate in clinics dealing with gems miners in Cambodia. This appalling state was reached because of the largely uncontrolled use of drugs and population movement. People could effectively purchase whatever antimalarial they chose in Thailand to take to Cambodia. They could then treat themselves in Cambodia if they became ill, and such self-treatment usually results in subclinical doses being taken. Even if people returned to Thailand for treatment they would return to Cambodia with antimalarial drugs persisting in their body. The movement to intense transmission areas in Cambodia with subclinical or residual amounts of drugs imposed an enormous selective pressure on the parasite leading to the incredibly rapid spread of multidrug resistance in this area. The migration from areas with malaria control to intense transmission areas with very little control over the drugs people were using caused this rapid spread of multidrug resistance. Fortunately, because most of this migration is just across the border area and transmission is not uniformly high across Cambodia and Thailand, these extremely drug-resistant *P. falciparum* strains are limited to the border area. However, large-scale migration throughout these countries could lead to the wider distribution of multidrug resistance.

3. Sociopolitical Disturbances and Natural Disasters

Wars, political unrest, and famines may cause increased risk of malaria. This can be due to either disruption of existing health care structures in endemic areas or the movement of people to new geographic locations creating new risk for migrants or the communities they cohabit. Malaria epidemics have been associated with military conflicts, social unrest, and natural disorders. The consequent movement of nonimmune people to malarious areas across borders or within countries (Kondrachine and Trigg, 1997) creates opportunities for large-scale epidemic increases in malaria transmission. The United Nations estimates that in 1993 there were 24 million internal refugees within their countries (United Nations High Commission for Refugees, 1995) The same report shows an increase in external refugees, mostly in Africa, Asia, and Latin America, from 2.5 million in 1970 to 20 million in 1995. These figures highlight the increased opportunities for epidemic malaria.

4. International Travel and Commerce

Air travel to the tropics for the purposes of business and tourism has increased tremendously over the 1980s and 1990s. The development of a global economy with markets in developed and developing countries will undoubtedly continue to increase business travel. Annually 30 million travelers from nonendemic countries visit malaria endemic countries. This short-term movement of nonimmune people to malaria endemic areas has contributed to the rise in imported malaria cases observed in developed countries. The lack of appropriate health advice and availability of safe, prophylactic drugs for certain areas has further increased the risk of imported malaria. Figures from the United States and United Kingdom indicate that over 1000 imported cases were reported for each country in 1991 (Kondrachine and Trigg, 1997).

D. Malaria Epidemics Due to Climatic Change

Increased rains in arid and semiarid desert areas with limited vector breeding and insufficient vector longevity, as well as abnormally high temperature and humidity in highland areas where *Plasmodium* spp. cannot complete the sporogonic cycle due to low temperatures, can lead to dramatic changes in malaria transmissions. Such climatic changes can lead to a sudden increase in anopheline densities and consequent malaria epidemics. Reports of climatic change causing malaria epidemics in Madagascar, Ethiopia, India, and Peru have been made (Kondrachine, 1996, 1997; Lepers *et al.*, 1991; Teklehaimanot, 1991). These epidemics were associated with high mortality and suffering.

E. Breakdown of Public Health Infrastructure

A major consequence of the cessation of the malaria eradication program was the dismantling of the manpower and infrastructure established in many endemic areas. Staff were not replaced by malaria control workers with an alternative agenda to deal with the increasing incidence of malarial disease. Indeed, the remaining malaria professionals were trained to implement residual insecticide vector control using standardized methodology from a centralized administration. Often old procedures were needlessly continued in the absence of effective management. The response time to adapt to the new era of malaria control and case management was as a consequence too long in many countries. The increased clinical workload from the 1970s onward has generally been absorbed by already overburdened health care systems. Malaria control programs were dismantled with consequent cost savings for health departments with no added budgets for malaria. The external sources of funds dried up for political and economic reasons (Gramiccia and Beales, 1988).

Individual governments now administer malaria health care funding in an economic climate where drugs to treat malaria are increasing in cost as resistance to chloroquine emerges; patient management costs are escalating as the need for transfusions increases with the consequent risk of human immunodeficiency virus (HIV) infection.

V. GLOBAL WARMING

The impact of human-induced global climate change poses an obvious threat to human health. The insect vectors of *Plasmodium* spp. thrive in warm climates of tropical countries. Global warming leading to increased temperature in temperate areas could provide a habitat suitable for the increased distribution of anopheline vectors. Whether the potential increase in vector populations will lead to a concomitant increase in malaria transmission is not clear (Rogers and Packer, 1993) and may depend on level of endemnicity (Lindsay and Birley, 1996). Increased temperature can both increase the mortality of the vector and the biting rate as well as affect the duration of the sporogonic cycle. Predicting the change in transmissibility (R_0, *see Section 6.3*) as a mosquito-transmitted pathogen such as *P. falciparum* moves into a new area is difficult, but a number of mathematical models have attempted to do this in the context of available data (Rogers and Packer, 1993). Entomologists have also turned their attention to measuring changes in the global distribution of vectors. This strategy has lead to the use of geographic information systems and satellite imaging to monitor vector populations. Detailed mapping of vector habitats and distributions will allow rapid detection of any significant changes in the possible risk of malaria transmission.

VI. THE FUTURE

A bleak picture has been painted for the future regarding the global malaria situation. Malaria is no longer a disease of developed and developing nations. The impact of this disease is being felt globally, but most seriously in Africa. The WHO Action Plan for Malaria Control (1995–2000) has estimated that approximately US$28 million per annum of external investment in malaria control is needed in Africa. Outside Africa, malaria control programs cost an estimated US$175–350 million a year. These sums will just maintain the status quo. Unless considerable resources are allocated to funding research and development of new tools, it is impossible to see how the situation will improve.

Four areas of research show promise for the future: vector control, chemotherapy, malaria vaccines, and genome molecular epidemiology studies. However, more financial input is needed to develop and/or implement appropriate interventions as well as to evaluate such interventions.

A. Vector Control

Insecticide-impregnated bednets and curtains have been evaluated as malaria control measures using both mortality and morbidity measures as end points. They appear to be promising tools when used in conjunction with disease management. Results of large-scale field trials of permethrin-treated bednets, organized by UNDP/World Bank/WHO Special Programme for Research and Training in Tropical Diseases in Burkino Faso, the Gambia, Ghana, and Kenya, demonstrated an overall mortality reduction in children aged 1 to 4 years of 15 to 33% (average 25%; Cattani and Lengeler, 1997). Efforts are underway to develop sustainable programs based on impregnated bednets. Further research is still required to enhance their effectiveness and sustainability in operational settings. Insecticide resistance in vector populations must also be assessed. There is also a need to monitor the long-term efficacy of impregnated bednets in areas of differing transmission intensity. Snow *et al.* (1997) have published findings from a multicenter African study that show that the incidence of severe disease, in particular, cerebral malaria, can increase as transmission intensity decreases (see Section 3.2). They conclude that bednets should be implemented with caution under conditions of long-term evaluation to determine if a rebound effect occurs as immunity in the population declines due to reduced exposure. The conclusions of Snow *et al.* (1997) are being actively debated in the malaria community at present. It is questionable whether comparisons between sites with different host genetics and environmental and socioeconomic factors are valid. Molineaux (1997) has discussed the paper in the context of lessons learned from the eradication era. He concludes that "these observations do not justify withholding preventative measures (vector control, reduction of man/vector contact and chemoprophylaxis) from anybody in any malaria situation." The debate will no doubt continue as will the implementation of vector control and, it is hoped, of "long-term evaluation" of the efficacy of insecticide-treated bednets.

It is clear that effective vector control depends on adequate taxonomic studies to define behavioural and genetic characteristics of local vectors. A resurgence of interest in vector biology has occurred during the 1980s. This has resulted in new taxonomic methods as well as detailed genomic studies of anophelines at both the individual and population level (Collins, 1994). The application of these new tools to field studies is now necessary. The technology has now been developed to transfect *Aedes* mosquitos with transposable elements carrying specific gene sequences (James, 1992). It is expected that this technology will soon be applicable to anopheline mosquitos. These advances encourage the view held by some that malaria control may be achieved by driving *Plasmodium* refractory genes through mosquito populations (Kidwell and Ribeiro, 1992).

B. Chemotherapy

The advent of resistance to all the known antimalarial drugs in current use has precipitated an urgent need for new antimalarial drugs. The increasing levels of chloroquine resistance in Africa, as well as emerging resistance to pyrimethamine/sulfadoxine combinations, point to the need for an inexpensive, safe, effective drug to replace chloroquine. Pyonaridine, a Chinese compound, is under international development by WHO/TDR as an affordable, possible replacement for chloroquine. Krogstad and colleagues (1996) have rescreened chloroquine analogs and found a compound which shows no cross-resistance with chloroquine. This discovery has failed to interest the pharmaceutical industry. Indeed the general lack of interest of the pharmaceutical industry in design and development of new antimalarial drugs stimulated scientists at the Malaria Conference in Dakar concerned with the malaria situation in Africa, held earlier this year, to propose to set up an African Drug Consortium to develop antimalarials for the African continent (Dakar Meeting, 1997).

The Chinese drugs artemisinin and its derivatives have become the mainstay of malaria treatment in areas of multidrug resistance in Southeast Asia and South America. They show no cross-resistance with known antimalarials. WHO/TDR has conducted randomized, multicenter trials with intramuscular artemether to support its registration outside China. Artesunate suppositories are also being screened for home management of clinical cases to reduce the incidence of severe disease and death due to malaria.

A combination of atovoquone with proguanil was registered in the United Kingdom in 1997 for the treatment of uncomplicated falciparum malaria. This combination is active against multi-drug-resistant malaria. It is to be donated to endemic countries through the Task Force for Child Survival and Development.

Although the drug situation is under control for the present, increased research activity is urgently required to develop new drugs to prepare for the inevitable evolution of resistance to even the most promising antimalarial drugs in the pipeline.

C. Malaria Vaccines

The development of malaria vaccines to reduce infection and disease would provide one of the most cost-effective approaches to malaria control. The 1980s and 1990s have seen a great deal of research on identification of candidate vaccine antigens using recombinant DNA technology to obtain purified antigens and more recently DNA vaccines. Vaccine research has focused on identification of conserved, immunogenic regions of surface antigens of different life cycle stages. Three types of vaccines are under development: (1) anti-sporozoite vaccines, designed to prevent infection; (2) transmission-blocking vaccines, designed to arrest the development of the parasite in the mosquito, thereby blocking transmission; and (3) anti-asexual blood stage vaccines designed to reduce the incidence of disease. There has been a great deal of research into the molecular and immunological aspects of malaria vaccine development. I will not cover this extensive literature but rather will briefly summarize the aims of vaccination and the results of recent trials to illustrate the considerable activity in this area which holds great promise for innovative control measures.

1. Vaccine Trials

Candidate antigens for all the above vaccine strategies have been identified. Because of the complexity and cost of malaria vaccine development, as well as limited commercial interest, relatively few vaccine candidates have so far progressed to human clinical trials (Kondrachine and Trigg, 1997).

The first malaria vaccine to reach population field trials (Phase III) was SPf66, a subunit synthetic peptide consisting of amino acid sequences from *P. falciparum* antigens. The sequences are thought to derive from the major merozoite protein (MSP1) and two undefined blood stage antigens, and are linked by NANP-repeat sequences from the circumsporozoite protein. The vaccine should target both the sporozoite and asexual blood stages of the parasite; however, its mode of action is still unknown. A number of phase III trials have been carried out to assess the impact of SPf66 on the incidence of nonsevere disease. The results have been conflicting, with a highly significant protective effect of 34% in Colombia (Valero *et al.*, 1993), a borderline significance of 31% protection in Tanzania, which was maintained (25%) 18 months postvaccination (Alonso *et al.*, 1994, 1996), and no protective effect in the Gambia (D'Alessandro *et al.*, 1995) or Thailand (Nosten *et al.*, 1996). A significant protective effect of 55% was shown in a trial in Venezuela (Noya *et al.*, 1994), but, as no placebo innoculation was given, vaccination status was known by the vaccinees and might have affected their treatment seeking behavior. The two trials in which no vaccine efficacy was detected had 80% power to detect an efficacy of less than 40% (D'Alessandro *et al.*, 1995). The lack of an effect is therefore likely to be real. Although less than encouraging results have been obtained to date, further SPf66 trials are underway. A great deal has been learned from these early field trials which will facilitate more effective vaccine evaluation in the future.

Several vaccines are under development after promising results in animal model screens and human trials. A recombinant circumsporozoite protein vaccine has shown protection against sporozoite challenge in human trials (Stoute *et al.*, 1997). DNA vaccine strategies offer many technical advantages, including stimulation of T-cell responses, and are currently being evaluated (reviewed by Doolan and Hoffman, 1997).

2. Vaccination Policy

The likelihood of eradicating malaria by mass vaccination with a transmission blocking vaccine is related to the transmissibility of the parasite. Transmissibility can be defined numerically for a microparasite such as *P. falciparum* by the number of new infections to arise from a single infection in a wholly susceptible population. This is defined as the R_0 or basic reproductive number for a pathogen and is specific for the transmission conditions in a particular endemic area. It is the maximum transmission potential of a pathogen.

To attempt to eradicate malaria by vaccination with a conserved vaccine active against all strains of the parasite, the fraction of the population P to be immunized to block transmission of malaria would be calculated from Eq. (1):

$$P > 1 - 1/R_0. \tag{1}$$

It is well understood that it is easier to achieve eradication of pathogens with R_0 values in the range 1 to 5 compared to pathogens with higher R_0 values due to the nonlinear relationship between P and R_0 as a consequence of herd immunity. R_0 values for a variety of pathogens given in Table I illustrate this relationship. Given that we have 100% effective vaccines for both measles and smallpox, it has been much easier to achieve eradication of smallpox due to its lower transmissibility. What is the R_0 for *P. falciparum*? How easy will it be to control malaria by vaccination?

During the era of vector control Macdonald calculated the R_0 for *P. falciparum* by the following equation:

$$R_0 = - \frac{ma^2bp^n}{rIn(p)} \tag{9}$$

in which m is the number of adult female anopheline mosquitos per person, a is the daily biting rate of an individual female mosquito on humans (accounting for meals taken on other hosts), b is the fraction of mosquitos with infective sporozoites that actually generate human infection (and infectiousness) when biting, p is the daily survival rate, and n is the number of days between mosquito infection and the production of sporozoites in salivary glands (the so-called extrinsic incubation period). Human infection is summarized completely in r, which often is said to be the rate of recovery from infection but which, strictly,

TABLE I Estimated Values of Transmissibility (R_0) of Various Pathogens and Critical Proportion of Population (P) to be Immunized to Block Transmission[a]

Disease	R_0 of pathogen	P (%)
Measles	11–17	90–95
Pertussis	16–18	90–96
Mumps	11–14	85–90
Rubella	6–9	82–87
Poliomyelitis	5–6	82–85
Diphtheria	4–5	80–82
Rabies	4	80
Smallpox	3–4	70–80
Malaria	?	?

[a]Data from Anderson and May (1991).

is the rate of recovery from infectiousness. This equation was summarized by Garrett-Jones as the product of vectorial capacity (daily rate at which future inoculations arise from a currently infective case) and duration of infectiousness. Typical values for R_0 calculated from vectorial capacity are 50 for Madang, PNG, and 200–1000 for Ifakara, Tanzania. These estimates of R_0 have been suggested to be high enough to preclude the eradication of malaria by vaccination alone.

Are the R_0 estimates accurate? It is well understood that calculation of R_0 values from vectorial capacity data is problematic (Dye, 1994). Equation (2) is an incomplete expression for R_0 since there is no parameter that allows for the efficiency of transmission from humans to mosquito. Components of vectorial capacity are notoriously difficult to measure. Duration of infectiousness is measured as the duration of infection, which may be a gross overestimate of this parameter. Macdonald's approach to assessing R_0 by vectorial capacity put an upper bound on this figure and was intended to be used comparatively in the context of vector control. How realistic are these values? For the purpose of vector control it was perhaps less important to understand than for the goal of eradication of malaria by vaccination (given the availability of an appropriate transmission blocking vaccine). Myself and colleagues (Gupta *et al.*, 1994a) suggested that the R_0 of *P. falciparum* may not be as high as previously believed if the malaria transmission system is a construct of independently transmitted antigenic types or "strains." In such a transmission system the risk of infection can be high as it is related to the number of "strains within the system," whereas the transmissibility of malaria may be low as it is the weighted average of the R_0 values of the constituent "strains" within the system. This would be extremely good news for malaria vaccination.

This "strain theory" of malaria transmission has met with considerable resistance. Although most malariologists would acknowledge that parasite diversity is of paramount importance in malaria transmission, they are not comfortable to develop a theoretical framework for transmissibility based on a "strain" structure. This has never been attempted before. The converse of the "strain" structure must be the view that all diverse parasite genotypes form one entity called collectively "malaria." This view of the transmission system is untenable with the proponents of "strain" theory. Several basic assumptions of the theory have been challenged. First, some do not believe that "strains" can exist in a population of organisms that recombine. This is an experimental question that remains to be answered by appropriate linkage disequilibrium studies. Second, the transmission system relies on a single exposure generating long-lived immunity that blocks *transmission* of a "strain" in a "strain"-specific manner. Many malariologists believe that immunity to malaria is short-lived. The only data available to provide an answer are those of Deloron and Chougnet (1992) showing that immunity to malaria is long-lived in the absence of ongoing transmission in Madagascar. This experiment adds weight to the "long-lived immunity" assumption. A third criticism of the theory was that the effective human host population size needed to maintain the strain structure as described is incompatible with reality. A small host population could harbor a large effective parasite population size. More experimental research is needed to explore the validity of the assumptions of "strain theory."

The incorporation of parasite genetics into a theoretical framework for control of malaria by vaccines is a new area of research. This area needs to continue to grow both experimentally and theoretically to determine whether malaria can be eradicated by mass vaccination given appropriate vaccines.

D. Molecular Epidemiology—Genome Studies

Evolution of both *Anopheles* spp. and *Plasmodium* spp. in the face of natural and man-made selection will inevitably occur. The development of molecular epidemiological approaches to monitor changes in the parasite and vector biology at a level of detection able to identify rare variants in populations may allow us to respond more rapidly to potential failure of drugs, vaccines, and insecticides. A fundamental understanding of the population biology of *P. falciparum* may also help design innovative control strategies in the face of social change and human migration. Support of *Anopheles* and plasmodial genome studies will aid this process as well as facilitate vaccine and drug development. If DNA vaccines prove effective, it is clear how useful the sequence information from the Malaria Genome Project will be to vaccine development.

E. Conclusion

The lessons of the past tell us that no single approach to the control of malaria will provide a long-term solution. Social change, as well as natural and man-made changes to the environment will all contribute to create complex global patterns of malaria transmission (Fig. 6). Multidisciplinary approaches to both

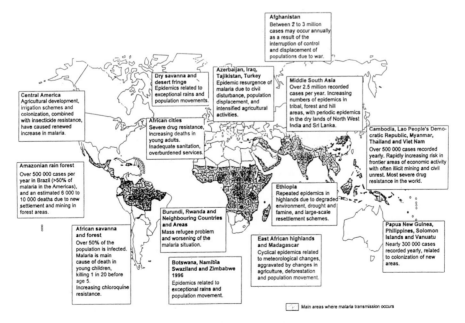

The designations employed and the presentation of material on the maps do not imply the expression of any opinion whatsoever on the part of the Secretariat of the World Health Organization concerning the legal status of any country, territory, city or area or of its authorities, or concerning the delimitation of its frontiers or boundaries.

Source WHO/CTD/MAL

FIGURE 6 Malaria distribution and problem areas, 1997.

basic and applied research, funded by adequate resources, hold the key to future improvements.

I conclude this review written in the year of the century of the discovery of *Plasmodium* parasites in mosquitos by Sir Ronald Ross in 1897 with a poem, "Indian Fevers." The poem was written by Ross while he was working as a medical doctor in India and attempting to describe how malaria was transmitted. It is regrettable that 100 years later the malaria situation in many parts of the world remains the same or worse than in the time of Ross.

Indian Fevers

In this, O Nature, yield I pray to me.
I pace and pace, and think and think, and take
The fever'd hands, and note down all I see,
that some dim distant light may haply break.

The painful faces ask, can we not cure?
We answer, No, not yet: we seek the laws.
O God reveal thro' all this thing obscure
The unseen, small, but million-murdering cause.
                                        Ross (1890–1893)

REFERENCES

Alonso, P. L., Smith, T. A., Armstrong-Schellenberg, J. R. M., *et al.* (1994). Randomised trial of the efficacy of SPf66 vaccine against *Plasmodium falciparum* malaria in children in southern Tanzania. *Lancet* **344**, 1175–1181.

Alonso, P. L., Smith, T. A., Armstrong-Schellenberg, J. R. M., *et al.* (1996). Duration of protection and age-dependence of the effects of the SPf66 malaria vaccine in African children exposed to intense transmission of *Plasmodium falciparum, J. Infect. Dis.* **174**, 367–372.

Amerasinghe, P. H., *et al.* (1992). Malaria transmission by *Anopheles subpictus* in a new irrigation project in Sri Lanka. *J. Med. Entomol.* **29**, 577–581.

Anders, R. F., and Smythe, J. A. (1989). Polymorphic antigens in *Plasmodium falciparum. Blood (Journal of the American Society of Hematology)* **74**, 1865–1875.

Anderson, R. M., and May, R. M. (1991) "Infectious Diseases of Humans: Dynamics and Control." Oxford Univ. Press, Oxford.

Babiker, H. A., Ranford-Cartwright, L. C., Currie, D., Charlwood, J. D., Billingsley, P., Teuscher, T., and Walliker, D. (1994a). Random mating in a natural population of the malaria parasite *Plasmodium falciparum. Parasitology* **109**, 413–421.

Baruch, D. I., Pasloske, B. L., Singh, H. B., Bi, X., Ma, X. C., Feldman, M., Taraschi, T. F., and Howard, R. J. (1995). Cloning the *Plasmodium falciparum* gene encoding PfEMP1, a malarial variant antigen and adherence receptor on the surface of parasitized human erythrocytes. *Cell (Cambridge, Mass.)* **82**, 77–87.

Bate, C. W., Taverne, J., Roman, E., Moreno, C., and Playfair, J. H. L. (1992). Tumour necrosis factor induction by malaria exoantigens depends on phospholipid. *Immunology* **75**, 129–135.

Biggs, B. A., Gooze, L., Wycherley, K., Wollish, W., Southwell, B., Leech, J. H., and Brown, G. V. (1991). Antigenic variation in *Plasmodium falciparum. Proc. Natl. Acad. Sci. U.S.A.* **88**, 9171–9174.

Boyd, M. F. (1949). "Malariology." Saunders, London.

Brewster, D. R., Kwaitkowski, D., and White, N. J. (1990). Neurological sequelae of cerebral malaria in children. *Lancet* **336**, 1039–1043.

Bruce, M. C., Donnelly, C., Walmsley, M., Lagog, M., Packer, M., Gibson, N., Narara, A., Alpers, M., Walliker, D., and Day, K. P. (1996). Dynamics of Plasmodium species and *Plasmodium falciparum* genotypes within multiply infected individuals from Papua New Guinea. *Br. Soc. Parasitol. 8th Malaria Meeting abstract.*

Bruce-Chwatt, L. J. (1965). Paleogenesis and paleoepidemiology of primate malaria. *Bull. WHO* **32**, 363–387.

Bruce-Chwatt, L. J. (1988). History of malaria from prehistory to eradication. *In* "Malaria: Principles and Practice of Malariology" (W. H. Wernsdorfer and I. McGregor, eds.), Vol. 1, pp. 1–61. Churchill Livingstone, London.

Carneiro, I., Donnelly, C., Cox, M., Austin, D., Ferguson, N., and Day, K. (1997a). Incidence and recovery rates for *Plasmodium falciparum* infection: Estimates using genotype markers. Submitted for publication.

Carneiro, I. A. M., Cox, M. J., Tavul, L., Kum, D. E., and Day, K. P. (1997b). Turnover of *Plasmodium falciparum* genotypes in symptomatic and asymptomatic children from rural Madang, Papua New Guinea. Submitted for publication.

Carter, R., and Graves, P. M. (1988). Gametocytes. *In* "Malaria: Principles and Practice of Malariology" (W. H. Wernsdorfer and I. McGregor, eds.), Vol. 1, pp. 253–305. Churchill Livingstone, London.

Cattani, J., and Lengeler, C. (1997). Insecticide-treated bednets and the prevention of malaria. *Rec. Adv. Paediatr.* **16**, in press.

Clark, I. A., and Rockett, K. A. (1994). The cytokine theory of human cerebral malaria. *Parasitol. Today* **10**, 410–411.

Collins, F. H. (1994). Prospects for malaria control through genetic manipulation of its vectors. *Parasitol. Today* **10**, 370–371.

Contamin, H., Fandeur, T., Rogier, C., Bonnefoy, S., Konate, L., Trape, J. F., and Mercereau Puijalon, O. (1996). Different genetic characteristics of *Plasmodium falciparum* isolates collected during successive clinical malaria episodes in Senegalese children. *Am. J. Trop. Med. Hyg.* **54**, 632–643.

Conway, D. J. (1997). Natural selection on polymorphic malaria antigens and the search for a vaccine. *Parasitol. Today* **13**, 26–29.

Conway, D. J., Greenwood, B. M., and McBride, J. S. (1992). Longitudinal study of *Plasmodium falciparum* polymorphic antigens in a malaria-endemic population. *Infect. Immun.* **60**, 1122–1127.

Cox, F. E. G. (1993). "Modern Parasitology: A Textbook of Parasitology." Blackwell, Oxford.

Creasey, A., Fenton, B., Walker, A., Thaithong, S., Oilveira, S., Mutambu, S., and Walliker, D. (1990). Genetic diversity of *Plasmodium falciparum* shows geographical variation. *Am. J. Trop. Med. Hyg.* **42**, 403–413.

Curtis, C. F., and Otoo, L. N. (1986). A simple model of the build-up of resistance to mixtures of anti-malarial drugs. *Trans. R. Soc. Trop. Med. Hyg.* **80**, 889–892.

D'Alessandro, U., Leach, A., Drakeley, C. J., Bennett, S., Olaleye, B. O., Fegan, G. W., Jawara, M., Langerock, P., George, M. O., Targett, G. A. T., and Greenwood, B. M. (1995). Efficacy trial of malaria vaccine SPf66 in Gambian infants. *Lancet* **346**, 462–467.

Dakar Meeting; Final Report of the International Congress on Malaria in Africa. (1997). Challenges and Opportunities for Co-operation. Dakar, Senegal.

Daubersies P., Sallenave-Sales S., Magne S., Trape J.-F., Contamin H., Fandeur T., Rogier C., Mercereau-Puijalon O., and Druilhe P. (1996). Rapid turnover of *Plasmodium falciparum* populations in asymptomatic individuals living in a high transmission area. *Am. J. Trop. Med. and Hyg.*, **54**(1), 18–26.

Day, K. P., and Marsh, K. (1991). Naturally acquired immunity to *Plasmodium falciparum*. *Parasitol. Today* **7**, A68–A70.

Deloron, P., and Chougnet, C. (1992). Is immunity to malaria really short-lived? *Parasitol. Today* **8**, 375–378.

Dietz, K. (1988). Mathematical models for transmission and control of malaria. *In* "Malaria; Principles and Practice of Malariology" (W. H. Wernsdorfer and I. McGregor, eds.), Vol. 2, pp. 1091–1135. Churchill Livingstone, London.

Doolan, D. L., and Hoffman, S. L. (1997). Multi-gene vaccination against malaria: A multi-stage, multi-immune response approach. *Parasitol. Today* **13**, 171–177.

Dye, C. (1994). Vector control. *In* "Parasitic and Infectious Diseases" (M. E. Scott and G. Smith, eds.), Academic Press, San Diego.

Dye, C., and Williams, B. G. (1997). Multigenic drug resistance among inbred malaria parasites. *Proc. Natl. Acad. Sci. U.S.A.* **7**, 277–281.

Felger, I., Tavul, L., Kabintik, S., Marshall, V., Genton, B., Alpers, M., and Beck, H.-P. (1994). *Plasmodium falciparum:* Extensive polymorphism in merozoite surface antign 2 alleles in an area with endemic malaria in Papua New Guinea. *Exp. Parasitol.* **79**, 106–116.

Genton, B., Al-Yaman, F., Mgone, C. S., Alexander, N., Paniu, M. M., and Alpers, M. (1995). Ovalocytosis and cerebral malaria. *Nature (London)* **378**, 564–565.

Gilles, H. M., and Warrell, D. A., eds. (1993). "Bruce–Chwatt's Essential Malariology." Arnold, Boston.

Gramiccia, G., and Beales, P. F. (1988). The recent history of malaria control and eradication. *In* "Malaria: Principles and Practice of Malariology" (W. H. Wernsdorfer and I. McGregor, eds.), Vol. 2, pp. 1335–1377. Churchill Livingstone, London.

Greenwood, B. M., Bradley, A. K., Greenwood, A. M., Byass, P., Jammeh, L. Marsh, K., Tulloch, S., Oldfield, F. S. T., and Hayes, R. (1987). Mortality and morbidity from malaria among children in a rural area of The Gambia, West Africa. *Trans. R. Soc. Trop. Med. Hyg.* **81**, 478.

Greenwood, B. M., Marsh, K., and Snow, R. (1991). Why do some African children develop severe malaria? *Parasitol. Today* **8**, 239–242.

Gupta, S., and Day, K. P. (1994). A strain theory of malaria transmission. *Parasitol. Today* **10**, 476–481.

Gupta, S., and Day, K. P. (1995). A theoretical framework for the immunoepidemiology of *Plasmodium falciparum* malaria. *Parasite Immunol.* **16**, 361–370.

Gupta, S., Trenholme, K., Anderson, R. M., and Day, K. P. (1994a). Antigenic diversity and the transmission dynamics of *Plasmodium falciparum. Science* **263**, 961–963.

Gupta, S., Hill, A. V. S., Kwiatkowski, D., Greenwood, A. M., Greenwood, B. M., and Day, K. P. (1994b). Parasite virulence and disease patterns in *P. falciparum* malaria. *Proc. Natl. Acad. Sci. U.S.A.* **91**, 3715–3719.

Haldane, J. B. S. (1948). The rate of mutation of human genes. *Proceedings of the Eighth International Congress of Genetics—HEREDITAS* **35**, 267–273.

Hastings, I. (1997). The biology and population genetics underlying the evolution of drug resistance in *Plasmodium*. Submitted for publication.

Hill, A. V. S. (1992). Malaria resistance genes: A natural selection. *Trans. R. Soc. Trop. Med. Hyg.* **86**, 225–226 and 232.

Hill, A. V. S., Allsopp, C. E., *et al.* (1991). Common West African HLA antigens are associated with protection from severe malaria. *Nature (London)* **352**, 595–600.

Hill, A. V. S., Elvin, J., Willis, A. C., Aidoo, M., Allsopp, C. E. M., Gotch, F. M., Gao, X. M., Takiguchi, M., Greenwood, B. M., Townsend, A. R. M., McMichael, A. J., and Whittle, H. C. (1992). Molecular analysis of the association of HLA-B53 and resistance to severe malaria. *Nature (London)* **360**, 434–439.

Hill, W. G., Babiker, H. A., Ranford-Cartwright, L. C., and Walliker, D. (1995). Estimation of inbreeding coefficients from genotypic data on multiple alleles, and application to estimation of clonality in malaria parasites. *Genet. Res. Cambridge* **65**, 53–61.

James, A. A. (1992). Mosquito molecular genetics: The hands that feed bite back. *Science* **257**, 37–38.

Joshi, H., Subbarao, S. K., Ragavendra, K., and Sharma, V. P. (1989). *Plasmodium vivax:* Enzyme polymorphism in isolates of Indian origin. *Trans. R. Soc. Trop. Med. Hyg.* **83**, 179–181.

Kemp, D. J., Cowman, A. F., and Walliker, D. (1990). Genetic diversity in *Plasmodium falciparum. Adv. Parasitol.* **29**, 75–149.

Kidwell, M. G., and Ribeiro, J. M. C. (1992). Can transposable elements be used to drive disease refractoriness genes into vector populations. *Parasitol. Today* **8**, 325–329.

Knudsen, A. B., and Sloof, R. (1992). Vector bourne disease problems in rapid urbanisation. *WHO Bull. OMS* **70**, 1–6.

Koch, R. (1990). Bericht uber die Thatigkeit der Malariaexpedition. *Dtsch. Med. Wochenschr.* **26**, 88–734.

Koella, J. C. (1993). Epidemiological evidence for an association between chloroquine resistance of *Plasmodium falciparum* and its immunological properties. *Parasitol. Today* **9**, 105–108.

Koella, J. C., Hatz, C., Mshinda, H., de Savigny, D., Macpherson, C. N., Degremont, A. A., and Tanner, M. (1990). *In vitro* resistance patterns of *Plasmodium falciparum* to chloroquine—A reflection of stain-specific immunity? *Trans. R. Soc. Trop. Med. Hyg.* **84**, 662–665.

Kondrachine, A. V. (1996). Mission report on malaria epidemics in Rajasthan, India. *Unpublished WHO Report.*

Kondrachine, A. V. (1997). Malaria in Peru. *Unpublished WHO Report.*

Kondrachine, A. V., and Trigg, P. I. (1997). Global overview of malaria. *Indian J. Med. Res.* **106**, 39–53.

Kremsner, P. G., Luty, A. J. F., and Graninger, W. (1997). Combination chemotherapy for *Plasmodium falciparum* malaria. *Parasitol. Today* **13**, 167–170.

Krogstad, D.-D. F. M., Cogswell, F. B., and Krogstad, D. J. (1996). Aminoquinolines that circumvent resistance *in vitro*. *Am. J. Trop. Med. Hyg.* **55**, 579–583.

Kwiatkowski, D. (1991). Cytokines and anti-disease immunity to malaria. *Res. Immunol.* **142**, 707–712.

Kwiatkowski, D. (1995). Malarial toxins and the regulation of parasite density. *Parasitol. Today* **11**, 206–212.

Lederberg, J., Shope, R. E., and Oaks, S. C., eds. (1992). "Emerging Infections: Microbial Threats to Health in the United States." National Academy Press, Washington, D.C.

Lepers, J. P., Fontenille, D., Rason, M. D., Chougnet, C., Astagneau, P., Coulanges, P., *et al.* (1991). Transmission and epidemiology of newly transmitted falciparum malaria in the central highland plateaux of Madagascar. *Ann. Trop. Med. Parasitol.* **85**, 297–304.

Lepes, T. (1981). Technical problems related to biological characteristics of malaria, as encountered in malaria control/eradication. *Unpublished WHO document WHO/MAL/81.934.*

Lindsay, S. W., and Birley, M. H. (1996). Climate change and malaria transmission. *Annals Trop. Med. Parasit.* **90**, 573–588.

Macdonald, G. (1957). "The Epidemiology and Control of Malaria." Oxford Univ. Press, London.

McGuire, W., Hill, A. V. S., Allsopp, C. E. M., Greenwood, B. M., and Kwiatkowski, D. (1994). Variation in the TNF-alpha promoter region associated with susceptibility to cerebral malaria. *Nature (London)* **371**, 580–511.

Maitland, K., Williams, T. N., Peto, Day, K. P. T., Clegg, J. B., Weatherall, D. J., and Bowden, D. K. (1997). Absence of age-specific mortality in children in an area of hyperendemic malaria. *Trans. Roy. Soc. Trop. Med. Hyg.* **91**, 562–566.

Marsh, K. (1992). Malaria—A neglected disease? *Parasitology* **104**, S53–S69.

Marsh, K. (1993). Immunology of human malaria. *In* "Bruce–Chwatt's Essential Malariology" (H. M. Gilles and D. A. Warrell, eds.), pp. 60–78. Arnold, Boston.

Missiroli, A. (1948). *Anopheles* control in the Mediterranean area. *Proceedings of the Fourth International Congresses on Tropical Medicine and Malaria* **2**, 1566–1576.

Molineaux, L. (1988). The epidemiology of human malaria as an explanation of its distribution, including some implications for its control. *In* "Malaria: Principles and Practice of Malariology" (W. H. Wernsdorfer and I. McGregor, eds.), Vol. 2, pp. 913–999. Churchill Livingstone, London.

Molineaux, L. (1997). Nature's experiment: What implications for malaria prevention? *Lancet* **349**, 1636–1637.

Neiva, A. (1910). Ueber die Bildung einer chininresistenten Rasse des Malariaparasiten. *Mem. Inst. Oswaldo Cruz* **2**, 131–140.

Nocht, B., and Werner, H. (1910). Beobachtungen uber eine relative Chininresistenz bei Malaria aus Brasilien. *Dtsch. Med. Wochenschr.* **36**, 1557–1560.

Nosten, F., Luxemburger, C., Kyle, D. E., Ballou, W. R., Wittes, J., Wah, E., Chongsuphajaisiddhi, T., Gordon, D. M., White, N. J., Sadoff, J. C., and Heppner, D. G. (1996). Randomised double-blind placebo-controlled trial of SPf66 malaria vaccine in children in northwest Thailand. *Lancet* **348**, 701–707.

Novelli, V. M., *et al.* (1988). Floods and resistant malaria. *Lancet,* 1367.

Noya, G. O., Gabaldon Berti, Y., Alarcon de Noya, *et al.* (1994). A population-based clinical trial with the SPf66 synthetic *Plasmodium falciparum* malaria vaccine in Venezuela. *J. Infect. Dis.* **170**, 396–402.

Ntoumi, F., Contamin, H., Rogier, C., Bonnefoy, S., Trape, J.-F., and Mercereau-Puijalon. (1995). Age-dependent carriage of multiple *Plasmodium falciparum* merozoite surface antigen-2 alleles in asymptomatic malaria infections. *Am. J. Trop. Med. Hyg.* **52**, 81–88.

Packard, R. M. (1986). Agricultural development, migrant labor and the resurgence of malaria in Swaziland. *Social Sci. Med.* **22**, 861–867.

Paul, R. E. L., Packer, M. J., Walmsley, M., Lagog, M., Ranford-Cartwright, L. C., Paru, R., and Day, K. P. (1995). Mating patterns in malaria populations of Papua New Guinea. *Science* **269**, 1709–1711.

Plowe, C. V., Djimde, A., Bouare, M., Doumbo, O., and Wellems, T. E. (1995). Pyrimethamine and proguanil resistance-conferring mutations in *Plasmodium falciparum* dihydrofolate reductase: Polymerase chain reaction methods for surveillance in Africa. *Am. J. Trop. Med. Hyg.* **52**, 565–568.

Reeder, J. C., Rieckmann, K. H., Genton, B., Lorry, K., Wines, B., and Cowman, A. F. (1996). Point mutations in the dihydrofolate reductase and dihydropteroate synthetase genes and *in vitro* susceptibility to pyrimethamine and cycloguanil of *Plasmodium falciparum* isolates in Papua New Guinea. *Am. J. Trop. Med. Hyg.* **55**, 209–213.

Rieckmann, K. H., Davis, D. R., and Hutton, D. C. (1989). *Plasmodium vivax* resistance to chloroquine. *Lancet* **2**, 1183–1184.

Robert, V. (1986). Urban malaria in Bobo-Dioulasso (Burkino Faso, Sudan). *Cahiers Office de la Recherche Scientifique et Techique Outre-Mer Serie Entomologie Medical et Parasitologie* **24**, 121–128.

Roberts, D. J., Craig, A. G., Berendt, A. R., Pinches, R., Nash, G., Marsh, K., and Newbold, C. I. (1992). Rapid switching to multiple antigenic and adhesive phenotypes in malaria. *Nature (London)* **357**, 689–692.

Rogers, D. J., and Packer, M. J. (1993). Vector-borne diseases, models, and global change. *Lancet* **342**, 1282–1284.

Rogier, C., Commenges, D., and Trape, J. F. (1996). Evidence for an age-dependent pyrogenic threshold of *Plasmodium falciparum* parasitemia in highly endemic populations. *Am. J. Trop. Med. Hyg.* **54**, 613–619.

Ross, R. (1897). On some peculiar pigmented cells found in two mosquitoes fed on malarial blood. *Br. Med. J.* **2**, 1786–1788.

Ross, R. (1911). "The Prevention of Malaria." Murray, London.

Saul, A. (1996). Transmission dynamics of *Plasmodium falciparum. Parasitol. Today* **12**, 74–79.

Service, M. W. (1993). Mosquitos (Culicidae). *In* "Medical Insects and Arachnids" (R. P. Lane and R. W. Crosskey, eds.), The Natural History Museum, London.

Sethi, N. K., *et al.* (1990). Role of migratory population in keeping up endemicity of malaria in metropolitan cities of India. *J. Commun. Dis.* **22**, 86–91.

Shahabuddin, M., and Kaslow, D. C. (1993). Chitinase: A novel target for blocking parasite transmission? *Parasitol. Today* **9**, 252–255.

Sharma, V. P. (1996). Re-emergence of malaria in India. *Indian J. Med. Res.* **103**, 26–45.

Smith, J. D., Chitnis, C. E., Craig, A. G., Roberts, D. J., Hudson-Taylor, D. E., Peterson, D. S., Pinches, R., Newbold, C. I., and Miller, L. (1995). Switches in expression of *Plasmodium falciparum var* genes correlate with changes in antigenic and cytoadherent phenotypes of infected erythrocytes. *Cell (Cambridge, Mass.)* **82**, 101–110.

Snounou, G., Pinheiro, L., Goncales, A., Fonseca, L., Dias, F., Brown, K. N., and do Rosario, V. E. (1993). The importance of sensitivity detection of malaria parasites in the human and insect hosts in epidemiological studies, as shown by the analysis of field samples from Guinea Bissau. *Trans. R. Soc. Trop. Med. Hyg.* **87**, 649–653.

Snow, R. W., Omumbo, J. A., Lowe, B., Molyneux, C. S., Obiero, J. O., Palmer, A., Weber, M. W., *et al.* (1997). Relation between severe malaria morbidity in children and level of *Plasmodium falciparum* transmission in Africa. *Lancet* **349**, 1650–1654.

Stoute, J. M., Slaoui, M., Heppner, D. G., Momin, P., Kester, K. E., Desmon, P., Wellde, B. T., Garcon, N., Krzych, U., Marchand, M., Ballou, W. R., and Cohen, J. D. (1997). A preliminary evaluation of a recombinant circumsporozoite protein vaccine against Plasmodium falciparum malaria. *New Eng. J. Med.* **336**, 86–99.

Su, X.-Z., Heatwole, V. M., Wertheimer, S. P., Guinet, F., Herrfeldt, J. A., Peterson, D. S., Ravetch, J. A., and Wellems, T. E. (1995). The large diverse gene family var encodes proteins involved in cytoadherence and antigenic variation of *Plasmodium falciparum* infected erythrocytes. *Cell (Cambridge, Mass.)* **82**, 89–100.

Teklehaimanot, A. (1991). Travel report to Ethiopia. Unpublished WHO Report.

Tibayrenc, M., and Lal, A. (1996). Self-fertilization, linkage disequilibrium, and strain in *Plasmodium falciparum. Science* **271**, 1300.

Trape, J. F. (1987). Malaria and urbanisation in Central Africa: The example of Brazzaville. Part IV: Parasitological and serological surveys in urban and surrounding rural areas. *Trans. R. Soc. Trop. Med. Hyg.* **81**, 26–33.

Trape, J. F., and Zoulani, A. (1987). Malaria and urbanisation in Central Africa: The example of Brazzaville. *Trans. R. Soc. Trop. Med. Hyg.* **81**, 10–18.

Trape, J. F., *et al.* (1987). Malaria and urbanisation in Central Africa: The example of Brazzaville. Part V: Pernicious attacks and mortality. *Trans. R. Soc. Trop. Med. Hyg.* **81**, 34–42.

Trape, J. F., *et al.* (1994). The Dielmo project: A longitudinal study of natural malaria infection and the mechanisms of protective immunity in a community living in a holoendemic area of Senegal. *Am. J. Trop. Med. Hyg.* **51**, 123–137.

United Nations High Commission for Refugees. (1995). "The State of the World's Refugees: In Search of Solutions." Oxford Univ. Press, Oxford.

Valero, M. V., Amador, L. R., Galindo, C., *et al.* (1993). Vaccination with SPf66, a chemically synthesised vaccine, against *Plasmodium falciparum* malaria in Colombia. *Lancet* **341**, 705–710.

Walsh, J. F., Molyneux, D. H., and Birley, M. H. (1993). Deforestation: effects on vector-borne disease. *Parasitology* **106**, 555–575.

Wang, P., Read, M., Sims, P. F. G., and Hyde, J. E. (1997). Sulfadoxine resistance in the human malaria parasite *Plasmodium falciparum* is determined by mutations in dihydropteroate synthetase and an additional factor associated with folate utilisation. *Mol. Microbiol.* **23**, 979–989.

Warrell, D. A. (1993). Clinical features of malaria. *In* "Bruce–Chwatt's essential Malariology," pp. 35–50. Arnold, Boston.

Warsame, M., *et al.* (1990). Isolated malaria outbreak in Somalia: Role of chloroquine-resistant *Plasmodium falciparum* demonstrated in Balcad epidemic. *J. Trop. Med. Hyg.* **93**, 284–289.

Wellems, T. E. (1991). Molecular genetics of drug resistance in *Plasmodium falciparum* malaria. *Parasitol. Today* **7**, 110–112.

Wernsdorfer, W. H. (1991). The development and spread of drug-resistant malaria. *Parasitol. Today* **7**, 297–303.

Wernsdorfer, W. H. (1994). Epidemiology of drug resistance in malaria. *Acta Tropica* **56**, 143–156.

White, N. J., and Olliaro, P. L. (1996). Strategies for the prevention of antimalarial drug resistance: Rationale for combination chemotherapy for malaria. *Parasitol. Today* **12**, 399–401.

WHO. (1991). Synopsis of the world malaria situation. *Weekly Epidemiological Record* **22**.

WHO. (1992). "WHO Expert Committee on Malaria," Nineteenth Report. WHO/CTD/92.1. Geneva.

WHO. (1996). "WHO Report 1996: Fighting Disease, Fostering Development." Geneva.

■ INDEX

Italicized numbers refer to figures.
"t" refers to tables.